"One of the icons of environmental criticism, Cheryll Glotfelty, engages in a conversation with a world-class photographer, Peter Goin. Together they give voice to the invisible density of radiation, the uncanny agency of climate and waste, the beauty of resisting places. A compelling journey across disciplines and artistic languages, *Peter Goin and the Photography of Environmental Change* enables its readers to access the stories that weave the 'humanatural' fabric of landscapes. This book provides a masterclass in visual studies and ecological humanities."

Serenella Iovino, *Professor of Environmental Humanities, University of North Carolina at Chapel Hill, USA*

"Of all the arts, photography has been the consistent guiding light during the past 150 years in helping us understand not only the larger American landscape but also the everyday places where we live, work, and play. And no photographer has done more to advance a new awareness of our contemporary landscapes than Peter Goin, who always seems to anticipate future directions of the art form. This new book thankfully confirms Goin's artistic efforts and the tremendous contributions he has made during his stellar career."

George F. Thompson, *Founder and Director of the Center for the Study of Place*

"*Peter Goin and the Photography of Environmental Change* is a powerful conversation between the beauty and mystery of our altered world, and an exciting way of reinventing our way of seeing and feeling Nature to pursue a better world. It is the most compelling book yet written about visual literacy."

María Antonia Blanco Arroyo, *Professor of Art, University of Seville, Spain*

PETER GOIN AND THE PHOTOGRAPHY OF ENVIRONMENTAL CHANGE

Peter Goin and the Photography of Environmental Change narrates the forty-year quest of award-winning and internationally exhibited contemporary photographer Peter Goin to document human-altered landscapes across America and beyond. It is a collaborative work between an artist and a literary critic, a retrospective of an accomplished environmental photographer, and an innovative education in visual reading.

Enduring howling wind, pounding rain, and blistering sun, Goin bears witness to radioactive landscapes, abandoned mines, simulated swamps, rechanneled rivers, controlled burns, overgrown ruins, industrialized agriculture, shrinking reservoirs, feral spaces in the city, architected wilderness, sacred wastelands, contested borderlands, and more. Based on more than seventy hours of taped interviews with the artist spanning over a decade, trailblazing ecocritic Cheryll Glotfelty narrates the arc of Goin's career, sharing excerpts from their conversations that reveal his brilliant mind and piquant personality while situating his work within the broader context of environmental thinkers.

This beautifully illustrated volume, with 200 images in color and black-and-white showcasing Goin's work, will be a fascinating and insightful read for upper-level students, academics, and researchers in photography, environmental history and culture, landscape studies, and environmental humanities.

Cheryll Glotfelty, the nation's first professor of Literature and Environment, enjoyed a twenty-eight-year career at the University of Nevada, Reno, before retiring in 2018. Her coedited *The Ecocriticism Reader: Landmarks in Literary Ecology* is a progenitive work in the environmental humanities. Her two coedited volumes, *The Bioregional Imagination: Literature, Ecology, and Place* and *The Biosphere and the Bioregion: Essential Writings of Peter Berg*, explore bioregional approaches to harmonizing culture with nature. Glotfelty's literary criticism and reviews have appeared in *ISLE: Interdisciplinary Studies in Literature and Environment, Women's Studies, ATQ, Southwestern American Literature, Western American Literature, Literature and Belief*, and edited collections. Her art criticism has appeared in *Material Ecocriticism* and *Environmental Criticism for the Twenty-First Century*. She is a cofounder, past president, and honorary lifetime member of the Association for the Study of Literature and Environment.

Routledge Environmental Humanities

Series editors: **Scott Slovic** *(University of Idaho, USA),* **Joni Adamson** *(Arizona State University, USA) and* **Yuki Masami** *(Aoyama Gakuin University, Japan)*

The Routledge Environmental Humanities series is an original and inspiring venture recognising that today's world agricultural and water crises, ocean pollution and resource depletion, global warming from greenhouse gases, urban sprawl, overpopulation, food insecurity and environmental justice are all crises of culture.

The reality of understanding and finding adaptive solutions to our present and future environmental challenges has shifted the epicenter of environmental studies away from an exclusively scientific and technological framework to one that depends on the human-focused disciplines and ideas of the humanities and allied social sciences.

We thus welcome book proposals from all humanities and social sciences disciplines for an inclusive and interdisciplinary series. We favour manuscripts aimed at an international readership and written in a lively and accessible style. The readership comprises scholars and students from the humanities and social sciences and thoughtful readers concerned about the human dimensions of environmental change.

Naturebot
Unconventional Visions of Nature
James Barilla

Trees in Nineteenth-Century English Fiction
The Silvicultural Novel
Anna Burton

Imagining Climate Engineering
Dreaming of the Designer Climate
Jeroen Oomen

Blue Ecocriticism and the Oceanic Imperative
Sidney I. Dobrin

Environmental Humanities in the New Himalayas
Symbiotic Indigeneity, Commoning, Sustainability
Edited by Dan Smyer Yü and Erik de Maaker

Subjects of Intergenerational Justice
Indigenous Philosophy, the Environment and Relationships
Christine J. Winter

Mosquitopia
The Place of Pests in a Healthy World
Edited by Marcus Hall and Dan Tamïr

Imagining Apocalyptic Politics in the Anthropocene
Edited by Earl T. Harper and Doug Specht

Cosmopolitical Ecologies Across Asia
Places of Power in Changing Environments
Edited by Riamsara Kuyakanon, Hildegard Diemberger and David Sneath

Peter Goin and the Photography of Environmental Change
Visual Literacy and Altered Landscapes
Cheryll Glotfelty with Peter Goin

For more information about this series, please visit: www.routledge.com/Routledge-Environmental-Humanities/book-series/REH

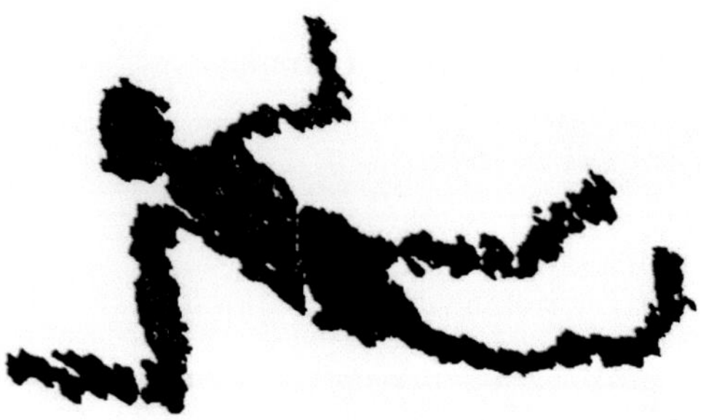

PETER GOIN AND THE PHOTOGRAPHY OF ENVIRONMENTAL CHANGE

Visual Literacy and Altered Landscapes

Cheryll Glotfelty
with Peter Goin

Routledge
Taylor & Francis Group
LONDON AND NEW YORK

earthscan
from Routledge

Cover image © Peter Goin

First published 2022
by Routledge
4 Park Square, Milton Park, Abingdon, Oxon OX14 4RN

and by Routledge
605 Third Avenue, New York, NY 10158

Routledge is an imprint of the Taylor & Francis Group, an informa business

British Library Cataloguing-in-Publication Data
A catalogue record for this book is available from the British Library

Library of Congress Cataloging-in-Publication Data
Names: Glotfelty, Cheryll, writer of added text. | Goin, Peter, 1951- photographer.
Title: Peter Goin and the photography of environmental change : visual literacy and altered landscapes / Cheryll Glotfelty with Peter Goin.
Description: Abingdon, Oxon ; New York, NY : Routledge, 2022. | Series: Routledge environmental humanities | Includes bibliographical references and index.
Subjects: LCSH: Landscape photography—United States. | Environmental degradation—United States—Pictorial works. | United States—Environmental conditions—Pictorial works. | Photographic criticism. | Ecocriticism. |
Goin, Peter, 1951-
Classification: LCC TR660.5 .P4725 2022 (print) | LCC TR660.5 (ebook) |
DDC 779/.36—dc23
LC record available at https://lccn.loc.gov/2021031596
LC ebook record available at https://lccn.loc.gov/2021031597

ISBN: 978-1-032-08032-1 (hbk)
ISBN: 978-1-032-08033-8 (pbk)
ISBN: 978-1-003-21260-7 (ebk)

DOI: 10.4324/9781003212607

Typeset in Bembo
by codeMantra

For Product Safety Concerns and Information please contact our
EU representative: GPSR@taylorandfrancis.com
Taylor & Francis Verlag GmbH, Kaufingerstraße 24,
80331 München, Germany.

CONTENTS

FIGURES

ACKNOWLEDGMENTS

I gratefully acknowledge the year of sabbatical support I received from the College of Liberal Arts, University of Nevada, Reno. Peter Goin and I appreciate support from the College of Liberal Arts Scholarly and Creative Activities Grant Program at the University of Nevada, Reno. This book was published with generous support provided by the Center for Art + Environment at the Nevada Museum of Art. At Routledge, we would like to thank editors Annabelle Harris, Grace Harrison, Rosie Anderson, Matthew Twigg, and Matthew Shobbrook, the production staff, the Environmental Humanities Series editors, and three anonymous outside reviewers. The staff at Special Collections, University of Nevada, Reno, cheerfully helped me access archival materials and rare books; special thanks to Donnie Curtis, Betty Glass, and Jacquelyn Sundstrand. Fond thanks to the UNR Department of English, especially chair Eric Rasmussen and my colleague Mike Branch. I received a Summer Research Assistant from the English Department and was fortunate to be matched with PhD student Will Lombardi, now an English professor at Feather River College, Quincy, CA. Shawn Sariti, UNR's Digital Multimedia Production Specialist, converted a dozen early cassette tapes to MP3 files. Goin's collaborators on previous books—Elizabeth Raymond, Paul Starrs, and Mary Webb—shared their time and perspectives in taped interviews with me. I presented work-in-progress at conferences of the Western Literature Association and the Association for the Study of Literature and Environment and received helpful feedback from my colleagues in those organizations. Special thanks to my long-time WLA friend Nancy Cook and to ASLE co-founders and dear friends Mike Branch and Scott Slovic. I deeply appreciate and am beyond indebted to Amy Anderson for transcribing upwards of seventy hours of interview tapes over a span of nearly ten years. My parents, Loren and Evelyn Acton, have been active supporters of this project. My husband, Steve, is always there for me through thick and thin. Finally, I would like to thank the star of the show, Peter Goin. Peter has been gracious and generous, patient and punctilious. He is a remarkable human being, and it has been a privilege to get to know Peter Goin and his work.

1

INTRODUCTION

Leave your shoes at the door

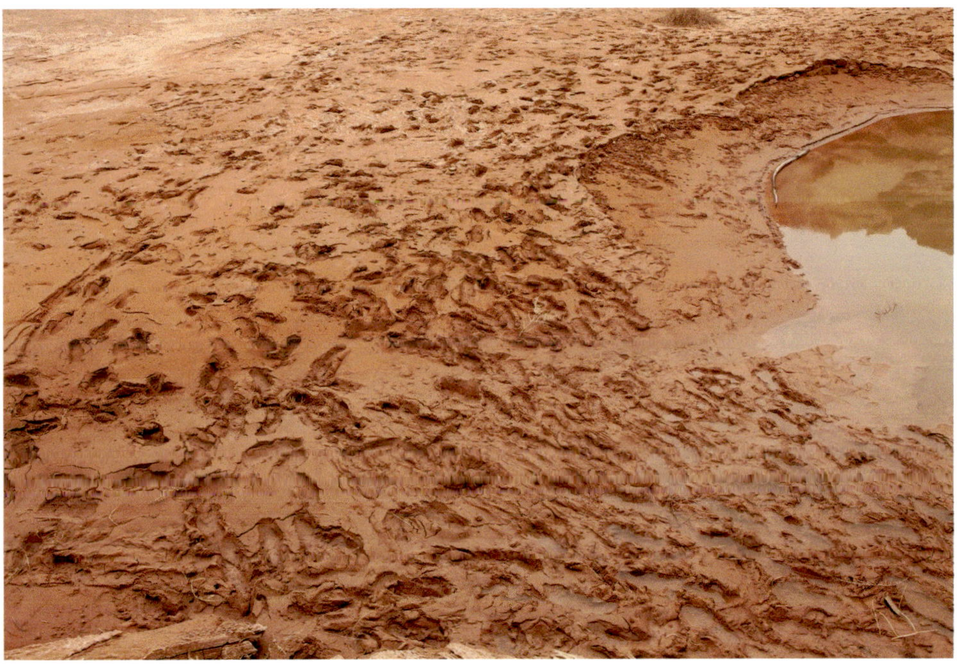

1.1 *Footprints, Scorup Canyon.*

Here is a photograph by Peter Goin entitled "*Footprints, Scorup Canyon.*" What would you say about it? One viewer comments dryly that this photograph shows footprints in mud by a puddle of water. Another viewer describes the photograph as brown and mucky, idly wondering where Scorup Canyon is. This is not a bad start, but unfortunately for these viewers, it is the end. After this mere glance, these casual viewers shrug and wander away. But wait!

DOI: 10.4324/9781003212607-1

There is so much more to be said. Contemplating this image can lead to some startling insights and may even turn your world upside down. (Hint: Note the inverted reflection in the water.) We will return to this image soon enough. But first, take off your shoes, remove your preconceptions, and prepare to enter *the mind behind the lens.*

Peter Goin and the Photography of Environmental Change: Visual Literacy and Altered Landscapes makes a case for the value of photography in offering constructive new ways to perceive environmental realities. However, to glean the insights that photography has to offer, one must become visually literate. Peter Goin is a contemporary American photographer who for more than forty years has documented human-altered, natural landscapes. Perhaps best known for *Nuclear Landscapes*, Goin has published nineteen books on topics such as arid waters, California agriculture, and post-mining landscapes, and in places that include Nevada's Black Rock Desert, Lake Tahoe, and the US-Mexico border.[1]

This long overdue study of Goin's work, informed by extensive interviews with the artist, collects Goin's most important images into a single volume, while offering a portrait of a brilliant thinker persistently engaged with the nature/culture interface in the United States and beyond. The approximately two hundred photographs featured in this book, culled from more than five thousand photographs that Goin has made, consist of a mixture of published and previously unpublished images from his career. The text serves as a knowledgeable guide to illuminate the trail of eco-consciousness that runs through Goin's work. We join Goin as he investigates American conceptions of nature and culture, documents a history of land-use practices, and challenges paradigms that structure the way we create and perceive landscapes, from wilderness to wasteland. With Goin, we gain special access to military bases throughout the American West, explore underground uranium "health mines" in Montana, thrash through possibly radioactive regrowth on Bikini Atoll, motorboat into secret coves on Lake Powell, watch bulldozers contouring imported sand at Virginia Beach, discover patches of wildness in San Francisco's Mission District, risk our lives in a high-speed car chase to evade Mexican drug lords in Juarez, and contemplate ancient petroglyphs off the beaten path in the Nevada desert. Through howling wind, pounding rain, and blistering sun, we travel into the mind of a photographer as he thinks about how ideas can be visually represented, makes decisions about craft, and experiments with new techniques and technologies.

This book will help readers become more visually literate. Educated people are trained to analyze the written word. We can identify parts of speech (nouns, verbs, adjectives, adverbs, prepositions) and elements of a sentence (syntax, diction, semantics). We are conscious of figures of speech (ambiguity, image, metaphor, hyperbole). We can outline an essay to lay bare its structure and organization. We can place a piece of writing into a genre (poem, drama, fiction, journalism) and a tradition (romanticism, realism, postmodernism). We are introduced to major writers and can talk about influence, how, for example, Stephen King resembles Edgar Allen Poe. We can identify the assumptions of a text as well as notice things that are *not* said. We have learned to analyze rhetoric, noticing how a piece of writing attempts to persuade. In short, required English courses have taught us to be literate.

But we receive little formal training in reading *images*. Unless we have spent time in art museums or taken an art history class, we will not know how to analyze art. When shown a painting, for example, most of us will say either "I like it" or "I don't like it." We won't be able to place it in a tradition, recognize influences, detect stylistic hallmarks of the artist, analyze its composition, or understand its intent. Complicating matters still further, photography is a particularly knotty case of visual representation as it is a relatively new medium, not even two hundred years old. While photography shares many of the pictorial properties of drawing and painting, it is also quite different. For example, photography bears an intrinsic connection to the "real"; the interpretation of any photograph is highly context dependent; photographs are easily reproducible; and the "hand of the artist" in photography is rooted in the visual kinetics of interpretation, rather than the obvious "skill" represented by drawing or painting. Furthermore, these days anyone with a smartphone can take a snapshot. The ease of making photographs tends to belie the layered complexity of the photographs made by artists.

Peter Goin and the Photography of Environmental Change: Visual Literacy and Altered Landscapes braids together three strands:

- *Eco-biography*: Chapter essays chart the evolution of Peter Goin's environmental thinking and share life stories of experiences that formed his character and shaped his outlook.
- *Environmental perception*: The text proposes new ways of perceiving the environment by surveying the arc of Goin's career as a documentary landscape photographer, interpreting his work, and analyzing selected images.
- *Visual literacy*: Accessible lessons discuss key developments and introduce selected practitioners in the history of photography, explain basic compositional principles, introduce the concept of "visual semantics," and demonstrate how to read a photograph.

I have written this book for the person I was before I wrote the book, environmentally conscious but visually naive. I was and am deeply concerned about environmental degradation but also naggingly aware of serious problems with the mainstream environmental movement, both politically and philosophically. Peter Goin's work critiques environmentalism from the inside, not to derail the environmental movement but to make it smarter. His work and ideas challenge and may possibly transform the way that we understand and, indeed, *perceive* and *experience* our environment. In a good way.

Like many people who enjoy landscape photography, I love nature. However, having read Bill McKibben's *The End of Nature*, I am aware that nature as we once knew it no longer exists. Semantically, what we mean and historically have meant by "nature" is itself ever-changing, as Arthur O. Lovejoy demonstrated in 1935 by documenting sixty six meanings of "nature" in the appendix of *Primitivism and Related Ideas in Antiquity*. More recently, entire books have been written about the shape-shifting concept of nature—see R. G. Collingwood's *The Idea of Nature* and Kate Soper's *What is Nature?* In other words, I accept that what we think of when we think of *nature* depends on who we are and in what culture we were raised. Nature is neither a fixed concept nor a stable, external reality. Nature is not simple. Natural systems are enormously complex, confounding our attempts to accurately computer-model them. And our ideas of nature are multifarious, tangled, and contradictory. Yet these ideas influence human interactions with and treatment of an always contested, ever-changing, and already complex nature.

Peter Goin and the Photography of Environmental Change explores new metaphors for understanding the dynamic mesh of nature and culture. Early in his journey, Peter Goin grasped a fundamental truth that he later termed "humanature"—the insight that nature is a human construction, both in the sense that the *idea* of nature is a mental construct and also, literally, on the ground in the form of human-altered landscapes. We fail to comprehend the extent of the human imprint on nature, because we so often don't *see* it. On the one hand, natural landscapes *appear* to be untouched, concealing the human decisions and actions that shape those places. On the other hand, severely altered landscapes are avoided, overlooked, or hidden from view—screened with trees, located in the boondocks, or fenced with No Trespassing signs. Goin's photographs first provoke us to notice the human mark upon the land. Second, his arresting stills paradoxically reveal that landscapes themselves are never static; they continuously evolve. Knowing that, and embracing our role in that ongoing evolution, we have reason for hope. Humans devastate landscapes, admittedly, but we also create them. In bearing witness to the inextricability of human culture and the environment, Goin advocates for the biotic community as a whole.

Coincidentally, Peter Goin's *Humanature* appeared in 1996, the same year that I put the term "ecocriticism" into wider circulation in my coedited anthology *The Ecocriticism Reader: Landmarks in Literary Ecology*. As the nation's first professor of Literature and Environment—at the University of Nevada, Reno—I have striven to inspire and enable the profession of literary studies, that is, academics in English Departments, to address the environmental crisis and contribute to ecological restoration and health. How? Simply put, literature is an expression of culture, and culture influences how we imagine and interact with the natural world. Literature and language are powerful levers with which we can critique, question, and revise how we think, talk, and treat our environment. To change culture is to change nature—and vice versa. Essentially, my work in literary studies as a critic complements Peter Goin's work in photography as an artist.

The idea for this book emerged incrementally. As colleagues at the University of Nevada, Reno, and faculty affiliates of the university's Academy for the Environment, Peter and I have known for years of the consilience of our work. We have been influenced by many of the same environmental historians and writers—William Cronon, Aldo Leopold, Patricia Limerick, Bill McKibben, Gary Snyder, and Donald Worster, to name a few. And we both have a long-standing commitment to and love for Nevada and the Great Basin. For many years, whenever Peter and I happened to encounter one another, we would say, "We should do a book together." Finally, one day we agreed to meet for coffee and discuss possible book projects. As I began to brainstorm for that meeting, I realized that I very much enjoy writing critical retrospectives of an author's career. I find it fascinating to trace the development of a writer's thinking, observe how one project leads to the next, discover patterns and detect persistent concerns in their work, view their work in light of their life, and use my own writing to help others more fully understand their work and appreciate its significance. I see my role as critic as grabbing my reader by the sleeve and saying, "Listen, let me introduce you to someone I think you should meet." In the case of living authors, my essays have benefitted enormously from the opportunity to interview my subjects during the research phase and to have them respond to my drafts. My subjects caught unwitting errors, sharpened the precision of my language, and even added details that would otherwise have been lost to history.

When Peter and I met for coffee, I therefore suggested that we collaborate on a retrospective of his work. Peter was aghast! He recoiled like a salted snail. He is a private person who prefers to stand behind the camera lens, not in front of it. He had intended to recruit me for a book or video project about an environmental topic in the American West, as he has done with several other coauthors. We parted ways that day inconclusively. But Peter agreed to think about my proposal, and eventually he came around. As he confided once the project was well underway, "One of the things that's very important to me is that, well, I don't want my ideas to get lost when I die." It is our hope that this book will sow the *ideas* that his work investigates into the wider world, where like seeds they may bear fruit.

There was only one problem. I am trained as a literary critic, not an art critic. I can write a five-page explication of a fourteen-line sonnet, but am at a loss when it comes to analyzing a photograph. Or at least I was before Peter and I embarked on our collaboration. However, I became convinced that my handicap was an asset. I could ask the innocent questions and represent the naive viewer, whom Peter's explanations could educate. (Fortunately for us, Goin is very articulate and has won several teaching awards.) Peter and I therefore chose an interactive and time-intensive methodology, namely, weekly one-hour taped interviews that took place intermittently over the course of a dozen years, skillfully transcribed by Amy Anderson at Solutions through Research. As months became years, the "interviews" evolved into conversations, and our professional collaboration grew into a friendship, which is why the text of this book alternates between "Goin" and "Peter." I hope that you, the reader, will likewise feel that you are getting to know Peter as a person. I myself developed a deep respect for him—the order in his life, his self-discipline, reliability, sanity and ability to weather the storms of life with grace, his high standards, and especially the extraordinary job he has done as a father raising young girls into beautiful, strong women in the wake of his wife's tragic death. But more about that later.

As we met to discuss his work, I was continually impressed—and at times exasperated—by how long Peter could hold forth about a single photograph. On more than one occasion we devoted the entire hour to a single image. As in Marcel Proust's *Remembrance of Things Past*, one little taste, one tiny detail, could trigger a seemingly endless chain of memories and thoughts. For example, take the "*Footprints*" photograph with which this chapter opens. That image is hitched in Peter's mind to a larger project—the Lake Powell of Glen Canyon—which is linked to memories of multiyear fieldwork—surviving frightful storms, exploring the lake with different houseboats and motorboats, watching the light, itself a memory linked to specific recollections of dancing reflections on the water, evening light warming red cliff faces, campfires at night, insects attracted to candle light, sunrise casting a sharp line of receding shadows. As is true of each project, Lake Powell is linked to thematic concepts—climate change, a new form of beauty, water in the West, wilderness, ruins which are related to specific photographic subjects and techniques—inverted reflections, signs of water even in its absence, evidence of humans, placement of the horizon, depth of focus, use of curved lines, and scarcity of the color green. Eventually, anxious that the tape recorder was piling up the minutes like a taxi meter compounds dollars and cents, I resorted to pantomime gestures: A swirling finger in the air to speed up the narrative, a hand slicing across my neck to cut it off, thumbs up for well done.

Throughout this book, Peter Goin's own words, lifted from our transcribed conversations, appear in quotation marks, and the authorial pronoun tends to be "we." An essay introduces each chapter and develops that project's unique ecological insights and artistic principles. Following the essay, selected photographs are interpreted by in-depth captions that reveal the artist's intent and craft. I have always liked attending poetry readings where the poet prefaces each poem with a story about that poem and then reads it with intonations that vastly help the audience to "get it." In that spirit, *Peter Goin* tells the stories, sets the context, and shares the explanations that grant readers privileged access to Goin's art.

In addition to interviewing Peter Goin, I audited his capstone course on the History of Photography. In both cases, certain phrases rang out again and again. I came to realize that these pet sayings are a key to understanding Goin's philosophy and photography. Here are my top five:

1. **It's not about if you like broccoli or not.**

Artists don't necessarily aim to cater to your existing tastes. Goin does not aspire to produce photographs that you will want to frame and hang above your mantelpiece as a feel-good picture. According to Goin, art is not about making beautiful pictures. Art is a path to insight. Art is one form that thinking takes. Art embodies ideas. Fundamentally, art is the profoundly human act of making meaning. When looking at Goin's work, you need not decide whether you *like* it or not. Instead, ask what the photograph might be trying to say. Ask *how* it communicates. Goin's photographs demand rumination and invite conversation.

2. **"Deliberately wrought visual poetry, disguised as plain prosaic fact."**

Critic Jerry L. Thompson described the "documentary style" of Depression-era photographer Walker Evans this way, and it has become one of Goin's favorite ways of characterizing his own work. As Goin explains:

> There's a presumption that the documentary photograph is without emotional shading. It's without intent. It's without creative interpretation. Of course, that's completely false. The allegations that it's scientifically factual versus lyrical is only a reflection of the accuracy of the illusion.

Peter regards his own work as *both* scholarship *and* art. It is scholarship because his projects require extensive research, reading, and fieldwork. It is art because Peter knows the history and traditions of art and can allude to them, pay homage to them, push them. Furthermore, he makes conscious, artistic decisions about how a given image is framed, composed, developed, and printed. Peter points out that in earlier days, the marriage of science and art was quite common. Consider Leonardo da Vinci. Scientist? Artist? Obviously both, employing reason and intuition, observation and insight, objectivity and subjectivity. Why is that integration such a difficult concept for the modern mind to grasp?

3. **Photographers are authors who use visual language.**

When he collaborates with a writer on a book project, Peter Goin insists on being listed as a coauthor, not as the photographer. To be listed as the photographer is to cede intellectual primacy to the writer and to endorse a reading that regards photographs as merely illustrative rather than asserting their intrinsic worth as primary information, artifacts, and interpretive historical documents. Goin views his collaborations as presenting complementary perspectives on a topic, text and image enjoying equal status and operating semi-independently. He regards photographs as similar to written text, in that the elements and structure of the composed image, like the semantics and syntax of a sentence, interpret the subject in layered and complex processes. Too, photography uses visual equivalents of nouns, verbs, adjectives, and adverbs. Having a heightened awareness of language, Goin is highly sensitized to the words we use to refer to photographs, and he is careful to avoid the language of aggression and appropriation. He trained me to say "make" a photograph instead of "take" one, explaining that this convention spells the difference between a colonialized view of photography and a decolonized view, that is, "taking" from rather than "creating." Along the same lines, Peter taught me to avoid the words "capture," "shoot," "snapshot," "trigger," "load," "aim," and "target," all of which carry violent associations. Preferable alternatives include "compose," "frame," "focus," "visualize," "picture," and "expose."

4. **It's a more complex narrative.**

Photographs are two-dimensional; nevertheless, Goin's images are complexly layered, a complexity that sadly goes unnoticed much of the time. Goin has studied art history. He is a well-read intellectual with a large personal library. He is engaged in the world of ideas. His work often pays homage to earlier photographers, pictorial traditions, painters, and paintings. His work reflects the influence of literature as well, particularly French farce, magical realism, and the dramatic innovations of Bertolt Brecht. "I feel it's important to represent my learning and my knowledge in the images I make," he says:

> but I don't want to walk around with a bat, hitting people over the head. It's not really about that. It's about continuing the dialogue. It's participating in a much broader discussion over issues that artists have dealt with in a long period of time, connecting my work to a rich vein of committed visionaries.

5. **What have you done to make the world a better place?**

After one of Peter Goin's slide shows a man raised his hand and asked, "Why aren't your photographs beautiful, like other nature photographers'? Why do you take pictures of bombing ranges, polluted rivers, chain-link fence lines, and mining pits?" Goin replied that as a citizen of a democracy, he feels an obligation to bear witness to the way people treat, trash, and transform the land, thereby provoking people to think about our place in the greater biotic community. In an interview with Julie Ramos, a faculty member at Université Paris 1 Panthéon-Sorbonne, Goin elaborates:

> Perhaps it is an essential element of the photographic practice, but there are times when bearing witness is a necessary and cathartic act. My children asked me, so many years ago, "Daddy, what have you done to make the world a better place?" And with my work documenting nuclear lands, interpreting the paradigm and construction of nature, and presenting the paradox of how a positive vein of hope resides in post-apocalyptical mining landscapes, and how desert regions are alive and rich and how spirituality resides in arid environments, and . . . exploring the new nature of agriculture, I hope that I offer some answers to that question. Perhaps these answers, these photographs, and the text I and my co-authors offer are not enough, and never will be. But for me, I wish to add my voice to the choir of like-minded voices who speak out, and sing, into and with the wind.

As artist and eco-cultural critic who employs the language of photography, Goin has covered a broad swath of North America in his work, specializing in landscapes of the arid American West, a West that has been bombed, mined, dammed, dumped on, and developed. While Goin's work pays homage to American nineteenth-century geological survey photographers such as Timothy O'Sullivan, William Henry Jackson, and Andrew Russell, from twentieth-century photographers he cites two major influences, Diane Arbus and the New Topographics. Diane Arbus may seem a curious choice, as she made portraits; however, the way Arbus represents people informs the way Goin represents landscapes. Arbus's work features society's misfits and marginalized people, people typecast as dwarfs, the disabled, homosexuals, cross-dressers, children, the elderly, and carnival freaks. Befriending rather than objectifying her subjects, Arbus focuses on the very people from whom polite society turns away; her images disturb viewers by exposing their own insecurities. She presents her subjects as comfortable in their own skin, looking straight into the camera; she dignifies them, revealing their beauty, and expressing the common humanity that binds the human family. Similarly, Goin has embraced and found eloquence in condemned and marginalized landscapes, photographing "wastelands" and places so toxic that they are literally off-limits to the public, America's sacrifice zones. And like New Topographics photographers Robert Adams, Lewis Baltz, Bernd and Hilla Becher, and others, Goin documents human-altered landscapes. Whereas renowned nature photographers such as Ansel Adams, Galen Rowell, and Eliot Porter celebrate the beauty of transcendent natural vistas, Goin and New Topographics photographers cast an unromantic eye on landscapes transformed by human use, implicitly raising uneasy questions rather than soothing frazzled nerves.

Other than the subject matter—altered landscapes in America—how might you be able to tell that a photograph is by Peter Goin? Does he have a signature style? The most obvious ways of identifying a photographer's work do not apply, for Goin has photographed in black-and-white and color; with 4x5, Polaroid, and digital cameras; close ups and panoramas; in bright daylight, night, and everything in between. Nevertheless, in Goin's work, foreground, middle ground, and background are almost always simultaneously in sharp focus. His work is architecturally adjusted for accuracy of perspective. Accuracy obtains in color as well, for he never enhances color saturation. Compositional hallmarks include deliberate use of triangles, circles, curves, and diagonal lines and highly conscious placement of the horizon. As mentioned, his work frequently alludes to artistic and literary traditions. Be alert for elements of tricksterish humor in Goin's work, even if it is so subtle as to require him to explicitly point it out, such as the green rubber snake that he placed inconspicuously in a "garden" of kudzu in North Carolina.

There is one compositional commonality to which Goin cleaves that ironically produces stylistic variability. Goin wants to become *of a place* before he photographs it and only then forge a specific visual approach to represent it. As he puts it, "each project demands its own solution." For example, he chose to photograph the Nevada Nuclear Test Site at high noon, creating a brightness that metaphorically connotes radioactivity. He experimented with a developing technique for his Mesoamerican ruins series that causes the edges of buildings to "bleed" unpredictably, conveying the humidity of those sites and symbolically intimating that architecture is melting back into nature. And he favored vibrant colors in his California agriculture project, evoking the sensuality of food.

We are now equipped to revisit Goin's "*Footprints, Scorup Canyon,*" approaching it as a portal to contemplation. We know that most of Goin's photographs are made within the context of a specific project and that each project explores a unique set of ideas and employs a custom-made visual strategy to express them. "*Scorup Canyon*" tips us off that this photograph belongs to Goin's Lake Powell series, which culminated in the 2016 book *A New Form of Beauty: Glen Canyon Beyond Climate Change*, a fuller discussion of which concludes this book. Lake Powell is a huge reservoir, which began to fill in 1963 when Glen Canyon dam was completed. Although the lake is artificial, people go there to enjoy nature, and Lake Powell is a prime example of the manipulated landscapes that Goin has named *humanature.* But due to years of below-average precipitation and above-average temperatures, by 2014 when Goin made this image, the lake level had dropped some 140 feet from its maximum, exposing once-submerged shoreline, which had accumulated a layer of silt, formed by the settling of water-borne particulates in the ongoing process of landscape change that this photograph documents. In the "*Footprints*" image the mighty Lake Powell appears to be a shrinking puddle, a metonym for climate change. The C-shaped curve made by the pattern of footprints and water's edge directs the viewer's attention to the water, which reveals the horizon in reflection, an inverted image that suggests the inverted landscape itself—what was once a canyon is now a lake.

The story behind this image is that Peter was exploring a remote cove of Lake Powell. In the wilderness all by yourself, who has not fancied that they are the first person to ever step foot there? The mob of human footprints in Scorup Canyon quickly dispelled the fantasy of first discovery. This image playfully riffs on the wilderness maxim to take only pictures and leave only footprints. The dead tumbleweed at the top of the image is another sign of humanature, as the tumbleweed is an introduced species from Russia that has become a rogue weed, ironically symbolizing the American West. A final clue to the ideas embedded in this image is that it appears in a gallery entitled "Flora and Fauna" opposite a similar photograph of bird and animal tracks. Goin upends the dichotomy we typically set up between humans and nature and reminds us that humans are animals—two-legged members of the biotic community, whose habitat we have had a hand in shaping.

Welcome, all, to this study of the art and environmental perceptions of photographer Peter Goin. As the first-ever retrospective of Goin, this research-level book contributes to scholarship and will appeal to academics. Chapters are project-based and chronologically arranged, following Peter Goin on his quest and tracing the evolution of his thought and art. Dates designate when each photograph was exposed in the field—namely, the year the shutter clicked—even if printing and publication occurred in a later year. Appendix A lists Goin's books, archives, and collections.

In the service of understanding Goin's art, this book offers lessons in visual literacy, introducing several key principles per chapter, covering topics such as text-image relationship, allusive photographs, conveying scale, fugitive color, and geometric patterns. In addition, we consider more than fifty themes related to environmental perception, including sacrifice zones, toxicity, borderlands, rewilding, and landscape zoos. My interpretative commentaries are concrete examples of *how* to critically read an image, while Appendix B lays out a step-by-step process for analyzing a photograph. These elements will be particularly useful to students.

As you may have deduced, *Peter Goin and the Photography of Environmental Change* is not an easy book to pigeon-hole. It is a new bird. It derives from literal conversations between a photographer and a literary critic. It can be assigned in classes as a textbook, while at the same time being attractive and artful enough to appeal to a general reader. It is a retrospective of an accomplished environmental photographer, while simultaneously teaching principles of visual literacy that can be brought to any photograph. Innovation inspires change, and we hope that this book will open up fresh possibilities for future scholarship and catalyze other cross-disciplinary, mixed-mode investigations in the environmental humanities.

Our exploration will be driven by intangible questions, such as what is the effect of a line imposed on a landscape? How does a photographer picture that which cannot be seen, such as radioactivity, water's absence, or the relativity of time? What happens when the romance of nature confronts the reality of large-scale extractive industries and industrialized agriculture? *Peter Goin and the Photography of Environmental Change* challenges conventional environmental thinking and causes us first to *see* altered, avoided, and condemned landscapes and then to view them in a new and ultimately hopeful light.

Note

1 Peter Goin's books have won awards and honorable mention in the Indie Book Awards, the American Association of University Press Awards, Graphic Design USA, the Addy Awards, the Association of American Geographers, the Natural Resources Defense Council Eco Award of Excellence, the International Photography Awards, and regional awards. His work has been shown in more than one hundred museums and galleries nationally and internationally; he is the recipient of two National Endowment for the Arts Fellowships; and his archives are housed at The Bancroft Library at the University of California, Berkeley; Duke University; the Center for Creative Photography at the University of Arizona; the Library of Congress in Washington, DC; and others. Goin's video work has earned him an EMMY nomination as well as the Best Experimental Video Award at the 2001 New York International Film & Video Festival. Peter Goin is currently a Foundation Professor of Art in Photography and Time-Based Media at the University of Nevada, Reno.

Works cited

Ramos, Julie. Interview with Peter Goin. *Écoplasties: Art et Environnement*. By Nathalie Blanc and Julie Ramos. Manuella Éditions, 2010.

Thompson, Jerry L. Essay. *Walker Evans at Work*. By Walker Evans. Harper & Row, 1982.

2

BIOGRAPHICAL SKETCH

It's not about me

"It's not about me," Peter Goin protested when we were debating whether to include a biographical sketch in this book. Decidedly *not* of the selfie generation, Goin is a private person who prefers to stand behind the camera lens, not in front of it. Although Peter has made thousands of photographs, he was hard-pressed to scrape together even a dozen pictures of himself to illustrate this chapter. Goin is much more interested in ideas than in what he calls "the cult of the individual." However, he agrees that knowing what drives a particular artist illuminates the work; therefore, with a nod to James Joyce, he reluctantly consented to sit for *A Portrait of the Artist as a Young Man*. This chapter traces the formative influences in Goin's early life that enable us to more fully understand his art, also highlighting the choices and guiding principles that will help the reader appreciate Peter as a person. Subsequent, project-based chapters reveal just enough personal information for the reader to track the co-evolution of Peter's photography and Peter himself.

"I was the rebel in the family," Peter recalls of his family dynamic. The youngest of three children, each spaced two years apart, Peter was thought of, in the family, as more willing than his elder sister and brother to challenge, to think differently and outside the box. Peter Jackson Goin was born in Madison, Wisconsin, on November 26, 1951, of Welsh, French, and Northern European ancestry. His father, Lauren Jackson "Jack" Goin, having earned a Master of Criminology from the University of California, Berkeley, in 1948, landed a job as Chief of the Microanalysis Section of the Wisconsin Crime Laboratory in Madison. Peter's parents met at UC Berkeley when they were both college students. His mother, Evelyn "Evie" Winn Goin, earned a BA in psychology.

For that era—the 1950s—the Goins ran a traditional household, Peter's father bringing home the paycheck, while his mother raised their three children as a stay-at-home mom. Peter describes his mother as gracious, elegant, attentive, and self-effacing, totally devoted to her husband. Jack Goin would spend long hours at work, returning home late, and then sitting down with Evie to a late supper. The children would have eaten earlier. After Madison, the Goin family spent a two-year stint in Pittsburgh, where Jack directed a crime laboratory, after which he became a career civil servant in the US State Department, working as

DOI: 10.4324/9781003212607-2

a public safety advisor in the International Cooperation Administration (ICA), a precursor to the Agency for International Development (AID). Jack Goin's foreign service career carried the Goin family to several countries, relocating every two to three years.

Peter's first sustained memories are of Jakarta, Indonesia (1955–1957). He recalls chasing chickens in the yard, sneaking rambutans from the fruit bowl on the dining room table, sitting at the small wooden table in the servants' quarters with Amah, the cook, and Darmo, the "tukang kebun" (the gardener), and eating rice as the servants did, a single bare lightbulb suspended from the ceiling. Too young to be put in school, Peter picked up Bahasa Indonesian from the house servants, and he became the family interpreter. Peter explains that:

> Part of my intrinsic nature is being within the environment, being of a place. I've always been that way. And you could say maybe it began in Indonesia, because I lived

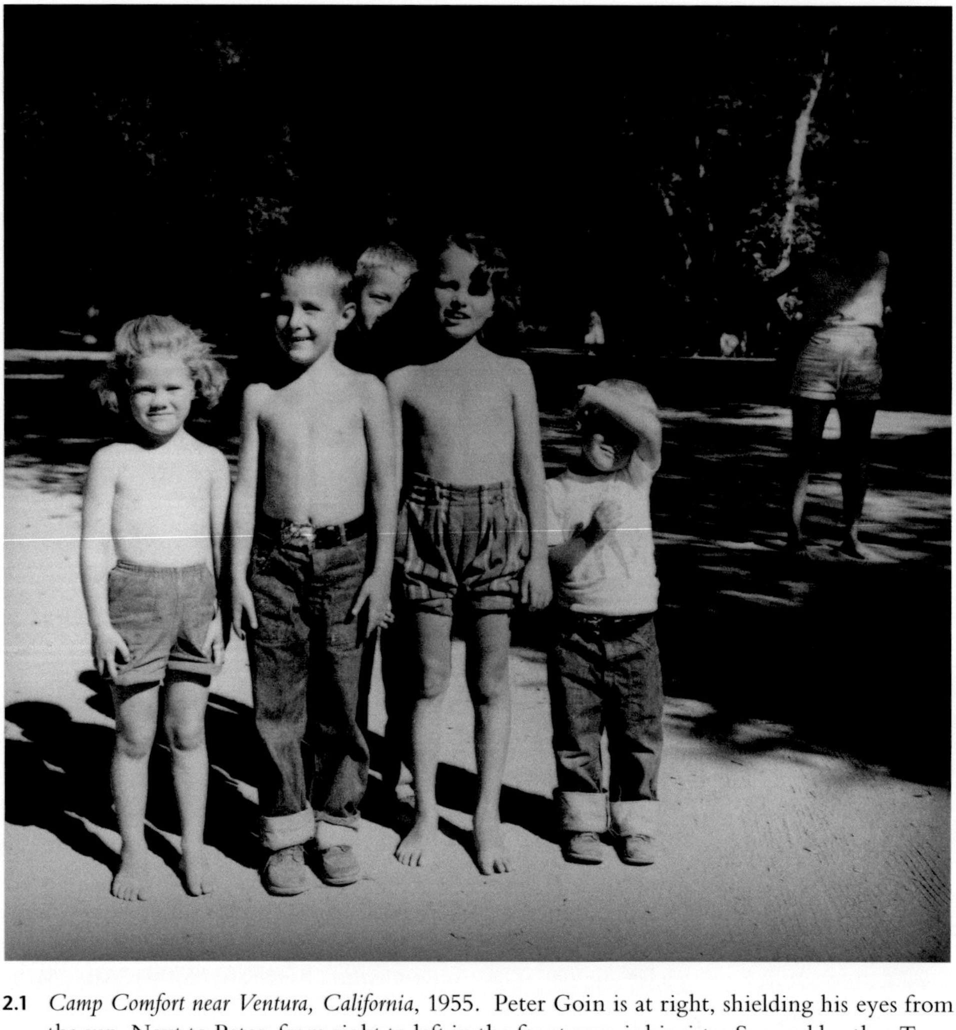

2.1 *Camp Comfort near Ventura, California*, 1955. Peter Goin is at right, shielding his eyes from the sun. Next to Peter, from right to left in the front row, is his sister Sue and brother Tom. The girl at far left and the boy in the back row are family friends.

> with the servants. I was out in the dirt floors and was there when the roosters would crow and the night was ever and always evident. The sky. . . I was never an interior person. And so, I think from that place-memory of Indonesia derives my interest in landscape.

In Indonesia Peter loved watching shadow-puppet shows, and that memory infuses his earliest contact-printed, narrative photogram projects, such as his piece "The Pursuit of Happiness," composed of 120 assembled photograph panels, samples of which are reprinted on the last page of this book.

After the idyllic years in Indonesia, lived mostly outside, the family's move to Ankara, Turkey (1958–1960), taught Peter harsher lessons of *being* an outsider. In Turkey Peter stayed mostly indoors in an apartment building, which did not suit him. Old enough then to go to school, Peter was bussed to an American Air Force school. He recalls those long bus rides, looking out the window at the landscape, but kept from it. The family lived in a Turkish neighborhood but was for the most part insulated from it, and Peter mostly played with his older brother, Tom. They would play kick-the-can for hours. One time when they were playing outside, they were shot at with BB guns. Peter recalls impressionistically:

> I remember not understanding that I'm being shot at, but the BBs are hitting the ground, and they hit a rock, *bee-ching*, and then there's dust kicking up and it's right at your feet and you look and you see them shooting at you and you kind of run. My brother got hit. He still has scars on his leg.

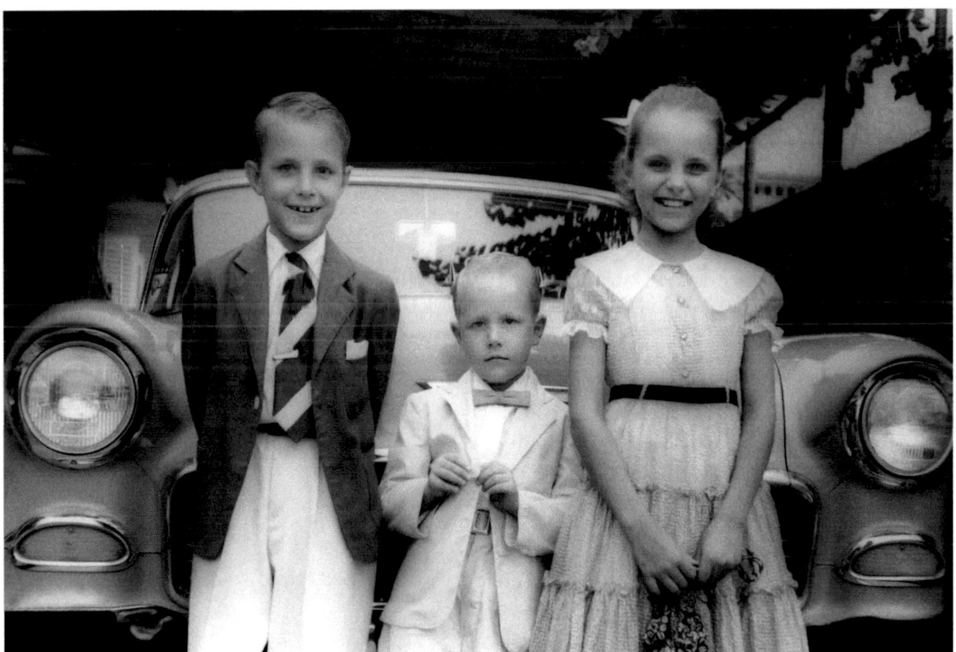

2.2 *Thomas, Peter, and Susan Goin, Indonesia, December* 1955, *ready to go to the US Ambassador's residence for a Christmas party.*

Another time, when the family cat died, Peter and his brother and sister gave the cat a burial and placed a little cross on the grave. The cat was dug up. They reburied it. It was again dug up. The Goin children were then advised to bury their cat in an unmarked grave. "We were so isolated," Peter reflects, "our connections and interactions with the Turkish and Muslim population were usually friction-based. As a kid, I did not understand why."

Brazil, the family's next residence (1960–1962), was a welcome breath of fresh air. Peter loved Brazil and Brazilians, and he happily embraced that culture. The Goins lived in the Vidigal district of Rio de Janeiro, near Morro Dois Irmãos—"the two brothers"—a pair of steep rock formations that tower above the Rio skyline. For a child who was in fourth grade through sixth grade at the time, Peter was given remarkably free rein. After dinner, he and his friend Chris would ride a bus into the city and explore, the only instructions being that they had to be home by 11:00 p.m. when the last bus ran. Peter remembers going to downtown Rio to watch Carnival, playing at the beach, and exploring in the favelas. He recalls:

> I liked the climate. I liked the ocean. I liked the Cariocas—that's what the people who lived in Rio were called. I loved the language. I loved the people. And I have behavioral habits now that come from being in Brazil.

In Brazil it is common to see friends or even mere acquaintances walking down the street arm in arm. While Americans greet each other with a formal handshake, Brazilians give each other a warm hug. Speaking of American mores, Peter observes:

> I realize this American culture has tremendously inbred fears and hostilities about sexual narratives. And that's what does not exist in Brazil, in the same way. I find the Victorian attitudes towards body image and sexuality to be an unfortunate legacy of the American identity.

Peter Goin lived abroad from preschool through sixth grade, roughly from age three to eleven—the developmental years. As such, he fits the profile of a "Third-Culture Kid," described by sociologist Ruth Hill Useem: A child who accompanies their parents into another society and has spent a significant period of their formative years in one or more culture(s) other than their own, integrating elements of those cultures and their parents' culture into a third culture. Peter first encountered the term *third-culture kid* in a 2004 newspaper article by Elizabeth Lowe in *The Orlando Sentinel*, analyzing democratic presidential nominee John Kerry, who, as the son of a Foreign Service officer, grew up in Paris, Oslo, and Berlin. When Peter read the article, he suddenly had a better understanding of himself. Third-culture kids, writes Lowe, must learn to adapt to alien cultures. They become both "a part of" and "apart from" whatever situation they find themselves in. They feel a general sense of not belonging, are prone to being misunderstood, and can be highly focused, energetic, and goal-driven. Peter reflects:

> A significant amount of my identity was formed by living in foreign countries as a child. Even though I came back to this country and went to middle and high school here, I was always an outsider. I've never been able to be an insider, because I don't feel I come from here.

Third-culture individuals have a tendency to have an international perspective and see the broader context, which explains in part why Goin's work examines cultural paradigms. Ironically, the sense of not fitting in and the need to adapt to an alien culture extend to Peter's dexterity. He notes:

> I'm left-handed. Left-handed people have to adapt to a right-handed world in a whole lot of ways that right-handed people don't even imagine. Place settings, where glasses go on a table, where the pour spout on a pitcher is relative to the handle, scissors [and, one might add, cameras]. All kinds of things are designed for right-handed people. At a certain point, you just adapt.

In terms of his career choice to photograph landscapes rather than exclusively people, Goin speculates that his itinerant childhood may be a factor. He explains:

> I have a very deeply curious mind about landscape, about the formation of landscape, about the structure of landscape, about how we interact with it, about sense of place. Maybe that comes from being ripped apart from so many different landscapes. You put a kid in Brazil, which I thoroughly loved and miss, and you say, "OK, bye." And you helicopter that kid to Southern California for three months. *Everything* is different. Then you take that kid and you move them to Fairfax, outside of Washington, DC, and everything's different. Everything's always been different. You're always having to negotiate, OK, where am I? What is this landscape? And you're going from Indonesia—that kind of tropical, intense environment—to arid, Middle Eastern Turkey, and then to Brazil. You just become so aware of the landscape—the stage—of how different it is.

After Brazil, Jack Goin's next assignment was to be Tanganyika, Africa. However, there was unrest in Africa, and the State Department was reluctant to send the Goin family into a conflict zone. Jack was promoted to a position in Washington, DC, in the Office of Public Safety. Peter thus went to public middle school and high school in Fairfax, Virginia. From these years, Peter remembers going on road trips with friends, especially to the Maryland shore, visiting Frank Lloyd Wright's *Falling Water* in Bear Run Nature Preserve, inviting his girlfriend on a day trip to the Blue Ridge Parkway, climbing the stairs inside the Washington Monument, sitting in the gallery during sessions in the US Senate and having lunch in the cafeteria amongst all those famous people, and biking along the C&O canal in the Georgetown district. As would be true throughout Peter's life, he was always out and about, enriching his experiences through adventure, friendships, and a curious mind. In high school, Peter was editor of the yearbook, supervising the photo editors and doing a lot of photography himself. His first camera, a Nikon 35 millimeter single-lens reflex, was a graduation present from his father.

In choosing a college, Peter was attracted to Hamline University in St. Paul, Minnesota, because Hamline had good foreign exchange programs that would allow him to travel. He chose Political Science as a major and thought of becoming a policy analyst for the UN, a job that would allow him to travel to other countries. At Hamline, one of Peter's dorm roommates in his sophomore year, Dennis Albrecht, became a good friend, and the two of them rented an apartment off campus together for the last two years of college. Dennis was a native Minnesotan, and he had a VW bug. The two friends enjoyed making day trips around Minnesota. The flat, farming landscapes intrigued Peter as exotic, especially the birthplace of Bob Zimmerman (Dylan), in the midst of the Iron Range. As they drove through the small Midwestern towns, Peter tried to imagine this rural lifestyle so entirely different than his own experience. Dennis enjoyed assisting Peter on photography, and the two of them have remained lifelong friends, making many trips together over the years.

Hamline University had a 4-1-4 semester system, where a student would take four classes in fall, four in spring, and just one class in January. Peter and Dennis availed themselves of the January block to do an independent study in Mexico. As he explains, "I am in Minnesota. It's freezing. We got to get out, go to a warm country, and explore." That first January, Peter and Dennis studied the structure of Mexican markets, Dennis taking sociology credits and writing a paper, Peter taking fine arts credits and producing a heavily illustrated, photographic documentary—a high-quality, photocopied and bound book that he thinks of as his first publication. Peter regarded his camera not as a hobby but as a research tool that complemented the intellectual inquiry he was undertaking, and he found photography to be uniquely suited for cultural analysis and critique.

Peter's maternal grandparents were Americans who emigrated to Guadalajara for an early retirement and became Mexican citizens. Growing up, Peter visited them from time to time, learned passable Spanish, became comfortable in and drawn to Mexico, and took Spanish courses in college. He has returned to Mexico throughout his life. In their January semester, Peter and Dennis began by staying with Peter's grandparents and studying the market in Guadalajara, from there riding busses to other parts of Mexico, observing other markets, and staying in small, local hotels. The following January, the two returned to Mexico, this time analyzing the development of Mexican parks, examining how those landscapes are designed and constructed. While Peter's friends ribbed him for earning college credit to spend Januaries in a vacation destination, Peter's own view, in his words, is that:

> The adventure-exploration aspect of a self-defined program of study was innovative and addressed from the very beginning a conceptual approach whereby paradigms of cultural design were investigated and documented, albeit from a "learning point-of-view." I was searching for more than "warmth," but a place under the sun, where my thoughts and ideas expressed visually would be valued as *scholarly*, as *important* as a reputable construct for the artist. These experiences also embedded the concept of fieldwork into my practice, which I consider a critical aspect of how I answer the question of the project.

Already one can find the origins of Peter's future work using photography to document human-altered outdoor spaces.

Peter was a serious, self-motivated student. Although he did try drinking tequila one time—"that wasn't fun," he says—he never drank beer or wine and did not smoke or even drink coffee, regarding them as unnecessary distractions. When his friends went out to a bar and bought a pitcher of beer, Peter, often the designated driver, would order a glass of cranberry juice at roughly the same price. He says, "I had a different point of view than most people. I did not subscribe to the party behavior," although he does acknowledge having a large stereo and a "party room," where he and Dennis entertained friends. During his college years, Peter had a deep and abiding concern over the direction of the United States, and he felt that the war in Vietnam was unjustified and deeply troubling, especially when colonialist histories are studied. He participated in anti-Vietnam War marches and he supported the environmental movement that gained visibility with the first Earth Day in 1970. Peter's growing eco-consciousness derived from early experiences and an intuitive empathy for damaged landscapes. During high school in Fairfax, he and a friend secretly cleaned up a section of roadside garbage, mostly cans, for no other reason than it was just sad to see so much garbage. They were stealthy because they thought that they might get in trouble if they were discovered. Then again, traveling in Mexico, Peter was saddened to see garbage littering the roadside habitats. Peter has a lingering memory from his college years of witnessing plumes of smoke billowing into the broad sky from a massive forest fire near today's Boundary Waters Canoe Area Wilderness. The scale of environmental destruction was, in his view, the ecological sublime, albeit weighted more on the "terror" than the "awe" side of the definition. On his day trips with Dennis to the Iron Range country of northern Minnesota, Peter witnessed firsthand the environmental distress resulting from the abandoned mining sites, entire landscapes treated as sacrifice zones and written off as waste places. Such sights left an indelible imprint in memory and planted seeds that decades later would grow into projects, such as Goin's 2004 book *Changing Mines in America*, for which he revisited and rephotographed sites in the Iron Range.

In summary, at Hamline University, Peter applied himself, was politically and ecologically conscious, and earned his BA in 1973 with highest honors. Although he now enjoys a cup of coffee and the occasional cocktail or glass of wine and hosts convivial dinner parties, Peter Goin remains at his core a self-motivated and industrious lifelong learner, dedicated, focused, and determined, sometimes to the detriment of the work-life balance.

In 1972, his senior year of college, Peter availed himself of Hamline's Washington Semester Program and enrolled in an internship in Washington, DC, at American University. Thinking to gain an inside perspective in what really happens in the halls of government, Peter worked as an intern for Congressman Michael Harrington, a democratic representative from Boston. It was a fascinating yet ultimately disillusioning experience. "Everything that was wrong with American society was projected and reflected in the power base," he explains. To the young idealist, the only thing that American politicians seemed to care about was raising money for their next campaign, and politics itself turned out to be less about ideas than about power. Furthermore, with the exception of Harrington, whom Goin found admirable, honest, and dedicated, the political system was offensively sexist. He wanted out. He needed to reinvent himself. Peter wanted to be in a profession that worked with ideas. He found photography engaging and relevant, a Fine Art that dealt with ideas and that was on center stage, as media became increasingly important in the telling of *culture*. As such, photography was a Fine Art practice that Peter felt he could embrace, one that was central to how he perceived the environment.

Goin decided to go to graduate school, and he applied to the University of Iowa's Master of Fine Arts program in photography, submitting some of the photographs that he made in Mexico as samples of his work. To this point, Peter was a self-taught photographer, who learned how to process black-and-white film and make prints from another student at Hamline who was one of the school's newspaper photographers. The University of Iowa was rigorous, regarding its MFA as equivalent to a PhD in the number of credit hours required, plus a thesis. MFA students first earned an MA degree, which Peter used as an opportunity to combine aesthetics with technical knowledge. For part of his MA research, Peter studied old

Kodak manuals and reports, extrapolating from the processes described therein to later experiment with alternative developing processes in his own work. Iowa's MFA focused less on craft and more on critical analysis of art. Accordingly, Peter produced a thesis on the history of representation of the sun in art, taking an international approach that encompassed the Egyptian relationship to the sun and interpretations of the sun in Mesoamerican culture. At Iowa, Peter became friends with Robert Cree, a Vietnam veteran and MFA Creative Writing student who was enrolled in the Iowa Writers' Workshop. In 1976, after graduate school, Goin was employed for eighteen months by the US Government, working in arts-and-crafts facilities for the Army. He then cashed in his retirement funds from that job, roughly $1200, and embarked with Cree in August 1977 on a road trip throughout Mexico and Central America, camping for nearly eight months, an adventure detailed in the next chapter.

We now know what we need to know to understand the major early influences on Goin's art—his independent streak, his international upbringing and third-culture-kid perspective, his abiding interest in landscapes, his sixties-generation eco-consciousness, and his political science and Fine Arts education. As a cultural critic, Peter wanted to use the language and form of photography like an essayist wields words—to evaluate cultural paradigms. A biographical thread will continue to weave its way through this study, divulging elements of Goin's life that help contextualize and illuminate his art. We will follow Peter on his postgraduate travels with Cree through Mexico, journey with him to San Francisco, where he invested six years learning the book trade, and watch him make the decisive career move to Reno, where he landed a professorship in the Department of Art at the University of Nevada, a position that enabled him to support himself, his family, and his photography for more than three decades. The remainder of this chapter will attempt to answer the question, "What do we need to know to understand Peter Goin as a person?" focusing on loved ones and outlook.

Peter Goin began dating Chelsea Miller in San Francisco in 1978 when he was twenty-seven years old. The story of their meeting and courtship is told in Chapter 6 "Urban Wild." For now, suffice it to say that Chelsea shared Peter's interest in art and was an early supporter of his photography. They married in April 1984, and Chelsea moved with Peter to Reno that same year so he could start his new job at the University of Nevada. They had two daughters, Kari, born in 1987, and Dana, born in 1990. It was a difficult time for Chelsea, in that her professional aspirations conflicted with the era's expectations that mothers be singularly devoted to hearth and home. For example, upon arriving in Reno, Chelsea was invited to join the Faculty Wives Club and help with teas. "Oh my," Peter cringes, "that didn't go over well." Peter supported Chelsea's career ambitions and shared equal responsibility for the children—diapers, dinner-making, doctor's appointments, and domestic chores.

Professionally, Chelsea encouraged Peter when he felt compelled to use photography to bear witness, while he supported her desire to go to graduate school and pursue a PhD in cultural anthropology at the University of Nevada. When Chelsea needed to live in Santa Fe, New Mexico, for a year for dissertation fieldwork, Peter helped her make contacts and assumed the primary care role for the children while she was away. Peter speaks enthusiastically of Chelsea's research project on the marketing of ethnic identity. As a foremost rug conservator, Chelsea was ideally positioned to document Turkish kilim designs in Navajo rugs that were marketed as being indigenous Native American. She also examined "Navajo" rugs being woven by the Zapotecs in the Oaxacan Valley of Central Mexico. Chelsea's research on the cultural diffusion of creative design, already under book contract, was poised to significantly challenge on an intellectual level the concept of originality. (Interestingly,

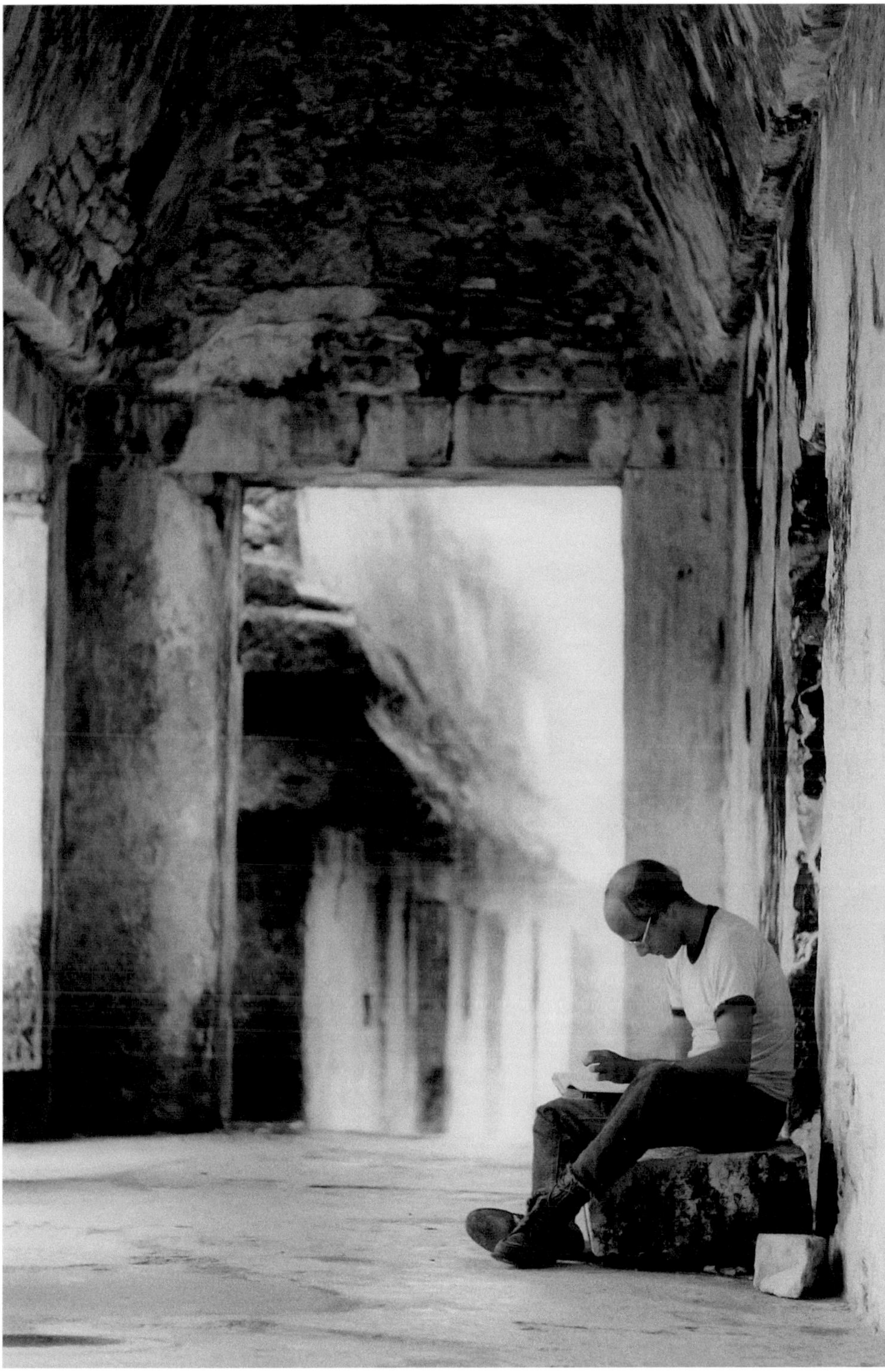

2.3 *Peter Goin, reading Thomas Pynchon's* Gravity's Rainbow, *Palenque, Mexico, on his road trip with Robert Cree*, 1977. Photograph by Robert B. Cree.

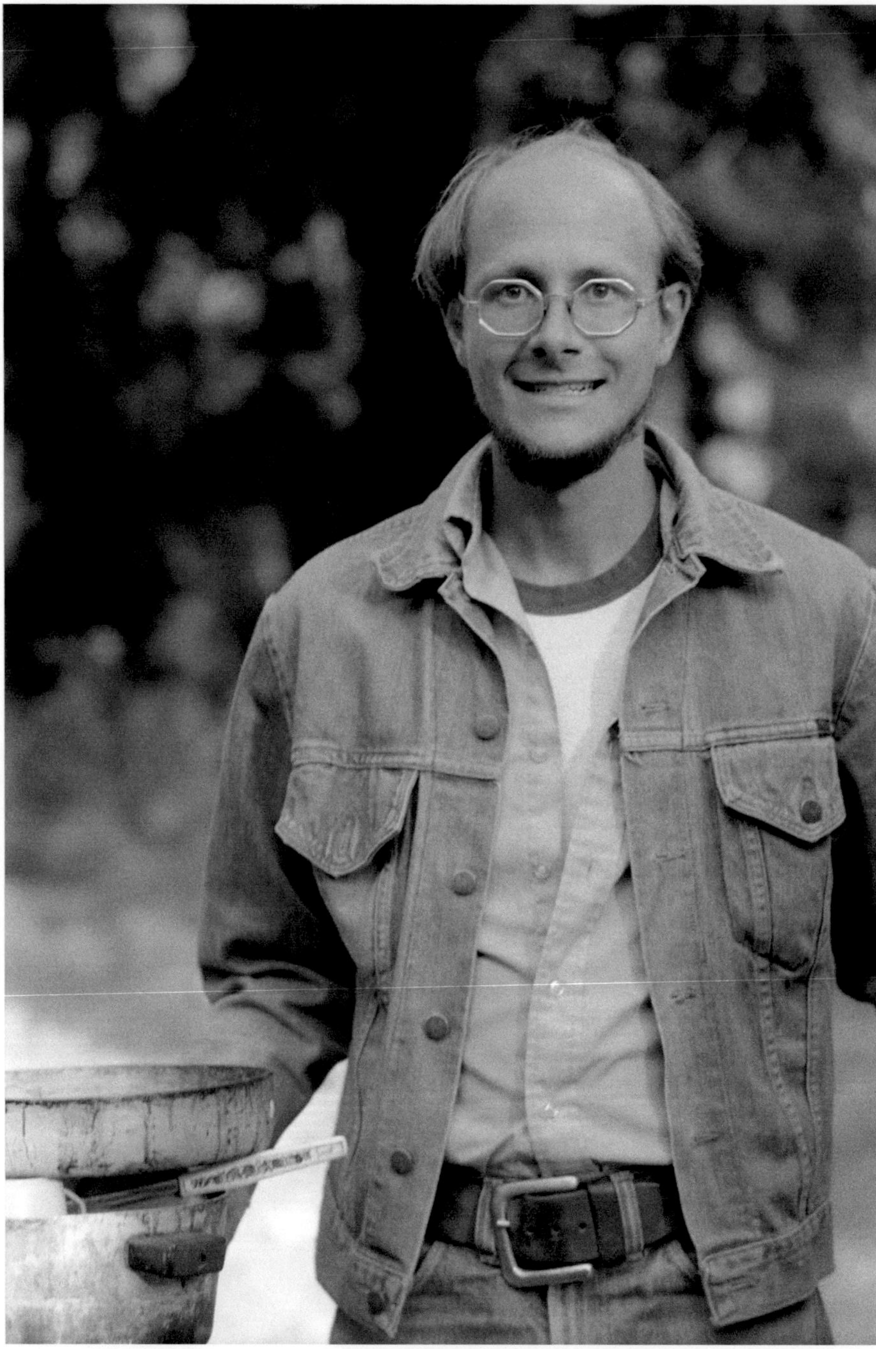

2.4 *Peter Goin, attending to dinner duties at a campsite near Izucar de Matamoros, Mexico, on his road trip with Robert Cree,* 1978. Photograph by Robert B. Cree.

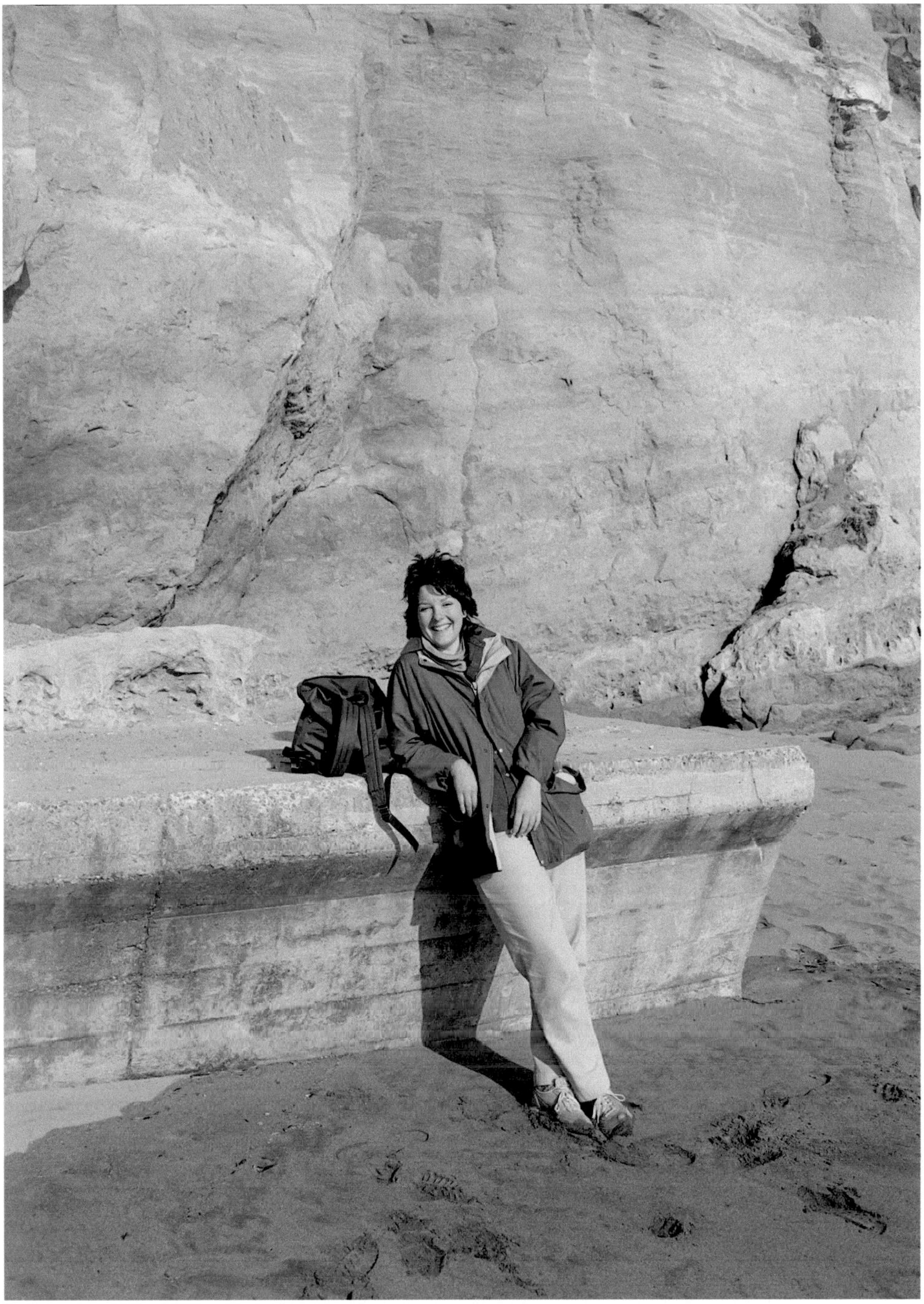

2.5 *Chelsea Miller, Fort Funston beach, near San Francisco, California*, c. 1982–1983.

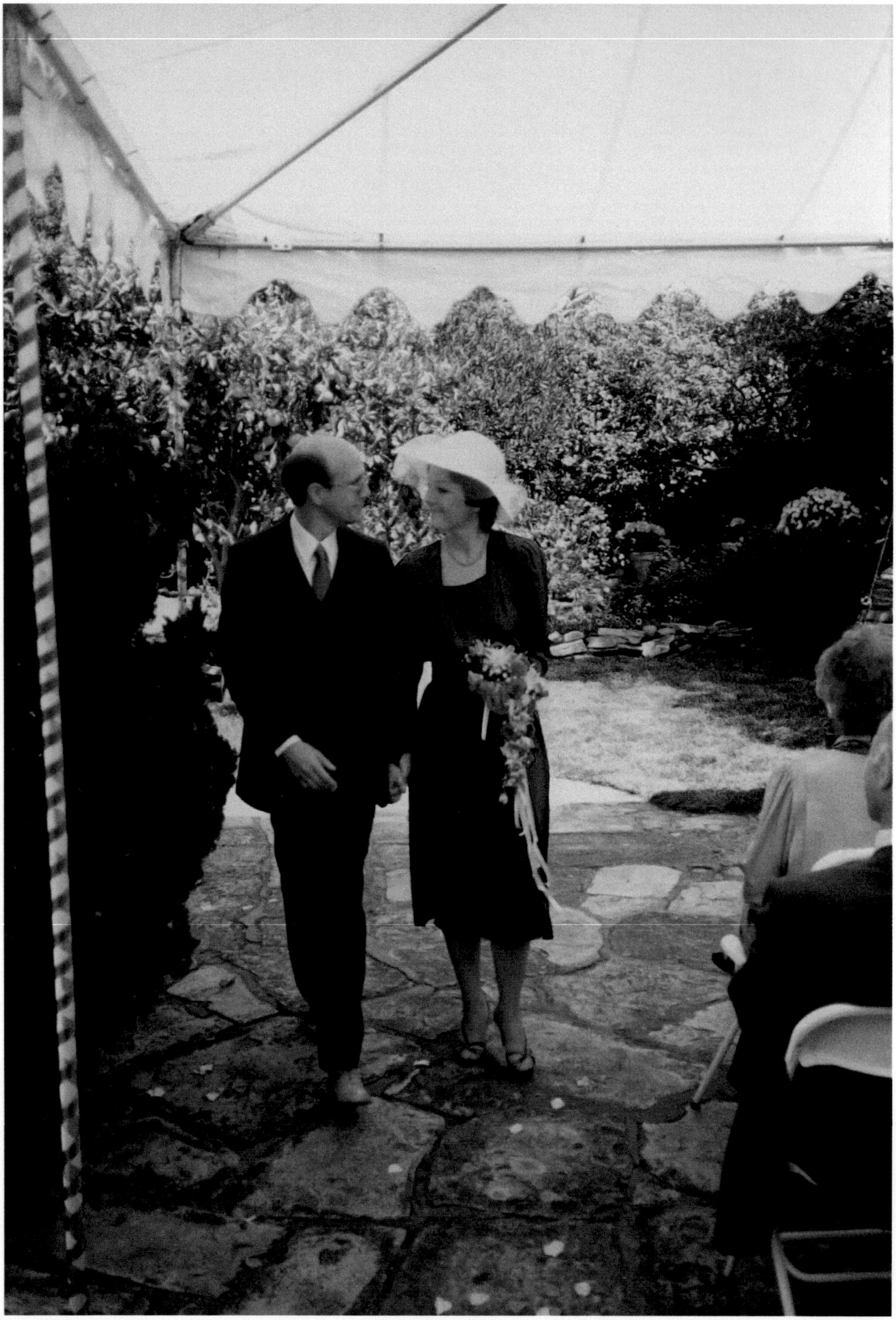

2.6 *Wedding of Peter Goin and Chelsea Miller at the home of Chelsea's mother, Carmel, California, 1984.*

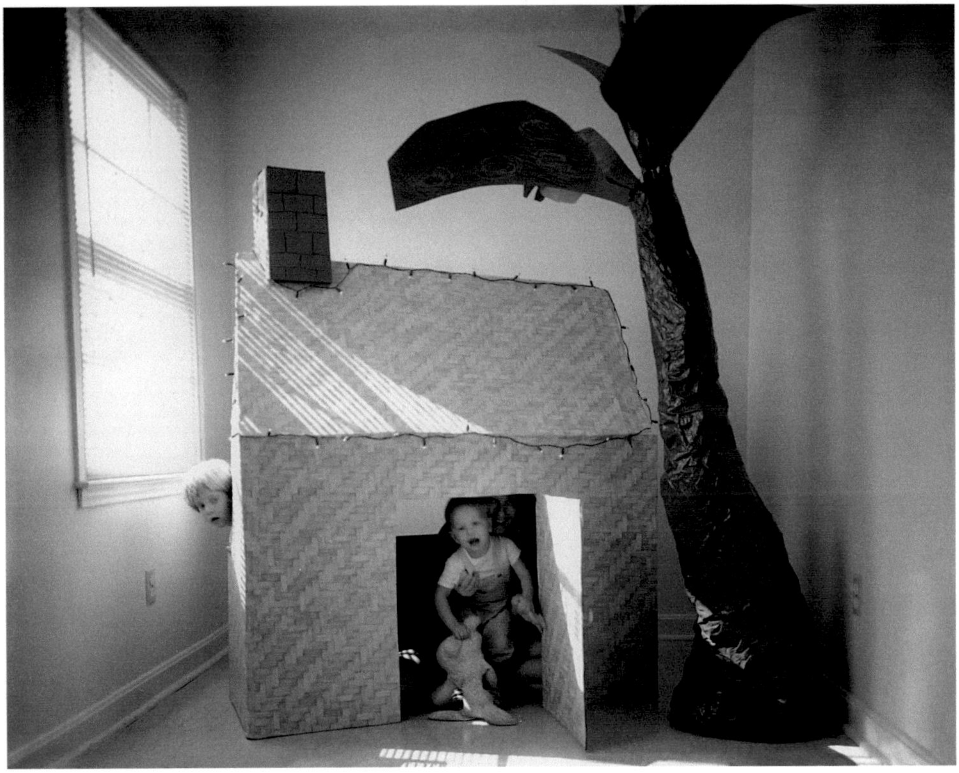

2.7 *Cardboard playhouse in the Goin family's sabbatical-year apartment in Durham, North Carolina,* 1991. Kari Goin at left, Dana in the doorway, Chelsea holding Dana.

as we shall see in later chapters, Peter's landscape photography also challenges prevailing notions of purity, originality, and being "native.")

Fieldwork completed, Chelsea was driving back to Reno to rejoin the family when she got into a fatal automobile accident near Holbrook, Arizona. The date was June 19, 1996. Kari had just turned nine, Dana was six, and Peter was in shock. He made a series of choices. Rather than quickly remarrying, as a well meaning relative advised, Peter chose to remain single and raise his girls himself. He preserved Chelsea's letters, diaries, artwork, and mementoes, so when the girls wanted to learn more about their mom in later life, there was a very detailed archive of objects and writing available. Together Peter and the girls installed a lovely fountain in their courtyard as a lasting memorial to Chelsea. They made tiles for the catchment basin out of clay and had them fired, choosing a glaze the color of Nevada Blue. Peter was and remains fully committed to being a good parent. He confides, "Being a good father is deeply important to me in part because my own father was essentially absentee. I wanted my children to have a dad who was always there, always loving and forgiving, always supportive and encouraging." Eventually he made two important decisions. First, he did not have the energy or time to manage what people thought of him. Second, while his girls were growing up, his social life would be child-centric. Although he did have lady friends, there was simply not enough time to cultivate relationships for himself alone. He enrolled Kari in swimming and Dana in gymnastics, and he would often find himself the

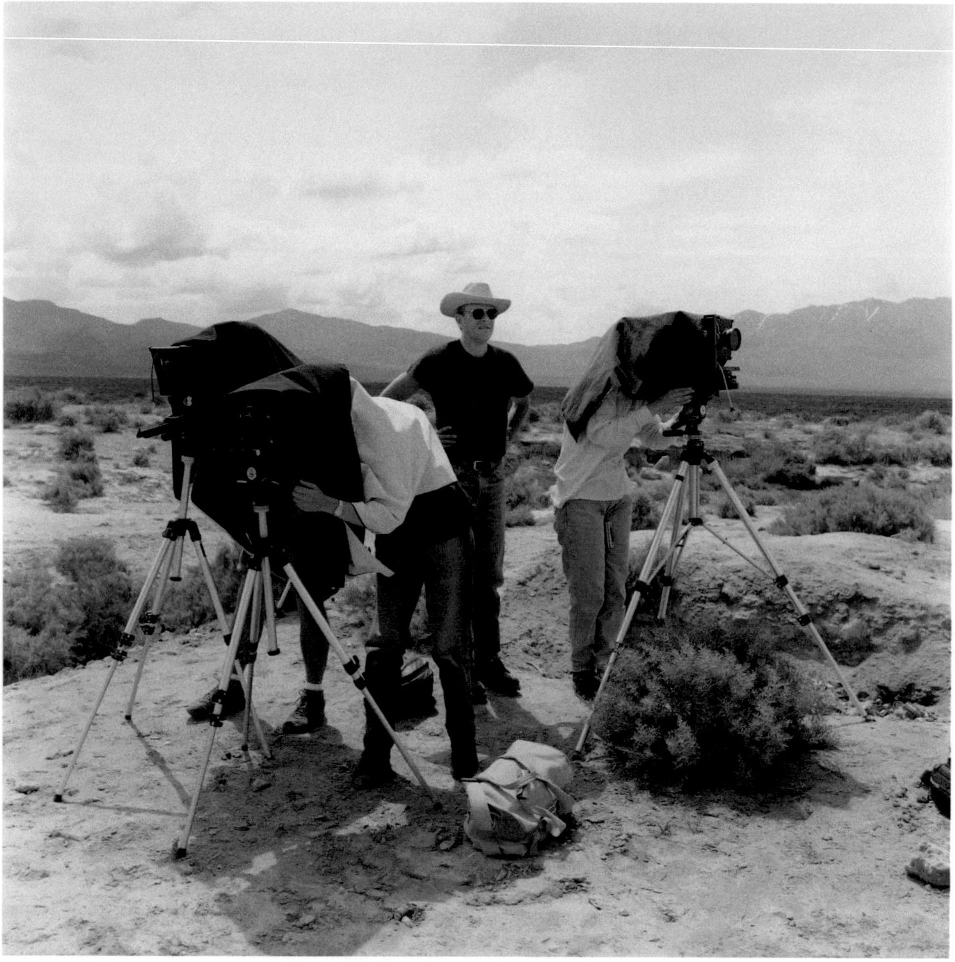

2.8 *Peter Goin, teaching students to use 4x5 cameras in the field, Nevada*, summer 1995. The workshop was entitled "Great Basin Experience: Large Format Photography." Once completed, the students' photographs were exhibited at the renovated historic Opera House in Eureka, Nevada.

only dad among the swim and gym moms, becoming an unobtrusive member of the coffee klatches and burger lunches.

In 1992 Peter agreed to sit for a portrait with daughters Kari (left) and Dana (right) for a photographic artist named Caroline Vaughan, who was making a series of photographs for her "Relationship Project." After Chelsea's death, this photograph took on a haunting quality.

Almost immediately after Chelsea's death, Peter realized he had to keep working. As he explains:

> Pacing around the house isn't a particularly good behavior. If you go to the zoo and see animals exhibiting non-purposeful behavior, it's not healthy for them. I think that animals go insane. I felt that I had to get going. I had to stay mentally and physically healthy for my girls.

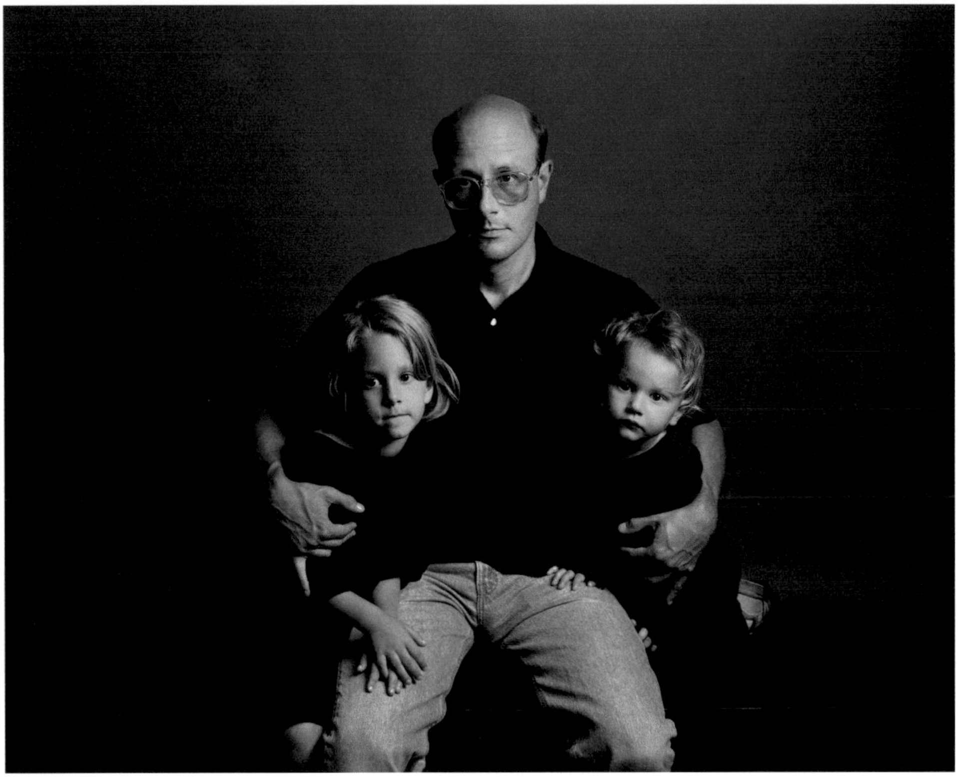

2.9 *Peter Goin and Kari and Dana Goin*, 1992. © Caroline Hickman Vaughan.

And so, Peter continued to work on projects and publish books, fitting in fieldwork around babysitters' schedules and gratefully accepting the offers of relatives to look after the girls for a week or so once or twice a year. His daily schedule was chopped up, but he made it work and became a consummate time manager.

"The life of a widowed, single dad is not all it's cracked up to be," Peter jokes. Nevertheless, he recalls his parenting years with joy and fondness. He loves children and loved being with his girls. He enjoyed the energy that would fill the house when his daughters invited their friends over for sleepovers. And he relished the "why?" conversations that many parents find exasperating. He tells this story:

> When a child says "why?" they're just trying to start and continue a conversation. "Why is that bird sitting there?" one of my girls would ask. I would say, "Well, because it's resting." "Why is it resting?" "Well, because..." I would love this! "It's resting because it's been flying and looking for worms to feed its babies all day, and it's tired, and even parents need to rest once in a while." (I always put it in a morality play.) "Well, why?" And I would say, "Well, because they've gone through a lot." "What have they gone through?" "Well, you know, birds used to be dinosaurs." "Really?" "Yeah, they used to be dinosaurs." "But dinosaurs were big." So now we're having a conversation. "Why were dinosaurs big?" "Well, because they had to reach these big tall trees and get the leaves on the trees, and they had to be big because they

> had to cover a lot of ground." "And why?" "Because they could." "Well, why could they?" "Because they had four legs." On and on and on. We could do this for an hour. To me it was great. I loved it.

To Peter's astonishment, a week or two after their mother's passing, Kari and Dana were having friends over and running around laughing. Granted, the girls had grown accustomed to their mom's absence during her year in Santa Fe, but Peter marveled at the seamlessness of their lives and the way that they lived in the now, a fundamentally healthy orientation to life. As Peter notes, "There is an interesting statement from Buddha, which I will have to paraphrase. But it goes something like this: 'Do not dwell on the past and do not dream of the future. Focus on the present.'" Children embody this principle, and Peter has consciously adopted it, explaining:

> That's a creed I live by. That doesn't mean that I don't make sure I have retirement, and it doesn't mean I don't have good memories, but it's where your consciousness resides. I use a different phrase. For me it's called 'being present.'

Peter acknowledges that Zen Buddhism is a guiding philosophy for him, although his training has been more informal than formal. "It's an unbelievable journey," he reflects, thinking back on the "ridiculous path of overwhelming circumstantial encounters" he has had:

> I would have someone that I would run into and find them very engaging, and they would talk to me in ways that I'm not used to being talked to, more in terms of someone whom I don't know who approaches me conversationally as if they were a mentor. They're giving me advice I hadn't solicited, but they never seem to care whether I follow it or not.

The first such encounter occurred in Mexico, when some Huichol people joined Peter and Robert around their campfire. Even though the two groups did not speak each other's language, through hand gestures and body language Peter absorbed from them "a certain kind of mindfulness and awareness of life forces—the power of the environment, the mindfulness of the air we breathe, the sounds of the environment around you." For example, Peter suddenly pivots mid-conversation from Mexico past to Reno present:

> a hummingbird just flew by. There's a hawk circling off in the distance. The birds in the neighborhood are paying attention to the hawk, and they're moving around. This whole little theatrical performance is going on at this very moment.

Another memorable encounter occurred when Peter was a young man living in San Francisco in Presidio Heights, next door to a Buddhist temple. Peter needed furniture for his bare apartment, and he noticed a comfortable chair that someone had abandoned across the street below his flat, free for the taking. As he and his friend were carrying the heavy chair, a saffron-robed gentleman asked them what they were doing. Thus ensued a pleasantly innocuous conversation about the chair and about Peter's peregrinations in the city. The next morning, Peter read in the *San Francisco Chronicle* that the Dalai Lama had visited San

Francisco as part of his US tour. Behold, Peter recognized in the newspaper's photograph the very gentleman who had engaged him in the street! To Peter:

> the essence of that meeting was the normalcy of two people who are completely different finding amazing common ground over the vernacular. There's something magical about living in the moment. And clearly, the Dalai Lama knows how to do that. That serendipitous meeting had an impact on me. I don't really interpret it in a kind of mystical sense. But if people believe that he is a god, he's a very approachable god. And isn't that great?

Several other Zen-inspired principles serve as guiding lights for Goin. "In the grand scheme of life," he says, "none of us truly matters. There's a larger, almost pantheist environment in which we live, and, in that context, we don't really matter." "We are of spirit," he continues:

> We have an energy within us. It's not just the corporeal identity, but we have a spirit within us. And the creatures who share our world—the hummingbird, the coyote, the raccoons that pass through my backyard—exist in the same spiritual spaces with us.

An important orientation for Goin, who has lived in the darkness of despair over what we are doing to our planet, is one he paraphrases from Buddha: "There are many troubles in the world. Now go plant a tree." "I do believe that we have an obligation to do good on some level," Peter elaborates:

> I think that we have an obligation to find a place of contentment based on how we negotiate our path through this amazing river of life, which is also a Zen sentiment. We bear responsibility as humans not just to take. What we need now are more makers.

To make is a good place to embark on an exploration of Goin's photographic career, the headwaters of which rise in Mexico, to which we now turn.

Work cited

Lowe, Elizabeth. "Understanding a Third-Culture Kid." *The Orlando Sentinel*, 29 July 2004, p. A21, http://unr.idm.oclc.org/login?url=https://search-proquest-com.unr.idm.oclc.org/docview/280170718?accountid=452. Accessed 21 November 2019.

3

ON THE ROAD

Our narrative—and Peter Goin's photographic career—begins with a journey. By age twenty-four, Peter had earned an MA and an MFA in art from the University of Iowa, majoring in photography. Upon graduation from Iowa in 1976, he had studied art history and had mastered the craft of making photographs, which included not only using a camera but developing film and printing pictures in a darkroom. Possessing both knowledge and skill, the question became *what next*? What was he going to do with his education? What ideas or questions would drive his art? In the words of Emily Dickinson, what would be his "letter to the world?" The idea of *travel* called, a siren song deriving from the Beats and Jack Kerouac's *On the Road*. To travel requires funds, which Peter earned by working as a cook in a residential facility for the disabled and finding employment with the US Government in Virginia, teaching crafts at a military base. By 1977 he had saved enough money to embark on a road trip with his graduate school friend Robert Cree. On this extended adventure through the United States, Mexico, and Central America, Peter took field notes and made photographs. But rather than taking snapshots of the I-was-there or we're-having-a-great-time ilk, Peter adopted a journeyman's attitude, using the trip to create a coherent body of work, which he later submitted to the National Endowment for the Arts, winning a prestigious grant in 1982. "I wasn't sure exactly what I was going to be photographing except that I was interested in the concept of prehistory," Goin recalls.

Of all the things to photograph, Goin chose to make pictures of mounds and ruins. These images record an epiphany of sorts, where he came to realize that what appeared to be natural hills covered in vegetation were, in fact, architecture. Even though they looked like nature, the mounds had been created by humans—they were burial sites or garbage piles perhaps, possibly even prehistoric cities, covered over with dirt and overgrown with plants until their original use was all but forgotten and their human-made origins nearly imperceptible. Similarly, the buildings—Mayan ruins—were being reclaimed by nature.

DOI: 10.4324/9781003212607-3

To Goin's imagination, the Mesoamerican ruins "dissolved back into the geomythology of the Central American landscape." *Nature and culture are intertwined.* This fundamental insight, grasped on the road trip with Cree, would become a signature theme, one that Peter would record in images and explore in diverse projects for the next four decades and beyond.

If you Google "Tulum, Mexico" you'll find that Tulum is a beach, an archaeological site, and a town located on the Yucatan peninsula, about eighty miles south of Cancun. "Tulum" means "wall" or "enclosure," the ruins being what remains of a thirteenth-century Mayan walled city. An earlier name for the area is "Zama" or "dawn," as the location atop Caribbean cliffs faces the rising sun over the sea. The literally hundreds of thousands of photographs of Tulum that are posted to the internet tend to fall into two varieties. The first are color images of impressive gray Mayan ruins, ancient and towering above the brightly clad, camera-toting tourists below, who walk on designated paths at the base of monumental ruins that loom above a park-like setting of green grass and the occasional tree. The second set of snapshots are gorgeous photographs of limestone cliffs sheltering a white sandy beach in the foreground, with the cerulean Caribbean lapping in the background and a blue sky promising a perfect day. In this vacation paradise, palm trees, beautiful women, and beachfront cabanas complete the image of a heavenly tourist destination.

In comparison to the vibrant and scenic images on the internet, monochrome photographs of Tulum ruins are, well, "interesting," as a polite dinner guest might remark upon being served a dish whose taste is unfamiliar and not entirely pleasant. *Why make images like these when it is possible to make images like those?* a viewer may ask when comparing the images in this chapter to the conventionally beautiful photographs on the web. Good question. That question of "why this way and not that way?" is one of the essential questions in understanding any artist's work. Goin's photographs of the Mayan ruins at Tulum and Uxmal are black-and-white, a bit smudged, and somewhat reminiscent of an etching or of the photographs that one might find in an old album exhumed from a dusty chest in an attic. But the photographs in the attic would almost certainly focus on people, whereas Goin's images of Mayan ruins are devoid of human presence. Furthermore, they are mostly devoid of background and setting. No Caribbean cliffs, no ocean, no sense of *where* these ruins are located. It is as if the viewer discovers the buildings quite by accident. They have not yet become a tourist destination, but rather a chance encounter with startling evidence of an ancient past. The photographs seem to be more about *time* than about place. But the style of the images, along with the absence of people and modern products, renders the age of the photographs ambiguous. Were the images made in the nineteenth century or the twentieth? The photographs themselves appear to be historical artifacts.

What we see in these representative images from the Mayan ruins series are frontal exposures of apparently small buildings, although their size is difficult to gauge because there are no familiar objects, such as a person or a car, against which to measure their size. In each photograph, the subject—the structure—is centered and invites contemplation. Taken together, the photographs create a typology, an inventory of sites that enables comparison and juxtaposition, variations on a theme. One building appears to be tipping to the right, its roof beginning to erode away, while another structure, perhaps more recent, stands perfectly rectilinear and seemingly impervious to decay . . . except for a fringe of weeds taking root on the flat roof, a subtle sign that nothing is immutable, and that given enough time nature will reclaim this building, just like the others.

In understanding the photographer's art, it is important to be aware of process, both the process of formulating ideas and the process of making the images. Goin is an avid reader, and most of the books he has published include a list of references and suggested readings. His work is thus informed by scholarship and engaged in intellectual debates. But the real *aha* moments that make his work stylistically unique occur in the field, in encounters with a place. It goes without saying of course that a landscape photographer must do fieldwork. One cannot photograph landscapes in a studio. Nevertheless, there is fieldwork and there is *fieldwork*. Some photographers zip around the world shooting pictures without ever staying anywhere long enough to listen to the places they visit. The photographs they produce are glances, composed within a rubric of formalism or another imposed aesthetic. However, Goin's alternative, preferred method is to let his interests guide him to particular places and then, through immersion of body and mind, to let those places transform preliminary interests into insights. While the "drive-by" photographers bring an aesthetic to the landscape, Peter first experiences the place and then develops an aesthetic to represent it. Go(in)'s mandate—to *go* there—has propelled him into some of America's most forgotten and forbidden landscapes. Of his 1977 trip with Cree to Central America, Goin recalls:

> You are bitten by the bugs, you're breathing that hot, humid, wet air, you're scratched by the bristles, you're walking over the same grounds that the Mayans walked. And you go and you say, "This is going to be interesting. What are we going to discover?" What I discovered was how architecture became landscape and landscape became architecture. In Mexico, a landscape that is all green and hilly *could* be a former city. By the same token, you look at these ruins, which are overgrown, and you don't say, "Wow that's architecture." You see architecture melting back into the landscape.

Goin and Cree camped for a week or so near the Tulum ruins. In this field camp, they slept on cots in a tall tent underneath a spectacular grove of palm trees. "The tent was translucent," Peter recalls:

> and I remember like it was yesterday the glow of the light from within the tent reflecting off the palm trees, ocean surf in the background, and a bright moon, casting shadows throughout the grove. Not a soul but us around.

At that time, the ruins were not yet a popular tourist attraction. There was nobody there. There were no signs telling you to keep off, no guards, no ropes. Peter brought along Thomas Pynchon's *Gravity's Rainbow* and Malcolm Lowry's *Under the Volcano* and remembers how he could sit up against a ruin and be absorbed in a book. He read Borges' *Ficciones* and Márquez's *One Hundred Years of Solitude*. Cell phones had not yet been invented, and the two travelers got no news and received very limited mail "general delivery" at major post offices in, for example, Guatemala City, and even then, only a few times in the nearly eight months in the field. They felt isolated in a way almost unknown today.

Once Goin has opened himself to the sensibility of a place and to the subtleties of a subject, he experiments with photographic processes that allow him to visually represent the ideas that he has intuited. Each project has its own visual language; therefore, to understand his art—or any art—one must pay attention to stylistic choices and their effects. In the case of the Mayan ruins, Peter's manipulation of the images during the printing process has an interesting back story, one that ties his work to that of earlier photographic methods and alludes to other practitioners in the history of photography. Goin wrote his MA thesis on alternative methods of toning or manipulating an image outside the straight gelatin silver print. He came across an early turn-of-the-century Kodak manual that talked about introducing a shift in the palette of how silver is treated. The process takes the silver nitrate and bleaches it, which makes the silver vulnerable to other manipulations, and then redevelops it. In the process of redeveloping, the silver will bleed, it will create shadows, it will shift color. This process was abandoned because it is very unpredictable, but Peter embraced the unpredictability and began to manipulate the Mayan ruins images to generate a palette that would reflect the atmosphere of that very intense historical landscape narrative and the climate, which was intensely hot and humid and sticky.

Goin explains, "If you look at a lot of photographs, climate is rarely imagined or presented. We don't look at an image and make a decision about its humidity." What he was trying to do was to use the bleaching process to generate a sensibility about the atmosphere. He observes, "If you compare them to a regular black-and-white print you realize that these images are significantly different. The process creates muted tonality. It gives that kind of age or that patina of a long-time era." The look of patina suggests that the photograph is less a window than an artifact. Goin's chemical manipulation of images when he printed them in 1979 anticipates the kinds of effects more easily achieved nowadays with Photoshop, and his darkroom experimentation takes part in a tradition of manipulations in the history of photography. Photographers such as Man Ray experimented with methods of redeveloping to create specific effects. Altered images such as these are a reminder that photographs are not an objective mirror of reality, but rather an artistic representation of the photographer's perception.

As you scrutinize the three photographs presented here, think about what narrative the work encodes. There are no people. There are no interpretive signs. The images of these ancient ruins with their dark and subdued portals provoke questions that have no answers. What do we really know about Mesoamerican culture? Who are these people? And how did they live? The Spanish conquistadors vigorously eliminated Mayan cultural history and writing in tablet form. These enigmatic photographs evoke that erased, mysterious Mayan past.

■

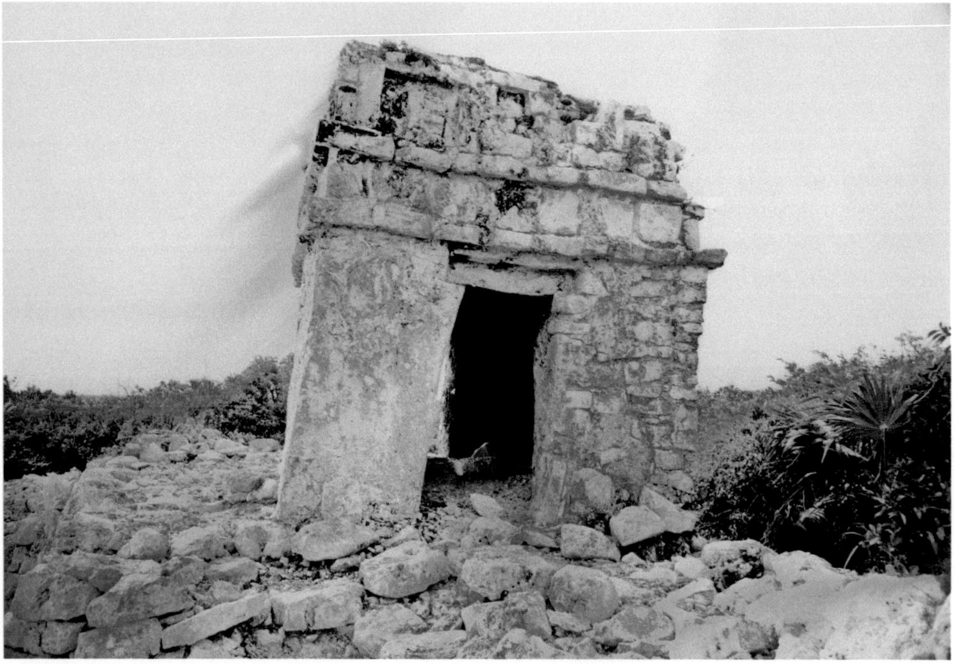

3.1 *Mayan structure, Tulum, Yucatan, Mexico*, 1977.

This black-and-white gelatin silver print shows signs of a manipulated printing process. The special printing process Goin used to make this image creates a downward smear from the upper left side of the building. This otherworldly smear imparts motion to the ancient structure, as if the photograph catches it in the act of tipping over. The smear, coupled with the washed-out look of the shrubs on the horizon, evokes the feel of the air in Central America; both are visual analogs of humidity and oppressive heat. Rising from the tumble of rocks in the foreground, the dark, rectangular portal in the center of the frame invites entry, perhaps into the shadows of the mythic past, into a sacred chamber backlit by a slit of radiance. The nondescript setting and absence of people or modern detritus lend a forlorn mood to this scene, as if this building, at one time bustling with human activity, has been long forgotten, left alone to crumble back into dust.

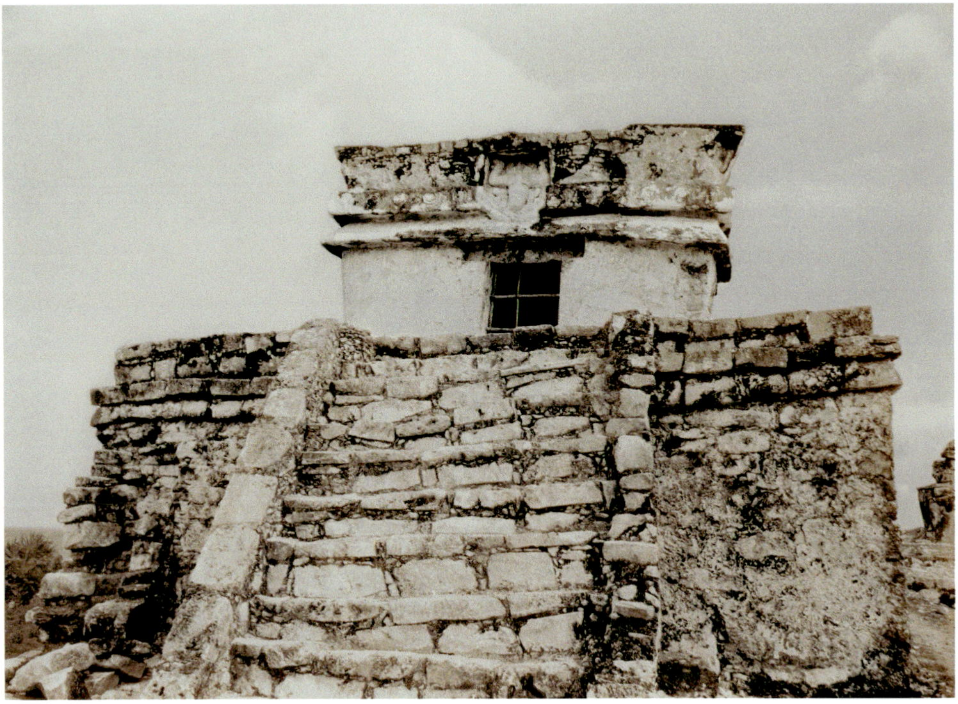

3.2 *Diving God, Tulum, Yucatan, Mexico*, 1977.

Black-and-white gelatin silver print manipulated printing process. Some buildings at Tulum, such as this one, depict a Diving or Descending God, whose meaning has been lost to time. The earthen tint and angular stones that dominate this image create a feeling of antiquity that contrasts with the youthful shapeliness of a bright, upside-down human figure, whose dive, chiseled in stone, is captured in suspended animation for all eternity. The photograph is composed and developed in such a way that the human figure is a visual anomaly, a whimsical intrusion of human shapeliness into an angular place of stone, solemn ruins long ago vacated by human inhabitants. The stark, frontal camera angle and the patina of age created by the unusual printing process of this and other photographs in Goin's Mayan series pay homage to the work of Francis Frith, a nineteenth-century British photographer who used an albumen printing process to document antiquities of the Middle East.

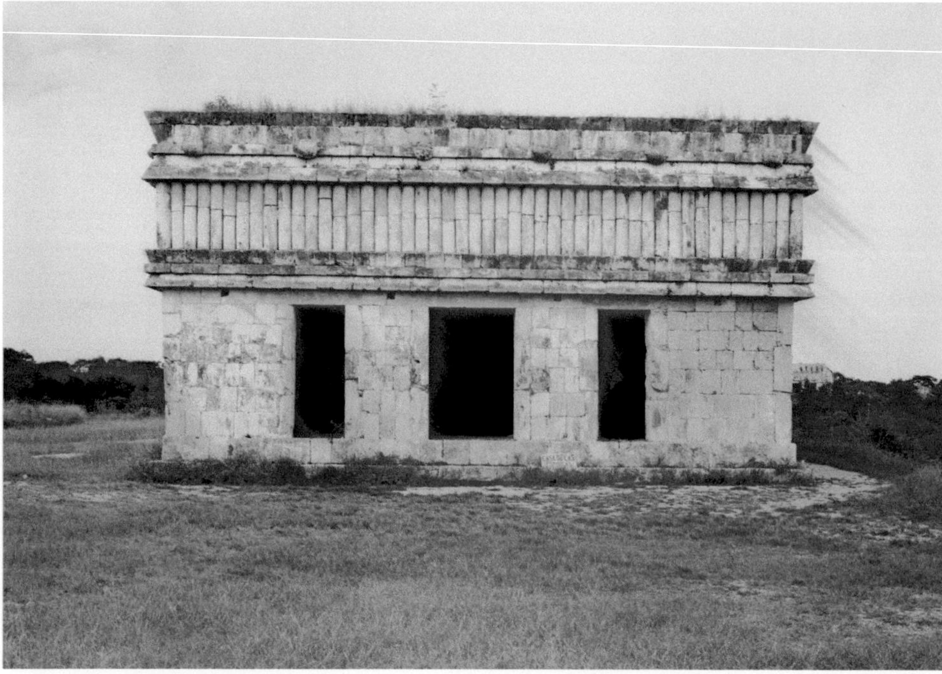

3.3 *Casa De Las Tortugas, Uxmal, Yucatan, Mexico,* 1977.

In the typology created by Goin's Mayan ruins series, this structure at Uxmal stands as relatively intact and precisely rectilinear. The Mayans did not discover the architectural possibilities of the arch, so massive walls were required to support the doors and roofs. Three dark rectangular doors become an element of contemporary abstract art, a counterpoint to the great age and photographically precise documentation of the building. The three black doors may symbolize the darkness that obscures our knowledge of Mayan culture, most traces of that culture having been forcefully obliterated by Spanish conquistadors beginning in the fifteenth century. Indeed, the Spanish name of this building, *Casa De Las Tortugas* (House of the Turtles), cannot be the structure's original name, which would have been in a Mayan language. However, the line of six rounded turtle figures on the cornice is a poignant reminder of the biological life that must have been a meaningful part of Mayan civilization in this region. Turtle shells appear throughout Mayan art, associated with water and with the earth, making them a fitting metaphor of the organic link between architecture and nature in Goin's Mexico and Central American photographs of mounds and ruins.

4

EVOLVING LANDSCAPES

Although Peter may not have recognized it at the time, the trip to Central America with Robert Cree became a point of departure for his life's work. Prior to his formative encounter with Mesoamerican ruins Peter's photographic career could have taken any number of directions. In graduate school Peter had worked in manipulating the physical construction of landscape images and had experimented with ways of representing cultural interpretations of light and symbols of light. Directly after graduate study, he worked with time-based portraits, grids, and atmospheric landscapes. But the Mayan ruins, experienced in that crucible of humidity, light, and heat, imprinted themselves in Peter's consciousness, so much so that when he returned to the United States, he began seeing and then seeking out places where human-made structures could be discerned in the process of dissolving back into nature. In the overgrown environments of the United States, nature's powers of regeneration became visible not in thousands of years—the time scale invoked by Mesoamerican ruins—but in a mere one hundred years or less. What might have once appeared as a static landscape, Peter now perceived as animated, an ever-evolving landscape, a dynamic interface between human-made structures and natural processes. From this point forward, Peter's interests propelled him out of the studio and into the field, where he became a student of altered landscapes.

Initially Peter did not have the time or money to pursue his studies in any kind of systematic way. His first priority had to be supporting himself. The Mexico trip exhausted his small savings, and he took the only job that offered itself, teaching a summer course at the University of Iowa. There he scraped together enough money to make it to San Francisco, where he stayed with his brother until he found a job that allowed him to live independently. The images collected in this chapter were taken over a five-year period, roughly 1978–1982, following his Central American road trip with Cree. During this time, Peter was able to take short trips and photograph sporadically. Peter made these images in pursuit of discovering and refining the syntax of environmental content, seeking signs and symbols that reflected inherent visual truths. He had no inkling that they might one day form a set. Given this half-decade time span and the scattershot geographical reach that encompasses

DOI: 10.4324/9781003212607-4

a wide range of environments—from the northern Minnesota lakes country, to the high deserts of the Southwest, to coastal sites in the San Francisco Bay Area, to the northern Baja Peninsula in Mexico—the photographs from this period evince a remarkable intellectual coherence, all contemplating human-made structures dissolving back into nature. With the benefit of hindsight, we can now label this series *Dissolving Edges*, or *Contemplating Regeneration*, or, as Peter himself thinks of them, *Evolving Landscapes*.

Considering these photographs as a set, we notice recurring images of abandoned roads, mounds, and vegetation. We see repeated patterns of notches, grooves, and contained but empty space. We may wonder where the people are. These places appear to be deserted. We register that these are black-and-white, not color prints. We observe that the subject of every photograph is centered declaratively so that the viewer knows unambiguously what the subject of the photograph is, even if the viewer may be puzzled by what the particular structure or space is. It would be difficult to label any of these photographs "beautiful" in the conventional sense. Rather than evoking beauty, they provoke questions. Why would a photographer take a picture of *that*? What happened here? What is the history of this site? What am I seeing?, the viewer asks.

All these observations and questions are important and provide keys to understanding Peter Goin's art. Abandoned roads are visual evidence of the land's human history. Traces, tracks, paths, scratches, marks of human origin. Who made the roads? Where did they go? Why were the roads abandoned? Without doing research, we cannot say for sure. Yet, like Mesoamerican ruins, abandoned roads are a remnant of human civilization. And, like Mayan ruins, the roads that capture Peter's attention are in the process of being reclaimed by nature. Although we tend to think of roads as places where plant life has been cleared, curiously, in the case of an old stage route in Colorado, the road bed is actually *lusher* than the surrounding land due to the way the road channels water to particular areas. The abandoned road photographs speak to the ultimate variability of human marks on the land, even lines inscribed by so indelible an ink as asphalt. Whether the transitory nature of cultural artifacts is to be interpreted as tragic or redemptive remains up to the viewer. The photographs assert that nature is an actor, not a mere stage for human action.

The overgrown mound in Cambridge, Minnesota (Figure 4.1) is a touchstone photograph in Peter Goin's oeuvre, and the alert viewer will notice allusions to this image in later work. Interestingly, the Cambridge, Minnesota mound image is treated with the same "bleaching" toning process as the images of the Mesoamerican ruins, linking the two experiences visually, and recreating an impression of a charged atmosphere, supernatural forces, and haunting ambiguity. How tall is the mound in this photograph? In the absence of a human figure to provide scale, it is hard to say whether this mound is three feet, thirty feet, or three hundred feet tall. It looms as big as the viewer's imagination is willing to entertain, its significance more about shape and form than about size. The close-up view, denying us the anchor of geographical context, abstracts the mound from the landscape, presenting it as a sculptural form. The composition of this picture metaphorically portrays the mound as voluptuous and feminine, revealing a sensual cleavage on the right-hand side, suggesting breasts or buttocks. Here is an early instance—repeated years later in Goin's Joshua Tree National Park series—where Goin depicts the landscape anthropomorphically, in sensual terms.

In this case, sensualizing the mound runs counter to a more orthodox environmental interpretation of this place. In reality, the mound is a waste heap, a tailings pile from the industrial mining of iron ore that took place in Minnesota's Mesabi Range in the late nineteenth and early twentieth centuries. A more conventional environmentalist photographic commentary on the Mesabi Range might select images that depict devastation—large-scale earth removal, visible scarring, unsightly slag heaps, and gaping pits, all of this a visual editorial decrying the environmental devastation of mining. Peter's photograph, however, shows the tailings pile in the process of rewilding, transforming from a waste heap into a hillock, a site for outdoor recreation and contemplation of nature. It's rather hopeful.

In the Cambridge, Minnesota mound as in the other photographs in the Evolving Landscapes series, vegetation is nature's active agent in rewilding human-altered landscapes. Shrubs take over an abandoned stage route in Colorado; grasses and perennials poke up through cracks in the pavement in Arizona; vines enshroud a vacant building in northern Mexico; and grasses soften the edges and cloak the former uses of defunct military structures at Point Pinole, California. Over longer time horizons, earthquakes and weathering join forces with vegetation to unmake human structures, as we see in photographs of the cracked and heaved up road at Thornton State Beach, California, and the disintegrating adobe wall at Fort Bowie, Arizona.

Given Peter's interest in the delicate balance between exploitation and the land's ability and tendency to regenerate, one might legitimately ask why he uses black-and-white film rather than color to make this series. Wouldn't greenery be an effective visual cue of nature's regenerative force, a symbolic contrast to the whites and grays of the built environment? Perhaps, but as Peter explains, there are both technical limitations and conceptual considerations that led him to choose black-and-white. First of all, in the late 1970s and early 1980s it was very challenging to work in color, not least of all because color printing was not commonly in the hands of the maker, but rather only possible through commercial printing businesses. At that time color film produced saturated colors, Kodachrome-like and almost garish. The colors at that time did not yield the subtlety that Peter thought the landscape required. But, to go to the heart of the question, why black-and-white?, it is because Goin's interest is about structure and form, not about how verdant and lush the landscape is. Black-and-white film is at its best when representing form. For example, consider the landforms made visible in the three photographs made at Point Pinole, California. What are we seeing?

Point Pinole Regional Shoreline was an abandoned munitions plant on the San Francisco Bay that today is a regional park managed by the East Bay Regional Park District. However, when Peter lived in San Francisco in the late seventies and early eighties, Point Pinole was not yet a park. It was an abandoned, post-industrial landscape, exactly the kind of no-man's-land and forgotten space that fascinate him as places to explore and sites to photograph. The park's interpretive brochure provides an answer key to the curious landforms documented in Goin's Point Pinole photographs:

> Scattered oddly-shaped foundations and raised earth berms, sunken bunker areas, half-buried railroad ties and wooden pilings give you a tiny glimpse of life at Point Pinole during the 80 years that it was home to four explosives manufacturing companies. From 1880 to 1960, 2 billion pounds of dynamite were manufactured here; the danger of explosions from the dynamite and nitroglycerine manufacturing processes kept the point isolated from the public and otherwise undeveloped.
>
> *("Point Pinole Regional Shoreline")*

The Point Pinole pictures favor muted tones in the middle values. This choice bucks a dominant trend in landscape photography at that time to produce dramatic images with rich blacks and bright whites, such as those made popular by Ansel Adams. Of the high contrast style, Peter observes, "It's almost like we need to say everything is lovely. Or everything is ugly. Everything is—forgive the pun—black or white." But he wants to look at subtlety. In the English language, we say "well, that's a gray area," admitting room for differences in judgment. Printing in a gray-scale palette here indicates a willingness to suspend judgment; furthermore, the gray-scale shows the detail and the richness of the sites and allows the structures and the information to be presented. In the same vein, the flat light, typical of the Bay's foggy shoreline, produces subtlety and enhances minute detail. The images record an ambulatory experience, walking through the landscape, reflecting, looking here, over there, and registering how we *see* the shapes, forms, and geometry.

The Point Pinole images comprise a typology, where the viewer is invited to contemplate the varieties of the ways that these landscapes have formal elements that are creating a visual edge, a dissolving between the human and the natural. Even though the landscape may be completely human-affected, there is an emergence of the natural. As Peter says:

> It's subtle. It's not four cloves of garlic. It's a pinch. I offer these visual poems, if you wish to call them that, as objects of contemplation. But there is no imperative that you actually contemplate them. By the same token, if you want dramatic color, well, you know, that's like asking for candy. This is not candy.

It is ironic that one of the Bay Area's historically most dangerous sites—a munitions plant—because it was off-limits to other kinds of development, has become one of the Bay's most "natural" shorelines. "You realize the tremendous irony of these industrial, war-based landscapes emerging as parks and restorative areas," Peter observes. "If *these* landscapes can be restored, there is some hope for the rest of what we do."

While the photographs taken after Peter's trip to Mexico may have been somewhat inchoate conceptually, they were nevertheless fully self-aware compositionally. Having earned an MA in art and an MFA in photography, Peter was by this time thoroughly steeped in art history and theory as well as in the history of photography. Furthermore, he had become an accomplished craftsman, skilled in camera use, principles of composition, and film developing and printing techniques. In this period and beyond, as he selects vantage points, chooses camera settings, and frames the view, Peter possessed a trained artist's heightened awareness and control of geometry. Geometry refers to the pattern of lines that comprise a picture. When you look for lines in the photographs from the Evolving Landscapes series, you begin to notice structural aspects of the pictures that would remain subliminal if your focus were fixed exclusively on subject matter.

Photography is a visual language, in which the lines in an image are analogous to the syntax of a sentence, forming the underlying structure that organizes the semantic elements in the message. Once you begin looking for them, you will find lines everywhere—lines along the edges of roads, lines of walls or bridges, lines along the corners of buildings, lines made by boards or stairs, horizon lines, telephone poles, vertical trees, cracks in a road, the curved lines of trails or tracks, pipelines, curbs, lines between planes of surfaces in the topography of the land, and, of course, the four lines that form the rectangular frame of the photograph itself. If you take a piece of paper and draw a sketch of the prominent lines in an image, you will begin to discover its underlying geometry, which in some cases becomes a kind of signature or compositional predisposition of a particular artist.

In the Evolving Landscapes series, particularly noticeable are triangles formed by the converging edges of a road that recedes into the distance, by the walls of an abandoned building or bunker, or by the pyramidal shapes of berms or hills. Berms and hills form what art critics call *positive* space, namely solid objects. In the lingo of electronics, one would call these *male* spaces, while *female* or *negative* spaces appear as vessels or enclosures. In the Evolving Landscapes series, the mound in Cambridge, Minnesota is a positive/male space, while the walls of the Fort Bowie Ruins in Arizona designate a negative/female space. As structural descriptors, "negative" and "positive," "female" and "male" are neutral terms, carrying no hierarchy of value or additional connotations. Recognizing positive and negative spaces in art helps to establish structural links between images; for example, notice how the Fort Bowie Ruins' negative space closely matches that of the third Point Pinole photograph in the gallery below (refer to Figures 4.4 and 4.7). Not only does negative space supply the structural foundation for both images, but in both cases there is an irony: The walls of the container-like spaces enclose *nothing*. The very fact of the presence of walls creates an awareness of absence, a ghost of whatever structure or life the space once hosted. Similarly, becoming conscious of Goin's use of positive space helps us see the close structural parallels between the mound in Cambridge, Minnesota (Figure 4.1) and the hill at Thornton State Beach Park in California (Figure 4.8). Once structural mirrorings bring the two photographs into juxtaposition, one can discern the vehicle tracks that crawl up both mounds—inviting reflection on the impacts of recreation on the land—as well as observe the forces of nature at work in the images, revegetation in the case of the Cambridge mound and landslide damage, "un"-paving the road, opening up cracks for future revegetation at Thornton State Beach.

Also salient structurally are intersecting lines, X's where the arcs and rays of a composition converge. These intersections focus energy and attention, structurally bringing together possibly divergent subjects, such as the place where the line of a wall meets the contour of a mountain, or where the line of a road finds a row of shrubs, or where a defunct military installation dialogues with the trunks of eucalyptus trees. As in a poem, in which word rhymes highlight thematic affinities or opposites, so, too, the arrangement of lines in an image can be used to underscore thematic concerns.

And so, this collection of seemingly miscellaneous images from apparently random locations, made within a low-budget, five-year period, is knit together by confident craftsmanship and by each image's conceptual resonance with Peter's memory of Mesoamerican ruins. But if the images are in this sense rooted in memory, they also become the seeds for future growth. In this Evolving Landscapes set we can identify many places, subjects, themes, and questions that will carry forward into future projects.

Peter will investigate the subject of ruins explicitly in a US context in his Erie Canal project in the next chapter. He will maintain photographic ties to California, making a series of neighborhood views as a pedestrian in San Francisco and, much later, co-authoring a field guide to California agriculture. He will return to Mexico many times in his life, arduously conducting a photographic survey of the two thousand-mile US-Mexico border, completed in 1987, and creating a whimsical *Árboles Urbanos* (urban trees) gallery twenty years later. He will remain fascinated with mining landscapes, culminating in *Changing Mines in America*, and with military land uses, which became his particular focus in *Nuclear Landscapes* and in his work on the Bravo 17 and Bravo 20 bombing ranges. And only hinted at in the Evolving Landscapes images of Colorado and Arizona, Peter will become a photographer laureate of the desert West, examining concepts of aridity and documenting the changing contours of the New West.

Even in this early portrait of the artist as a young man, we can begin to characterize Goin as an American landscape photographer whose images document the imprint of human land use even as they refrain mostly from picturing people. As an artist and a thinker, Peter gravitates toward altered landscapes, post-industrial spaces, and abandoned places. The fact that these places are abandoned means that they are for the most part unseen. Peter's photographs invite us to *see*, to notice, and to contemplate the dynamic interface between nature and culture. While he may opt to evoke the sensuality of his subject, he rarely chooses to aestheticize it in a conventional sense. Rather, the images are "alternatively aestheticized," in style and grammar consonant with the post-New Topographics' visual design and concept, discussed further in the next chapter. Rather than catering to ideas of conventional beauty, Peter wants to provoke questions. His work gathers visual evidence of human history written into the landscape and of natural forces reclaiming human structures. In the age of the Anthropocene, where signs of human impact are everywhere to be discerned, Goin asks, "What happens when you put a line upon the land?" "How can we picture time?" "What is wild?" "Can you commune with nature in an altered environment?" "What do ruins say?" The work examined in this chapter, then, hints of tantalizing visual investigations still to come.

■

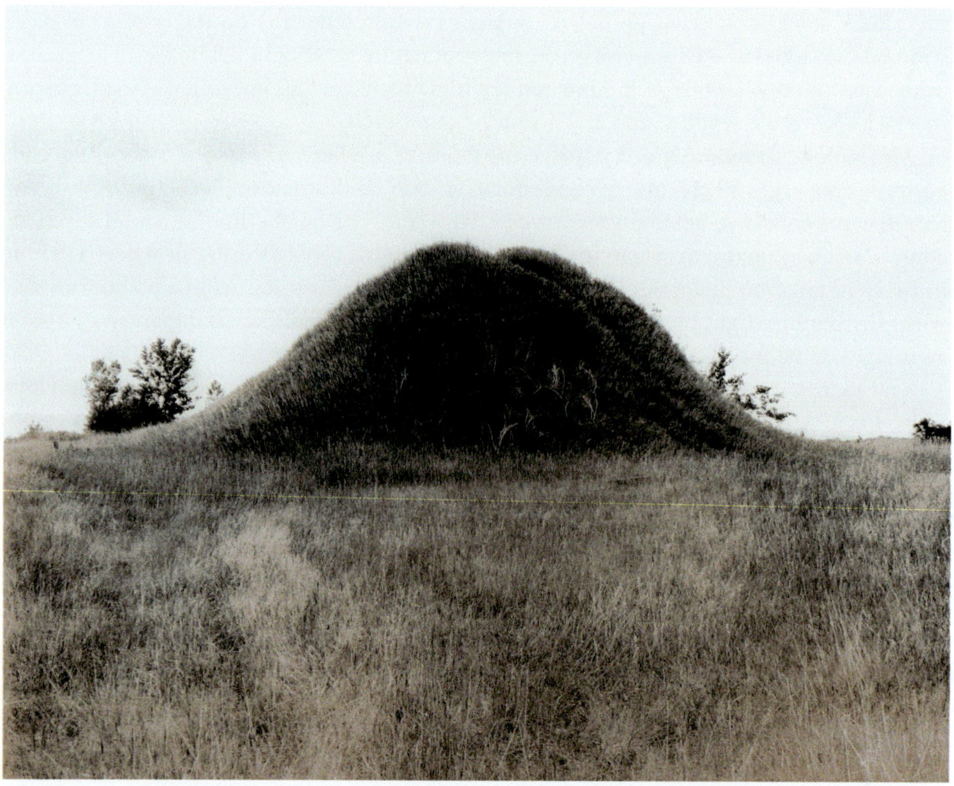

4.1 *Landfill mound near Cambridge, Minnesota*, 1981.

The unusual "bleaching" printing process used to make this image produces a subtle, earthen, muted red tonality, creating a sci-fi quality, as if there is a supernatural force at play just beyond the visible range. We might name this animating force Revegetation, an energy vibrating between nature and culture. Like Wallace Stevens' "Jar in Tennessee," this Mound in Minnesota becomes a point of reference in Goin's imagination, a site he would return to and rephotograph a quarter century later.

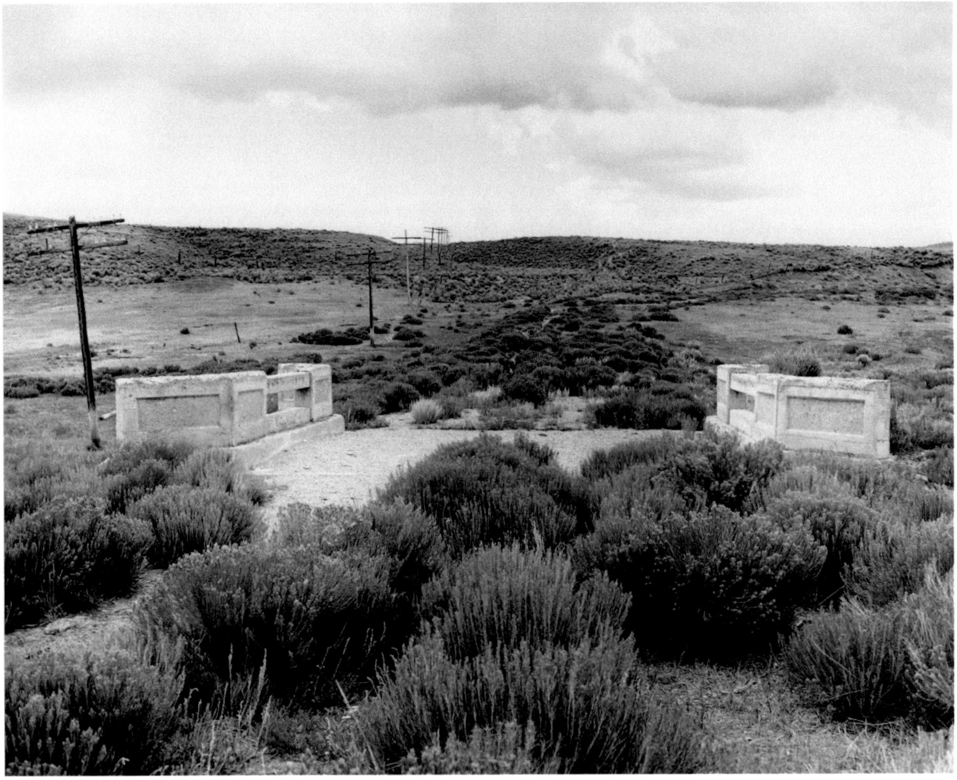

4.2 *Overgrown road, originally a stage route, Colorado,* 1978.

Ironically, this roadbed sports denser shrubs than the surrounding undisturbed land, perhaps because of the way the road channels water drainage. Note the bridge to nowhere, inviting speculation, and the row of remnant power poles, marking an abandoned electrical line. The relics in this photograph suggest past movements of people, water, and power. "I am trying to deal at some level with the assumption that human presence is irrevocable," explains Goin. "Well, it may be. But it may not be. You have to start talking concepts of time."

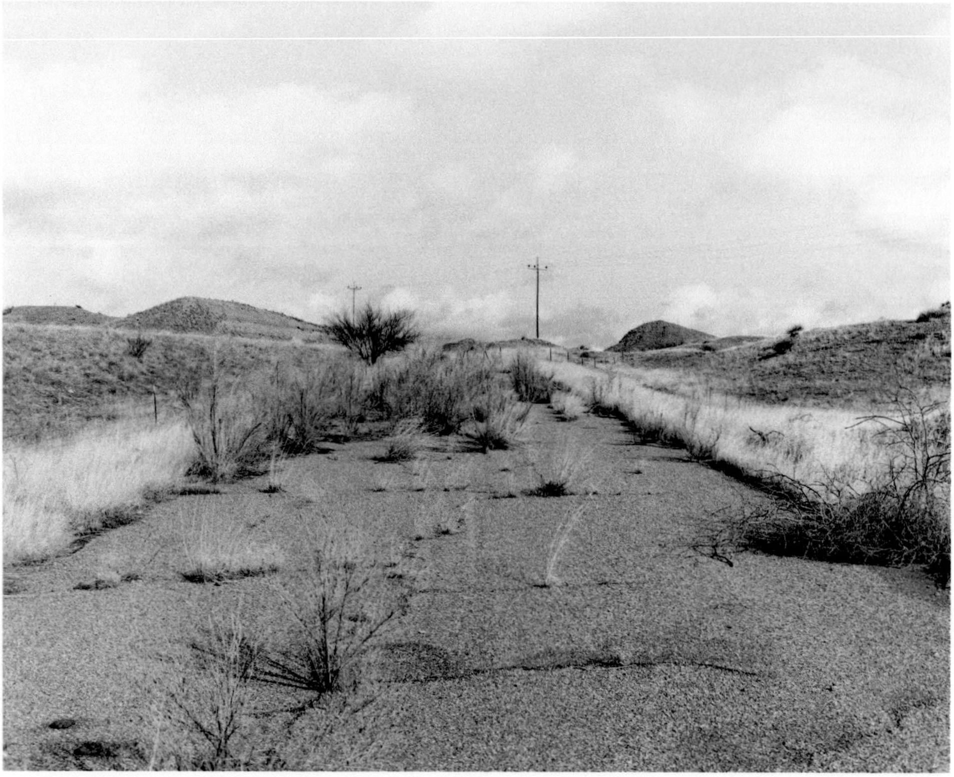

4.3 *Overgrown highway, old Hwy 83, south of Tucson, Arizona*, 1982.

This forgotten highway is arrested here in an in-between state, its visible past written in pavement, its future an open question. Note the geometrical similarity of this image to the photograph of the overgrown stage route in Colorado (Figure 4.2), both pictures creating a triangle, whose base is the lower edge of the frame and whose sides, paradoxically, are the parallel edges of a road. Vertical lines in the image—fence posts and power poles, so ubiquitous throughout the West that we hardly notice them—are further evidence of human designs on the land.

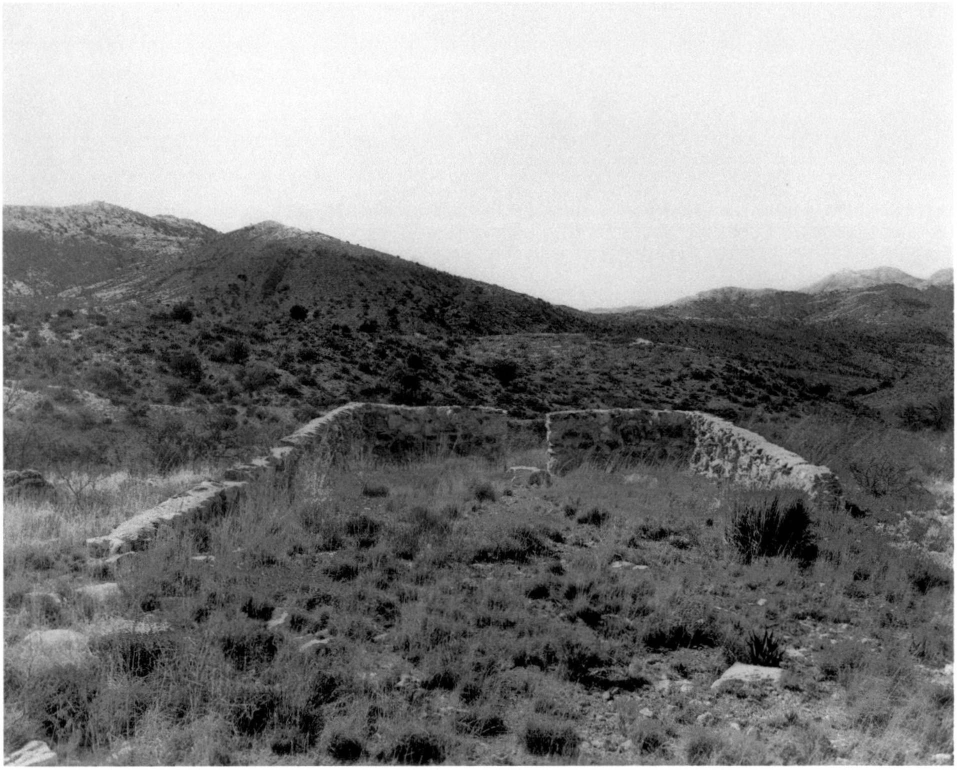

4.4 *Fort Bowie ruins, Arizona*, 1982.

The Fort Bowie ruins in Arizona belong to the same visual stream of consciousness that originated in Goin's contemplation of Mesoamerican ruins, discussed in the previous chapter. At Fort Bowie, Peter became intrigued with the way that adobe "melts." We don't usually think of walls as melting, but unless these walls are preserved, one day they will entirely dissolve into the landscape, dust returning to dust. Fort Bowie was a base for the US Army's actions against the Chiricahua Apaches during the Civil War era, forcing the surrender of Geronimo in 1886 and the banishment of the Chiricahua Indians to Florida and Alabama. Today, Fort Bowie is a National Historic Site. The geometrically negative space of the three walls creates an enclosure that encloses . . . nothing, a poignant reminder that the history of the West is not only one of conquest, but of serial abandonment.

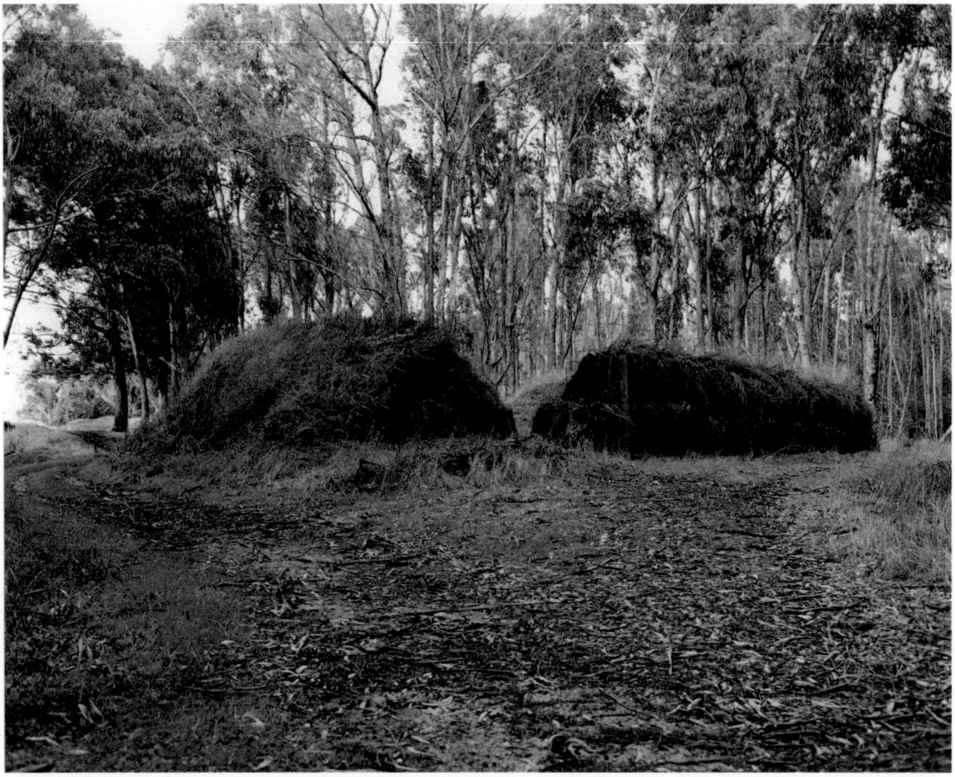

4.5 *Overgrown berms, Point Pinole, California*, 1979–1980.

Point Pinole, now a Regional Park near Richmond, California, on the East Bay, used to be a dynamite manufacturing plant. When Goin visited there in the late 1970s and early 1980s, the site was a virtual no-man's-land, no longer a munitions factory and not yet a park. For Goin:

> This is an affirmation of spring, of the regeneration of the human-nature world. Here you have this landscape that is essentially there to produce weapons of destruction, and yet by virtue of this, in some weird, twisted way, it's emerging as this environmental refuge where people can find solace and can find open land, where these human industrial landscapes are being overgrown. People go out here with their dogs. They go out there to have picnics. The walk itself becomes an interesting tale.

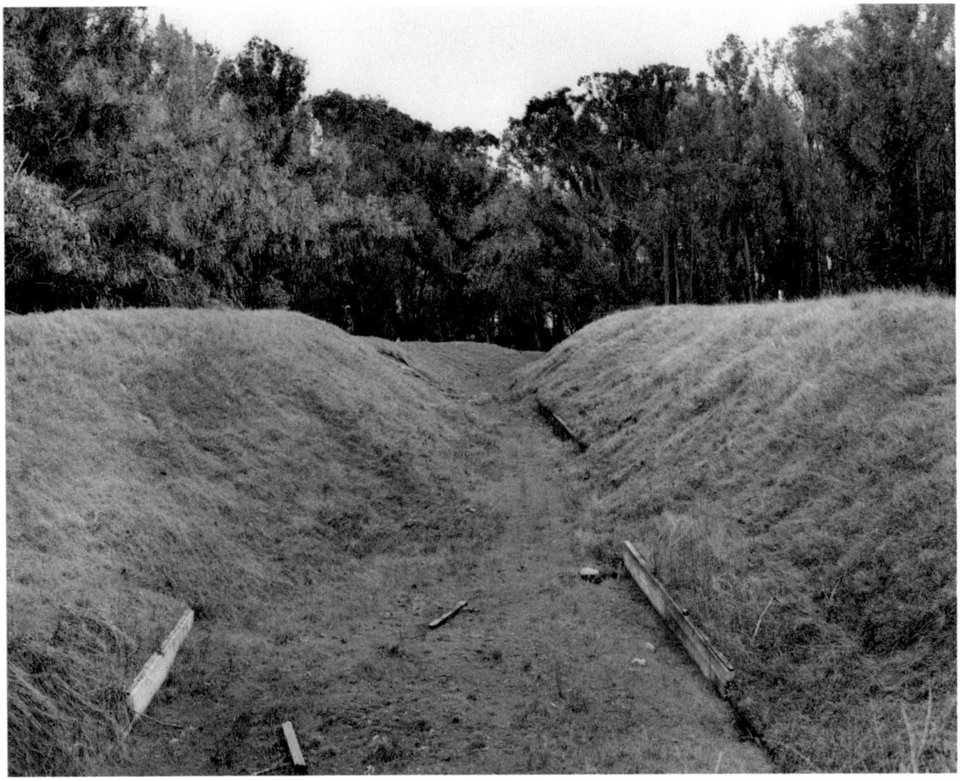

4.6 *Berm structure, interior, Point Pinole, California*, 1979–1980.

At Point Pinole, one is at liberty to walk through the landscape and try to guess the past uses of human-created landforms. When Goin visited Point Pinole, Mesoamerican landscapes were in his mind's eye; in both places, architecture is dissolving back into nature. In this and other Point Pinole images, you may recognize eucalyptus trees, which are exotics (native to Australia) but have become strongly associated with a California sense of place, posing yet more puzzles about the nature of nature.

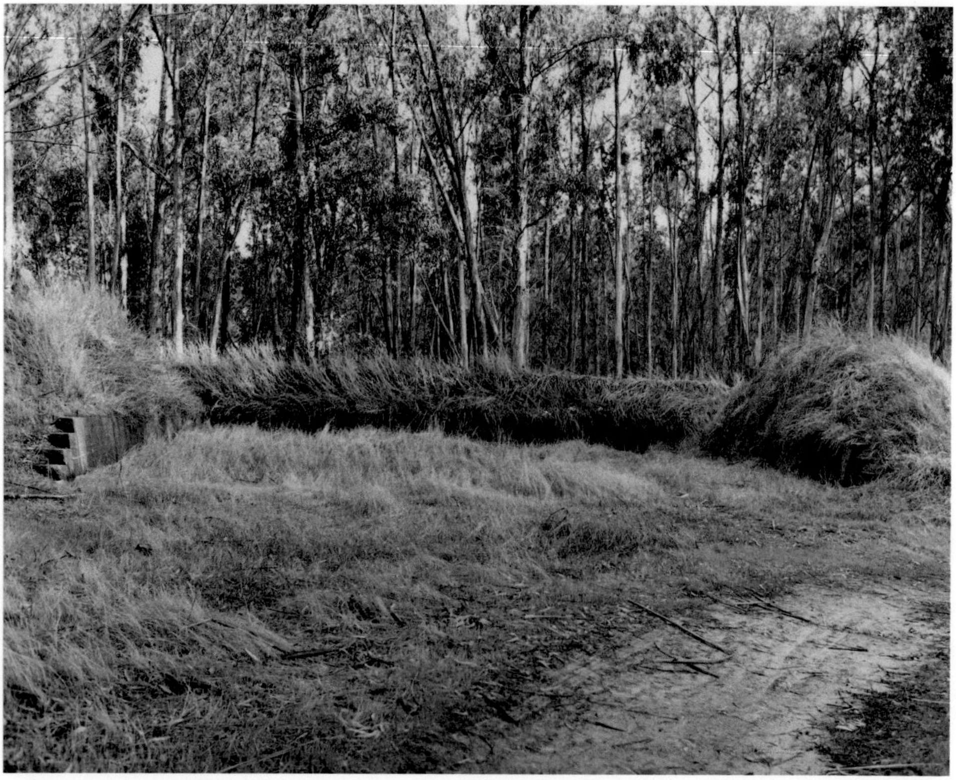

4.7 *Three walls, berm structure, eucalyptus forest, Point Pinole, California*, 1979–1980.

The muted, gray-scale in this photograph invites contemplation, a bit like a Zen koan. Observe that the shape of this enclosure—the foundation of a former building—mimics the negative space delineated by the crumbling walls in the Fort Bowie Ruins photograph in Figure 4.4. In both cases, the photographs evoke buildings and times past. When Goin made these photographs, these landscapes were simply "anonymous" and did not have plaques, trailside notes, or any interpretive information. For those walking or hiking, unless they were previously informed of the site's historic dynamite plant, Point Pinole appeared to be a generic industrial site, abandoned and left to nature.

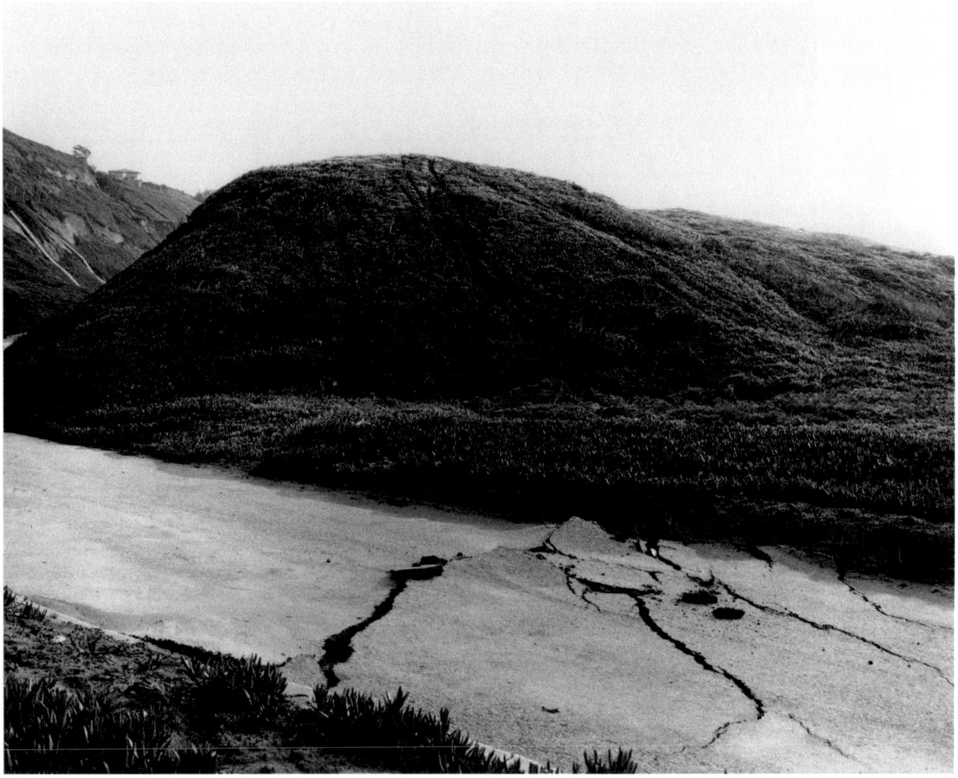

4.8 *Thornton State Beach, California*, 1980–1981.

Thornton State Beach at Daly City, California, is an extraordinarily dynamic landscape, defying human efforts to establish permanent structures. The high bluffs and slumping landforms are the result of a massive landslide in the 1906 San Francisco earthquake. In the 1930s access roads and a picnic area were built, but the 1982 El Niño storms created more landslides and upheaval, buckling the road and forcing closure of the state park. By 2018, there was no longer beach access due to erosion. As of 2020, the property is operated by Daly City as Thornton Beach Vista, an overlook platform with views out over the Pacific. Signs warn visitors that the area is unsafe and not to go beyond the cordoned off pavement. Goin's photograph, whose dark mound references his image of the Cambridge, Minnesota mound (Figure 4.1), documents a volatile coastal landscape, becoming overgrown by ice plants (native to South Africa), while pipelines at left are stranded and exposed and a house sits precariously on the verge of un-solid ground. If we were to set this scene to music, its high contrast exposure coupled with the jagged crack lines in the foreground might call for something dramatic, such as Beethoven's Fifth Symphony.

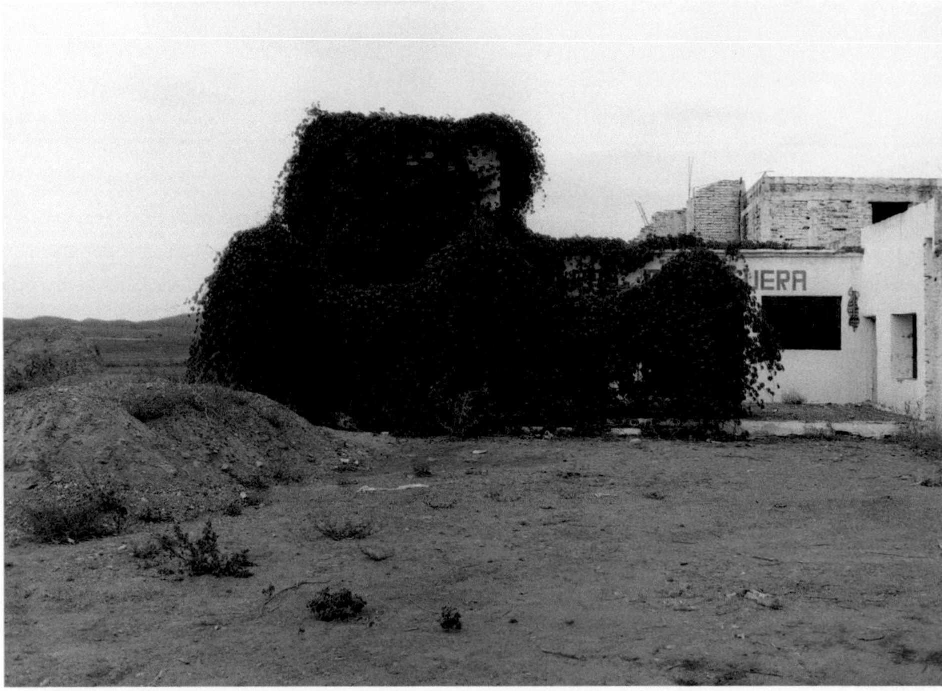

4.9 *Overgrown building, Mexico*, c. 1982.

When Peter made this photograph of an overgrown building in northern Mexico in the early 1980s, Mexican laws allowed people to begin construction projects before full funding was in place, and a building was not taxed until its completion. These factors tended to result in unfinished buildings. Peter observes that in the United States, an abandoned, unfinished building is a symbol of failure, but in Mexico it represents hope. It was started, and it could one day be finished.

Work cited

"Point Pinole Regional Shoreline." Brochure. East Bay Regional Park District, 2004, https://www.ebparks.org/parks/pt_pinole/. Accessed 9 January 2020.

5

AMERICAN RUINS

The year is 1981. Peter is nearing thirty. He is a photographer, yet his job managing a large bookstore in San Francisco leaves him little time for trips. Furthermore, the dominant modes of landscape photography leave him little room to maneuver, or so it seemed to him. At that time, mainstream work in landscape photography included sublime wilderness paeans by Ansel Adams and his followers, breathtaking Alaskan vistas by Robert Glenn Ketchum, exquisitely sculptural forms by Edward Weston, and emotionally evocative scenes by Minor White. All of these deservedly famous artists created highly aestheticized renderings of their subjects. In a word, in the 1980s, landscape photography was mostly about creating images of great beauty. While not averse to beauty himself, Peter was much more interested in exploring other ideas, in using the language of photography to ask questions about our relationship to nature. He did not feel compelled to photograph beautiful wilderness spaces, but rather to use the camera to document and invite contemplation of waste places, abandoned or overgrown structures, no-man's-lands, and ruins.

The Mayan ruins were still on Peter's mind when he was granted an artist's residency at Light Work in Syracuse to spend a month photographing upstate New York in general and the Erie Canal in particular. Light Work, founded in 1973, is an artist-run, nonprofit photography organization that supports emerging and underrepresented artists and those "who are not a part of photography's mainstream." The fit was a good one. According to records kept by Light Work:

> During his stay as Artist-in-Residence Goin photographed the evidence of the changing landscape, making images that present the unusual balance between man's constructions and nature's adaptations. Goin concentrated on scenes that had been altered by man and were in the process of being reclaimed by nature.

Peter's work on the Erie Canal is the first example of what would develop into a disciplined, project-oriented career. It is as if Light Work's characterization of his residency became internalized as a mission statement that defined Peter's focus and carved out a space to move within the field of landscape photography. Goin was not alone; in fact, the kind of work he wanted to do had already been named, and he knew a few of its practitioners.

DOI: 10.4324/9781003212607-5

New Topographics is by now a well-known term among photographers and art critics, but that label may be unfamiliar and perhaps even misleading to the lay public. The term was coined by William Jenkins in 1975 to name an exhibition of contemporary photographs at the International Museum of Photography at the George Eastman House in Rochester, New York. Jenkins was then Assistant Curator of the Museum, and he organized a show of ten photographers who worked independently, but whose photographs nonetheless seemed to him of a school, signaling a new movement in landscape photography, away from beautiful images of pristine scenery and toward a kind of clinical documentation of the "man-altered landscape," as the exhibition book was subtitled. For hikers and hunters, *topographics* will call to mind a US Geological Survey topo map, a pale green and white map of the land's natural features, showing vegetation zones, including brown contour lines that indicate elevation gradients, and marking roads and trails in black. Jenkins, however, drew upon an older definition of topography: "The detailed and accurate description of a particular place, city, town, district, state, parish or tract of land." For Jenkins, the key word is *description*. He believed that the photographs of the New Topographics exhibit represented a new generation of documentary landscape photography, one that adopted an intentionally "style-less" style, more interested in observation than judgment, and whose viewpoint, in his words, "is anthropological rather than critical, scientific rather than artistic" (7). Jenkins's purpose in organizing the New Topographics exhibit was "to postulate, at least for the time being, what it means to make a documentary photograph" (7).

The photographers whom Jenkins included in the exhibition, all born in the 1930s and 1940s—and all but one male—subsequently became strongly associated with the New Topographics movement: Robert Adams, Lewis Baltz, Bernd and Hilla Becher, Joe Deal, Frank Gohlke, Nicholas Nixon, John Schott, Stephen Shore, and Henry Wessel, Jr. While curator Jenkins envisioned the exhibit as above all about an emergent new *style*—straightforward, plain, matter-of-fact—to many viewers, what is most striking about the collection of photographs is their seemingly non-photogenic *subject matter*. Why would one go to an art gallery to be inflicted by images of a derelict, coal processing plant, the flat fronts of generic office buildings in a southern California industrial park, a random parking lot, or an airport terminal? Or to be asked to contemplate photographs of a water tower, cheap motels along Route 66, run-of-the-mill intersections in obscure towns, TV antennas, and power lines? These questions persist. Even today, these photographs challenge the viewer to expand his or her conception of art. Their flat, documentary style, rather than being opinion-less, seems to imply a point of view that the viewer can't quite pin down. Collectively, the New Topographics photographs direct our view, give us no choice but to stare at aspects of modern-day life that are so mundane, so taken for granted, so quotidian that we never pause to think about what they may mean, never stop to see our culture as a moment in an historical era, as the infrastructures of our lives might appear to an ethnographer or an extraterrestrial.

When Peter went to Syracuse, New York, in 1981, he was only a couple of hours away from Rochester, where the influential New Topographics exhibit had shown just six years earlier. Peter would, of course, one day come to know and acknowledge an affinity with New Topographics, but at this time he felt very much on his own. He honed a documentary style similar to theirs, but he was drawn to slightly different kinds of landscapes, still operating under the influence of Mayan ruins. Although "man-altered landscape," or simply "altered landscapes," fits both kinds of work, the New Topographics artists tended to depict buildings encroaching onto the land, transforming a natural setting into a cultural space; in contrast, Peter Goin became intrigued by American ruins, human structures scarcely one hundred years old that were already in the process of being reclaimed by nature. If New Topographics mapped the translation of nature into culture, Peter recorded instances of nature responding. And if New Topographics drew the viewer's attention to the built structures of contemporary culture, Goin's work reveals the sometimes hidden, decaying structures of a bygone era and documents the uses to which these places are being put—or not—in our own time.

A selection of Peter's Evolving Landscapes and Erie Canal photographs appeared in the 1982 issue of *Landscape* magazine, one of his earliest publications. *Landscape* magazine was founded by cultural geographer J. B. Jackson in 1951, a fitting choice of venue on Goin's part, given that his Erie Canal photographs were informed by Jackson's writings. Peter's essay, "Old Landscapes Overgrown," which introduces the gallery, begins with a paragraph that serves as a fitting introduction to the Erie Canal series:

> Is the landscape natural? Is it artificial? What forces mold it? Should the land be used or preserved? The interaction between people and the land is complex and intricately woven. Land use changes over time, determined by political, social and economic forces. Policies range from complete protection to benign neglect. New environments are created, almost by accident, as new sites are constructed, abandoned, and overgrown. Vegetation reclaims the land, slowly.
>
> *(22)*

Here we may note several signature predispositions of Peter Goin, including his inquisitive nature, posing questions that have no easy answer, some of which become ethical conundrums. We see his awareness that landscape—a seemingly static concept—is, in fact, always dynamic and that both natural and human forces co-create landscapes. We note a keen awareness of complexity in human-nature interactions, recalling one of Peter's frequent, insistent remarks, "It's a much more complex narrative." We see an educated interest in policy, knowing that government policies make both anticipated and unanticipated impacts on the land. We observe a subtle, ironic sense of humor—"benign neglect?" Peter's prose, like the images of American ruins presented in this chapter, ultimately encourages us to notice overlooked places and to ponder the effects of policy and the agency of natural processes such as vegetation.

The Erie Canal photographs that are reprinted in this book were made on the site of the abandoned Lock 51 near Jordan, New York. In this small sampling of the extensive Erie Canal series, we can begin to see imaginative traces of the Mesoamerican ruins images. In photographs one and two, for example (Figures 5.1 and 5.2), walls and block structures of abandoned locks are centered declaratively in the rectangular frame of the photograph, requiring inspection. Like the Mayan ruins images, these topographical studies document plants growing in the crevices of building blocks, weeds beginning to get a toe-hold on the top of the structures, and surrounding vegetation growing up around and beginning to conceal the stone structures, which look to be forgotten in the process of merging back into nature. The structures are mute and abandoned, reflected in or casting their shadows on stagnant water, silent remnants of some past civilization. In essence, the processes that Peter observed in Mexico—ancient structures melting back into nature—are already beginning to happen in even so young a country as the United States. Images three and four in the Erie Canal Series (Figures 5.3 and 5.4) are compositional cousins to the abandoned roads photographs in the previous chapter, centered views of a road or in this case a canal, where the parallel edges of the road or canal form perspectival diagonal lines, creating a triangular design, a spatial negative to a pyramid, signifying absence and vacancy. These images show nature moving in opportunistically with such rank growth that it begins to obliterate the underlying walls of the canal. In images four and five, only the oddly angular contours of the vegetation might tip one off that the scene is not an untouched natural setting, but rather an altered landscape, an architected space.

When Peter photographed this stretch of the Erie Canal in 1981, it was an abandoned landscape, the working canal having been enlarged and rerouted in 1918, leaving some segments of the smaller canal built in the 1820s to fall into benign neglect. In his book *Real Places*, cultural geographer Grady Clay has named such places "opportunity sites," where the past is clearly past, the present condition is one of disuse, and the future awaits an entrepreneurial imagination (196). As Goin teaches students in his History of Photography classes at the university, landscape photographs inscribe history by capturing a dynamic landscape at a particular moment in its evolution. In analyzing a landscape photograph, therefore, it is always a good idea to ask how history informs the image. Ask not just *where* a photograph was made, but *when*, and think about how time makes its appearance in the image. The Erie Canal series invites precisely this kind of historical speculation. We ponder not only the age and purpose of the original structures and the probable length of time they have been neglected, but also what their future may hold. Interestingly, some of the sites that Peter photographed in 1981 are today more manicured and tidy than they were forty years ago when he made the photographs. Stretches of the old canal are today city parks, state historic parks, national historic sites, and museums, with interpretive signs, brochures, and websites that commemorate an important engineering feat and economic era in America's past.

Goin is intrigued by how certain places such as the Erie Canal become memorialized, preserved, and sometimes even restored as historical monuments. *Why is that?*, he wonders and then proceeds to speculate that "we are a culture that immigrated from other deeply historical lands, and we lost our sense of tradition and sense of place, and so to establish and to create our own history, we create our own ruins." He recalls reading a book by the influential landscape interpreter J. B. Jackson entitled *The Necessity for Ruins*, wherein Jackson discusses how in some cultures public monuments celebrate famous political leaders, but in the United States an alternative, democratic tradition can also be observed, in which monuments celebrate a "vernacular" past, "a sense of the way it *used to be*, history as the chronicle of everyday existence" (94–95). Commemoration of the Erie Canal fits Jackson's idea of celebrating a vernacular past, as labor is the dominant motif in Erie Canal lore—"I've got a mule, her name is Sal, fifteen miles on the Erie Canal," goes the folk song many Americans know by heart.

More than a few of the abandoned landscapes that Peter has photographed throughout the years later became reconstructed and memorialized as official, historic sites or parks. The project of historical preservation tends to favor one historical moment, restoring and then maintaining structures to their appearance at that chosen time period, sometimes preserving structures in a curious state of arrested decay, such as crumbling adobe walls or rickety wooden boardwalks of a mining ghost town. The topic launches Peter into a favorite disquisition:

> Why do *we* get to say when the most interesting romantic moment is in the history of this canal, of this place, of this site, of this ruin, of this culture? Why do we get to say, *this* is *the* most important moment? And it's not even this moment now. Because sometimes it's too far gone, too overgrown, so we resurrect it. We recreate it. We hire people who are academically trained and we say, "Well, we'd like it to look like it looked twenty years after it was abandoned." We don't choose it when it was

> functioning. Except at Williamsburg and simulated history. We *love* the ruin. But we want a certain *kind* of ruin. Well, what's that cultural value and meaning, and where does that come from?

Peter speculates that Euro-Americans yearn to have a history in the New World as a way of naturalizing their claim to the land and forging an American identity. Descendants of Old World immigrants severed roots in Europe, thereby necessitating the invention of a historical and mythological past in the United States. Ruins conveniently offer visible proof of the Eurocentric colonizers' coveted antiquity.

For his part, Goin tends to prefer the dynamism and open-endedness of nature's rewilding process to the fixity of a neatly preserved and scripted historical site. Exploring and photographing the Erie Canal satisfied his quest for North American ruins. In the next chapter, we will join the roving photographer on his rambles through San Francisco, seeking not ruins but irruptions of wilderness in the urban environment.

■

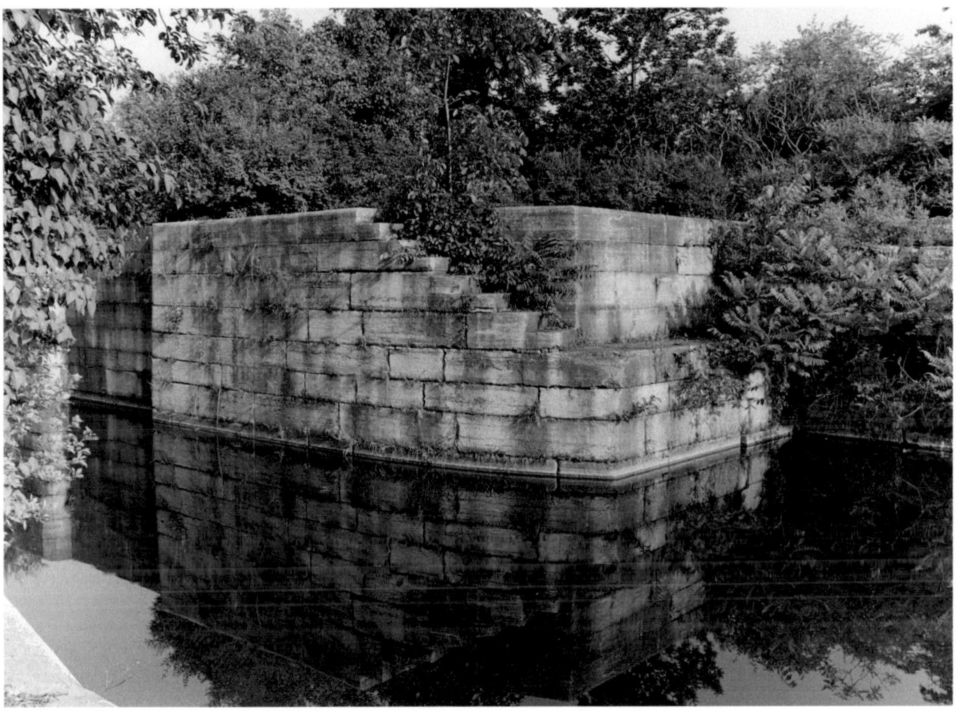

5.1 *Erie Canal, Lock 51, near Jordan, New York*, 1981.

This image resembles the photographs of Mayan ruins discussed in Chapter 3, complete with vegetation taking root between the cut stone blocks of an abandoned structure. The visual allusion to Mesoamerican ruins is intentional; Lock 51, built in 1847 and abandoned in 1918, furnishes a good example of "ancient" ruins in a US context. The framing of this image, using vegetation to mask the left and right sides of the picture, invites the comparison with Mayan ruins, in that this lock segment is made to look like a stand-alone building, when in reality it is a corner of a continuous wall. The reflection of the structure in the water creates an illusion of greater height, while the black-and-white print directs attention to details of the structural form without the distraction that the palette of greens of a color print would create.

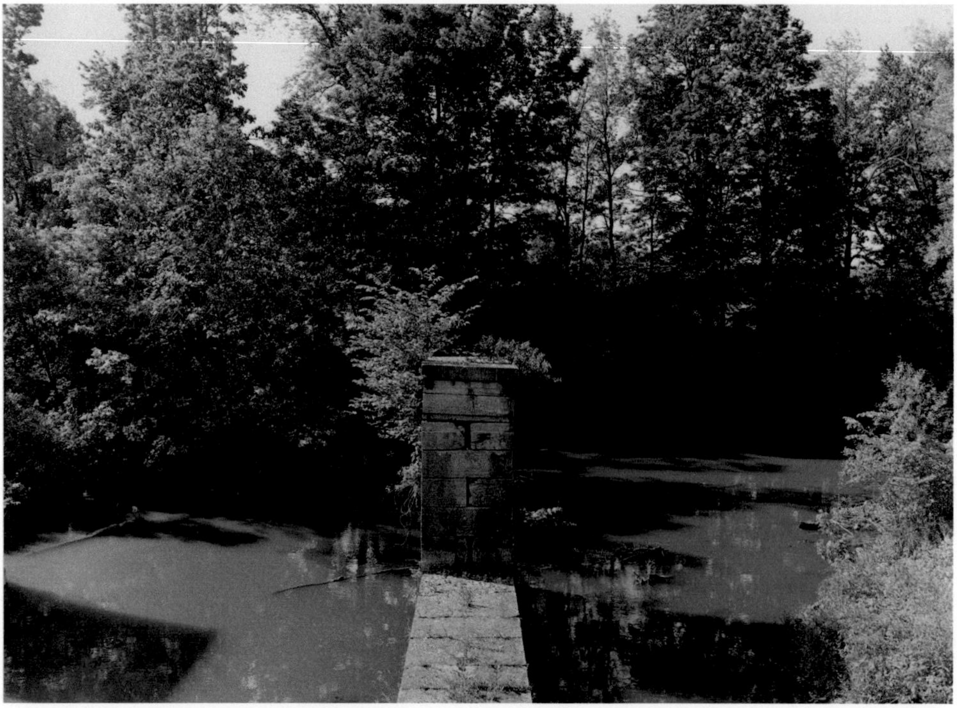

5.2 *Erie Canal, Lock 51, near Jordan, New York*, 1981.

Compositionally, the wall dividing the space in this picture creates two rooms in nature, even as the blunt centering of the wall states declaratively that the wall itself is to be the principal object of contemplation. The standing water suggests a waxed floor of the rooms and is an aqueous reminder of the working lock that once occupied this quiet glade.

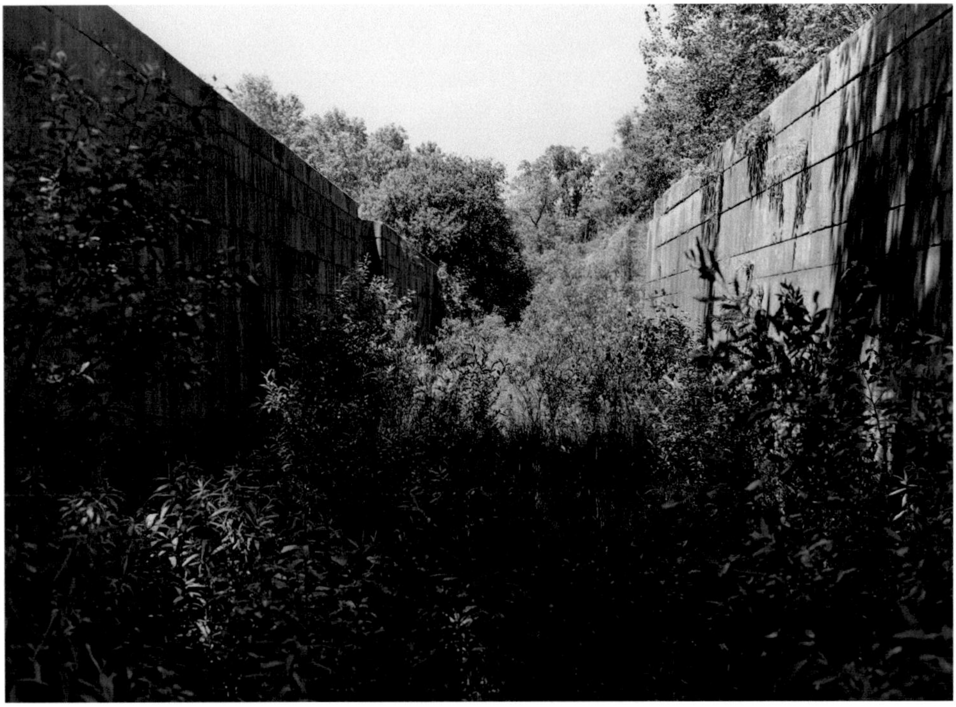

5.3 *Erie Canal, Lock 51, near Jordan, New York*, 1981.

The park interpretative sign currently located at Lock 51 reads:

> Looking at it today, it's hard to imagine that less than a century ago Lock 51 was a busy thoroughfare with traffic jams, schedules to meet, and the sound of lock gates clanging shut. But until 1918, the year that the old Erie Canal closed, the view from here was a line of barges, excursion boats, and maintenance scows being towed by mule teams.

Goin's vantage point—on the ground, *inside* the lock—creates perspective lines along the top and bottom edges of the walls that form an X, the point of intersection being a dark, vegetated area a shade higher than center, a kind of earthbound black hole that draws human-made artifacts into its leafy maw.

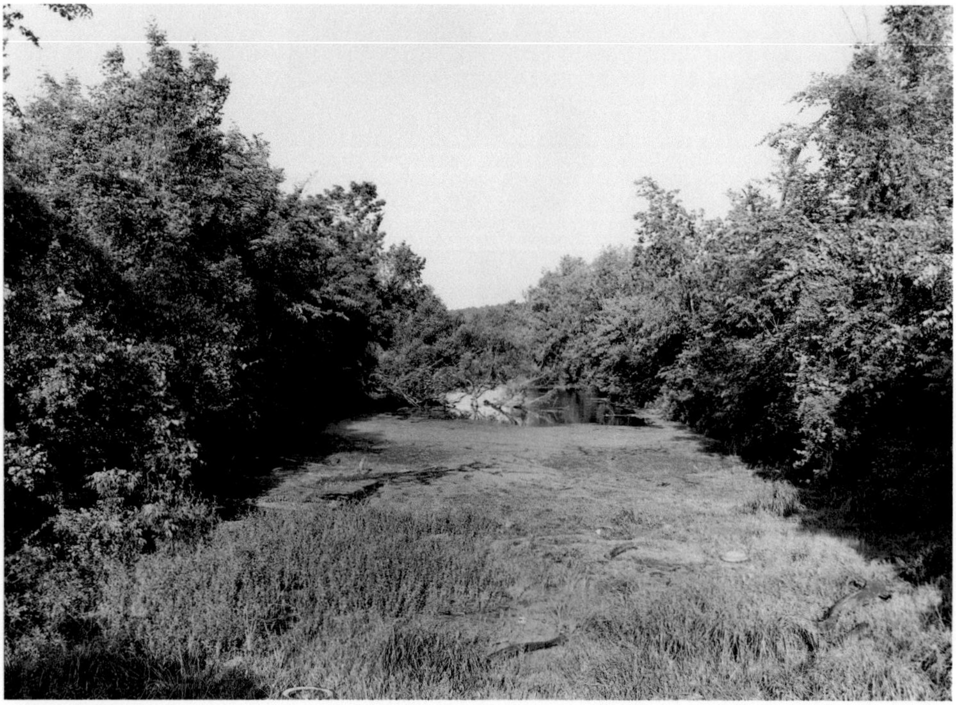

5.4 *Erie Canal, Lock 51, near Jordan, New York*, 1981.

The preternaturally straight-line sides of this swampy watercourse—which like the previous image create a central X pattern—provide a subtle clue that this site might be an altered landscape, representing the remains of an old canal rather than a stretch of untouched river. The angular lines of the shadows in this image support the interpretation of this site as human-architected. Goin calibrated the timing of this photograph so that the compositionally significant shadows would be most pronounced.

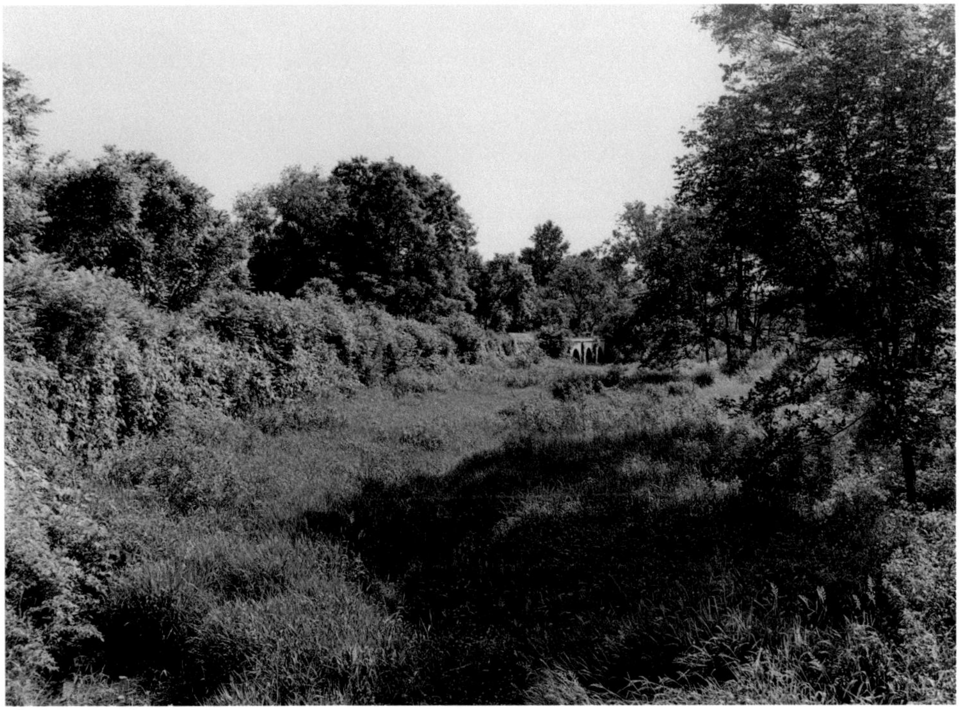

5.5 *Erie Canal, aqueduct near Jordan, New York*, 1981.

The diagonal line of shrubbery that extends from the lower left corner of this image to the bridge in the background looks suspiciously unnatural in structure. Might there be a human-made wall hidden under that wall of vegetation? Indeed, there is, this being yet another photograph of the old Erie Canal. Goin waited to make this photograph until a broad shadow fell onto the foreground, the shadow a cue to the viewer *not* to look there, but rather to shunt one's focus to the brighter vegetated wall at left, whose mysterious angularity poses a riddle about biomorphology. The New York State Canals interpretive sign erected at this site years after Goin did fieldwork there explains:

> Less than a century ago, what is now a park was the busy Erie Canal—Jordan's principal artery of commerce. The stone aqueduct standing nearby was the mule towpath, and canal barges floated in an adjacent seven-foot deep wooden trough (now gone) that was supported by large stone piers.

Works cited

Clay, Grady. *Real Places: An Unconventional Guide to American's Generic Landscape*. University of Chicago Press, 1994.

Goin, Peter. "Old Landscapes Overgrown." *Landscape*, vol. 26, no. 1, 1982, pp. 22–25.

Jackson, John Brinckerhoff. *The Necessity for Ruins, and Other Topics*. University of Massachusetts Press, 1980.

Jenkins, William. "Introduction." *New Topographics: Photographs of a Man-Altered Landscape*, Robert Adams, et al. International Museum of Photography at George Eastman House, 1975.

"The Jordan Aqueduct, Jordan, N.Y." *The Erie Canal*, Frank E. Sadowski, Jr., 2019, https://www.eriecanal.org/JordanAqueduct.html. Accessed 16 January 2020.

Light Work. Light Work, 2015, http://photography.cdmhost.com/cdm4/about.php; http://photography.cdmhost.com/cdm4/item_viewer.php?CISOROOT=/p4023coll6&CISOPTR=849&CISOBOX=1&REC=1. Accessed July 2015.

6

URBAN WILD

When Peter moved to San Francisco in 1978, he was an energetic young man with very little money, no job, no girlfriend, and no clear future prospects as a photographer. Similar to many others flocking to the Bay Area at that time, he was looking to restart, renovate, and innovate. He knew that he wanted to continue making art and that he wanted to work within the rubric of photography, and he was willing to take a chance and move to San Francisco without a job lined up in advance. When he left the City by the Bay six years later in 1984, he had gained valuable professional experience in the book business and in videography, held an artist-in-residency position, accrued an engaging body of work, published two articles in a respected journal, married Chelsea Miller, and was offered a tenure-track professorship in photography and videography in the Department of Art at the University of Nevada, Reno, a position he holds to this day. Before we turn to the photographs that Goin made in San Francisco, let's take a moment to fill in some biographical details, giving Peter's life story a chance to catch up to narrative threads that were emerging in his new work.

When Peter was earning his MFA at the University of Iowa, some unethical behavior that he observed there was profoundly disillusioning. One of his professors, whose decisions about graduate student funding favored sexually compliant co-eds, soured Peter's taste for academia. At that time, there was no viable means of grievance—and who would have believed him, anyway? The system seemed in some respects structurally unjust, and Peter wanted nothing to do with it. MFA in hand, Peter taught art and crafts at Fort Myer Army base in Virginia, left that job for extended travels in Central America, and, when his money ran out, returned to Iowa, where he taught a course at the university while plotting his next move. Given his antiestablishment attitude at that time, where better to sojourn than to the epicenter of the counterculture, San Francisco? Peter's brother was living there and allowed the disillusioned young idealist to stay with him until he could support himself.

How to support himself? Peter knew that he wanted one day to publish his photographs in books, so he decided to pursue work in a bookstore to gain inside knowledge of the publishing industry. Fortuitously, Books Inc./Tro Harper's was hiring. Books Inc., whose business in San Francisco dates back to 1859 and once employed Bret Harte, was a San Francisco

DOI: 10.4324/9781003212607-6

institution. Peter worked at the Powell Street location, formerly Tro Harper's Books, at Union Square in the heart of downtown. As far as the book culture goes, Books Inc., along with City Lights Bookstore, was a happening place for authors, readers, connoisseurs, and collectors. To work there was to have privileged access to culture and ideas. Books Inc. did very well selling remainders as well as Eastern philosophy and foreign imports, which at that time were rarely available elsewhere in the United States.

When Peter became manager, he built up the store's art book section, including photography books by Edward Weston, Ansel Adams, *Rolling Stone* photographer Annie Leibovitz, and all the other recognized names. He also noticed the robust sales of what he calls "gratuitous photographic books of scantily clad nubile females," which alerted him to the financially perilous divide between the kind of photography books that sell well and the projects that he wanted to undertake. The idea of ever being able to support himself from sales of his work was highly unlikely. If he wanted to be a serious photographer—and not a commercial photographer—he would have to find a career that supported that work. Managing a bookstore was not the answer: His fifty hours a week in Books Inc. ate up all the daylight hours and precluded extended field trips. Peter began to reconsider university teaching, a job where publishing one's work was expected and where summers could be dedicated to fieldwork. He began to harbor dreams of being the kind of professor he wished he had and of building up a premier university photography program. In order to improve his qualifications, Peter worked with a documentary filmmaker in San Francisco on his time off from the bookstore, learning skills that not only enhanced his competitiveness on the academic job market but also prepared him to later make documentary films of his own.

During the five years he spent as a carefree young man in San Francisco, Peter was rather systematically laying the groundwork for a future career as a distinguished professor and published photographer. Peter's driving ambition—the fire in the belly—was already kindling, and it would soon burn off the veneer of a Bay Area bohemian. Even as the young man was enhancing his professional credentials, he did not neglect his personal life, and he began to spend many enjoyable hours with a new acquaintance named Chelsea Miller. He had first met Chelsea briefly back at the University of Iowa, but that was about it. He and some college friends had organized an apple pie baking contest as a way to meet women. Much to their surprise, more than fifty people showed up, bringing thirty-two pies, to each of which they awarded a prize, including a ribbon for the largest number of thumb edges on the crust. The opportunity to meet Chelsea arose again when a mutual college friend encouraged Peter to contact Chelsea when he arrived in the Bay Area. Not long after getting settled in San Francisco, Peter telephoned Chelsea to ask if she'd like to meet for coffee. She agreed to meet him at Café Trieste, an Italian-owned café in North Beach. Peter arrived early, sat down along the windows, and began watching the door. In strolled two women, one a petite brunette, the other an oversized Wagnerian diva who carried an opera helmet. Not knowing which woman was Chelsea, Peter began schooling himself to be open-minded. Chelsea's commanding bodyguard was named April, a lovelorn friend whom Chelsea planned to set up with Peter. As it turned out, April and Peter had little to talk about; Peter's knowledge of opera was limited to having once seen *La Bohème*. But Peter and Chelsea had a lot to talk about, and they began to meet regularly for coffee, unchaperoned.

Chelsea had received her BA from the University of Iowa and then, like Peter, gravitated to the San Francisco Bay Area, where she earned an MFA in textiles from the California College of Arts and Crafts in Oakland. They made a great match. When Peter met her,

Chelsea worked for a Persian family restoring rugs, but she later managed a trend-setting San Francisco fine art gallery named the Grapestake Gallery, the first gallery in the city to exhibit photographs concurrently with painting and sculpture. Grapestake opened its doors in 1974 with an Ansel Adams retrospective and later became a virtual who's who of American photographers, showing the work of major artists such as Paul Strand, Harry Callahan, Joel Meyerowitz, Berenice Abbott, Imogen Cunningham, Jerry Uelsmann, Judy Dater, William Eggleston, Lewis Baltz, and Richard Misrach. As the years went by, Chelsea developed keen insight into what Peter was doing and became a trusted advisor. For Peter, the Point Pinole photographs, which we contemplated in the "Evolving Landscapes" chapter, evoke memories of Chelsea, as the two took many walks together in that derelict dynamite plant, which was for them a secluded arbor for quiet talks and solitary picnics.

Prior to his employment at the bookstore, Peter set up a darkroom in a small office that he rented in downtown San Francisco, and he began experimenting with photograms, a camera-less process where you place objects on light-sensitive paper and then expose it to light. The paper underneath the objects remains white, while the surrounding areas turn black, producing silhouettes and interesting black-and-white designs. Peter was familiar with the photogram art of Man Ray and László Moholy-Nagy, early practitioners of using the process to make abstract art. His own photograms leap to life with a swirl of human vitality, featuring dancing, twirling, falling, vaulting, cavorting human forms, inspired by the Indonesian puppets which he recalled from his childhood in Jakarta. At once whimsical, joyful, and surreal, these creative products of what Peter calls "photographer's theatre" appear on the last page of this book.

Between the bookstore, learning videography, and spending time with Chelsea, Peter found little time to devote to photography. Sunday mornings were virtually the only daylight hours available for making photographs. Peter made the most of that little window of opportunity. By this time, he and Chelsea were renting the top floor of a house in the Mission District, and Goin conceived the idea to make a detailed photographic survey of that neighborhood—his own backyard. As North American landscapes go, the Mission District has a relatively long history of change. Yelamu Indians lived in the area for two thousand years before Spanish-Mexican people began colonizing it in the eighteenth century. Built in 1776, Mission San Francisco de Asís in the eponymous "Mission" District stands today as the city's oldest surviving building. When Peter lived there in the early eighties, the Mission District was an ethnically mixed neighborhood with Victorian-style houses, the majority of which post-date San Francisco's massive 1906 earthquake. Compared to the sightseeing hotspots of San Francisco, such as Fisherman's Wharf, Chinatown, and Golden Gate Park, "the Mission," as it was called, was below the cultural radar at the time and was not much photographed or celebrated. In other words, it was exactly the kind of overlooked landscape that Peter gravitated to as ideal material for artistic reinterpretation.

In 1983 Goin made more than four hundred photographs detailing the Mission District. He converted a bedroom of the apartment he shared with Chelsea to a darkroom and printed the photographs late at night. Selected images from this urban survey appeared in *Landscape* magazine in 1984 under the title "Anonymous Places." The article is a visual tour de force, demanding that the reader assimilate an archive of thirty-two photographs—two photographs on the first page, followed by five consecutive pages of six images each, stacked in two columns of three identically sized horizontal-format photographs. It is as if the reader is invited to take an early Sunday morning stroll through the neighborhood of San

Francisco's Mission District while the residents are still sleeping and the streets are deserted. The early morning, fog-encrusted light is flat, perfect for achieving subtle variations in middle-value tones with black-and-white photography. We pause now and again to register an odd assortment of unremarked places and peculiar spaces of the built environment. We are forbidden to focus on prime real estate—the beautifully restored Victorian showpiece houses or well-kept gardens or historic churches of the district. No, the insistent camera tells us to become aware of what landscape theorist J. B. Jackson terms the "vernacular" landscape, the unselfconscious spaces of the lived-in environment. So, we notice exactly those things that would never appear in a promotional brochure of the area—clotheslines, weedy lots, untrimmed hedges, chain-link fences, cracked sidewalks, alleys, delivery trucks, boarded windows, graffiti, an oil-stained parking lot, patched streets, a net of electric tram wires, nondescript corner buildings, brick walls, stucco walls, shiplap wooden walls, garage doors, staircases, curtained windows, telephone poles, and "no parking" signs.

It is a cognitively demanding photo-essay, asking the reader to interpret an uncomfortably long series of photographs devoid of captions. In casting about for a photographic tradition that might help to contextualize this copious urban inventory, we might recall Eugène Atget, who in the first decades of the twentieth century took more than ten thousand photographs documenting Parisian architecture and street scenes in a compulsively encyclopedic effort to capture every nook and cranny of that great city. However, while Atget seemed driven by love and nostalgia to record and thereby fix in the collective memory every minute detail of Paris before his subjects became obliterated by time and change, Goin's work seems entirely dispassionate, less interested in preserving sites than in studying them, presenting them for visual analysis and interpretation. In their choice of subject and their emotional distance, the photographs bear an affinity with New Topographics and, indeed, may have been influenced by that increasingly prominent movement, conveying information-rich, topographical detail delivered in an unemotional, patently styleless style.

But unlike the New Topographics artists, whose work was exhibited in a gallery and was to be experienced as art, Goin published his urban survey in *Landscape* magazine, where it would be encountered as scholarship, alongside essays with academic titles such as "A Scottish Counterpoint: Urban Design in Princes Street," "Woven Landscapes: Saris of Orissa," and "The Burden of the Past: Rethinking Historic Preservation." Goin's photo-essay presents the findings of his visual research, while his brief introduction to the gallery speaks to a cultural geographer's interest in the spatial relationships of primary elements in urban landscapes, even as he asserts the value of photography in landscape study:

> As a scarce commodity in cities, space is a crucial ingredient. Understanding how space is handled allows us to identify the difference between neighborhoods. It can also clarify how foliage, architecture, and land are organized. As we build houses (or tear them down), and as we plant bushes (or dig them up), space is created, occupied, or divided. The geometry we produce in the urban environment leaves pockets of anonymous, empty space—either overlooked or ignored. Conversely, open spaces can be energized with tension, or they can inspire meditation. People exercise great freedom and imagination in shaping the space around them. Adults enclose yards, cement lawns, grow flowers, construct unusual houses, and erect barriers. Children make playgrounds out of vacant lots. Teenagers scrawl graffiti.

> The photograph is a sensitive tool for articulating the character of the urban landscape. The photograph can document who lives in a city, how it looks, how it is arranged, and how it changes. Photographs record not only architecture, foliage, and land, but the spatial relationships between them. These photographs of the Mission District in San Francisco are a sample of the visual evidence that can help us recognize and evaluate patterns in the use of space.
>
> *(24)*

"Understanding," "clarify," "sensitive tool," "document," "record," "evidence," and "evaluate." Hmm. These terms are more often found in scientific discourse than in art circles, right? Herein we stumble headlong into The Problem of Photography, a debate that has raged since the announcement of the daguerreotype in 1839. In a nutshell, people have been unclear about the status of photography, which in its literal translation means "light-writing." Since photography is a mechanical process, requiring skill in the use of specialized equipment, early practitioners were called operators, and photographs were generally taken to be two-dimensional copies of nature—much like we might think of a photocopy today—rather than kin to more obviously subjective works such as painting. As technology advanced and cameras became widely available and easy to use, anybody could make a picture, and it isn't always easy to tell which photograph was made by an amateur and which was made by an artist. The artistic status of *documentary* photographs is even more ambiguous, as they run the gamut from the standardized mug shots of a criminal line-up to the evocative work of Depression-Era photographers Dorothea Lange, Russell Lee, and Walker Evans, whose original prints today command princely sums in the fine art market. Ultimately, we will have to live with ambiguity and differences of opinion about photography and about the artistic status of any given photograph or body of work. Photographs that first appeared in one context—such as a nineteenth-century government survey report or a newspaper—may later or even simultaneously be collected and interpreted as art works. Photography light-writing—like writing itself is a tool that can be used to make anything from ephemeral memos to enduring masterpieces.

For his part, Peter Goin thinks of himself as an artist who does visual research. He reminds his university students that science and art have not always been compartmentalized as separate activities; in the Italian Renaissance, for example, science and art are married in the work of Leonardo da Vinci. For the most part, as his initial choice of publication in *Landscape* magazine and later books published by university and scholarly presses signals, Peter has chosen to collaborate with the humanities and pursue visual research and cultural criticism rather than developing affiliations with galleries. On the flip side, however, Goin holds a Master of Fine Arts degree and a faculty position in an Art department, and every one of his images can be interpreted as an aesthetic object. Straddling these two worlds—remarrying scientific inquiry and artistic expression at a time when our culture keeps them divorced—is fraught with conflict, to say the least. Peter Goin's work carves out its own creative and intellectual space between high disciplinary walls, a visual metaphor that operates in the Mission District urban survey as well.

Viewed aesthetically, Goin's San Francisco images are an artistic exploration of the way that we perceive space. He became interested in how the little, nondescript plots tucked between houses and sprinkled throughout his immediate neighborhood define space and create particular places, each one unique, each easily overlooked by virtue of their very "in-betweenness." Culture influences perception. Goin explains that in Western society, we generally look at objects, whereas in other cultures—Asian cultures in particular—people tend to look at the spaces between objects. By the same token there are different ways of reading. Not every culture reads left to right. Some texts—a Chinese handscroll, for example—are meant to be read from right to left, with their text, comprised of Chinese characters, linearly arranged from top to bottom. At this time in his life, Peter was beginning to build his own art library, which included books about Asian art. The different ways of perceiving and reading not just text but image provoked him to think about the use of space in artistic composition. In the "Anonymous Places" photographic series Peter was playing off the notion of the center, using the center not so much to feature a declarative *object* of contemplation—as he did in the Mayan Ruins series—but, instead, to create a visual *space*. This is why the photographs in this chapter may look vaguely foreign to an American audience, like they are empty at the center. "What are these pictures *of*?" a viewer might ask in mild irritation. Goin has framed many images to include high walls at the edges, directing the viewer's attention not to an object, such as a building or a sculpture, but exactly to the shape and characteristics of the *space* created between those walls, a little space with its own identity. It is the *nature* of that identity that so intrigued Peter and that planted the seeds for later book projects such as *Humanature*.

While Goin emphasizes geometry and spatial relations in his article for *Landscape* magazine, another thread runs through the series that, for this ecocritical study, becomes the more salient and in retrospect may, in fact, be the more conceptually important. That thread is the human-nature interface. The four images featured in this chapter not only document how humans shape space geometrically—the subject of Goin's article —but they also focus on what Goin calls the "urban wild." Is *urban wild* an oxymoron? To understand why *urban* and *wild* may initially seem like incompatible or mutually exclusive conditions, it helps to analyze the cultural assumptions packed into the idea of "wilderness." Historian William Cronon ignited a heated debate among environmentalists when he declared in his 1995 essay "The Trouble with Wilderness" that "[t]he time has come to rethink wilderness" (69). The problem, he says, is that Americans tend to think of wilderness in terms of pristine nature, untouched by man and uncontaminated by the taint of civilization. If we want to escape modernity and experience the restorative tonic of nature, we assume that we must get out of the city—the place where most of us actually live—and sojourn to the sanctity of the wilderness to renew our spirit. Any number of popular quotes by John Muir lend eloquence to our urge to flee the city for the wilderness; for example:

> Thousands of tired, nerve-shaken, over-civilized people are beginning to find out that going to the mountains is going home; that wildness is necessity; and that mountain parks and reservations are useful not only as fountains of timber and irrigating rivers, but as fountains of life.

According to Cronon, this way of putting wilderness on a high mountain pedestal, far removed from the grime of the workaday world, tends to create a "dualism that sets humanity and nature at opposite poles" (81), creating an equation where nature = good and the human world = bad. Cronon argues that there are many historical and conceptual flaws with this concept of wilderness, not the least of which is that if "untrammeled" wilderness is our image of a healthy, natural ecosystem, then people have no place in the natural world. Furthermore, if we imagine wilderness to be a place apart from human habitation, we run the risk of being unaware of the wildness in our own neighborhoods.

Peter Goin's San Francisco urban survey, made in the mid 1980s, anticipates the trouble with the dominant conception of wilderness that Cronon would so controversially articulate a decade later. Goin's photographs ask us to notice the presence of wilderness in the nooks and crannies of the city. In images of the rank growth of unpruned shrubs, the hummocks and stone outcropping of a weedy lot, the cock-eyed angle of a spindly juniper tree growing in the soil of a curbside strip, and the botanical tumult in a tiny front yard garden, one senses the wild force "that through the green fuse drives the flower," to quote a line from poet Dylan Thomas. In his Sunday morning photographic expeditions through the Mission District, Goin recorded instances of nature's exuberance, flinging sprays of dark foliage against the built environment's bright, hard-angled planes and straight lines. Open to paradox, Goin offers humble portraits of everyday emergences of wildness in the city. His images implicitly redefine wilderness as the nature on the street.

■

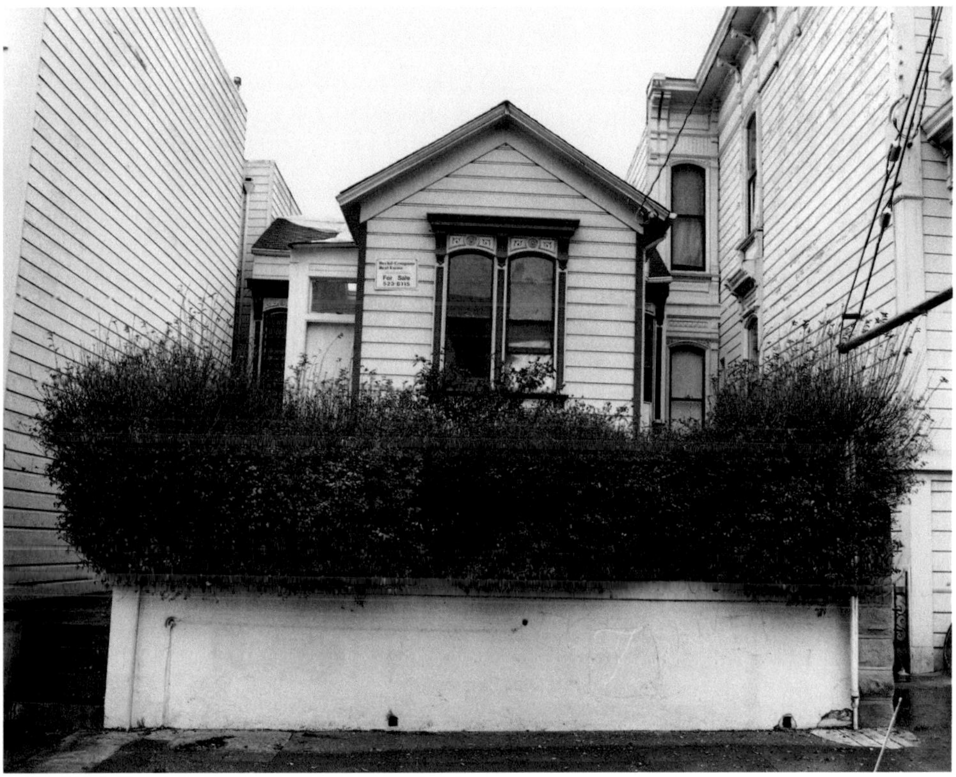

6.1 *House for sale, 312 Sanchez Street, San Francisco*, 1983.

This little house emerges within the open space created by monolithic walls on either side. Rather than cropping the frame to exclude the towering walls and focus on the house, Goin chose to include the walls as dominant geometric delimiters that define a space. A decorative hedge gone feral hides the elevated front yard and dwarfs the small For Sale sign on the house. The camera's low vantage point exaggerates the impression that the foliage is out of control—in a word, wild.

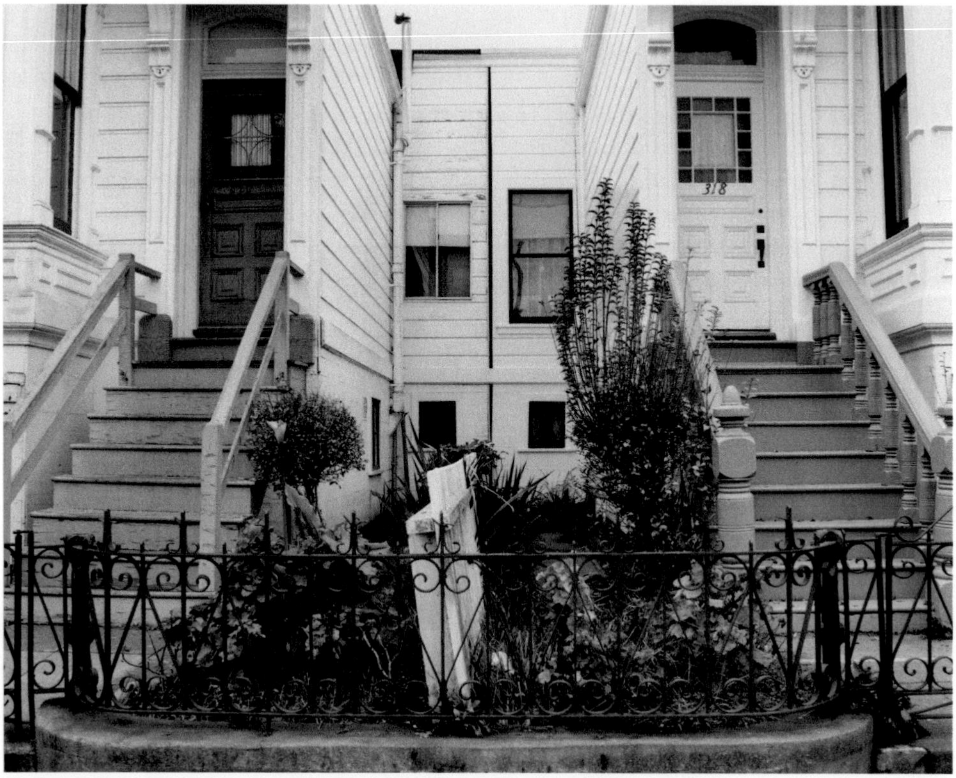

6.2 *Split yard between 318 and 320 Bartlett Street, San Francisco,* 1983.

"It's so San Francisco," Goin says of this split yard, "I got my house. I have a white picket fence. I got my front yard. I have stairs leading up to my front door." The white picket fence, with its vernacular tilt, divides a space that was small even before being split in two. The irony, of course, is that both yards look just about the same, the plants responding to similar conditions of soil, water, and climate.

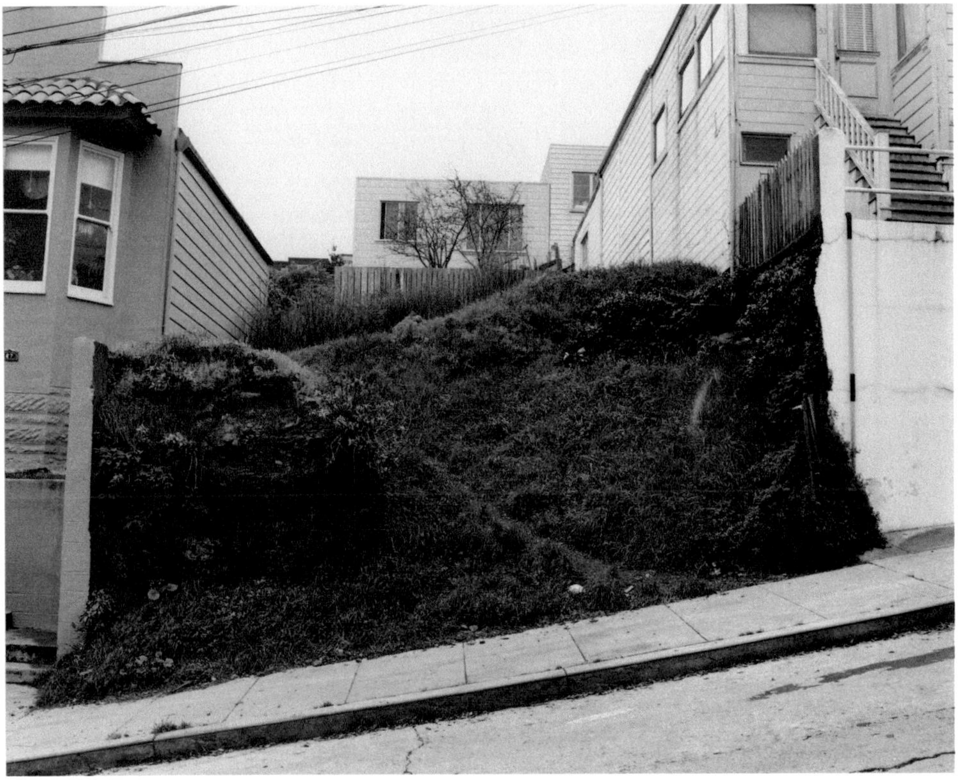

6.3 *No Man's Land. Lot between 47 and 53 Winfield Street, San Francisco,* 1983.

From a distance, this little abandoned lot looks completely natural, a patch of wilderness in the city. If one were to crop the houses out of the frame, this anonymous place could be the English countryside or the Scottish Highlands. While some may regard this undeveloped plot as overgrown or ugly—not properly managed—Goin fancies that this bit of wildness may be a refuge for urban wildlife.

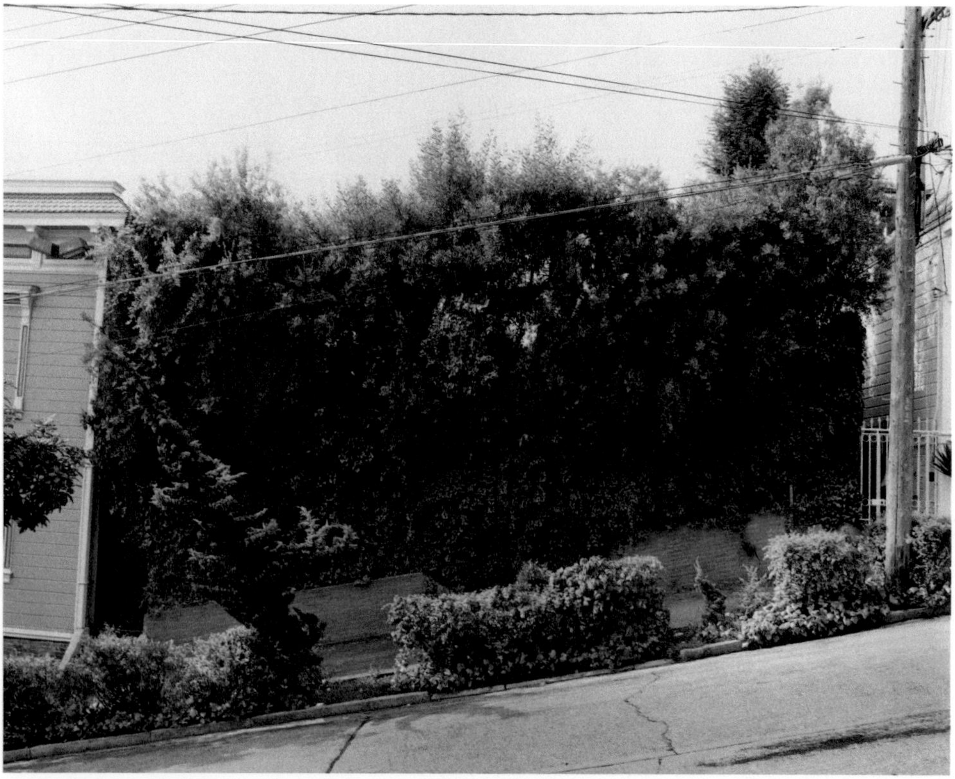

6.4 *Overgrown wall, 3647 22nd Street, San Francisco,* 1983.

What if this were a color photograph? One might be overwhelmed by the rich, vibrant, saturated greens. Goin, however, wants the viewer to look at forms. Note, for example, the crazy angle of the juniper tree in the foreground, expressing a wild will of its own. Observe, too, the contrast between the boxlike, pruned hedges along the street and the less disciplined growth of the vines, shrubs, and trees in the background, the former an example of human-managed nature, the latter of nature less contained.

Works cited

Cronon, William. "The Trouble with Wilderness: Or, Getting Back to the Wrong Nature." *Uncommon Ground: Toward Reinventing Nature*, edited by William Cronon. Norton, 1995, pp. 69–90.

Goin, Peter. "Anonymous Places." *Landscape*, vol. 28, no. 1, 1984, pp. 24–29.

Muir, John. *Our National Parks*. Houghton Mifflin, 1901.

7

A LINE ON THE LANDSCAPE

Lines were on Peter's mind. When he and Chelsea crossed the California-Nevada state line, they had a powerful feeling that they were entering another state of mind. In April 1984 Peter and Chelsea Miller had married. In May the University of Nevada, Reno, offered Peter a job as an Assistant Professor of Art. He would teach beginning through advanced photography courses, and he would also have an opportunity to build a photography program within the undergraduate art degree. Peter was eager to start a career that would give him more time to photograph.

Reno, the Biggest Little City in the World. In July 1984, when the newlyweds arrived, the population of Washoe County, which encompasses the Reno-Sparks metropolitan area, was just shy of 200,000. Compared to San Francisco, Reno was a cow town, an all-you-can-eat, steak-and-eggs kind of place, with jacked-up pick-up trucks, slot machines in grocery stores, cheesy neon lights by night yielding to blinding sun by day, skin-cracking aridity, and nowhere to get a good cup of coffee. In just a few short months the intrepid couple had crossed a decisive line, closing the pages of a California chapter—the Golden State, already tinged with nostalgia—and commencing a Nevada chapter whose narrative line, like the state itself, was a big unknown.

Now that he was at a research university, Peter was keen to embark on a major photographic research project. *Tracing the Line: A Photographic Survey of the Mexican-American Border* became that project, a massive undertaking that demanded Goin's allegiance for the next three years. Given how serious this work became, its origins are whimsical. Having crossed the state line into Nevada, Peter began playing with the idea of the significance of lines. One day Peter asked Chelsea, "What happens if you put a line right here? And we say, 'that side of the line is yours and this side is mine.' Well, what happens?" He pressed on, "What happens if I begin encroaching over the line into your space? You'll say, 'Hey, you crossed the line!'" Peter's point is that we all have lines. We have body space. We have personal space. We have property lines. We have county lines. We have school zoning lines. We have state lines. We accept these lines without question. But what happens if we put a line between two cultures? Peter's Mesoamerican and Evolving Landscapes projects documented

DOI: 10.4324/9781003212607-7

ruins and mounds. His Erie Canal and San Francisco work investigated the spaces made by architecture. Now what happens to the landscape if you put a line across it, a simple, little mark?

This freewheeling conversation soon moved beyond state lines and lines defining personal space to lines dividing two nations and two cultures—to the US-Mexico border. Peter had ties to Mexico. His grandparents had retired early to Guadalajara, Mexico, and had become Mexican citizens. Peter visited them periodically from childhood through his college years. He studied Spanish in college and twice enrolled in January mini-term independent studies through Hamline University. He became fluent enough in Spanish to function on the street, and his trip with Robert Cree to Central America several years later further deepened his understanding of and feelings for Mexico. He sensed that most people in the United States dismissed Mexico as a Third World country. America's southern neighbor was looked down on with condescension and misunderstanding in a way that Canada to the north was not. In short, Mexico was exactly the kind of ignored, neglected, no-man's-land type of place that attracted Peter like a moth to a flame—and would continue to attract him for the rest of his career.

Having long been intrigued by the concept of a "line" dividing cultures, Peter hatched an ambitious plan to make a photographic survey of the Mexican-American border. He became committed to the idea of furthering knowledge, bearing witness, and answering that deceptively simple question—what happens when a line is imposed on the landscape? He first undertook diligent (pre-internet) library research, which he supplemented by writing letters to the Library of Congress, his US Representatives in Washington, DC, and museums and immigration offices along the US-Mexico border. He discovered that a survey had been done in the nineteenth century, but that it was shoddy and incomplete, with amateurish photographs. Apparently, no one had heretofore attempted to make a serious photographic survey of the entire length of the 1,952-mile US-Mexico border.

Now that Peter was a professor with summers off, the plan was to spend his first summer "vacation" in the field, driving from the Gulf of Mexico to the Pacific Ocean, making photographs of "the line" on the landscape at notable sites. He would need support. Taking a chance, he applied for a Junior Faculty Research Award from the university; his proposal characterized his project as "a visual photographic study of the border landscape from Brownsville, Texas to Tijuana, Mexico." He made a strong case for the project's significance, explaining that:

> This study will introduce new visual imagery of a landscape previously avoided, neglected, or fenced with "no trespassing." . . . [T]his border has never been photographed from beginning to end. Within the field of photography, there have not been any significant projects that seriously investigate the relationship of the border-line on the landscape.

Goin's proposal took care to differentiate his work from that of Carleton Watkins, Ansel Adams, Minor White, and others, who worked in a tradition of American landscape photography that relied on the presence of "harmony, beauty, and awe-inspiring drama" to incite reverence for the "natural" landscape. In contrast to the photographic school of natural beauty, Goin allied his own work with that of the New Topographic photographers Robert Adams, Lewis Baltz, and Joe Deal, whose work had begun to "question the role of beauty within the photograph" and whose choice of imagery "reflects the character of exploitation, desperation, solitude and/or neglect of the land." For Goin, their work signaled that landscape photography had entered a "new, dynamic phase," one that provided viewers with a new vision of the relationship between "human-manipulation and the environment." It was in this spirit that, continuing the trajectory of his past projects, all of which investigated "evolving landscapes," Goin proposed to "document the altered landscape and to provide critical visual evidence of that interaction."

The US-Mexico border survey project provided an opportunity to study how a human-made line on the land affected the way that people used and interacted with that landscape. Those *uses*, in turn, created *changes* in the landscape, changes that could be visually documented by photographs that would record and perhaps influence human *perceptions* of the landscape. Perceptions affect use, and use creates change, creating an endlessly recursive process—a "complex narrative," in Goin's words—in which landscape, use, perception, and expression coevolve. As he stated in the proposal, the US-Mexico border survey project would use field research to document how a boundary line divides the landscape, what differences an arbitrary line creates within a previously "unified" landscape, what evidence of land use is created by the border line, and how a line changes the visual space within the landscape.

Surprisingly, the earnest young art professor made the first cut. The university's administration congratulated him for being a finalist for a Research Award and instructed him to appear on a designated day to defend his proposal. When Peter was summoned, he entered a narrow room with a very long table, at whose sides were seated eighteen men in suits and ties, equally spaced with nine per side. Peter was seated at one end, while the Dean of the Graduate School presided at the far end. The Dean began, "We have basically one question. How is it, Professor Goin, that artists *do* research?" The question blindsided Peter, but he ad-libbed an answer. "It's simple," he said with a show of confidence:

> Art is a visual language. The focus on language is paramount in education. We use language to comment on history, to reinterpret history, to investigate history, and to provide documents. Throughout scholarly interpretation, the photograph is used in many occasions to support, to identify, to provide artifacts that are used by scholars to justify a particular thesis in the humanities and even the sciences.

He explained that he perceived his own role as professor of photography at a research university to integrate with the humanities and, to the degree possible, with the sciences. "Throughout the history of landscape surveys," he continued, "photographers and artists were always included. Our representation of the American West comes from visual identity, and it is very important to continue that identity." There were no further questions. Peter was awarded the grant, the first artist in the history of the university to receive research funding, paving the way for other faculty in art, who now commonly receive research grants.

Placed on the spot, with critical funding at stake, Peter had done a passing job of making a case for the value of photography, and indeed of art itself, in research. But although his rhetoric was concocted, his answer was not disingenuous. He remains adamant that artists and photographers contribute to science. In his view, the idea that to do research one must produce statistical data and have quantifiable investigations is too narrow an interpretation of research. He argues for a broader view of what research is, favoring a return to an earlier integration of science and art. "If we go back to Leonardo da Vinci, science and art were the same," he contends:

> They were woven together in a fabric. In the Cartesian evolution, we split out the art because it became slightly suspicious because of its relationship to sentiment. We did not trust it any longer for science, which is the pursuit of objective knowledge. Cast aside was the value of *intuitive perception*, and artists have been undervalued ever since.

If we look to the United States for examples of the marriage of art and science, we may recall that during the nineteenth century, photographers participated in scientific surveys. Granted, photographers were not thought of as artists at the time; however, they *were* artists and, as such, used the artistic language of formalism and composition. One linguistic clue that government survey photographers maintained an aesthetic sensibility while documenting scientific expeditions is that their captions frequently include the line "Best General View." Those early photographers determined which places and features of the land were fit subjects for photographing and then chose the "best" place to stand and time of day to make the photograph. "In that phrase, 'best general view,'" Peter explains:

> is embedded some kind of genetic aesthetic code. There's this presumption that those photographers (they were called 'operators') working with a camera were not fine artists in the traditional sense. Yet if you look at their methodology and how they envisioned landscape, they are doing the same thing that landscape painters did.

His point is that a creative origin is implied in the very notion of best general view. However, considered critically, the idea of *best general view* is presumptuous. How many best general views can you have of a place? Only one. That is a not a pluralistic interpretation of the western landscape. It rather reflects nineteenth-century Euro-American conceptualizations of the landscape. In titling one of his own images "best general view" (see Figure 7.2), Peter not only pays homage to the artistry of scientific survey photographs but also highlights the irony of the notion that there is only one aesthetic frame through which to represent a given landscape.

Once he was in the field, along the border, Peter's professional challenges were threefold—physical, technical, and artistic. Physically, he had to find a way to function under conditions of relentless heat and personal danger and to navigate the extremely remote, unmarked roads and trails of the borderlands. Technically, he had to effectively manipulate the variables that a 4x5 view camera afforded, process 4x5 film on the road, and record the subject and precise location of each image. Artistically, he had to decide where, when, and of what to make pictures. If a photographer were to document the border today, she would likely use a digital camera and take thousands of pictures, later selecting the best ones to print. But in 1985, before the digital camera, film and developing was expensive, 4x5 film especially so. The work was time-consuming and often unpredictable, and Peter was unable to see the result often for a week or more at a time. He had budgeted approximately two hundred images to document a border nearly two thousand miles long. Therefore, each photograph must convey more than what appeared in the frame of the rectangle; each image must be representative, highlighting a salient aspect of the border landscape. As a visual researcher,

Peter needed to find a way to use the language of photography to communicate information and his intuitive perception of the border line. And as an artist, Peter wanted to make each photograph visually compelling; however, the landscape itself was often minimal and spacious, seemingly devoid of visual interest.

One of Peter Goin's maxims is that "each project requires its own solution." By that, he means that each of his projects takes on a different look, a result of adapting his photographic technique to meet the unique conceptual and visual challenges of a particular subject. But in south Texas, the most basic challenge was not aesthetic. It was how to function. "In 115- or 120-degree heat," he recalls, "you don't feel like eating. You have no appetite. Drinking water that is 100 degrees can make you throw up. It's not pleasant." The solution is counterintuitive. Heat the water still further and make hot tea, a recognizable beverage. Peter also heated canned foods on his little stove, "really dreadful." Soups, bread, and oranges rounded out his rations. He lost a lot of weight. "It is so desolate," he shakes his head at the mere thought, "You couldn't stay alive. It's like Lawrence of Arabia, crossing the desert." But there were compensations:

> The nights were spectacular. It might be ninety, but it feels cooler. If it's 120 in the day, that's a thirty-degree drop. Ninety degrees and you're wearing a sweater and sitting by a little campfire. Can you believe that?! It's ridiculous.

Along the border corridor, frequent lightning storms in the evening created unforgettable night views.

Film processing presented especially vexing problems. As the name implies, a 4x5 view camera requires that the photographer work with individual sheets of film, each measuring four inches by five inches. Each sheet of film must be loaded and unloaded into the rectangular film holder in complete darkness. At regional stops, Peter could load film at night in his darkened motel bathroom, using towels to block the light, but in the field, loading had to be accomplished by feel in a black bag. "It was such a challenge to keep film clean," he recalls. Dust was pervasive, invasive, and abrasive. Using color film was out of the question at that time because the heat caused a chemical reaction that created a color imbalance, and color film cannot be processed in a hotel room. Being restricted to black-and-white film was not a problem, though, because Peter wanted to pay homage to the nineteenth century tradition of black-and-white survey photography. Once a week he would get a motel room, process film, eat fresh food, take a shower, and replenish supplies. At each regional stop, Peter turned his motel room into a photo processing lab, setting out tubs of developing fluids and stringing up lines to which he pinned the negatives to dry. In a field notebook, he recorded the number of each negative and its corresponding photograph subject and location. With characteristic self-discipline, Peter persevered, stayed organized, and focused on the task at hand, becoming an indentured servant to a mission that demanded his complete devotion, sixteen hours a day, seven days a week, determined to make the most of the long-light days of summer.

As became true of many subsequent projects, Peter felt a weight of responsibility. Since no one else had made a photographic study of the entire length of the US-Mexico border, Peter felt a duty to the historic record to make images that not only documented the effect of the line on the landscape, but that created a visual record of precise locations that future researchers could use as a benchmark to study change. Until May 2000, Global Positioning System (GPS) readings were intentionally degraded by the US government as a national security measure, rendering GPS coordinates unreliable for civil and commercial users. Peter thus brought along a full set of US Geological Survey topographical maps for the entire length of the border and precisely marked every photograph location onto the appropriate map. A penciled dot on the map indicated vantage point, while an arrow designated the direction the camera was pointed. Often, he would write field notes directly onto the map, recording the local vernacular name for a site or jotting down the name of the Border Patrol agent who guided him along a particular stretch of border.

Border Patrol agents proved to be an unexpected boon. At first, Peter started out on his own. When a Border Patrol agent asked him what he was doing and whether he had obtained permission, Peter said that he didn't realize that he was on lands that were off-limits. He asked if there was a law that prevented him from being there. Once Peter explained who he was and what he was doing, the agent invited him to do a ride-along. It didn't take Peter long to realize that riding in a 4-wheel-drive, air-conditioned government vehicle with a guide who knew the area was a huge improvement over picking his way alone in his hot little pick-up truck. Soon, ride-alongs became an integral part of Peter's field methodology. Although Peter worked alone for some sections of the border, in a number of situations one Border Patrol agent would call ahead to the next office along the border to set Peter up with his next ride-along.

But which places to photograph and how to create a photograph that would convey his intuitive perception of the place? Peter's first artistic choice was to travel—and arrange the photographs in the survey—from east to west, Texas to California. His choice recapitulates the westward movement of emigrants in US history, from the Atlantic to the Pacific. In making the final prints, Goin occasionally used the same bleaching process used in his Mesoamerican work to evoke a sense of an almost surreal atmosphere of heat and humidity. Of the two hundred images that he made, only sixty-six were included in *Tracing the Line*, sixteen of which are reprinted here. When we discuss the images that made the final cut, we will trace the line of Peter's thinking about the significance of these particular images as exemplars of the border landscape.

When Goin presents this work to audiences, someone will invariably ask, "What was it like?" Because *Tracing the Line* pays homage to nineteenth-century expedition volumes, which omit personal information and subjective perceptions, Peter followed suit. However, the experience was intense, and every image triggers vivid recollections from his epic adventure. Advance correspondence with Border Patrol officers sobered him into the reality that the border zone was very dangerous due to the combination of heat, remoteness, rugged or barren terrain, and crime. One Border Patrol agent taught Peter how to read tracks. The "final exam" was for Peter to decipher a set of prints on the ground. Peter studied the prints and took a guess that they were made by cows. The agent gave him an "F" and said that they were the tracks of a person (probably an undocumented worker) wearing shoes with cow hooves pointing backwards glued to the soles. The gait of a cow would have been at a different distance, pacing, and weight than that of a human, the agent explained.

A Border Patrol officer out of El Paso invited Peter to cross the border into Juarez with him for dinner in Mexico. When Peter showed up to meet him after work, the officer was wearing plain clothes, and he handed Peter a derringer, telling him to tuck it into his boot. At the restaurant, after they ate, the officer whispered to Peter that he had been recognized and that they had to scram. A mad car chase ensued, straight out of an action-adventure movie. Apparently, the Mexican drug lords wanted to kill this officer of the law. The two narrowly made it back into the United States, but Peter was spooked and had a painful lesion from the derringer rubbing against his ankle.

"Where are all the people?" is the other most common question that viewers ask about the project. Goin has answers. First of all, he did not see very many people in that whole no-man's-land of the Rio Grande. And if people saw him, they would hide. Furthermore, if one were to photograph people, the question arises: Which people? Which stories would you represent? For example, he says:

> If you have Maria, who crossed here and was raped, her story is going to be radically different than someone who came across and all of a sudden provided for their family and got married and their life became better. And you have both of those narratives in the same place. So, which is the true story? It's a slippery slope. Historical revisionism comes into play in choosing stories. You would really need a Studs Terkel approach to represent them all. *Tracing the Line* isn't that. It is a survey of the landscape. I asked the question, 'What happens when you draw a line upon the land?'

Peter wanted to document the stage upon which human drama takes place: "I am working with the interpretation of space. I am trying to capture a sense of that place." So many things happen along the border line: Crossing, smuggling, hiding, fighting, eating, playing, working, trashing, driving, building, irrigating, ranching. These actions create the altered landscapes that Peter set out to record. These places can later be rephotographed to study change, whereas people cannot be so easily found again. "The landscape is verifiable," Peter explains, "You can go back to these exact same places." While Peter is keenly sensitive to the human plight, documenting that subject would have been an entirely different project.

Concerning people, Peter quickly perceived that the border line creates its own culture. "That bleeding of cultural identity is really interesting," he remarks:

> It's almost as if you were to take eye liner and put it on a page and then mist it, and it bleeds. The border line has created this culture that is such a blend of confusion. It could almost be its own country in a way. Think about Spanglish. It's very colloquial, very slang, derived by the identity of the people crossing. It's not going to be your physicians crossing; it's not going to be your college professors. And they come not just from Mexico, but from Guatemala, Nicaragua, and El Salvador. You can walk in El Paso and see so many people who are not citizens, who are not in this country legally. They might be workers, in which case they are undocumented. And it might make you do things like lock your car when you are sitting in traffic. It's a different *feeling* there than people might have from where they normally live. You walk more carefully. You hold onto your possessions more carefully. You are uncomfortable where you park your car. All of that is part of this grand narrative.

The border joins as much as it divides.

The border survey project resulted in a remarkable book with an interesting story behind it. As a recently hired Assistant Professor of Art, Peter envisioned the book being published as a Fine Art book by the University of Nevada Press. However, the press turned down the project. Naively, Peter assumed that if one press was not interested, none would be. Rather than approaching other publishers, Peter decided to take this setback and turn it into an opportunity. He recalled the work of nineteenth-century photographers such as William Henry Jackson, Carleton Watkins, and Timothy O'Sullivan who accompanied government surveys.

Their photographs were published in government reports, but they also published their work in illustrated, limited-edition volumes that were worth phenomenal amounts of money that only rich people could afford. Peter once had an opportunity to see a William Henry Jackson volume, and it made an impression. He agreed to collaborate with the university's Black Rock Press, which was interested in working with him to publish a limited-edition art book that pays homage to those earlier volumes while itself becoming an important historical document. As he recalls, "It really was a way of saying, OK, based on the value of the production of this book these volumes will be preserved in institutions dedicated to archival preservation."

The resulting volume, entitled *Tracing the Line: A Photographic Survey of the Mexican-American Border*, required extreme dedication and meticulous attention to detail. Peter commissioned the Schuberth Bookbindery, a German company based in San Francisco, to bind fifteen copies of the book. Each bound book cost him $350. It took a year to print the photographs and process them archivally, and then Peter himself hand-mounted the photographs to the pages—sixty-six images per book, fifteen books, for a total of 990 images. The university owned an antique letterpress printer, and Peter's friend Robert Blesse, director of Black Rock Press, and student Kathi Rick generously typeset and printed the book. The book is housed in its own box and both are covered in gray-and-white linen. The impressive tome measures 10½ inches tall, fourteen inches wide, and two inches thick. The ultra-thick pages, Rives BFK 280 mg, are a creamy off-white with a soft texture like flannel. Jet-black letters in centaur typeface, pressed into these thick pages, connote great care and expense. The book begins with a black-and-white, line-drawn map of the border, from San Diego/Tijuana to Brownsville/Matamoros, depicting political jurisdictions, place names on both sides of the border, and major rivers.

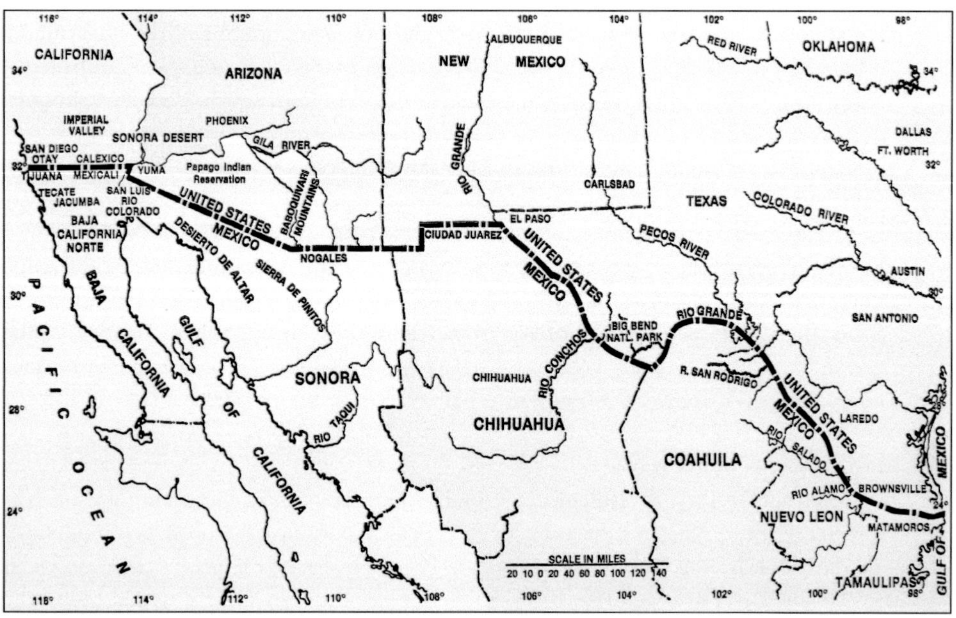

7.1 *United States-Mexico Border*, map by Kathi Rick, reprinted from *Tracing the Line.*

As the book lies open, on each right-hand page, centered, is a black-and-white, semi-gloss photograph, 6½ inches tall and 8½ inches wide. The corresponding caption appears on the left-hand facing page. Pages are not numbered. The book seems like it is at least one hundred years old, an anachronistic packaging given that some of the images include modern-day buildings, industrial sites, and cars.

Peter wrote to libraries and collectors, offering a pre-publication discount of $2000. Once published, the book sold for $2400, and today a copy would cost $10,000. The irony is that even though *Tracing the Line* does not have the circulation of his other books, Peter has made more money from it than from all of his other books combined. The book was purchased by the Library of Congress, Whitney Museum of American Art, the New York Public Library, the Bancroft Library in Berkeley, the Beinecke Library at Yale, the Texas Humanities Research Center, the Center for Creative Photography in Tucson, the University of Nevada, Reno library, the Sierra Nevada Museum of Art (now the Nevada Museum of Art), California State Library, and private collectors. El Paso Museum purchased a set of forty eleven inches × fourteen inches exhibition prints. After *Tracing the Line*, publishers began approaching Goin, and his subsequent books have appeared from some of the best scholarly presses, affirming his decision to pursue making books.

Goin's brief preface introduces the project and explores how a simple-looking line on a map becomes complex when it divides two countries and runs across the landscape. The preface subtly reveals a poetic sensibility as the words evoke the political, spatial, and metaphysical aspects of the US-Mexico border line:

> The most basic political element in any landscape is the boundary. Landscape is defined by boundary, and that dividing line is the first step to organizing space. Every landscape is a web of boundaries. The boundary line is a point of contact, presumes a separation, and represents an enclosure. . . . [I]t has length, width, height and construct. It is arbitrary, and at places, apparently random. It is meant to isolate, identify, and protect as much as it is to contain. Rather than just a "line of authority," a border creates a district, section, or sphere of influence. And like all structures, this line affects how the landscape is formed, utilized and perceived.
>
> The boundary line encourages and creates a frontier with completely different economic, political and sociocultural characteristics than would be expected given the natural resources of the area. The pervasive influence of the line extends beyond the physical appearance of its structure. The geometry of the boundary conflicts with the natural order of the landscape. Instead of being simply an anonymous stretch of arid land, the landscape derives identity from the structure imposed on it.
>
> This line is not the simple connection from point A to point B. It is a dynamic fusion of how pattern is created in the landscape. Line becomes direction, as the walker follows the line to that mountain peak. . . . Converging lines indicate depth while the horizon line implies distance. . . . Paths, roads, bridges, and fences with barbed wire become line. Line is implied. Lines intersect each other. The line creates tension by dividing space, both visually and culturally.

Although *Tracing the Line* presents itself as an artist book, and although the preface reflects an aesthetic and philosophical understanding of the line, photograph captions are strictly factual and objective, much like an official report or textbook. For example, here is a caption that explains the monument numbering system for the land boundary west of the Rio Grande:

> Without water, the landscape cannot sustain the urban or agricultural development typical of the Rio Grande Valley. The topography is entirely arid desert. The border is marked by 276 permanent monuments covering 698 miles from Monument No. 1 to the Pacific Ocean. The last monument, No. 258, reflects the numerical system adopted in the Boundary Survey of 1889–1896. The eighteen monuments added since 1906 were not assigned greater numbers, but were designated 2A, 2B, 84A, and so on.

Goin's captions provide geographical and historical information, in keeping with a photographic survey *research* project and a book designed to be preserved in archives.

On one level, in the tradition of the New Topographics, the photographs may appear similarly documentary and objective, not images that one would frame and hang on their wall as art. The scenes would seem baffling and nondescript without the captions that clarify the significance of the site as a representative example of the border landscape. Among the most common subjects are riverscapes, oxbows, canals, bridges, agricultural fields, vegetation (bushes, cane, grasses, cacti), seemingly endless stretches of barren plain, sand, rugged hills, eroded canyons, human-made structures, boundary monuments, urban edges, railroads, paved roads, dirt roads, cattle trails, and footpaths—and fences of all types. Indeed, *Tracing the Line* is already becoming increasingly useful as an anthropological inventory of fence types used along the border in the mid 1980s. We see three-, four-, five- and six-strand barbed wire fences, chainlink, chainlink with barbed wire top, chainlink with razor-wire top, metal mesh fence, breached fences, a wooden cattle fence suspended from a cable over a dry wash to flex with flooding, steel walls, steel cable fences, a line of big rocks, tilted fences, upright fences, reinforced fences, patched fences, and enormous elevated baffles to act as a heat shield from jet blasts at the Tijuana airport.

Beyond their obvious documentary and historic value, however, the border photographs may be read as visual poetry. The captions in this chapter aim to explain the ecocultural significance of the sites depicted and interpret their visual language and artful composition. Ultimately, Goin's art provokes an ongoing debate about the border. Goin's commitment to bearing witness and his engagement with perceiving and visually narrating landscapes of meaning reflects his passionate investment in the scholarship of photography.

■

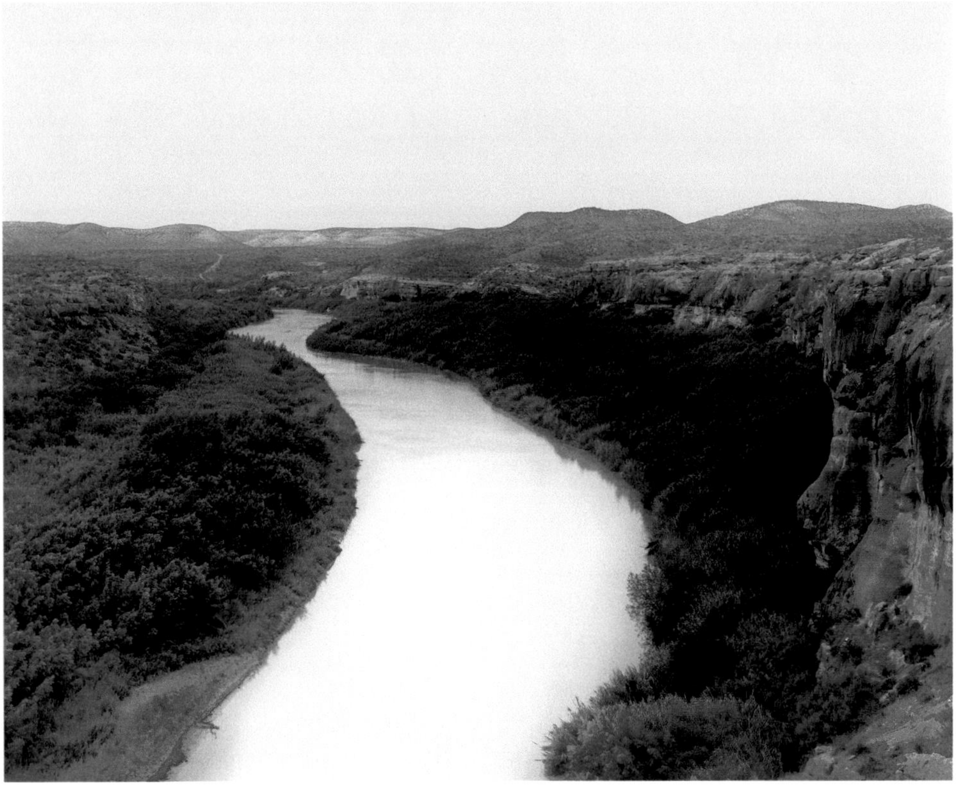

7.2 *Best general view of the Rio Grande looking west from Pump Canyon*, 1985.

One might expect a border study to begin at the Gulf of Mexico; however, Peter found that area to be water, water everywhere, impossible to discern a border. Instead, this chapter opens lyrically, with an image of the Rio Grande, a river boundary running the length of the southern border of Texas. Peter needed to include the Rio Grande, known in Mexico as *Río Bravo* or *Río Bravo del Norte*, bravo meaning, among other things, "furious" or "agitated." However, photographing the river posed a problem: "It's covered with weeds. It's blocked. You can't get down there to see it." Near Langtry, Texas, roughly halfway between Brownsville and El Paso, he finally found an elevated escarpment, revealing a spectacular S-curve of the river, a view that matched people's expectations of the symbolic river. The phrase "best general view" alludes to nineteenth-century expedition photographers such as William Henry Jackson and Timothy O'Sullivan, who accompanied government surveys of the American West. Fittingly, Goin's high-contrast, black-and-white image betrays no evidence of modernity in the undisturbed scene or photographic style. This image at the beginning of *Tracing the Line* thus initiates a temporal arc from the seeming pre-history of this pristine Rio Grande image to the industrial-age effluents polluting the New River in San Ysidro, California, toward the end of the book.

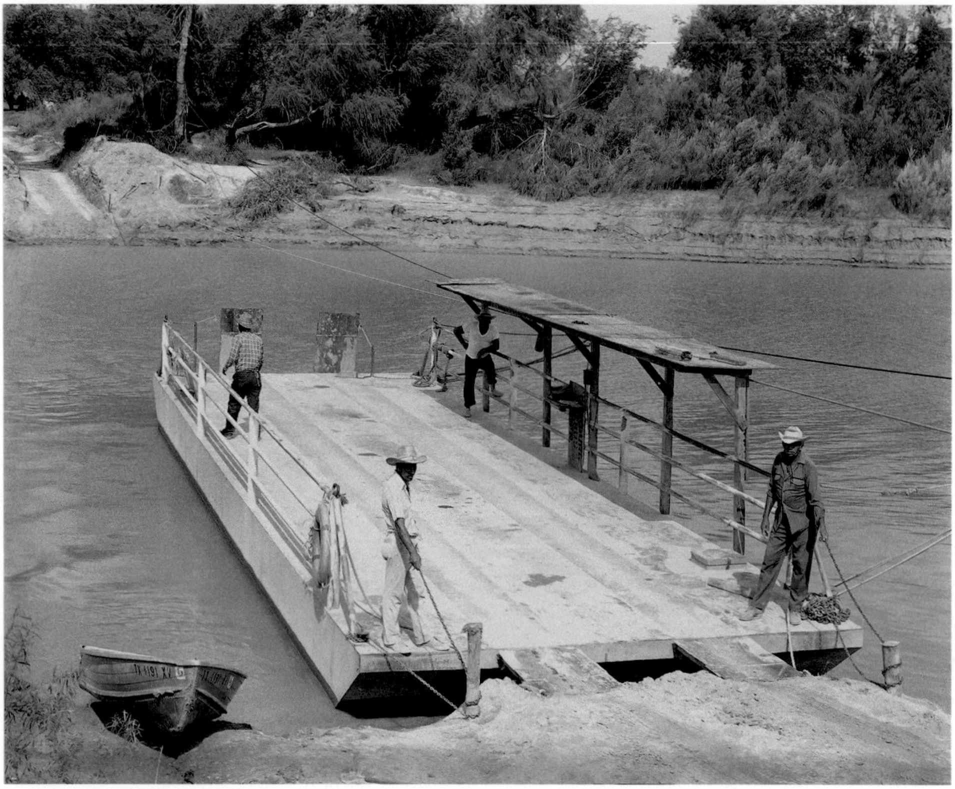

7.3 *"International Ferry" at Los Ebanos*, 1985.

Crossing the border between Texas and Mexico entails crossing the Rio Grande. In the mid 1980s, there were fifteen bridges where vehicles could cross, monitored by customs checkpoints on each side of the river. The humble "International Ferry" at Los Ebanos, just west of the sister cities of McAllen, Texas, and Reynosa, Mexico, was the last hand-pulled ferry along the river. Peter was told that this ferry was going to be discontinued (it no longer exists today), so he decided to document it as a land-use practice along the border line. In Goin's *Tracing the Line*, this image was printed in a light sepia tone to give it a tinge of nostalgia for simpler times when manual labor was used, never mind the irony that the ferry carries motor vehicles. Because there are people in the photograph, the viewer's attention automatically centers on them, and the landscape becomes of secondary interest. The most effective way for a photographer to guide viewers to contemplate the character of the human-altered landscape is to leave *humans* out of the picture, a practice almost necessitated in this project since the border is most often a no-man's-land where there aren't any people to be seen. If our attention were not fixed on the ferry workers, what lines would we notice?

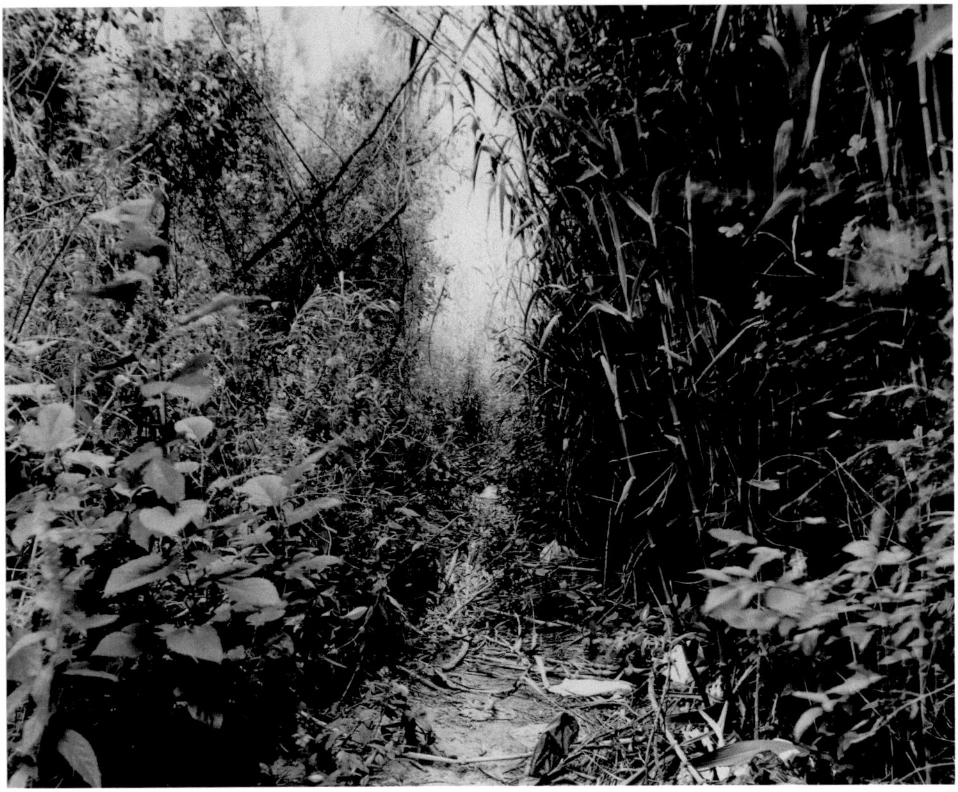

7.4 *River cane and salt cedar on land fronting the Rio Grande*, 1985.

This image is devoid of people, yet their traces imply a human story if one can decipher clues, such as the discarded garment in the foreground. Peter was told that undocumented workers illegally crossing the border, strip down to their underwear and wade across the Rio Grande, carrying a full set of clothing in a plastic bag on top of their heads. Reaching the US side, they hide in the cane and change into dry clothes, leaving their wet underwear in the bushes. When they sneak out of the canebrakes, they hope to pass for legal workers and not be apprehended by the Border Patrol as obvious "wetbacks," a derogatory slang term that references wading or swimming across the Rio Grande. This photograph evokes a feeling of chaos associated with the abstract expressionist painters. Peter explains, "What I am trying to communicate is that mood, that mood of being there. It's kind of scary. There are plants in the foreground that are out of focus. There's darkness; you can't see beyond it." Although the centered pathway draws the viewer into the picture, a vague sense of danger sets up counter-tension, making the viewer hesitate, wary of proceeding.

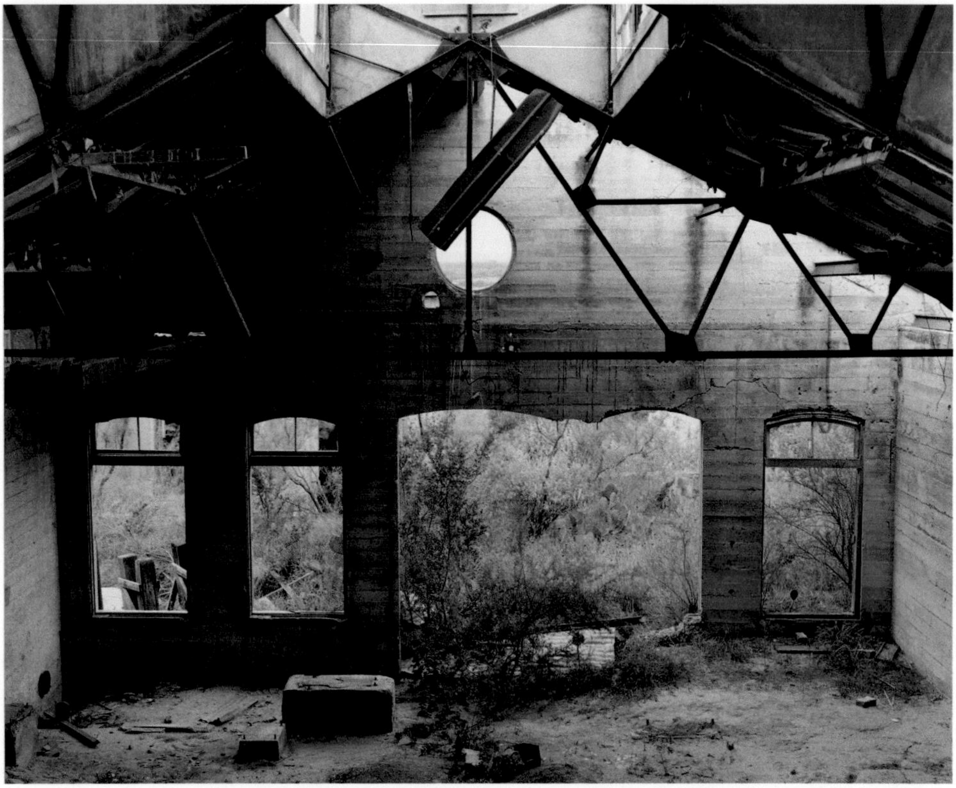

7.5 *Pump station of an abandoned mining operation near Kinney, Texas*, 1985.

Although not literally on the border line, this image is an important element of Goin's complex mapping of the line. This way-station on the *immigrado* trail—one of many such shelters—reflects an understanding that where people hang out is an extension of a line that creates a distinct border culture and border zone, extending for several miles on both sides of the line. This photograph, made from inside the building looking out, represents the immigrants' viewpoint and creates an emotional space. The ramshackle building belies the careful crafting of the photograph itself, arranged so that there is no distant view out of the lower windows, no way to orient oneself spatially in the landscape. Ah, but look through the circular window, centered in the upper half of the picture. *There* is the horizon line, *there* is Mexico, as if seen through the crosshairs of a rifle. Implied in this image is the border line, the horizon line, the structure lines, and the pathway lines. This picture, then, is not just about the literal, graphic nature of the border line. It's about the conceptual nature of the line.

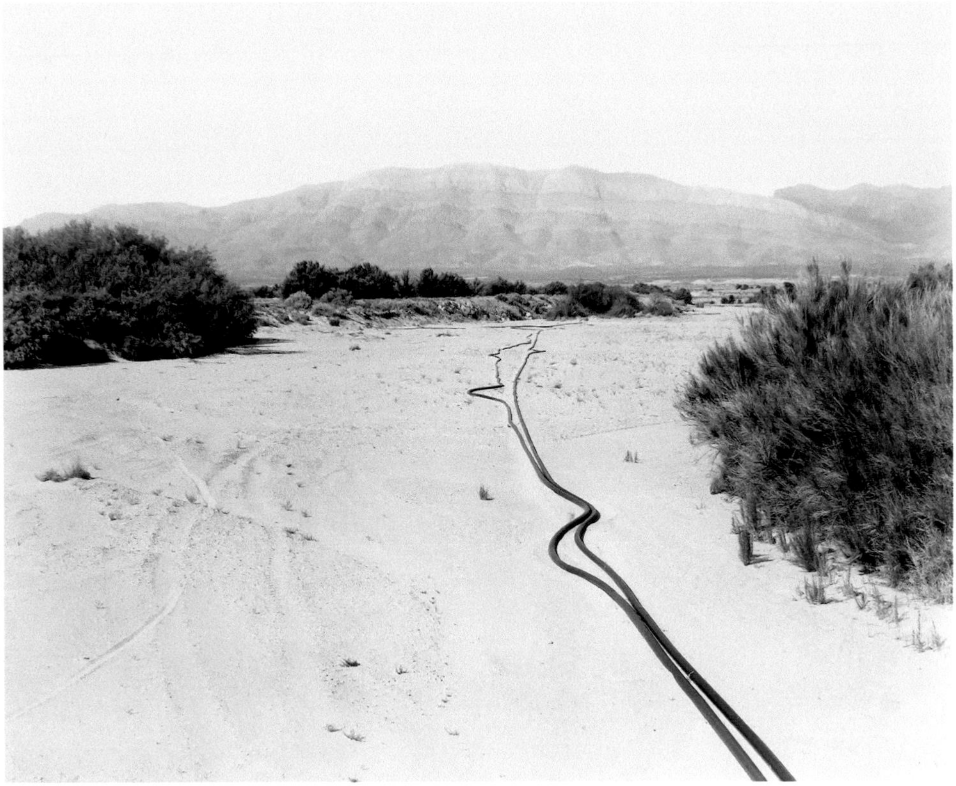

7.6 *Alamo Arroyo, near Ft. Hancock, east of El Paso and Ciudad Juarez*, 1985.

While the three images discussed above focus on the complex story of people crossing the border, this image implies an environmental narrative. According to treaties negotiated between Mexico and the United States, shared waterways must maintain specific water flow. Instead of losing water to evaporation and seepage, hoses may be used to carry the required water across the border. The hoses in this image run through the Alamo Arroyo toward the Rio Grande. Near El Paso, the Rio Grande becomes channelized and the landscape increasingly managed. If the first image in this series depicts a timeless river of light moving lazily through a romantically dark landscape, this image flips the values, the landscape now blindingly exposed and the lyrical river reduced to two functional black hoses whose random bends are an ironic reference to the river's noble curves. Visually, this photograph tricks our sense of depth, creating an optical illusion in which the two black hoses appear to be in a vertical plane, almost like an abstract metal sculpture or perhaps a line drawing on white paper.

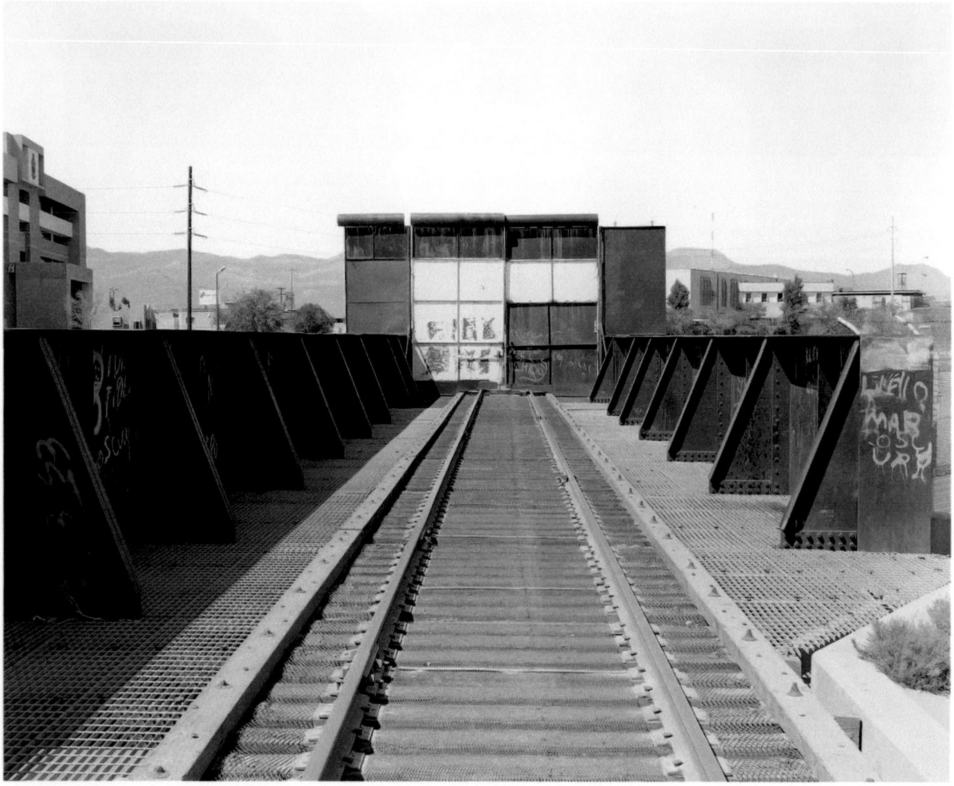

7.7 *Train yards, El Paso*, 1985.

This photograph of "East Railroad Bridge," also known as "Black Bridge," shows a border gate blocking the railroad tracks between Ciudad Juarez and El Paso and documents the repeating geometric shapes, grids, and lines of the engineered railroad bridge that spans the Rio Grande. While the photograph is geometrically compelling, it is more than simply an interesting graphic design. Its lines express insights about the border. For example, while the converging railroad tracks create forward momentum, propelling the reader into the scene, the square, static gate abruptly halts the viewer's forward movement, blocking not only progress but line of sight. The arrangement of lines in the photograph symbolizes the tension created by the border, which simultaneously creates a desire to cross while at the same time denies passage. The irregular bumps and dips of the horizon line in the distance contrast with the straight lines of the railroad bridge. The deliberately low vantage point from which this photograph was made exaggerates the imposing specter of the steel gate, making it dominate the view and appear taller than the hills in the background.

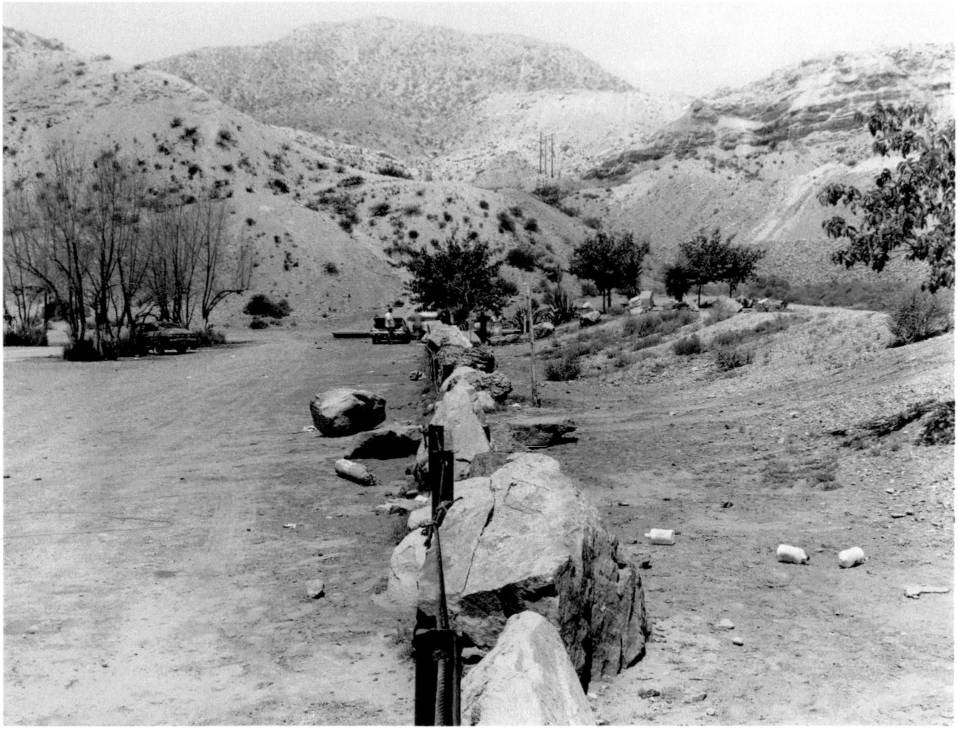

7.8 *Sunland Park, Monument No. 1*, 1985.

The headwaters of the Rio Grande are above Albuquerque, New Mexico. A traveler tracing the border from east to west begins by following the curves of the Rio Grande (also called the *Rio Bravo del Norte*) that flows between the two countries. This photograph of Sunland Park, west of El Paso and Ciudad Juarez, represents where the land border line begins, a straight-line boundary drawn on the land with a surveyor's precision. This image, looking west, depicts a fence dividing Mexico on the left and from the United States on the right. Monument No. 1 is hidden among the trees at the end of the rock fence line. In this first segment, the border lies east-west, a direct line for 104 miles, marked by forty-one monuments. The photograph provokes questions: Why is there a cable fence at a park? Why are there rocks on one side of the fence and not the other? Why is this arbitrary line in this place? Constructing the border was an effort to build a straight line in a topography that doesn't adapt to straight lines. The cable and boulders are to prevent vehicle "drive-throughs," and while Mexicans use their park, the park on the US side is comparatively vacant. The image thus registers a social relationship in which the people of the United States seem to want to keep Mexicans at an uneasy distance, hinted at by the differential engineering and use patterns.

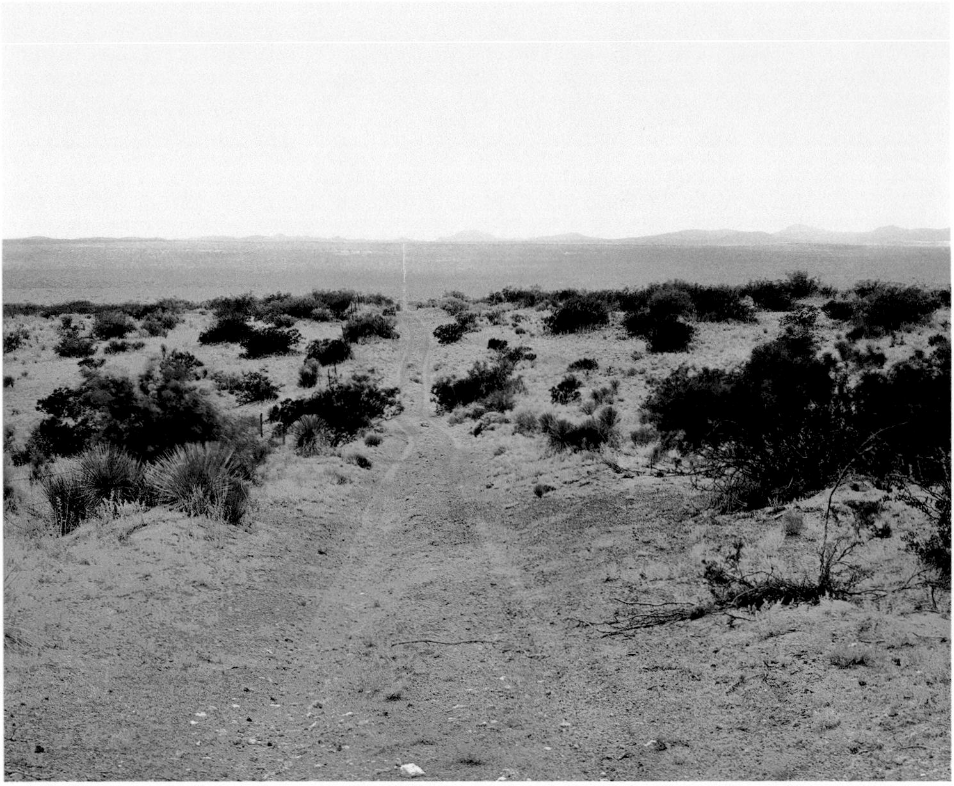

7.9 *Border road, near Monument No. 9*, 1985.

The border is marked by 276 permanent monuments covering 698 miles, from Monument No. 1 on the west bank of the Rio Grande—the eastern start of the land boundary—to the Pacific Ocean. This view looks west at the border road immediately beyond Monument No. 9. Without water, this arid stretch of the borderlands cannot sustain the urban or agricultural development typical of the Rio Grande Valley. This photograph communicates scale, depicting the border region as enormous and vast, challenging and difficult to transect. The image also playfully explores the presumption that a line transforms a space. The straight, vertical line of the road receding into the distance visually takes that landscape and flips it up vertically to look like a bar, flattening that whole space. It becomes a visual illusion (is it the ocean?). The road looks like it is a pole sticking up, very much in the tradition of the work of John Pfahl, an American photographer best known for his "Altered Landscapes" series, in which he uses vertical and inserted elements to create a completely different visual effect, an optical illusion. Aesthetically, then, Goin's image of a straight-line road alludes to what a line does optically.

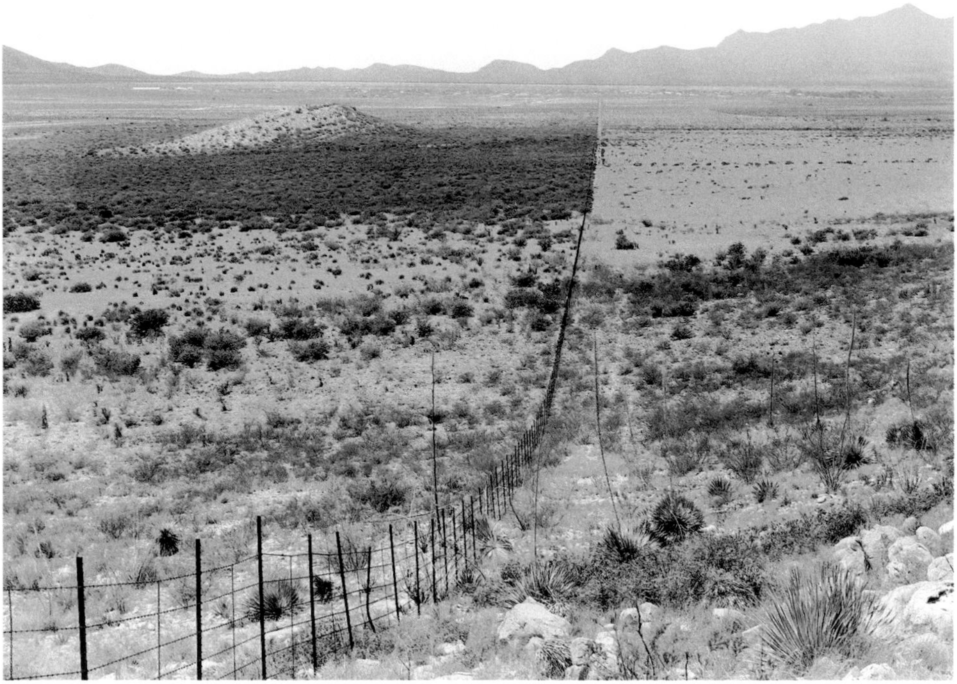

7.10 *Animas Valley, looking west from Monument 91*, 1985.

In the previous image, near Monument 9, the border line on the land carried no apparent land-use consequences; however, the line in this image has resulted in strikingly different biota. The hard edge of this fence line illustrates the results of different land management practices. While the untrained eye may assume that the land on the left side of the border (Mexico) is lusher because of the darker vegetation (which would show up green in a color photograph), in fact, paradoxically, the opposite is the case. Overgrazing on the Mexican side has depleted the grasses, giving way to hardier, grazing-resistant species such as creosote bush. The knee-high dry grasses on the right side of the fence—in Texas—grow on private land. The even more trained eye will note that the barbed wire fence in this image has six strands. It only takes a three-strand fence to keep out cows, so six-strand fences usually indicate a government subsidy, with fence construction complying with federal standards. Fences can be "read" to reveal clues about ecological, political, and economic patterns.

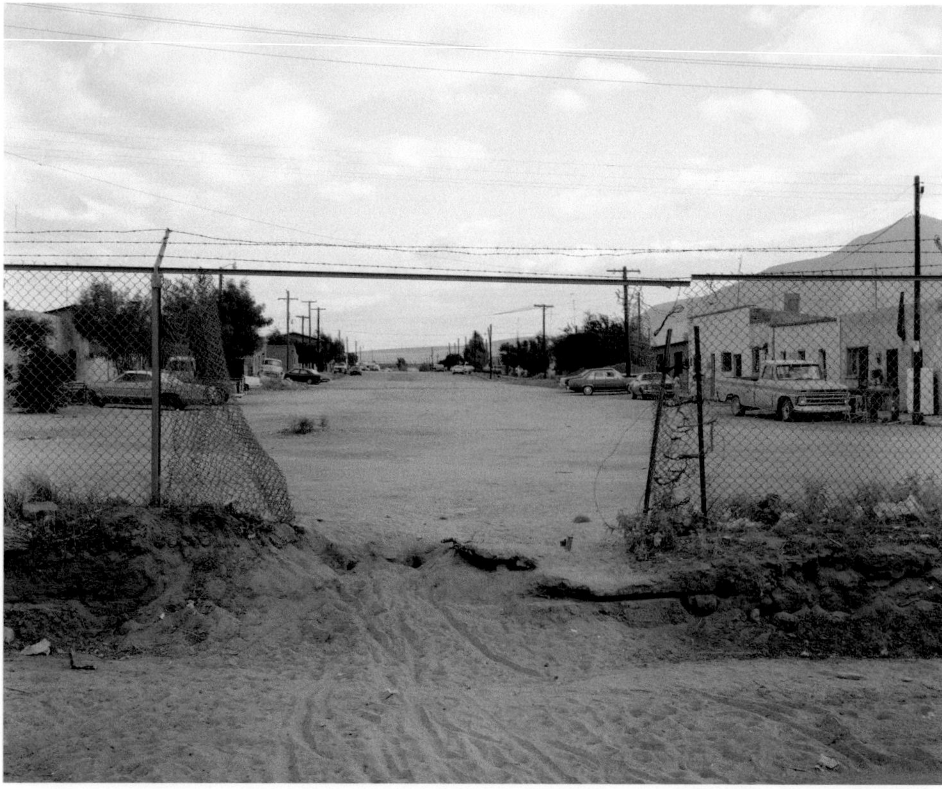

7.11 *A "drive-through," 1/4 mile west of the port-of-entry at Naco, Arizona and Sonora, looking into Mexico*, 1985.

A Border Patrol agent told Peter that this fence had been repaired multiple times, but kept getting ripped open. On this site, a theatrical narrative is enacted daily. The single tracks and many footprints in the foreground are traces of motorcycles and people on foot, crossing into the United States illegally. Unbeknownst to the Mexicans, the United States installed motion detectors and set up a perimeter about a mile into the United States. The border crossers suffered through the intense heat and across the desert before being apprehended. Rather than arresting or detaining them, US Border Patrol agents returned apprehended individuals back to Mexico, in what became a merry-go-round, a game of "cowboys and Mexicans," as one agent put it. In this photograph, the fence lines, motorcycle tracks, telephone poles, power lines, road edges, and horizon line create a layered taxonomy of lines. "The border is a line," Peter observes, "but a line intersects; it crosses, it divides, it separates, it can be violent, it can be divisive."

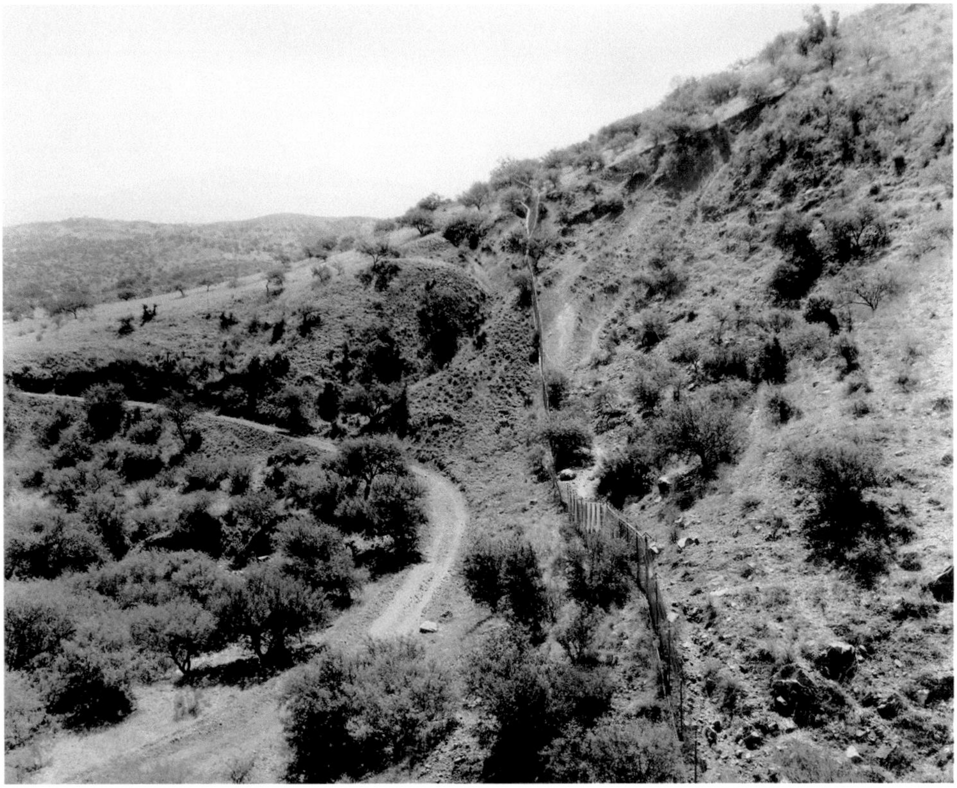

7.12 *View from Monument No. 120, looking east into "Smuggler's Gulch," near Nogales, Arizona and Sonora*, 1985.

Mexico is on the right. This rugged landscape looks like the setting for a John Wayne film, and the name "Smuggler's Gulch" that *La migra*—immigration agents—give to this area plays into that Wild West mythology. A confusing tangle of paths cuts the hillside slope on the Mexican side, a mix of cow paths and drug smugglers' trails, the latter going all the way to the fence. Although difficult to discern in this picture, the chain-link fence is topped with razor wire. However, smugglers simply throw their contraband over the fence to a waiting car. Smugglers know the times that the Border Patrol agents patrol the area, and they evade detection by driving at night with their headlights off. It's an interesting and dangerous place, and, with a little imagination, the trees and shrubs in this image seem animated, like disguised people crossing. In *Tracing the Line*, Goin varied his camera direction in deference to the alternating points of view held by people in the border region. Although the order of the photographs in the book proceeds from east to west, the point of view of every image varies, this one looking east.

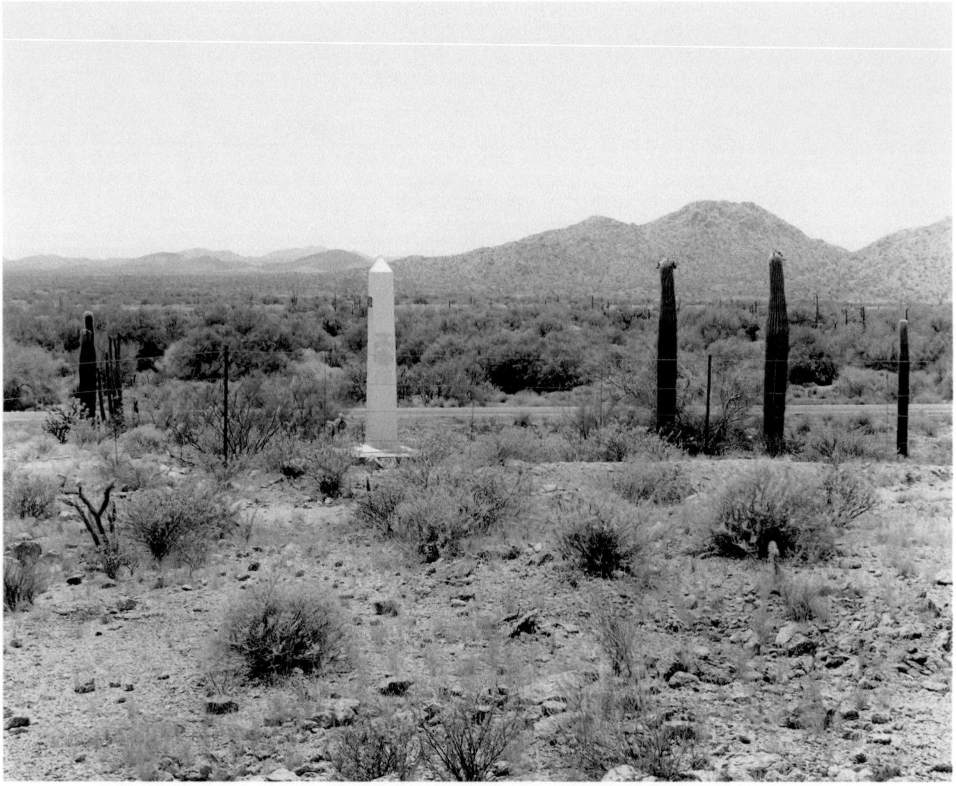

7.13 *Monument No. 172 and sentry cacti near Aguajita Spring west of Lukeville, Arizona, and Sonoyta, Sonora*, 1985.

Notice how the monument, nestled into the topography, tips the space up vertically and flattens it. "In a way, this is a very oblique reference to the processes by which we 'otherize' Mexican society, culture, and wishes," Goin remarks. "We do this in terms of our language and policies. We reduce them to a flatness. And we reduce the landscape, seeing it without complexity or layers of context." Note, too, how the tip of the monument connects to the horizon line. The horizon appears to be balanced—or precariously imbalanced—on the fulcrum of the monument point, creating a subtle sense of psychological unease. The saguaro cacti are lined up, "just trying to behave," adding a touch of playfulness to the scene. What if the photographs were made from a vantage point where the monument and cactuses were not pictured? The landscape would look different and the image would have a different meaning. Vantage point and compositional structure of the elements in a photograph are not random but meaning-bearing, and the visually literate reader will consider such elements when analyzing an image.

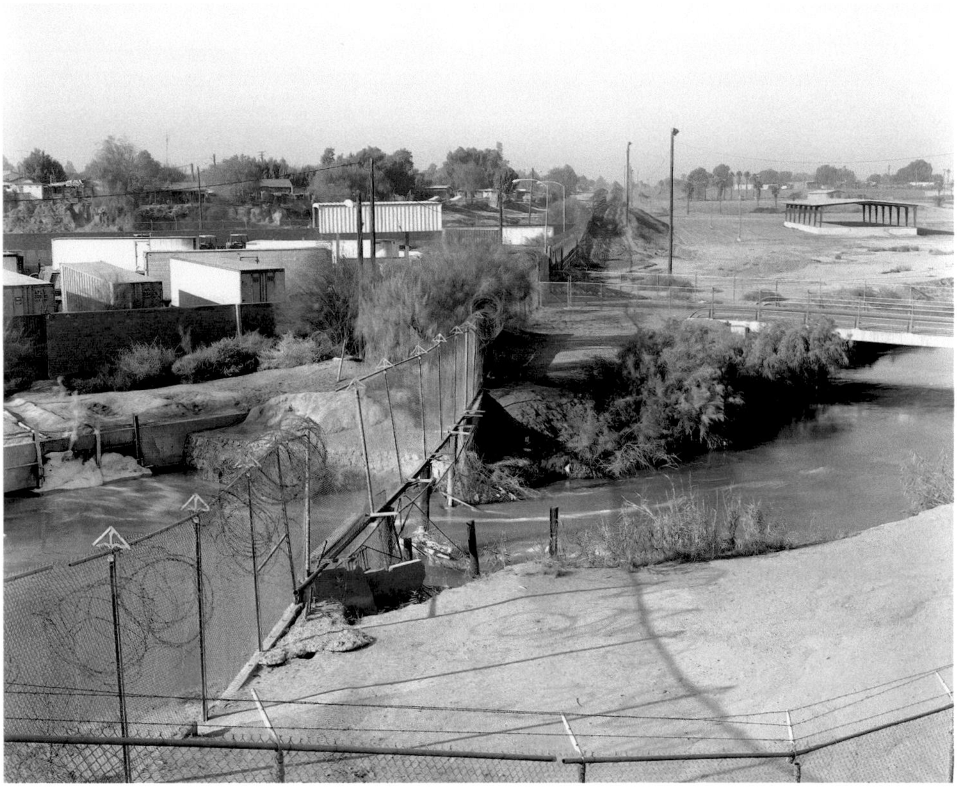

7.14 *Looking west at the entry point of the New River into the United States at Calexico and Mexicali,* 1986.

The New River is severely polluted, posing a major health hazard. Factories in Mexico, on the left side of the fence, discharge chemical waste into the river, which flows north to the Salton Sea in California. The white suds and streak in the water are effluent. Although water flows across the border unimpeded, people are forbidden to cross freely. The photograph depicts the *structure* of the concentration camp style management of this industrialized part of the border. Note the shadow of the fence, the boundary of the fence, and the roll of concertina wire. "Border patrol officers will be stationed off at a distance with binoculars, and they will fly over in helicopters," Peter recounts. "There is a lot of supervision because this is a very heavily urbanized area. It's a contested area." In May 2005, the New River was designated an environmental justice water quality pilot project for the State of California; the problem of raw sewage in Mexico was addressed and some sections of the river were covered over with a new thoroughfare, but according to the *L.A. Times* in 2018, the New River remains "a cesspool of pollution."

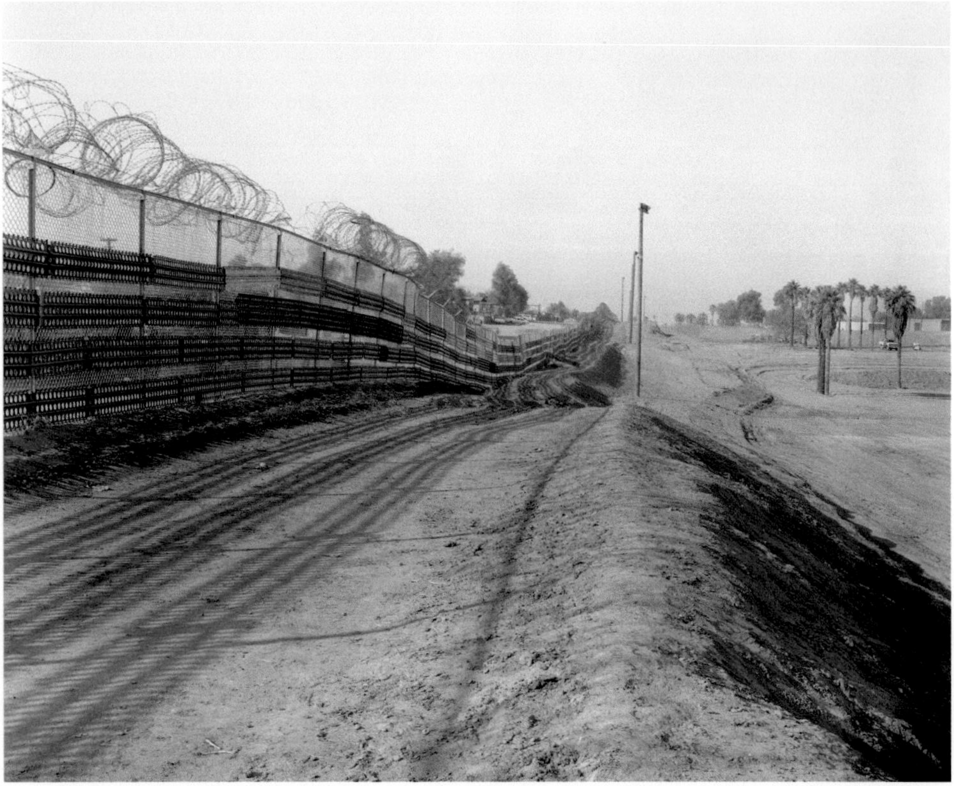

7.15 *"Tortilla Curtain,"* 1986.

In the previous photograph—of the New River—locate the dirt berm on the right side of the fence in the background and a dim grouping of palm trees in the upper right corner. The present image of the border fence was made from a vantage point on top of the dirt berm, narratively linking these images, as if the viewer is walking into the New River photograph to trace the line. Peter Goin's caption in *Tracing the Line* explains:

> During President Carter's administration, the Immigration and Naturalization Service constructed an 'impenetrable' fence at selected areas near El Paso, Calexico, and San Ysidro, among others. It is twelve feet high and constructed of metal webbing (much like chain link) topped with barbed concertina wire. It is called the "Tortilla Curtain" and does little to stem the flow of illegal crossings into the United States. The metal sections along the fence were installed to repair holes. Mexico is on the left.

Lines form a dominant design element—perspective lines, light-pole lines, concertina lines, fence lines, and shadow lines. This image has been reprinted many times, perhaps because it graphically conforms to people's understanding of the border.

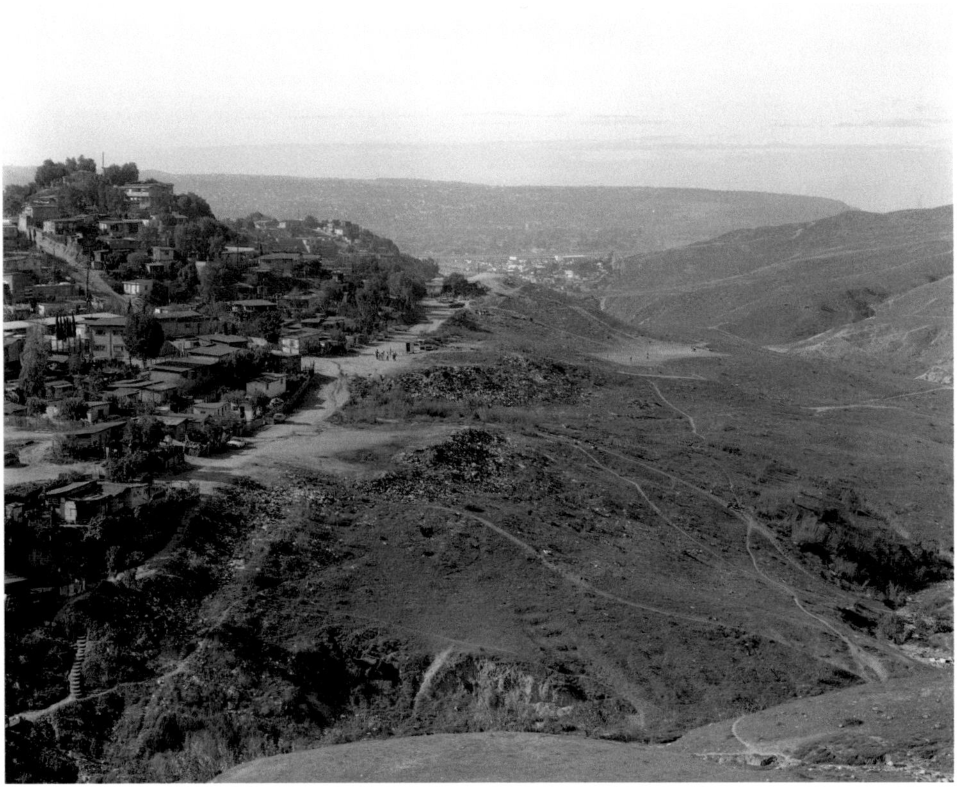

7.16 *"Soccer Field," Colonia Libertad, Otay Mesa,* 1986.

This photograph of Otay Mesa, Colonia Libertad, nicknamed the "Soccer Field," triggers hair-raising memories. A Border Patrol agent, acting as guide, took Peter to this overlook. He said they could not get any closer because they would endure the risk of gunfire, aimed in their direction. Every evening, anywhere from two hundred to two thousand potential immigrants await nightfall to cross from the populated Mexican side (at left) into the United States. The space between the actual border, marked by the road at the edge of the town, and the Border Patrol "defended line," at the north side of the canyon, forms a no-man's-land that is one of the most dangerous areas along the entire frontier. "During 1984," Goin writes, "these canyons were reported to be the stage for 176 robberies, 8 murders, 15 rapes, and more than 501 victims." In the background is the marine layer over the Pacific Ocean. "It is a phenomenal theatrical performance," Peter marvels, "You can see all of these paths. You'll see the people start to gather up here, waiting. Once it gets dark, whoosh, a mad rush. It's *Waiting for Godot*—with danger."

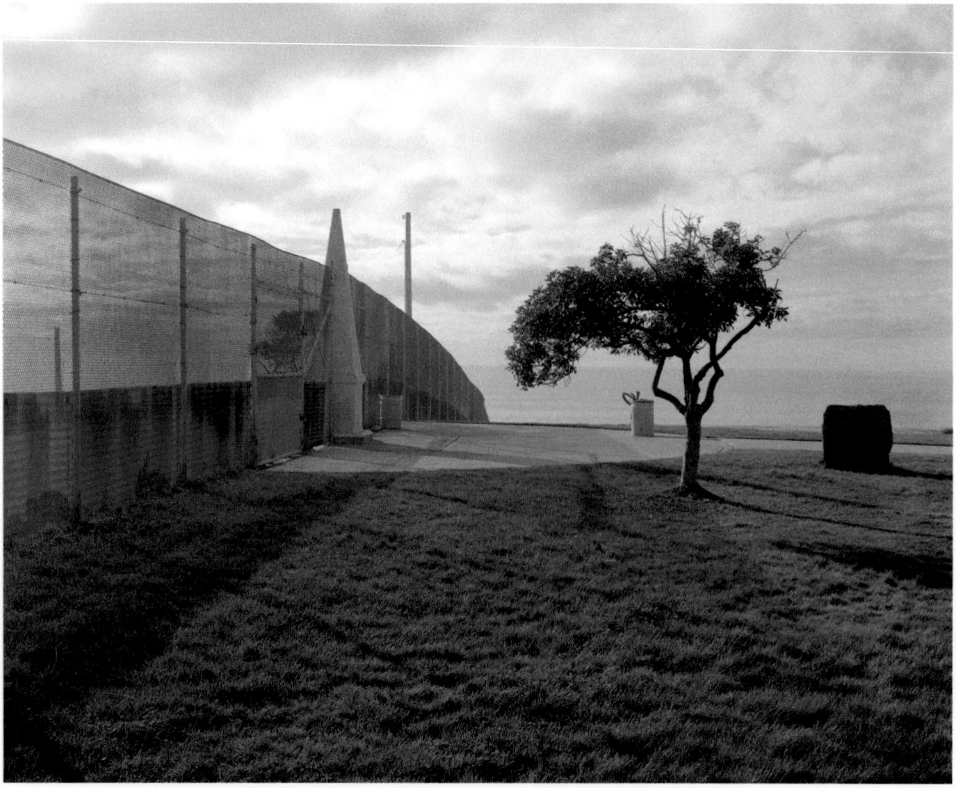

7.17 *Westernmost marker, Monument No. 258, Border Field State Park and Pacific Ocean,* 1986.

In this photograph, San Ysidro, California, is on the right, Mexico on the left, behind the fence. When Goin made this view in 1986, you could see through the border fence, creating a porous feel. Today, however, the fence is a sixteen-foot corrugated iron wall. This photograph features a number of design elements that by now should be recognizable as the syntax and grammar of Goin's art. There are lines—the fence's uprights and diagonally sloping top, the shadow line, the monument, the horizon, the telephone pole, the concrete plaza edges, the tree trunk—there are triangles and squares. And there is the natural element of the tree, looking relatively free-form, unkempt in its leafiness, in contrast to the hard edges of human-made structures and the rigidity of border policies. Because of the tree trunk, the ocean appears visually to be flipped up vertically like a bar. The soft sunset light, illuminating the collection of objects, creates a still life. The border's terminus, at the Pacific Ocean, lyrically concludes *Tracing the Line*'s westward moving survey. We have reached the end of the line.

Works cited

Goin, Peter. "Proposal: Junior Faculty Research Award, 'The Mexican-American Border from Brownsville, Texas to Tijuana, Mexico, A Photographic Project of the Border Landscape.'" Unpublished manuscript. University of Nevada, Reno, 1984.

———. *Tracing the Line: A Photographic Survey of the Mexican-American Border.* Reno, Nevada: Peter Goin, 1987.

8

LANDSCAPES OF FEAR

Tracing the Line sought to research what happens when you draw a line on the land. In the case of the US-Mexico boundary line, what does the border landscape look like? How has the line affected use? What are the sociocultural ramifications of the international border? Peter Goin's next project—*Nuclear Landscapes*—continued the visual investigation of bounded landscapes, in this case, bombed landscapes—national sacrifice zones.

Having moved to Reno in 1984, Goin enjoyed exploring Nevada's basin-and-range country. Since Nevada has so much federal land—nearly 80% by some estimations—people have access to camp, hike, photograph, and explore freely without encountering very many "Private Property, No Trespassing" signs. However, the state's natural beauty and biotic diversity are somewhat of a best-kept secret, in that much of the nation views Nevada as a barren wasteland. Besides having 314 mountain ranges, the most of any US state, Nevada is the most arid state. Goin became interested in how arid environments are the origin places of many of the world's major religions, yet Nevada's desert environment has led to its being perceived—and treated—as a godforsaken wasteland. Between 1951 and 1992, 928 nuclear tests have occurred at the Nevada Test Site, approximately 100 of them above ground, making Nevada the most bombed state in the nation. In the 1980s—and some might argue to this very day—Nevada has been relatively lacking in artistic interpretation, its deserts deemed less photogenic than the red-rock canyonlands of Utah or the saguaro-studded slopes and grand canyons of Arizona.

As an art professor at the University of Nevada, Peter began to think about ways to be a voice for Nevada. As luck would have it, he was invited to submit work to an art exhibition that was being planned about the Nevada Test Site, where atomic testing took place during the Cold War. Since the Test Site was off-limits to the general public, the exhibit's organizer asked artists to *imagine* what it was like and to produce art based on their imaginary visions. Although that planned art exhibition never materialized due to lack of funding, a seed was planted. Rather than producing imaginary work, why not gain access to the Test Site, cross the line into the restricted area, and document what it *really* looked like, Peter wondered. Why not photograph the effects of nuclear testing on the landscape? He was told that artists were forbidden to enter the Test Site. Not easily dissuaded, Goin began a campaign to be

DOI: 10.4324/9781003212607-8

granted access, contacting myriad state and federal agencies, including two Nevada congressional representatives and ultimately the Director of the Department of Energy. He described his project as visual research for educational purposes, intended for publication by a nonprofit, university press. Eventually, after more than six months, thanks to the help of Nevada's congressional representatives, he got a yes, the first artist to be granted access to the Test Site.

Once Goin became immersed in the project, it began to be less about Nevada specifically and more about the effects of the nuclear testing era on the landscape. The project furthered his ongoing inquiry concerning how we form landscapes and how the land itself reflects our culture. He decided to also document the Trinity Site in New Mexico; Hanford, Washington; and the Marshall Islands in the Pacific. Gaining access to each of these sites demanded rhetorical prowess and tenacity, and permission was granted on the condition that Peter be accompanied by a guide at each site save the Marshall Islands, where he was a solitary figure.

Goin's method of fieldwork had by this time become well established. First, he did extensive research to understand the history, character, and issues related to the place. He procured funding, arranged access, and made meticulous plans for a trip. And then, he spent time within, around, and immersed in the locale. In essence, the method required Peter to *be there*, to be fully present with all his senses, immersed in and keenly attuned to the landscape, allowing the place to make an impression on him, almost as if he himself were photographic film.[1] Once he not only saw but viscerally *felt* the place, he made decisions about how to use the visual language of light, color, framing, and perspective control, among others, to document the views in a way that conveyed his interpretation of the place. This process involves synesthesia, the use of one sense, such as sight, to convey the impressions of another sense, such as feeling. The *Nuclear Landscapes* photographs reflect the experience of heightened blood pressure, elevated heart rate, and health anxiety triggered by being at sites that are radioactive.

The *Nuclear Landscapes* project brought its own set of challenges, foremost among them that radioactivity made the work dangerous—dangerous in an uncanny way in that one cannot see, smell, hear, or feel radioactivity. Some sources said that people may get a metallic taste in their mouth when exposed to radioactivity, hardly reassuring. Goin purchased a pocket-sized Geiger counter and relied on his guides to take him to places that were relatively safe. But often, the guides would stay in the vehicle with the windows rolled up, while Peter was alone outside with his tripod and 4x5 camera, normally requiring ten minutes to make a photograph. He decided to use color film for the project, not only to document the color of nuclear landscapes but because color prints were at that time "fugitive," meaning that the image would decay in about twenty-five years. The idea of prints that had a half-life, so to speak, resonates conceptually with the subject of radioactivity. In 1987, when Peter made the photographs, the most popular color film was positive slide film, Ektachrome or Kodachrome. To his taste, however, the color of the "chromes" was too saturated, producing excessively vibrant hues, so he opted for 4x5 color negative film, actually designed for portraiture not landscape photography, which he could develop and print himself, retaining more control of the color palette.

In the summer of 1987 Peter crossed the line onto the 1,350-square-mile Nevada Test Site, where he lodged in the town of Mercury, sleeping in barracks and dining in the cafeteria for Test Site workers. Immediately his Geiger counter became an unnerving companion. He recalls:

> Now when you go to a site like this and you're standing there and everything is fine and you move *three feet* and your Geiger counter starts going 'TRRTTTTT,' you think, 'What the heck was that?!' And then you step back to the first spot, and it kind

> of dies down, and you go, 'Well what's *there?*' Your palms get sweaty. You can't see it. So you step over three feet, and it goes off again. As lame as this sounds, that Geiger counter was just a *scare machine*. It scared the bejeezus out of me. It informs you.

The Geiger counter went off without warning at the Nevada Test Site, Hanford, and the Marshall Islands, causing Peter to characterize nuclear landscapes as *landscapes of fear*.

A formidable artistic challenge is that radioactivity is *invisible*—how do you show it?—and yet it is *the* most salient aspect of nuclear landscapes. Another challenge posed by the Nevada Test Site is that, for the most part, it did not conform to dramatic, post-apocalyptic visions of a bombed landscape. While we may imagine a deeply pockmarked area, perhaps resembling the surface of the moon, what Goin actually encountered was a fairly standard Mojave Desert landscape. He explains, "You cross into this vast territory, and you realize that it looks like any other Southern Nevada creosote region." How to convey the *feel* of that landscape—radioactive, off-limits, fenced, guarded, and somehow alien? The solution lies in *what* to photograph and *how* to use the limited variables of photography to interpret the site. One decision Peter made for the book *Nuclear Landscapes* was to include government photographs of the 1950s bomb tests, showing magnificent mushroom clouds, enormous orange fireballs, and roiling landscapes. These fiery images evoke awe and terror—the contemporary sublime. When Goin visited the Test Site, the Limited Test Ban Treaty of 1963 barring atmospheric testing was in effect, and the tests had moved underground; therefore, he was photographing the aftermath of the above-ground testing era, documenting the visual legacy.

Sixteen of Goin's Nevada Test Site photographs appear in *Nuclear Landscapes*, accompanied by captions that help the reader understand the significance of each view. For example, one image features a very rocky canyon sparsely vegetated with juniper, piñon pine, and sagebrush. To a hiker's eyes, it looks like a fun place to scramble. The image is entitled "Accelerated Erosion," and the caption explains:

> This canyon is in the South Silent Canyon area on the grounds of the Nevada Test Site. Although the area is not used for testing it demonstrates the accelerated erosion caused by nuclear testing nearby. Vibrations from underground testing fracture the rock cliffs, breaking loose huge boulders.
>
> *(52)*

Another image that looks like a bunchgrass-dotted cut-bank is revealed by the caption to be a subsidence crater that "resulted from the collapse of earth into the cavity created by an underground nuclear detonation" (58). For most of the vast acreage of the Nevada Test Site, it takes a trained eye to read the subtle clues that indicate landscape change due to nuclear tests. Other images, however, while perhaps baffling, are more dramatic. Sedan Crater, for example, is an *enormous* hole in the ground. Other images include human-made detritus and structures, such as abandoned oil drums at Jackass Flats, two remnant drilling towers at Yucca Flat, the base of a disintegrated tower at ground zero of an above-ground test, a stranded segment of an aerial railroad, experimental cement structures erected to test resistance to nuclear explosions, a paintless wooden house with blown-out windows, a radioactive military tank, a collapsed hangar, powerline poles, a fence enclosure once used to contain animal subjects exposed to an atomic blast, sections of a destroyed road, and a variety of engineered earthworks.

The look of the Test Site images reflects Goin's interpretation of landscapes irradiated with fear. He photographed during the brightest, hottest time of day, providing a subtle clue to the emotional response to the landscape. A viewer may unconsciously squint or feel an impulse to wear sunglasses, so effectively do the images communicate an almost intolerable intensity of light. In printing the photographs, Goin kept them as high-key as technically feasible, making the whites the highest possible value of white without becoming the same as the border of the white photographic paper. "Light is a verb," Goin likes to say. In this series, bright light metaphorically alludes to radioactivity.

As the *Nuclear Landscapes* project evolved from its origin in Nevada, Peter decided to visit several more significant nuclear-era sites. Although not open to the public, these places are essentially open-air museums where land alteration and remnant objects are silent witnesses to the nuclear age and where one can study "The Nuclear Past in the Landscape Present," as one of Goin's essays is entitled. An important stop is the Trinity Site in New Mexico, where the first atom bomb was exploded on July 16, 1945. As Peter observes, "Trinity is *the* spot on this planet that represents the true clear dawn of the nuclear generation." The site is open to the public two days a year, one day in April and one day in October, but Goin received special permission to visit in July 1988. He was escorted by an Air Force colonel. An obelisk at ground zero commemorates the explosion and provides visual interest in an otherwise relatively featureless, desert plain. The obelisk reminded Goin of a sundial, and he patiently waited at the site to photograph when the sun was directly overhead so that the obelisk cast no shadow, a subtle reference to time standing still. He originally printed the image as a triptych, a reference to the trinity, later reprinting it as a panorama with the obelisk centered and intersecting the horizon, creating the feel of a teeter totter, a visual cue to the precarious atomic age ushered in by this blast. In the distance stand the Jornada del Muerto mountains, literally "journey of death," and one can imagine sound waves from the explosion reverberating off of them. The colonel invited Peter to collect shards of Trinitite, a mildly radioactive translucent greenish glass formed from sand melted during the blast.

Another important investigation for Goin's project was Hanford, Washington, where the world's first full-scale plutonium production nuclear reactor began producing plutonium in September 1944. Plutonium from Hanford was used to make the first atomic bomb, detonated at the Trinity Site. Before the 570-square-mile Hanford Nuclear Reservation was selected by the government, this area in southeastern Washington along a bend in the Columbia River had been an agricultural community. The region is high desert, but abundant water from the Columbia River allowed farmers to grow fruit orchards that produced peaches, apricots, and apples. The government condemned the area and relocated 1200 residents in order to convert the land use to a nuclear fuel production facility. As Peter recalls, "The town was itself sacrificed. It is eerie to walk the streets and see the 'ruins' of Hanford, as those remnants still exist. It's a revisionist creation of the boom-and-bust western narrative, although with different story lines." During the Cold War, nine nuclear reactors and five plutonium processing plants operated at the Hanford site, supplying plutonium for the US' nuclear weapons. Even though the weapons production reactors were decommissioned at the end of the Cold War, the area remains vastly contaminated.

In an interesting process of reverse rephotography, Goin first photographed the Hanford nuclear site and subsequently located archival photographs of the area prior to its withdrawal by the government for plutonium production. Where Goin encountered a field of tree stumps, an archival photograph shows a farmer sitting on a horse-drawn wagon among his fruit trees. Goin's photographs of the Hanford site reflect a depressingly drab, barren, and industrial wasteland, signaling a condemned landscape not only legally but visually—a national sacrifice zone. Some images document contaminated waste ponds, retention basins, and overflow drains, while others depict dumpsites for radioactive waste such as cement blocks, metal barrels, crates, dumpsters, and even two huge nuclear reactor compartments from a decommissioned submarine. Some images show what the blocky, gray reactors look like, while others reveal trenches where contaminated waste is deposited for later burial. Perhaps the most haunting photographs are of what look to be vast, gravelly, barren plains nearly devoid of vegetation. One learns from the caption that this area, originally called a "tank farm," "holds massive storage tanks containing hundreds of millions of gallons of liquid radioactive waste" buried in leaking tanks beneath the surface and is "one of the largest nuclear waste burial grounds in North America" (82).

It was at Hanford that Peter had the worst experience of the entire project. He learned that there was an area called the Burial Gardens, and the name was so evocative that he asked the guide to take him there. Peter had his tripod set up and was getting under the black cloth to make the photograph when he heard a rapping on the window of the van where the guide was waiting. Peter recounts the story in the present tense:

> I take off the cloth and I look, and he's yelling at me, 'Get the hell in the van!' And I pretend I don't hear him. My heart is pounding and I'm thinking, 'Why is he doing that? I'm busy.' I made the photographs, got the job done. He rolls down the window and he says, 'We're not supposed to be here. This is a plutonium-contaminated area. So you need to get your shit in the van.'

The driver then took Peter to a building where they impounded all of his gear and told him to strip down to his underwear and submit to a full-body Geiger counter search. Two women, dressed in complete white suits with little visors, booties, gloves, and probes, said, "'We need to see if you've had any radioactive contamination.' So they have me stand up in a Leonardo Da Vinci pose, and they go over every inch of my body." If they found radioactivity on any part of his body, he would need to submit to a wire brush scrub in the shower, down to the second layer of skin. (Think of the film *Silkwood*.) For this reason, he was told, many workers at the Hanford site wear bathing suits to work under their clothes to preserve their modesty in the event of a mandatory shower scrub. "That was not pleasant," Peter recalls, "That was the low point of everything. I'd had other low points, but that one was very scary to me." The sinister specter of contamination informs the Hanford images.

Final stop, Marshall Islands in the Pacific Ocean, where extensive Cold War atomic testing took place immediately following World War II, before the atomic testing program was moved to Nevada. After World War II, the United States was granted authority to oversee the Marshall Islands under the UN trusteeship system. Located 2400 miles southwest of Hawaii, the thirty-one sparsely populated atolls that comprise the Marshall Islands were remote from fishing and shipping lanes. They were therefore considered a safe place to test nuclear weapons. Bikini Atoll was selected in 1946 as the site for Operations Crossroads. The 167 people who lived on Bikini Island were asked to let their homeland be used to preserve world peace. They were relocated and told that they could return after the testing program was over. In 1948 nuclear testing was expanded to include Enewetak Atoll, whose 142 residents were relocated to facilitate Operation Sandstone. When the Bikini residents departed, 42,000 US military personnel and civilian support staff arrived to prepare the island for testing. Islands were bulldozed, cement runways were poured, and nuclear bunkers were erected to house monitoring instruments and to test how well different structures and materials withstood atomic blasts. By 1963, when the Limited Test Ban Treaty went into effect,

106 nuclear weapons had been detonated in the Marshall Islands. Despite costly restoration efforts, the islands remain too contaminated for human habitation, and the food chain still registers unhealthy levels of radioactivity. The former residents return for brief visits only.

Goin's stay on the Bikini and Enewetak atolls was unnerving to say the least. As he recalls:

> I was dropped off on Bikini Atoll, where there was a quadrunner and one Marshallese boatman who was required to stay in the research station camp. The island is half a mile wide and three miles long, and I rode around the island with my camera by myself all day with my Geiger counter, gear, and water. The concept of no one there is not truly understood until you're in a place like that. It felt like no one was around you for literally five thousand square miles. You go there and look and say, "I want to photograph here," and you get out, walk over, and set up the tripod, and then all of a sudden "TTTHHHCCCCHHHHTTT," that high staccato sound from the scare machine. And it's hotter than hell, felt like a hundred degrees and ninety-five percent humidity. But the wildlife was beyond your imagination. And tame. Different types of birds would land on your head and on you. They were curious. They did not have human predators, so they had no fear.

To photograph one ruin that was surrounded by water except at low tide, Peter waded through shark-infested water for a thousand yards to gain a vantage point. Just as he made his way through a rising tide back to the boat, a nine-foot gray shark whacked against the side of the boat and bit the propeller. On another occasion, when he was setting up his camera for a timed exposure of the dark interior of a bunker, Peter began to hear a CLCLCKTCLCKT scratching sound. When he shined his tiny flashlight beam around, he espied an enormous crab scuttling around his feet, with others around the perimeter. He was surrounded by coconut crabs, *Birgus latro*, the largest terrestrial arthropods in the world. He thought, "Yikes! They are mutated crabs from hell! It's the size of a basketball! I've been transported into a really bad science fiction movie. I'm going to be eaten alive by mutated crabs!"

The profuse wildlife on the uninhabited islands may have exhibited no fear of humans, but their cheeky proximity heightened Peter's own sense of a *landscape of fear*, a feeling that permeates the images he made there. While one might expect that photographs of a tropical island would evoke paradise—gently waving palm trees, white sandy beaches, bikini-clad beauties, a turquoise ocean, and bright blue skies—the Bikini/Enewetak Atolls series instead creates a sense of unease, of oppression. The twenty-two photographs in *Nuclear Landscapes* mostly center on decomposing military detritus being overgrown by aggressive green plants under hazy or cloudy skies. These unpeopled landscapes seem like *The World Without Us*, to quote the title of Alan Weisman's best-selling 2007 book imagining a post-human Earth. A rusty water tank tilts off-kilter toward the ocean; a bird nest perches atop a heap of electrical cable; and random collections of twisted and jagged metal litter the intertidal zone. Black splotches creep over strange-looking, abandoned concrete hulks, whose purpose can only be surmised. These modern-day ruins from the nuclear testing era bear an uncanny resemblance to the overgrown Mesoamerican ruins that Goin photographed in the 1970s. Each civilization leaves behind architectural elements and detritus that provide clues to its culture. How will our culture be interpreted a thousand years hence?

Having been personally exposed to and recorded on film the effects on the land of the above-ground nuclear testing era, Goin's next step was to disseminate this work, to make visible and raise public consciousness of places that have been "out of sight, out of mind." *Nuclear Landscapes* was published by Johns Hopkins Press in 1991 as a beautifully designed, large-format book, whose title in the paperback edition is printed in the yellow used for radioactivity warning signs. The book sold out and circulated widely, cropping up in unexpected places, perhaps the craziest being an Italian fashion advertisement for Valentino, in which a sexy model wearing a revealing black minidress reclines seductively in a lounge chair reading *Nuclear Landscapes*! In addition to *Nuclear Landscapes*, the book, Peter also made a video entitled *Protest Theater*, which documents the live, collaborative theater that took place on January 5, 1991, at the gate to the Nevada Test Site, performed by creative protestors as a demonstration against the testing of nuclear weapons.

The aesthetic of the *Nuclear Landscapes* photographs aroused controversy, anticipated and addressed in the prologue of the book. Goin explains:

> In contemporary landscape photography, there is a critical dialogue that centers on how photographers will address the role of beauty. How does a photographer convey *content* and still make the photograph appealing to the eye? Should a photograph appease and soothe the viewer, or challenge the viewer to think about the subject? Formal beauty can be a contradictory element in a photograph that comments critically on land use and land management.
>
> *(xxii)*

His solution was to adopt a stark, declarative, and solemn documentary style that uses "the visual language of photography to articulate [his] ideas about these landscapes of fear" (xxii). There are numerous choices a photographer can make in composing a view and in the printing process. Goin agrees with Ansel Adams here, whose musical metaphor states that the negative is the score and the print is the performance. Specifically, Goin employed "hot, bright, white light, the intersection of the subject with the horizon, the use of scale to demonstrate the pervasive effects of nuclear testing, subtle bleached colors, [and] the absence of people" to "capture on film a severely impacted environment" and to "communicate the irony and effect of subliminal fear these landscapes provoke" (xxii).

In a still more controversial move, for the exhibition prints Goin printed text in all caps onto each image, identifying its subject. The font is Helios bold, reminiscent of the text used on television of the Cold War era to label scenes on the screen, creating and containing a context for the viewer. The letters superimposed on the exhibition prints are red-tinted black with subtly smoldering edges, as if they have been burned onto the image. The labeled photographs allude to playwright Bertolt Brecht's *Verfremdungseffekt* ("estrangement effect"), in which he used a variety of techniques, including labeled stage sets, to remind the audience that they are watching a play, a *representation*, not reality itself. Brecht did not want his viewers to lose themselves emotionally in the play, but rather to maintain a critical distance that provokes rational reflection. Rather than creating catharsis or emotional release, Brecht wanted to goad people to think and take action against injustice. While Goin's strategy of post-aestheticizing the prints by aggressively stamping labels over the images offended some artists and critics, he won a National Endowment for the Arts fellowship for this work, vindicating his artistic vision.

By 2007, the color film used in 1987–1989 for the *Nuclear Landscapes* project was breaking down, meaning no new prints could be made. Scott Hinton, Coordinator of Digital Labs and Faculty Research in the Department of Art, University of Nevada, Reno, worked under Peter's supervision to laboriously scan the prints and create three digital portfolios of forty-five images each to make this work available in a two hundred-year archival history as a baseline documentary record of the nuclear testing era. Peter describes the 2011 portfolio as a "reinterpretation," because not only were the images technically cleaned and repaired, but the printing changed from dyes to ink on paper. To the trained eye, the resulting prints will look different, especially since instead of printing on color photographic paper, they were now printed on high-quality watercolor paper, which has a different feel and appearance.

In 2018, thinking back on the *Nuclear Landscapes* project from a thirty-year remove, Peter says that some of the memories still make him break out into a sweat:

> When you go to these places and you look at what we are doing it is a bit like going through Dante's *Inferno*. It is seeing consequences that will exist *far* beyond the conceptualization of humanity. We're talking *a quarter of a million years*. We can barely envision ten thousand years, much less a quarter million years into the future when these legacies will still be around.

Why was Peter driven to see this depressing and personally dangerous journey through to the end? He replies, "There's a compelling obligation, I felt, to bear witness to these sites that other people did not have access to." He was carrying out the responsibilities of a citizen in a democracy. Publishing this work contributes to the debate over how we relate to the biotic communities upon which we depend. As he argues in *Nuclear Landscapes*:

> The landscapes at the Nevada Test Site, the Hanford Nuclear Reservation, the Trinity Site, and the Marshall Islands sites of Bikini and Enewetak atolls have been radically altered. The perception of the arid western landscapes as wastelands uniquely suited for massive contamination neither alters nor justifies the damage. . . . There are many characteristics of these landscapes that give them beauty and value beyond any potential for human habitation. Their dramatic geological areas, arresting vistas, and unique vegetation are found nowhere else. Many birds, animals, plants, and insects thrive in the desert ecosystem. . . . Now, these areas will remain severely contaminated for hundreds of years. The operating principle 'out of sight, out of mind' cannot justify this lack of care for the environment.
>
> *(26)*

Peter Goin gave many talks throughout the United States on the *Nuclear Landscapes* project. Audiences would ask him what it was like to visit these places and whether he was personally harmed by being exposed to radioactivity. He would tell some of the stories recounted above and then sometimes slyly lift his hat and say, "well, I used to have hair." Humor, an antidote to fear. Often, though, the questions took a sober tone. "What have we wrought?" people wanted to know, and Goin would suddenly find himself thrust into the role of a postmodern preacher of optimism. His answer:

> I believe now that what we've ultimately wrought in the nuclear era is fundamental evidence that we must exercise civil action and responsibility to protect our habitat for our children. That's the lesson. There is a tremendous opportunity for us as a people to learn the importance and sacrifice and risks of inactivity and ignorance. War comes

out of ignorance and fear. If we invest in education and have a critically-thinking, intelligent, informed, civic-minded population, we will be peace-seekers. So here's the challenge. It's up to you.

■

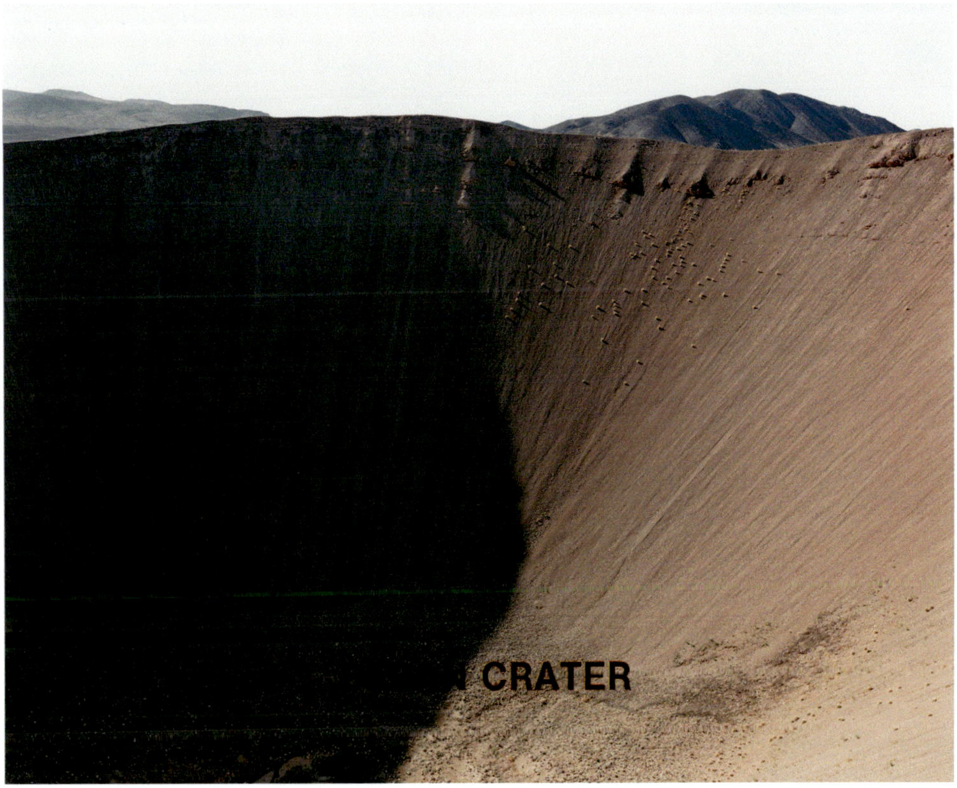

8.1 *Sedan Crater*, 1987.

Sedan Crater was created from an atomic blast in the Ploughshares program, which sought to find peaceful uses for atom bombs, such as large-scale excavation. Whereas much of the Nevada Test Site looks like a relatively undisturbed desert peppered with creosote bushes, Sedan Crater conforms to people's notions of a post-apocalyptic landscape. The challenge was how to convey the *scale* of this crater (1280 feet in diameter and 322 feet deep) in a rectangular photograph. Standing on the Test Site's designated viewing platform, Goin adjusted the camera's field of view so that the edges of the crater exceed the borders of the rectangle, and he chose a camera angle that evokes the feel that the whole earth has fallen except for the horizon. He chose a time of day and camera direction such that a shadow divides the image and severs the two words of the superimposed text, SEDAN CRATER. The dark shadow creates dimensionality in what otherwise might appear as a flat wall and becomes a divisive graphic element that also carries metaphoric and linguistic meaning, alluding to phrases such as "lurking in the shadows" or "the shadow of death." Nuclear detonations are the contemporary sublime, and this image conveys that combination of awe and terror.

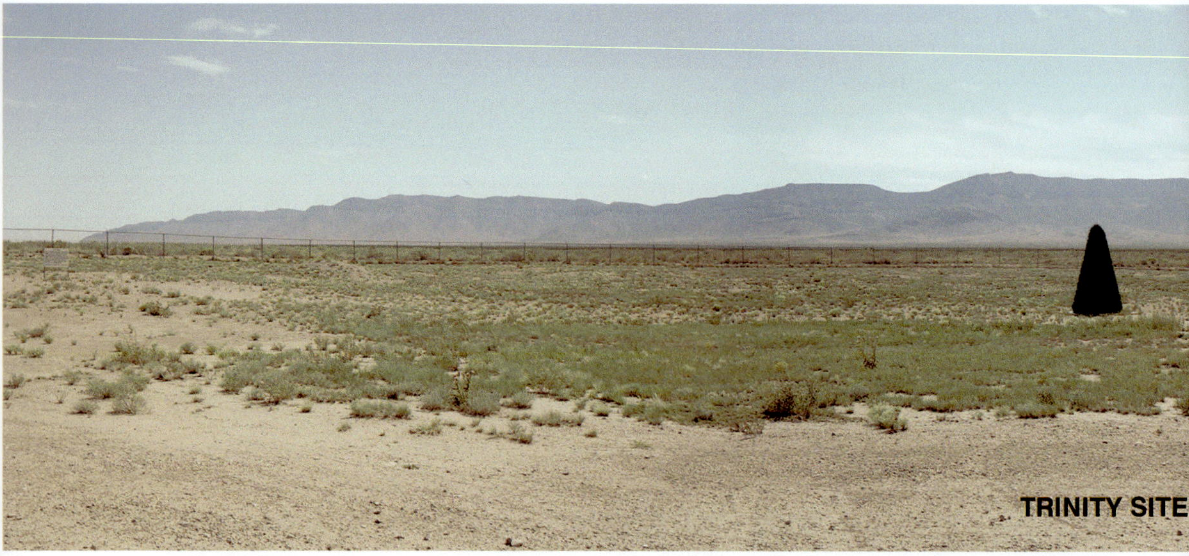

8.2 *Trinity Site*, 1988.

This panorama raises questions: What is that black triangle that looks like it was set down by aliens from outer space? What is that delicate fence that tracks the flat horizon? Why is there no one here? And why is the triangle casting no shadow? The triangular monument marks the Trinity Site in New Mexico, where the world's first nuclear explosion took place on July 16, 1945, on the Alamogordo bombing range in the Jornada del Muerto desert, ushering in the nuclear age. This is the spot. Although a National Historic Landmark, the site is fenced off and closed to the public, except on two days per year. Goin was granted

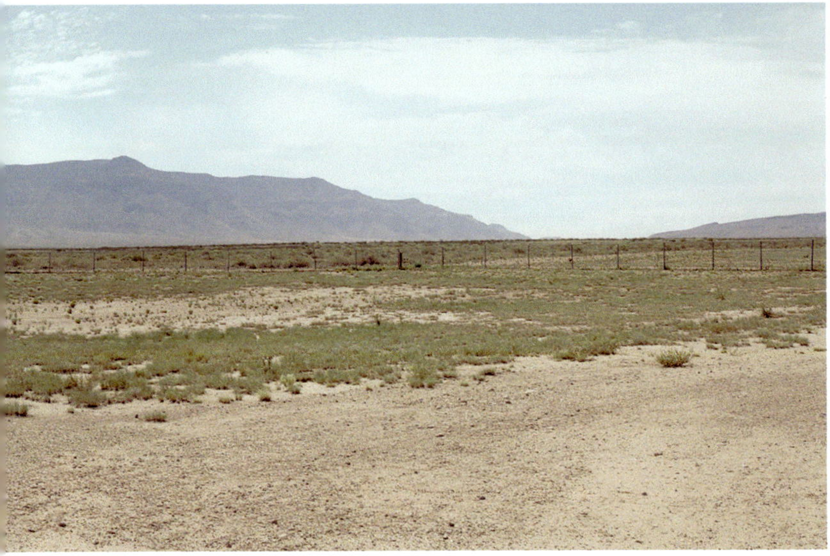

special access. For this panorama, the vantage point shows the road swinging off to both left and right, creating the look of a concave lens, a subtle reference to a bomb crater. Goin deliberately waited to make the image until the sun was directly overhead, aesthetically the worst time to photograph. He explains, "If you think of the obelisk as a sundial, its shadow implies the passage of time. Without a shadow, there is no time—time stands still. And that's the point."

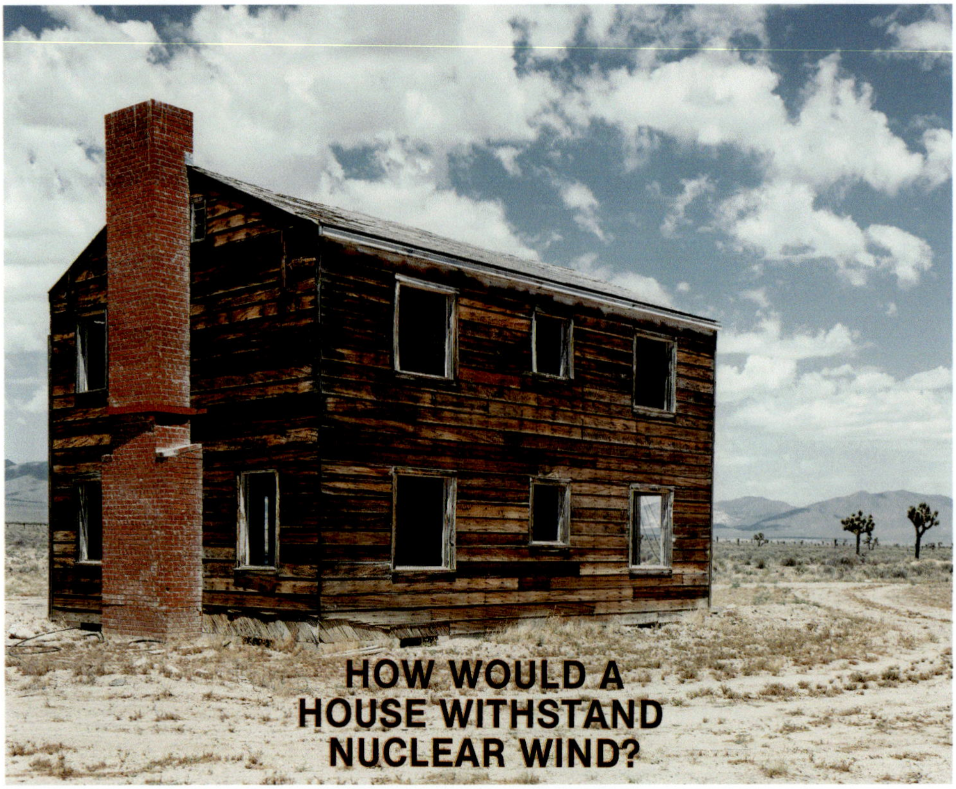

8.3 *How would a house withstand nuclear wind?*, 1987.

The military built a simulated town to see how a middle-class American house would survive an atomic blast. There are no surviving remains from the original "Doom Town," but portions of a second mock town, "Survival Town," still stand, seen here. Inside the frame of this photograph looms the dark, hulking shell of a two-story wooden house, whose paint has apparently burned off. To create an ominous sense, the camera was set low, looking up, with adjusted perspective to render the vertical walls parallel. Goin's use of jet black in the windows invokes Hitchcock, while the lower right window—where we can see all the way through the house—signals abandonment. The image ironically contrasts the cultural meaning of home—shelter, security, and safety—with the military's twisted idea to explode it and see what happens. (Inside the house, clothed mannequins of a family were also subjected to the blast.) How would a house withstand nuclear wind?" was one of the questions that researchers investigated in Operation Teapot in May 1955. In the context of the nuclear era, the burnt-on caption of this image asks *us* to consider the dramatic consequences of our actions: We are affecting the weather.

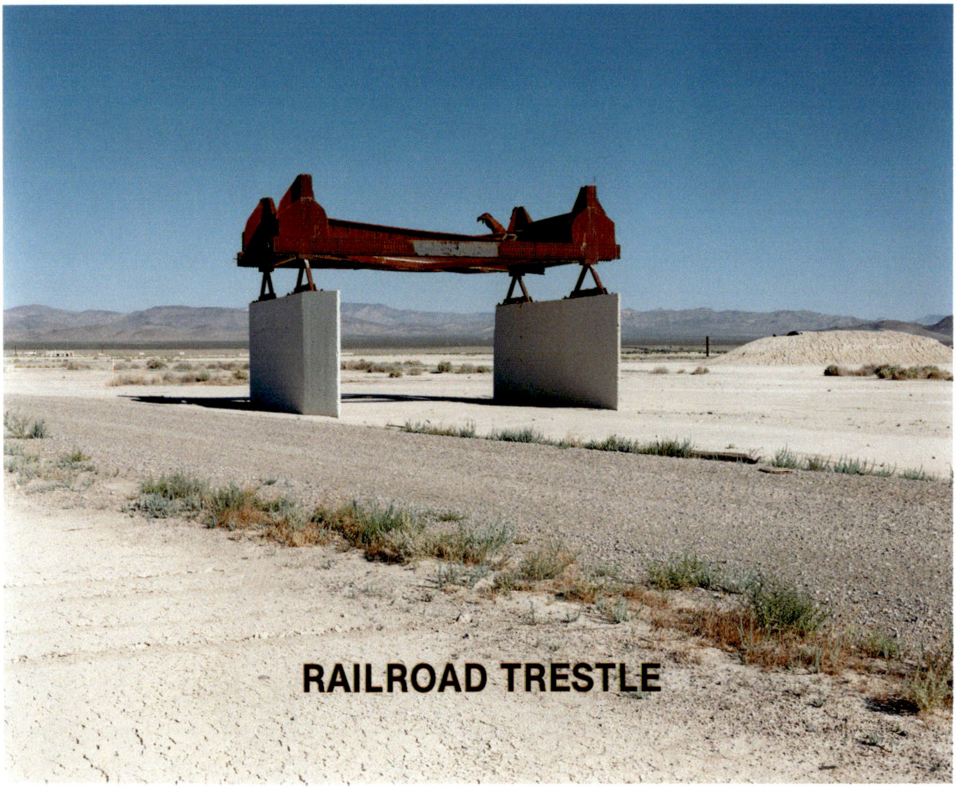

8.4 *Railroad Trestle*, 1986.

This surviving trial railroad trestle stands in Frenchman Lake, Yucca Flat. This bridge, the only remaining section of an elevated railroad, was located 1800 feet from ground zero and received overpressures exceeding 450 pounds per square inch. The blast blew sections of the structure off the foundation and bent large I-beam girders. This detonation, code-named "Priscilla," had thirty-seven kilotons of force. By comparison, the blast at Hiroshima measured thirteen kilotons. When Goin made this photograph, the Nevada Test Site was off-limits to the public, but today one may take a bus tour, which passes by this trestle. Nuclear tourism has been born. The trestle is radioactive, so tourists are not allowed to get off the bus. The experience of taking a picture through the window of an air-conditioned tour bus is a mediated view, viscerally different than being outside, alone in the heat, exposed to radioactivity. Although a tourist's photograph may closely resemble this image by Peter Goin, will their meanings be the same?

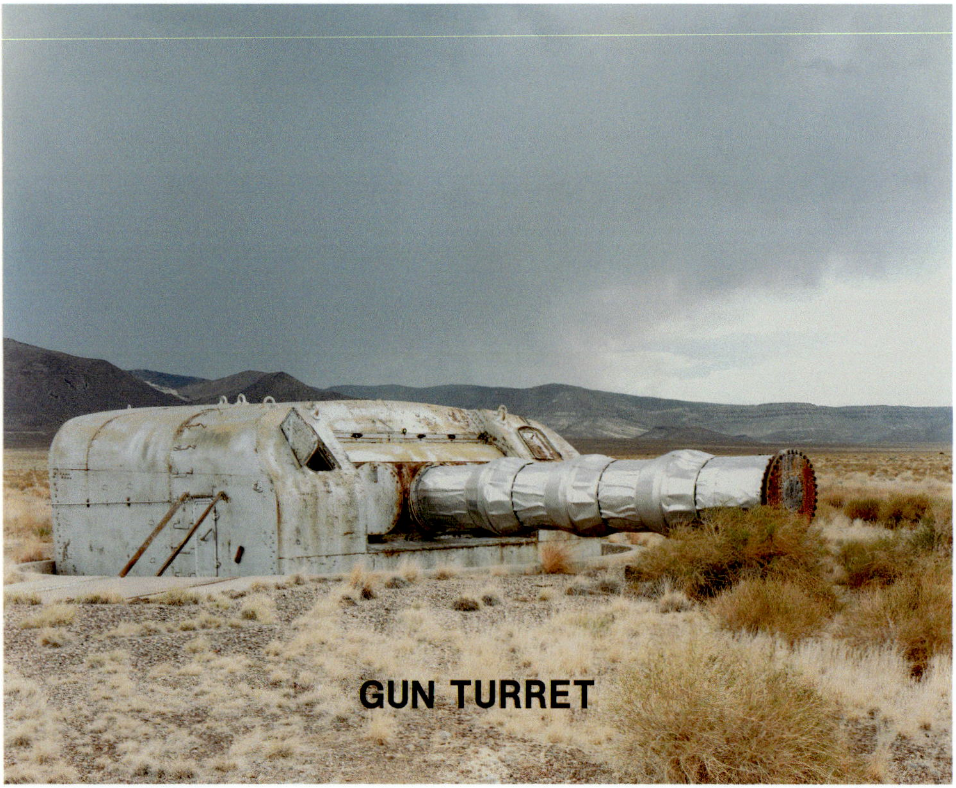

8.5 *Gun Turret*, 1987.

How often do you expect to encounter a heavy cruiser gun turret in landlocked Nevada? This rusty gray and silver component from a naval vessel seems wildly out of context amidst scrubby bushes, gravel, and bunch grasses in the desert, raising questions that the image does not answer. Where is the rest of the ship? How did it get here? *Why* is it here? The looming gray cloud above and blue-gray barren mountains in the background underscore the sense of foreboding that infuses this scene, dominated by a derelict metal war machine that resembles the head of a robotic elephant. There is a lot that we don't know and haven't been told and don't see, because we are not allowed onto restricted ground or into classified documents. Just what is the military *doing* out here? Because we are denied information, we piece together a story based on rumors. An informant told Goin that after World War II, bombing tests on the Marshall Islands also targeted ships brought to the area. This gun turret was apparently salvaged from a bombed ship and hauled all the way to the Nevada Test Site. Who knows why? It is radioactive.

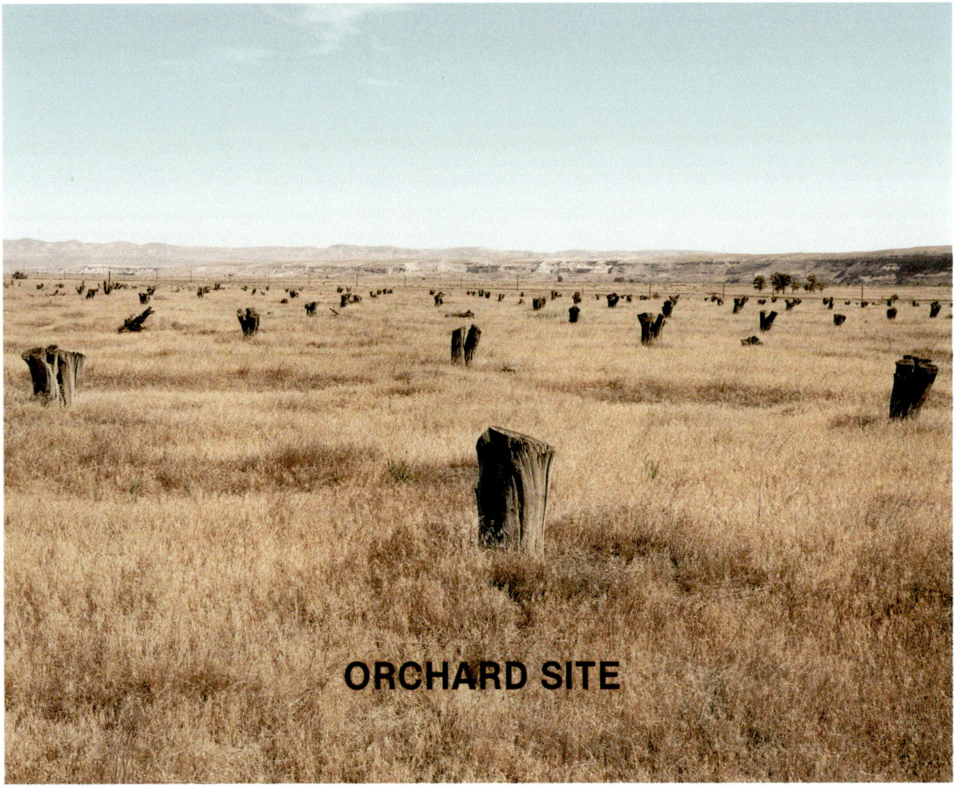

8.6 *Orchard Site*, 1988.

Here, mimicking the structure of a cemetery, countless dark stumps stand out against yellow fields. The Orchard Site at Hanford, Washington, is where the government cut down fruit trees to prepare the area for nuclear reactors. The logic of clearing the site seems to be that officials did not want the fruit from these trees to be contaminated. This sad image thus raises concerns about safety. In fact, pollution from Hanford's nuclear reactors became so serious that, at one point in time, drinking a glass of water from the Columbia River was the radioactive equivalent of having a chest X-ray, and Hanford became known for cancer clusters. Apparently, in the name of national defense the government was willing to sacrifice citizens. Biblically, this image alludes to the destruction of the garden. We have turned a plentiful garden of nourishment into a sacrifice zone. "Orchard Site" is of course an ironic label given that there are no trees, and this denatured view asks us to contemplate the cost of sacrificing environmental and human health.

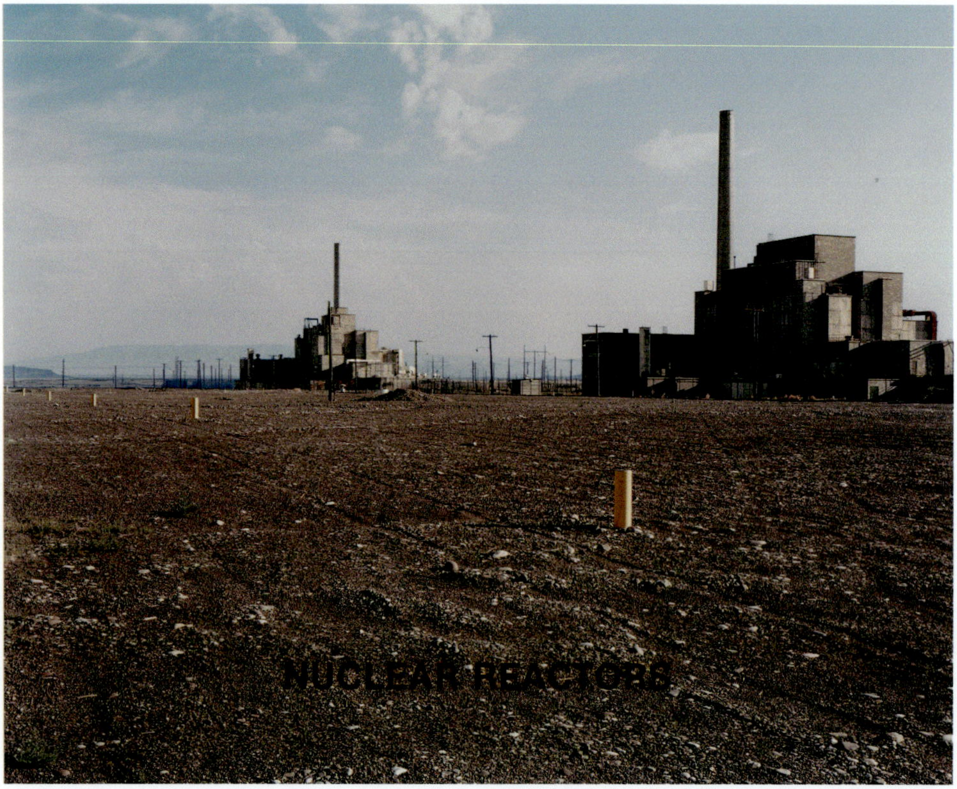

8.7 *Nuclear Reactors*, 1988.

This photograph depicts Nuclear Reactors D—one of the three first nuclear reactors ever built—and DR, which stands for D-Replacement. Located on the former town site of Hanford in southeastern Washington, these reactors produced plutonium for atomic tests during and after World War II. The nuclear reactors at Chernobyl share a similar design to these Hanford reactors, reflecting an industrial sensibility of pure utility without aesthetics. However, the design of this image is symbolic. The posts in the foreground—by now, yellow has become a ubiquitous color-code for danger—warn of the radioactivity buried beneath, which seeped from the two reactors. The lines in the photograph—the road track lines, the line of yellow posts, and the horizon line—open out to the right from an origin point on the left edge. This left-to-right, acute geometric angle suggests a cornucopia, a sinister basket of plenty that spreads out, creating a broader sensibility of the radioactive contamination of the planet. Today, because the fenced-in Hanford site is off-limits and out of sight, people take false comfort that toxicity is localized and contained. But this image conveys a more pervasive nuclear contamination in space and time.

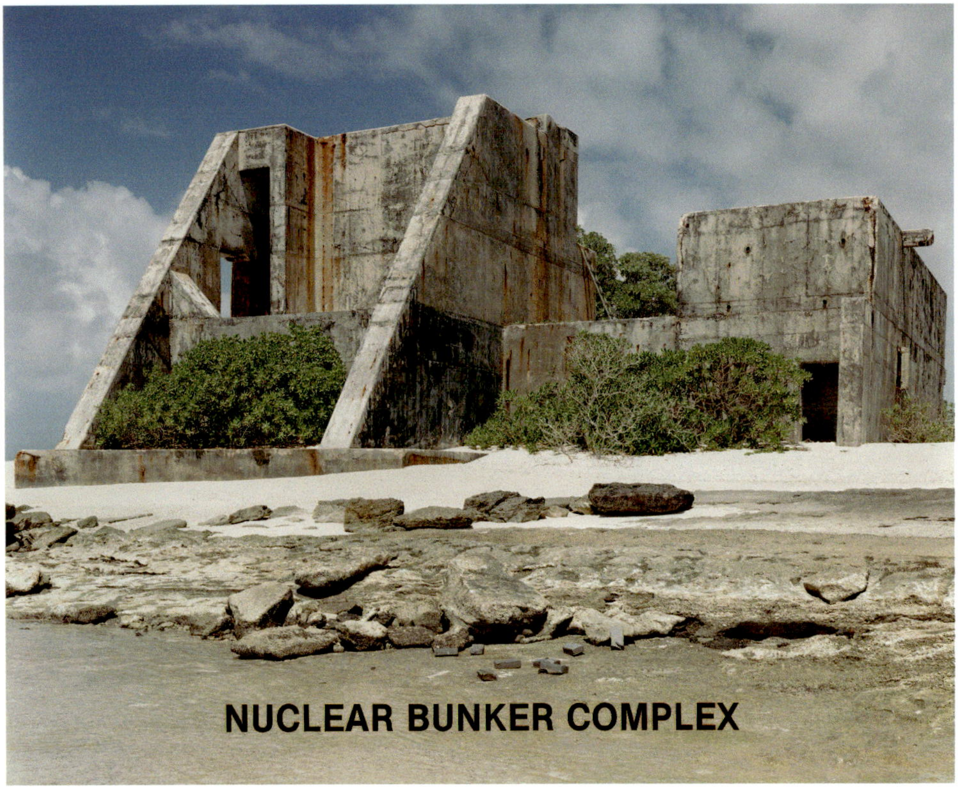

8.8 *Nuclear Bunker Complex*, 1989.

This bunker complex on Aomen Island in Bikini Atoll was a photographic station used during Operation Redwing in 1956. The lead bricks on the sand in the foreground were used to construct radiation barriers. The shape of this bunker is reminiscent of the Meso-american ruins that Goin photographed in the 1970s. And the wooden ladder that spans the two buildings conjures memories of yet another ancient civilization, the ancestral Pueblo communities of the American Southwest. These are the ruins of *our* civilization. Whatever we may think is our legacy, what we are really leaving behind are these bunkers, because they are designed to withstand a nuclear blast; therefore, they withstand time. Invoking the futuristic suspense genre, we could imagine future explorers exposed to this site's persistent radioactivity, whose half-life is 10,000 years, an almost unimaginable timespan. Perhaps the adventurers die, and a new myth is born about the curse of the tombs.

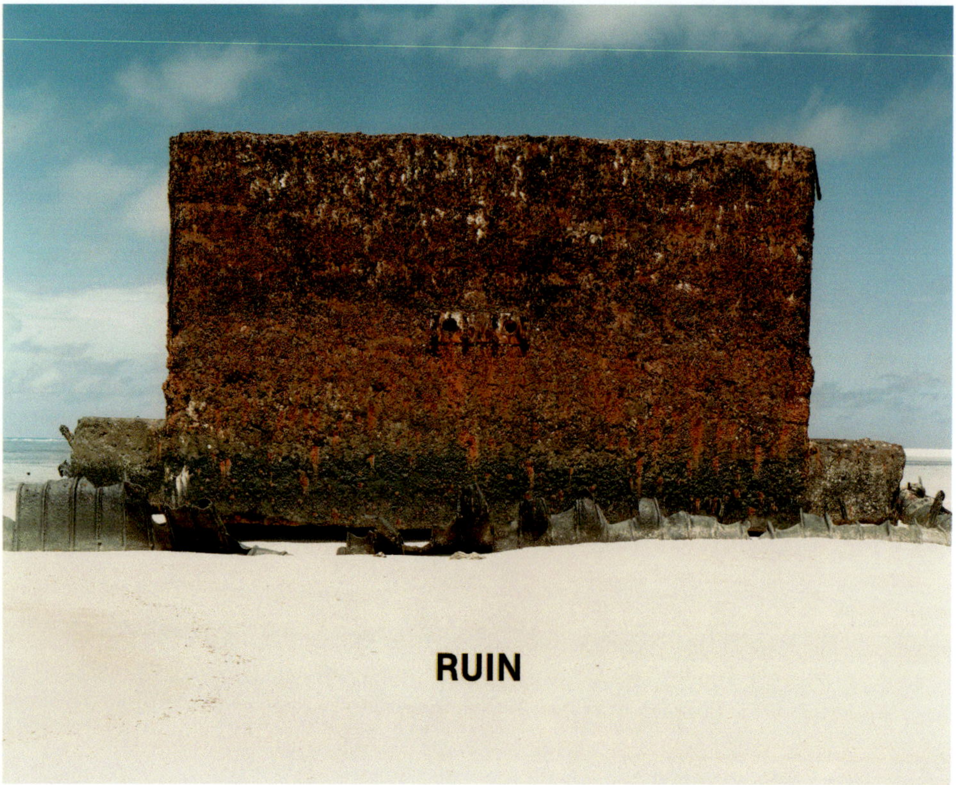

8.9 *Ruin*, 1989.

Rusting ruin near the Bravo Crater, Bikini Atoll. Abstract art is what comes to mind with this beautifully colored image: Dark box, centered, with two eyes, aquamarine blue above, narrow band of ocean, pure white below. Low, jagged mouth, perhaps the bent remains of a ripped-open pipeline. Clean lines. A declaratively simple image. Horizontal bands of color on the bunker indicate the life-forms of different intertidal elevations. At high tide, the lower five feet of this bunker are submerged under water, so this image was possible to make only at low tide. The bunker is where cameras were set up to record a test blast, so encoded in the image is the irony of a photographer making a photograph of a photographic bunker, the viewer's eyes looking at the eyes of the test cameras. This particular island in the Marshall Islands was lost in the bombing, literally bombed out of existence.

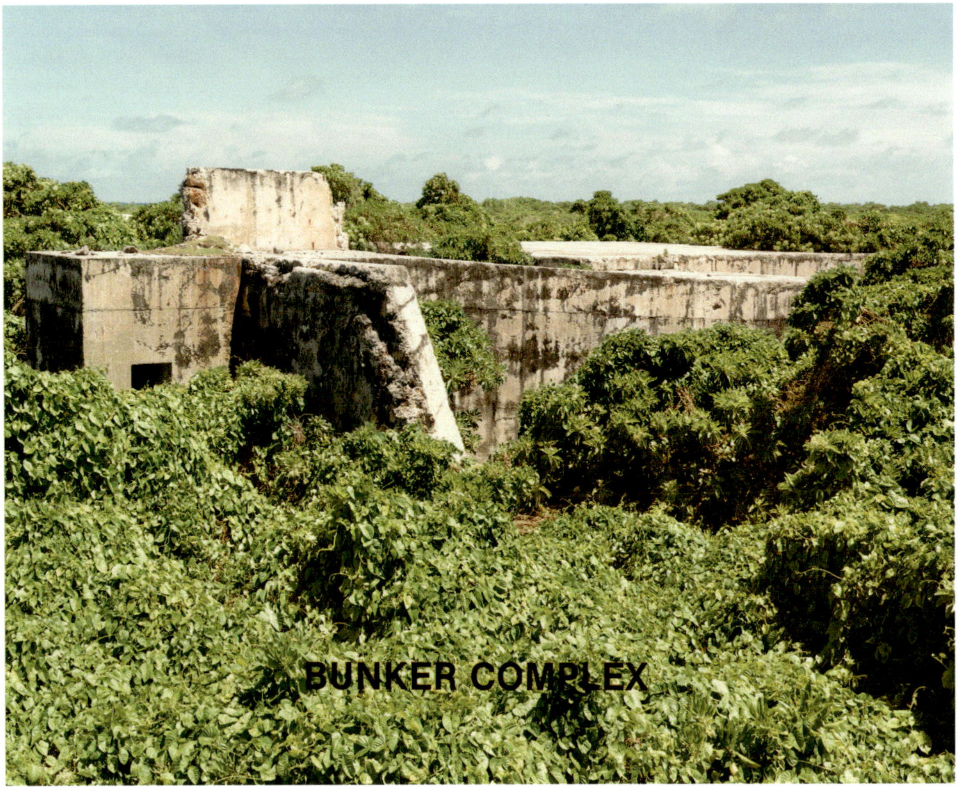

8.10 *Bunker Complex*, 1989.

This Bunker Complex on Nam Island in Bikini Atoll was built for the "Cherokee" test in Operation Redwing. "Cherokee" was detonated directly overhead, so the bunker, which was unmanned but contained sensitive instrumentation, needed to be massive, with walls eight feet thick. By 1989, when this photograph was made, radioactive vegetation was beginning to reclaim the site, reminiscent of the Mayan ruins that Goin photographed in the 1970s. He was struck by a thought. The Mayans did not understand the architectural principle of the arch; therefore, they needed massive walls and thick roofs to support their temples. In some ways, the ruins here similarly reflect a lack of understanding. In our fail ure to understand the equitable distribution of weight—of sharing and living within our means—we create massive bunkers. Rather than using our technologies to integrate us to our environment and live in harmony with our fellow humans and other species, we manufacture ever more powerful weapons, necessitating ever thicker defensive walls. In contrast to the crumbling concrete, the super-vibrant plant growth—exaggerated by the photograph's composition—raises the specter of genetic mutation. Perhaps radioactivity will produce super-organisms of the future.

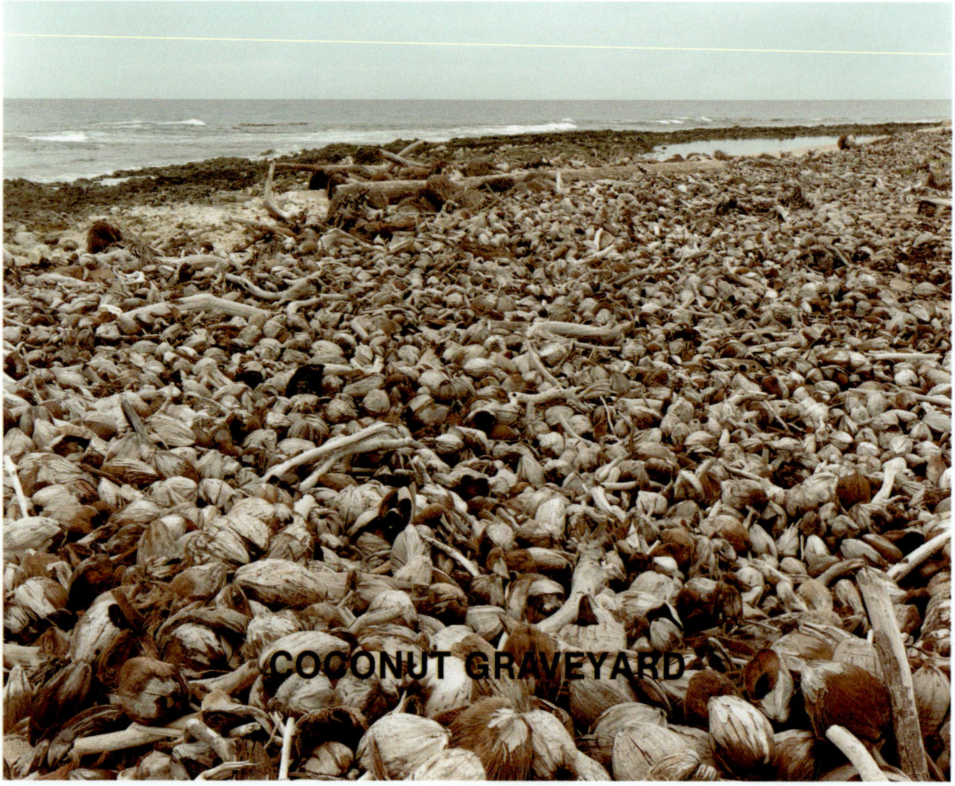

8.11 *Coconut Graveyard*, 1989.

These coconuts, on the shore of Eneu Island in Bikini Atoll, are too high in cesium 137 to be safely consumed. Goin recalls, "I walked there and saw this and thought, 'Oh no! It's *The Killing Fields!*'" referring to the acclaimed 1984 Roland Joffé film about the war in Cambodia. "Being there was just overwhelming," Goin remembers:

> Many nuclear sites are cerebrally overwhelming. But this one was emotionally overwhelming. You just think about *the consequences of our actions*. I was viscerally so impressed with the isolation of this place and knowing it's contaminated and knowing that there's this tremendous history of detonating sixty-six nuclear bombs and that these people have been completely displaced from what was their homeland.

A photograph of a beach of washed-up coconuts on a tropical island seemingly has nothing to do with nuclear landscapes, but Goin's burned-in caption, COCONUT GRAVEYARD, signals death, and in that context the coconuts and driftwood suddenly become metaphors for human skulls and bones, underscoring the power of a caption and of context—a Nuclear Landscapes exhibit or book chapter—to direct interpretation.

Note

1 The interpretive lens of immersive fieldwork discussed in this chapter was originally explored in Glotfelty's essay, "Corporeal Fieldwork and Risky Art: Peter Goin and the Making of *Nuclear Landscapes*" in *Material Ecocriticism*. Glotfelty would like to thank the editors of that volume, Serenella Iovino and Serpil Oppermann, for permission to reprise some of that material in this chapter.

Works cited

Glotfelty, Cheryll. "Corporeal Fieldwork and Risky Art: Peter Goin and the Making of *Nuclear Landscapes*." *Material Ecocriticism*, edited by Serenella Iovino and Serpil Oppermann, Indiana University Press, 2014, pp. 221–238.

Goin, Peter. *Nuclear Landscapes*. Johns Hopkins University Press, 1991.

Goin, Peter. "The Nuclear Past in the Landscape Present." *Atomic Culture: How We Learned to Stop Worrying and Love the Bomb*, edited by Scott C. Zeman and Michael A. Amundson, University Press of Colorado, 2004, pp. 81–100.

Goin, Peter, director. *Protest Theater*. KNPB Channel 5, Reno. Peter Goin, 1991.

9

MILITARIZING THE AMERICAN WEST

> Well, here's to Nevada, the leave-it state. . . . You want to gamble your money, leave it here. A wife to get rid of? Get rid of her here. Extra atom bomb you don't need? Just blow it up here and nobody will mind in the slightest. The slogan of Nevada is, 'Anything goes, but don't complain if it went!'
>
> —*Thelma Ritter as Isabelle Steers in* The Misfits *(1961)*

In 2005, in an article entitled "Bravo 17: Militarizing the Great Basin," Peter Goin ponders the "leave-it state" lines from John Huston's classic film *The Misfits*:

> Perhaps this quote best represents the mentality of conquest so prevalent in our human understanding of watersheds, history and geography; a broad assumption that the open space of aridity and basin and range allows for whatever one wants there to be. *(23)*

In conversation, Peter elaborates:

> The fundamental reason for this mentality is *aridity*. Water is a spiritual element—we baptize infants in water, bathe in water, take sacraments with water, and share a fundamental understanding of the human need for water to survive. We equate aridity with the absence of water. When water evaporates, it somehow takes with it the value of the space, so that value is evaporated too. We equate aridity with lifelessness. If the landscape is dead—so our thinking goes—then we can do whatever we want.

Peter himself would counter that arid landscapes are *not* dead. In an area that most people would write-off as lifeless, a fascinating, exquisitely adapted biotic community exists, and there is an animating spirit palpable to those who pay attention. This chapter concerns the uses of public lands in the arid West. "This land is your land," Peter's "Bravo 17" essay concludes, and he questions why we so readily accede to sacrificing arid lands to the military.

DOI: 10.4324/9781003212607-9

Goin's essay appeared in the inaugural issue of *High Desert Journal*, a literary and visual arts magazine "dedicated to further understanding the people, places and issues of the high desert east of the Cascades and into the Great Basin" (Quinn, back cover). This vast inland desert has been relatively neglected by scholars and artists, off the radar of most Americans, and marked on nineteenth-century maps as *Terra Incognita*. Editor Elizabeth Quinn states that the journal's goal is "to examine the layers of place through story, poetry, people and art." She thanks the contributors of the first issue for "chart[ing] the changes" and answering the question "what's out there?" (3). Goin's work aligns well with Quinn's mission. He writes:

> For more than twenty years, I have been photographing the landscape of the Great Basin. . . . My photographs are . . . about living a life of creative and dedicated citizenship, of bearing witness to our time, and of providing visual evidence of that, which we often choose to ignore, neglect or evade.
>
> *(24)*

The previous chapter's *Nuclear Landscapes* project, pursued in a spirit of dedicated citizenship, culminated in a published book and international traveling exhibit, both of which helped to raise public consciousness about the effects of the nuclear testing era on the landscape. However, related photographs on the militarization of Nevada and the Great Basin that Peter made during his first twenty years in Nevada—1984–2004—were not collected in a book. Many of the images from that ongoing project are published here for the first time. But lest you think that the issues raised by these images are safely relegated to the past, consider that according to Nevada historian James Hulse in *Nevada's Environmental Legacy*:

> by the 1990s, the military had claimed more than four million acres of Nevada land, mostly from the Bureau of Land Management but taking some of it from citizens by condemnation, purchase, or harassment. In addition, the armed forces had entered the airspace above the desert-and-range province, disturbing with the roar of jet-powered engines, much of the sky above and the wilderness below.
>
> *(72)*

Furthermore, as of 2018, the US Navy proposed to withdraw *an additional* 604,789 acres of public land and acquire approximately 65,160 acres of nonfederal land to expand and "modernize" the Fallon Range Training Complex, which includes Bravo 16, Bravo 17, Bravo 19, and Bravo 20 ("Fallon Range," 4). The National Defense Authorization Act for Fiscal Year 2021 extended the existing withdrawal of public lands for another twenty-five years, but did not approve the expansion and modernization as proposed by the Navy. The Navy will continue to work with stakeholders "to secure a mutually-agreed upon expansion" ("Fallon Range Training Complex Modernization Update").

What troubles Peter is that a defining element of the national identity of the United States is militaristic. The size of the US defense budget is nearly the combined total of the military spending of all the other countries on the planet. The Myth of the West, which is a powerful aspect of American identity, is fundamentally a conquest narrative. Vast amounts of land in the Great Basin are dedicated to military uses. According to an article in the *Las*

Vegas Review-Journal, in Nevada alone, there is Nellis Air Force Base, boasting more Air Force squadrons than any other base; Creech Air Force Base, the primary training, maintenance, and operations base for unmanned aerial vehicles; Fallon, a Navy tactical air warfare training center; and the Hawthorne Army Depot, a 147,000-acre ammunition storage facility (Martin). Hawthorne proudly boasts that it is "The World's Largest Depot." As Peter sees it, the militarization of western lands is a continuation of the conquest narrative.

In addition to the troubling issue of national identity, militarized lands raise questions about how we relate to the environment in the arid West. Conceptualizing an area as a bombing range reflects an entirely different relationship to the land than thinking about it as a complex biotic community or as a sacred space. As Peter states bluntly, "our militaristic nature trumps environmentalism." Militarized landscapes, such as the ones documented in this chapter, may have toxic waste, contaminated water, geomorphological interruptions, unexploded ordnance, radioactivity, and detritus. Why do American citizens so unprotestingly accept the notion of sacrifice zones in the name of national defense? Peter speculates that perhaps our apathy is because we do not feel "of the land." The prevailing Euro-American *dis*-identification with the land stands in sharp contrast to the sensibilities of native peoples. At a public meeting convened by the Navy in Gabbs, Nevada, to discuss the proposed Fallon Range Training Complex Modernization plan, Johnnie Bobb, the Western Shoshone National Council Representative, was asked what he thought about the Navy's massive land expansion. Bobb placed his hand on his chest, "In the heart," he said. "We don't call this the desert. These are our traditional lands" (quoted in Benti).

Goin began photographing the military's use of western lands soon after he arrived in Nevada in 1984. The sites represented in this chapter are Bravo 20 and Bravo 17 in northern Nevada and the Desert Training Center (DTC) near Indio in southern California. Peter's *High Desert Journal* article provides some helpful context on Bravo 20 and Bravo 17, both of which are in the Fallon Range Training Complex. He explains that this complex is:

> located in western Nevada with an operating area of 6.5 million acres and more than 10,000 square miles of air and land space. The Fallon range is the focal point for all Navy, and some Marine graduate level aviation strike warfare training. Of the 1,310 Navy aircrew who participated in the original Desert Storm, 1,298 flew at Fallon.
>
> *(26)*

Peter made many photographs at Bravo 20, three of which are represented here. Belying the lifeless appearance of the landscape during the day, wildlife make their homes on the range, active mostly at night. A 1980 federal government report on Bravo 20 found an impressive diversity of biota near Lone Rock, a preeminent bombing target, whose height was reduced by repeated shelling, bombing, and target practice. Dominant plants were Greasewood and Seep-weed, while small mammals trapped in the study included desert kangaroo rat, antelope ground squirrel, and deer mouse. Other wildlife found to be present in the sand dune habitat near Lone Rock include black-tailed jack rabbit and kit fox. Birds that were identified include Rough-legged Hawk, Mourning Dove, Common Flicker, Horned Lark, Common Raven, Western Meadow Lark, Brewer's Blackbird, and Sage Sparrow. The study was conducted in October; at that time of the year, no reptile and amphibians were active on the surface and therefore are not mentioned in the report. A biological investigation of a nearby undisturbed area on the eastern edge of the Carson Sink found a much greater number of plant species than in the Lone Rock area, comprising a mix of shrubs, forbs, grasses, and fungi ("Final Environmental Assessment," 26–30). Peter discovered Bravo 20 when he was photographing at the Stillwater Wildlife Refuge, which ironically is located adjacent to and immediately south of the bombing range.

In addition to making photographs at Bravo 20, in 1988 Peter made a haunting seven-minute video entitled *Bravo 20 Bombing Range* for public access television. Instead of voice narration, the story unfolds through a sequence of sounds choreographed to visual images. The arc of the narrative moves from innocent child and flourishing nature to military take-over and relentless bombing, a successful operation that pleases those in command. "What are we allowing to happen? What legacy are we leaving our children?" the film seems to ask.

Bravo 20 garnered public attention through the protests of citizen groups and the well-known photographs of Richard Misrach, whose book *Bravo 20: The Bombing of the American West* appeared in 1990. Like Goin, Misrach photographs human-altered landscapes to raise public awareness and provoke debate about land use. In his work on Bravo 20, Misrach and his activist friends staged ingenious publicity stunts, picked up by the media. For example, he purchased a share of a mining claim filed by his friends Dick Holmes and Dr. Richard Bargen when Bravo 20 was in the public domain, and for eighteen months he proceeded to "work his claim." He photographed Dick and Doc planting an American flag on the summit of Lone Rock, drawing attention to that high-point, which Navy bombers had already reduced in height by more than a third. And he proposed a brilliantly audacious plan to set aside this bombing range as America's first environmental memorial, Bravo 20 National Park. As a result of the attention drawn to the area by Misrach and other protestors, along with the continuing danger of unexploded ordnance, the Navy fenced off the area and posted No Trespassing signs. Peter began photographing at Bravo 20 in 1986 when these lands were still being accessed by the public. However, Goin moved on to other areas, such as Bravo 17, when he became aware that Misrach had staked an artistic claim to Bravo 20.

The Bravo 17 bombing range is located on the south edge of highway 50, a highway nicknamed in the 1980s by *Life* magazine "The Loneliest Road in America," a slogan later embraced and made official with road signs by the Nevada Department of Tourism. Bravo 17 is an alkali basin about fifteen miles east of Fallon, flanked on the west by the Sand Spring Mountains and on the east by Fairview Peak. According to Goin's essay, "Of all the Fallon ranges, Bravo 17 is the most frequently used" (27). Peter toured the site with a military guide on a day when there was no bombing scheduled. It was a remarkable sight, sort of like a giant game board for a war tournament. He saw life-sized mock tanks and missile launchers made out of plywood, cardboard tents, and miscellaneous wooden structures cloaked in camo-netting as well as real, retired airplanes perched on pedestals made of railroad ties. All these objects are bombing targets for Navy pilots, who learn how to spot and strafe them from the air. Peter noticed a yellow school bus. It was unclear whether the bus was intended as a target or as an object to recognize and avoid, but the bus was perforated with huge bullet holes. There was an area where old, retired tanks painted in desert camouflage were stored as if in a junkyard. These tanks were no longer operable but could be dragged out to a chosen site and bombed.

To Peter, a civilian with no military training, Bravo 17 aroused mixed emotions. On the one hand, he found it all very interesting to explore an area and observe the props of a restricted military training ground. On the other hand, having worked in theater, Peter couldn't help thinking of the stage setting for a French farce. His photographs of Bravo 17 have a double-edged quality of straight-faced documentary realism tinged with subtle hints of the absurd. You have to admit that flying a high-performance jet and blowing up tanks and strafing faux encampments must be really fun. At the same time, however, these exercises epitomize our militaristic culture and prepare our young men and women to perpetuate a legacy of conquest.

The third site represented in this chapter is the DTC in southern California, east of the small town of Indio. Also known as "Patton's Camp," this area in the Mojave Desert was used in World War II under the supervision of General Patton to prepare troops for the North African campaign. According to an article by Sarah McCormick Seekatz, published in conjunction with a public arts project named *Incendiary Traces* in 2015, the DTC, created in spring 1942, is the "largest military training ground in the United States": "During its two years of operation over a million troops trained in its desert heat, in an area roughly 18,000 square miles, a district larger than the states of Maryland and Delaware combined" (Seekatz). At its zenith, almost 190,000 men at a time trained at the DTC (Ruhge). A 1942 newspaper article heralded the project as follows:

> Somewhere in the California Desert, under a molten sun and in a country where the very earth feels like fire, American armored vehicles are training. . . . It is this force that will someday leave death in its wake in the sandy places of Libya, or wherever it may be sent.
>
> *(quoted in Seekatz)*

Seekatz explains that:

> The vast expanse of land provided enough room for multiple battalions to train in situ, to create living spaces from the ground up, and experience maneuvers that mimicked actual warfare. Engineers outlined camp roads, signal corpsmen laid telephone lines, and army air corpsmen took advantage of year-round clear skies to obtain crucial flying skills and practice. Thus, troops gained actual practice in the roles they would hold abroad in a landscape and climate similar to what they would encounter.

The US National Park Service, which manages the Mojave National Preserve to the north of Patton's Camp, notes the environmental impact of the military's use of these lands:

> The army's training exercises left permanent scars on the desert. Tank tracks remain visible in many places inside and outside the Preserve. Rock alignments laying out huge tent camps dot the desert, especially in the southeast corner of the Preserve, and military debris still litters the desert floor in some places. Military use compacted the soil and changed the natural ecology of the desert.
>
> *("General Patton's")*

Perhaps even more damaging to the fragile desert ecology are the off-road vehicle use and motorcycle races that now chew up the Mojave Desert, again reflecting an attitude of macho conquest. The Park Service points out that:

> The experiences of soldiers at the wartime desert camps translated to increased recreational use of the desert after the war. War surplus jeeps and camping gear enabled former soldiers and their friends to explore the desert in four-wheel drive, as they had done during their training, with new maps of the Mojave to guide their travels. One US Army photo, dated 1942, shows a soldier in full battle gear steering a huge military Harley-Davidson motorcycle down a Mojave sand dune - a stunning prelude of recreation to come.
>
> *("General Patton's")*

Peter had heard of General Patton's World War II training camp while photographing at Bravo 20 and Bravo 17, but he encountered the site almost by accident when he was working on another project at Joshua Tree National Park. As he recalls, on the DeLorme Southern California Atlas that he was using to explore the park:

> I noticed the road matrix in what was called 'Patton Camp' and found the grids amusing, at first assuming it was some large real estate project gone awry. But that was *at first,* because once I started investigating the site, I knew that it was an abandoned military site, and that piqued my interest. What's interesting to me about the site is that it is this arid landscape, but there are white rocks surrounding a creosote bush, or two, or three, and pathways to nowhere, and a kind of presence-absence that was intriguing. The white paint on the rocks amused me as it revealed the military's pattern of order, even within the wild of abandonment. It's the funniest thing. Creosote is probably not offered anywhere in any nursery. Who would buy it? So, here's this plant that's fundamentally poorly integrated into human society, and you have the military decorating it. There's something quirky. I'm not an expert on French farcical theater, but if I were, this is an element. This is the evolution of the landscape-equivalent of the farce.

Like so much of Goin's work, the photographs in this chapter invite us to contemplate the uses and significance of landscapes that are out of sight, out of mind; in this case, public lands withdrawn from public access for use as military training grounds and bombing ranges. As Goin's *High Desert Journal* piece remarks:

> The question of what shall be in the landscape is one of the most persistent and least resolvable themes in western politics. One inescapable presence remains in the Great Basin, and perhaps will always remain—that of the United States military.
>
> *(23–24)*

As you consider the images collected here to document the militarizing of western lands, be open to subtle touches of humor and playful optical illusions. Peter uses humor in these photographs not only to lighten an otherwise alarming topic, but to deconstruct the paradigm by alerting the reader to amusing ironies, thereby metaphorically disarming our militaristic culture.

■

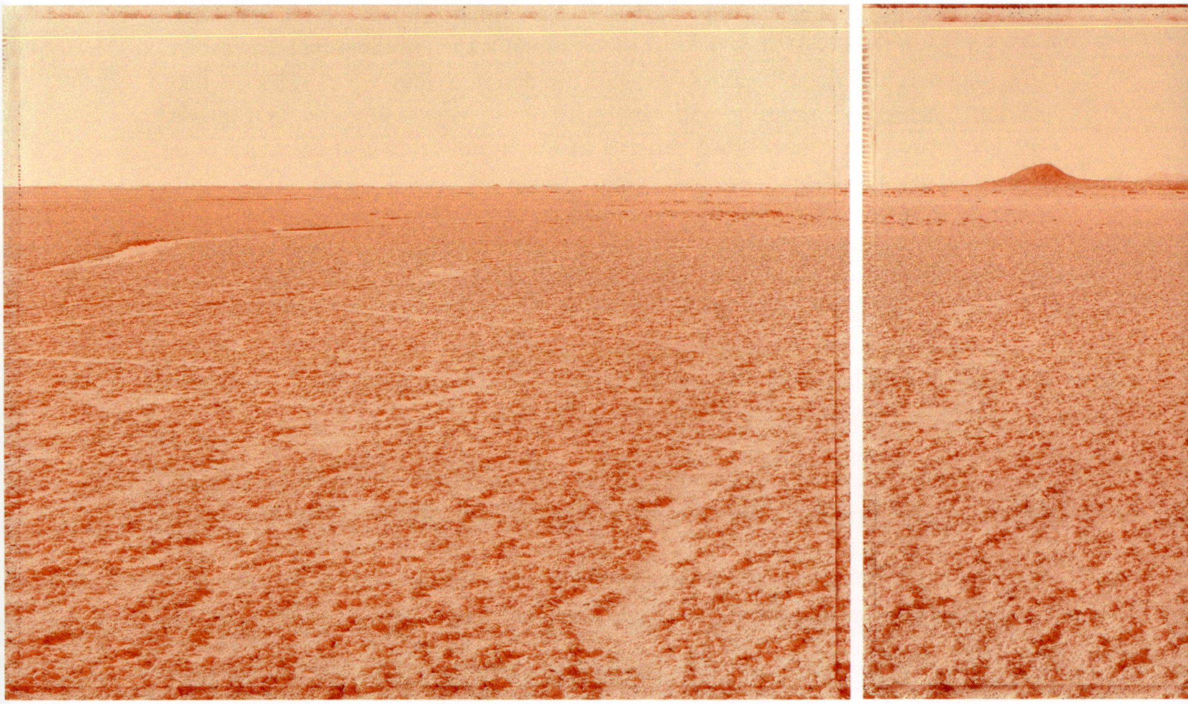

9.1 *Lone Rock, Bravo 20 Bombing Range, Nevada*, 1986.

When Peter became a professor at the University of Nevada, Reno, in 1984, the Polaroid Corporation gave him one of their photographer's grants, providing free film that he could use however he wanted. This 1986 Polaroid panorama provokes a discussion about bomb testing and the military's relationship to the landscape. The triptych is composed in such a way as to bring the horizon in close, highlighting the texture of the playa and the prominence of Lone Rock, the high point on the horizon, a designated target of military pilots. This photograph is a black-and-white image that Goin put into a color enlarger and

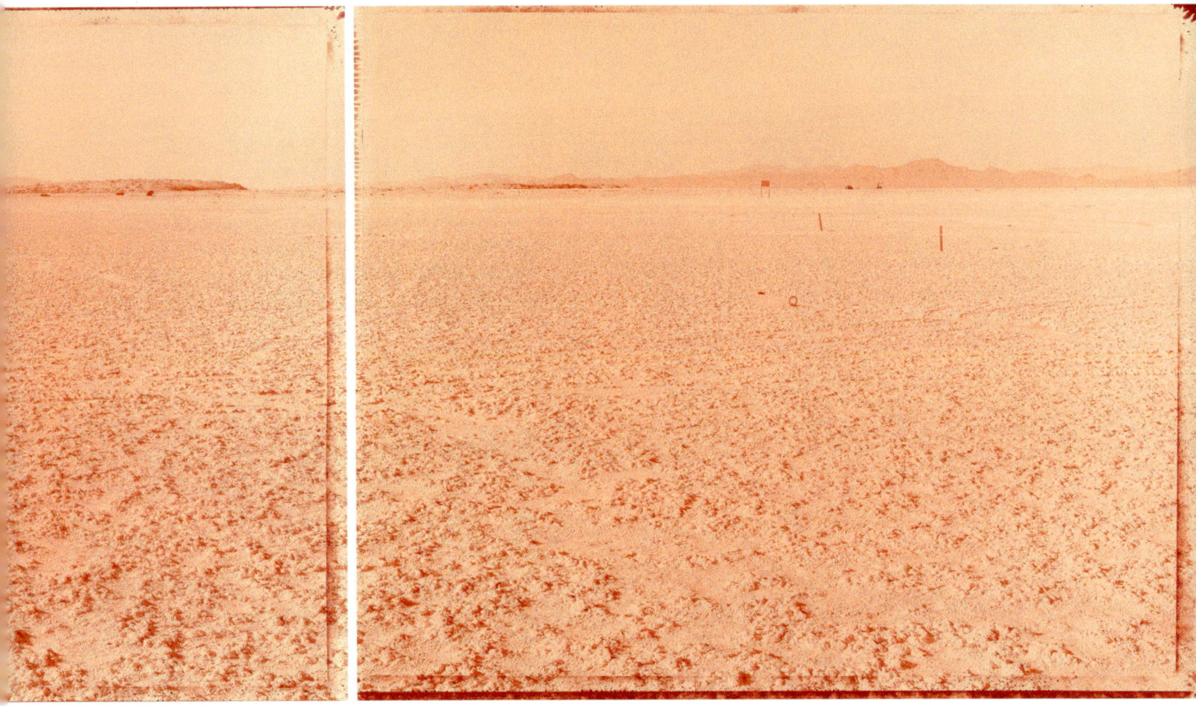

contact-printed in a red tone reminiscent of the blood-red color of the standing water in some of the bomb craters at Bravo 20. Like photographer Mark Klett, whose prints often show the edge of the film, Goin here retains the rough edges of the Polaroid negative, a self-referential reminder that the photograph is an artifact. While we tend to think of photographs as windows through which we see an unmanipulated view of the world, in fact, photographs are crafted objects.

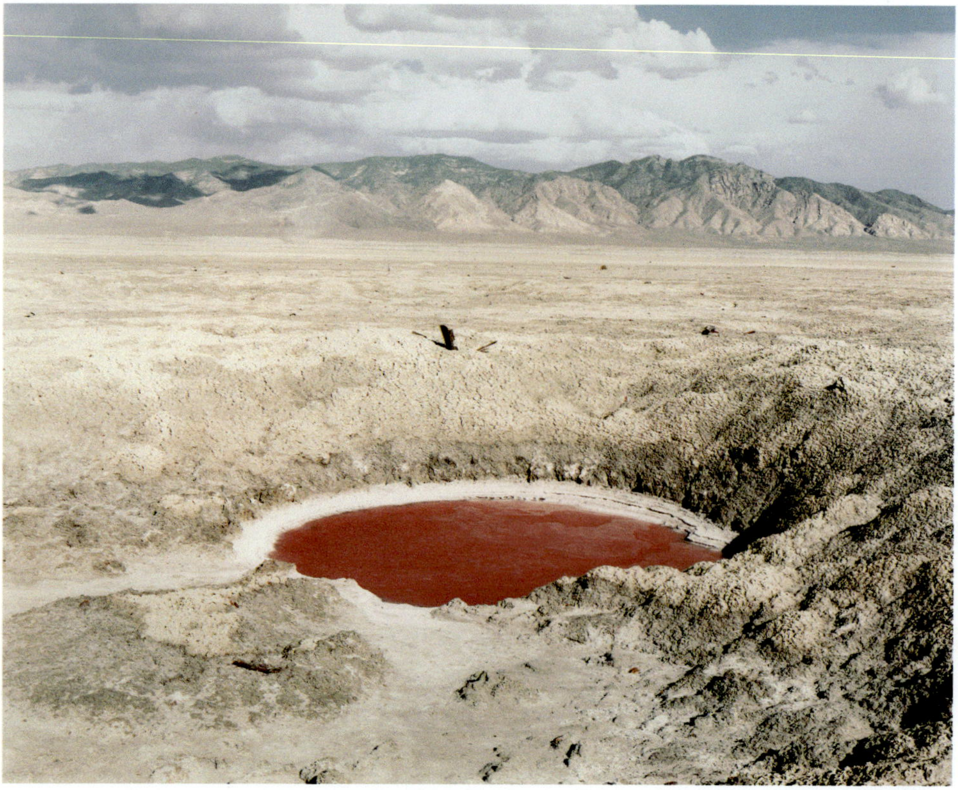

9.2 *Crater, Bravo 20, Nevada*, 1987.

Why would there be a red pond in the middle of the desert? There is something unnatural-looking about this weird pond. Ah, we are looking at a bomb crater on the Bravo 20 bombing range, northeast of Fallon, Nevada. To the south of Bravo 20 is the Stillwater Wildlife Refuge. It is an irony in the management of public lands that adjacent to a bombing range—an area sacrificed for military target practice—we preserve land for wildlife. The water table under this playa, or dry lakebed, is fairly high, dramatically revealed by the blood-red water that lies exposed in the crater. The color of the water likely results from red algae, but it functions here symbolically for the blood of a wounded landscape. In this composition, note the centered red circle—a gaping mouth—above which sits the nose-like shape of the tail fin of a bomb, with the furrowed brow of a low mountain range in the distance, capped by brooding clouds. Despite the artfulness of the arrangement of geometric elements, the effect is one of devastation.

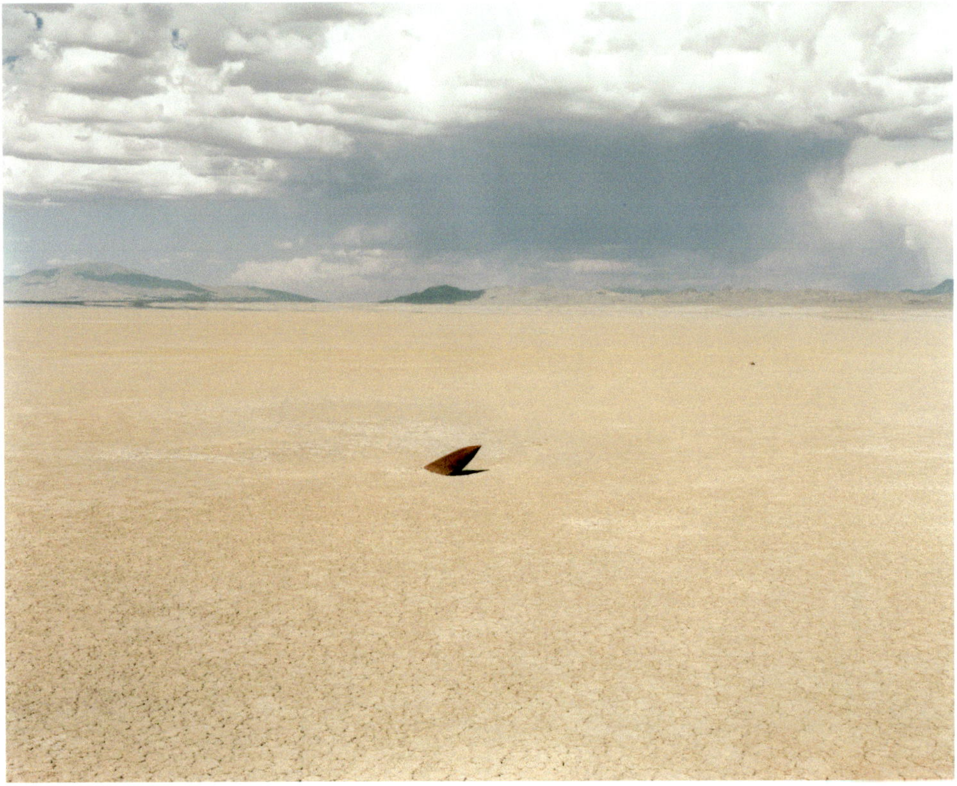

9.3 *Missile, Bravo 20, Nevada*, 1987.

Before the Navy closed Bravo 20 to the public—a decision made after photographer Richard Misrach's work and actions aroused increased interest in the area—it was possible to freely wander across the landscape of the bombing range. Wander at your own risk, however, as the site is littered with unexploded ordnance. The muted colors, uncluttered space, and minimalist style of this image create a sense of peace and tranquility that exists in tension with the explosive associations of the missile, curiously breaching from the playa like the nose of a whale or dolphin. Playas, of course, are made from the dried sediment of former lakebeds, so the watery metaphor carries a certain degree of prehistoric cachet. The storm brewing in the background animates this placid landscape, faintly suggesting that in geological time it could once again be under water. This photograph appears on the cover of John Beck's book *Dirty Wars: Landscape, Power, and Waste in Western American Literature.*

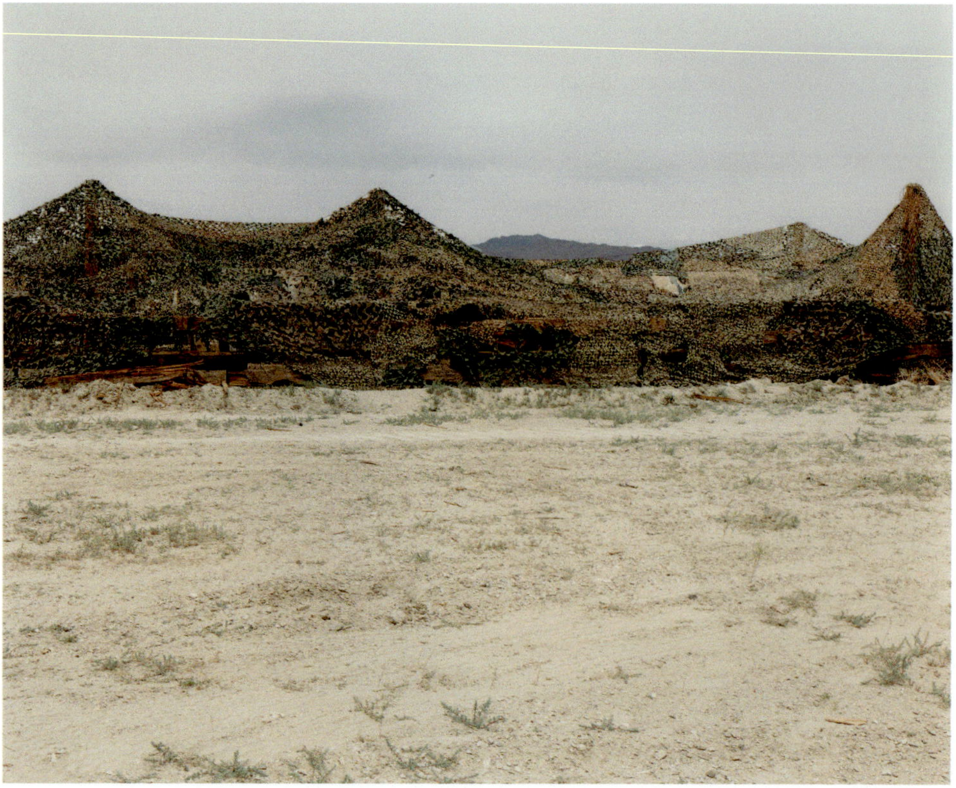

9.4 *Bravo 17, Nevada*, 1997.

What do you see in this photograph? It is something of a visual trick. What appears to be a low range of pointed dark hills, upon closer inspection resolves into a large camouflage net concealing wooden structures and cardboard tents, with upright poles forming peaks. The horizon line is much more iconic in landscapes of the American West than in those on the East Coast, and it functions here to give a sense of space and to naturalize the scene. However, the tire tracks in the foreground and blue-gray real mountains in the background confirm this range of pointy low "hills" to be human-made. Camouflage was developed in World War I to conceal weaponry and troops as they appeared on aerial photographs. On the ground and in person, the dimensionality of structures is apparent, but a photograph flattens depth perception, making camouflage an effective tool of deception.

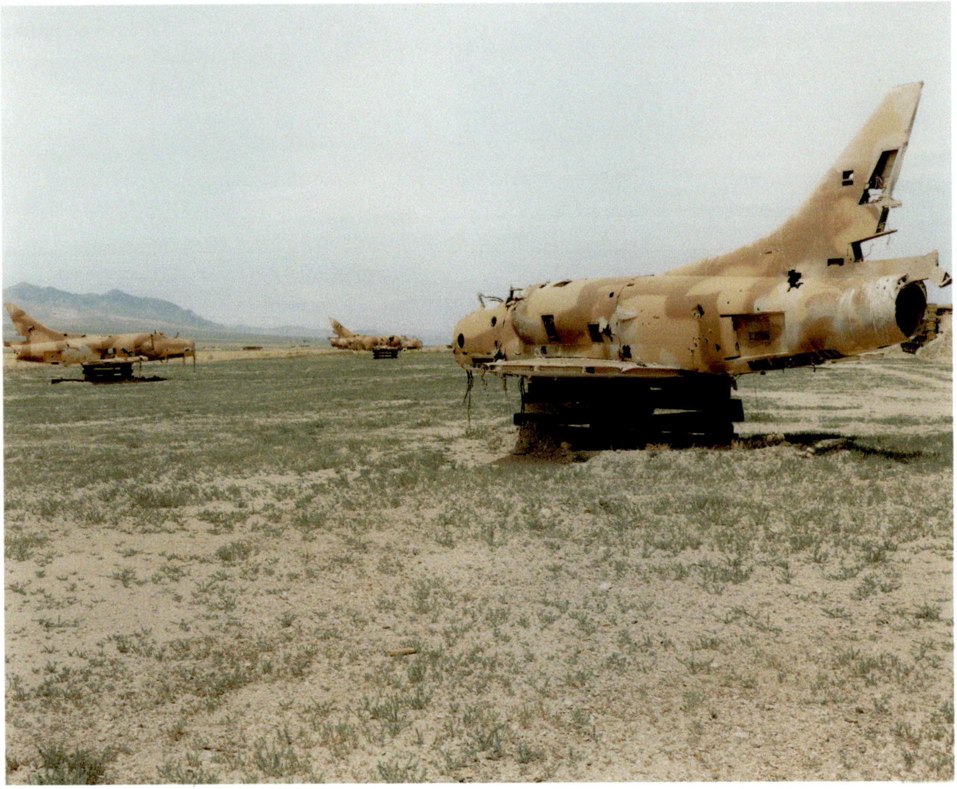

9.5 *Desert-brown camouflaged airplanes, Bravo 17,* 1997.

Peter was given permission to photograph at Bravo 17 with a military escort on a non-bombing day. The many holes ripped into the fuselages of these old, sitting-duck war planes, perched like birds on wooden bases in a simulated airport, attest to successful strafing. The military's practice runs are videotaped so that pilots can evaluate their performance. In the context of this chapter on militarized landscapes, these planes are quickly intelligible as practice targets, but if this same image were to appear in a photography exhibit of Nevada basin and range country, these prehistoric-looking planes would seem bizarrely out of context. While this photograph reveals the crazy things that hunker out of sight in the arid West, it also somewhat playfully reminds us that military operations are, at some level, war games.

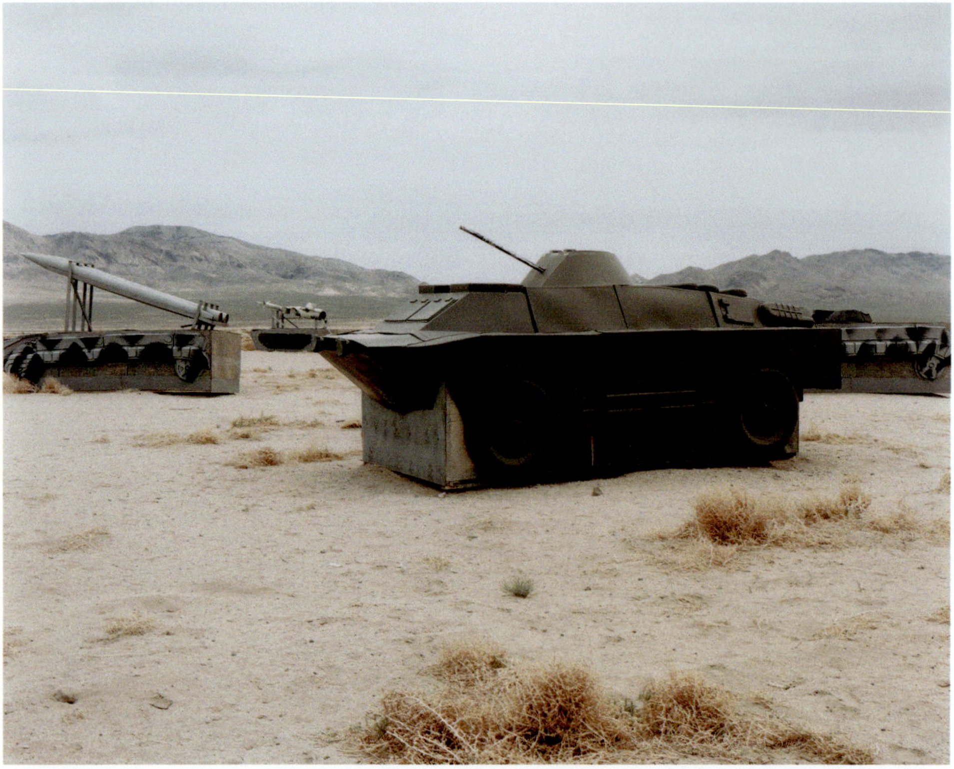

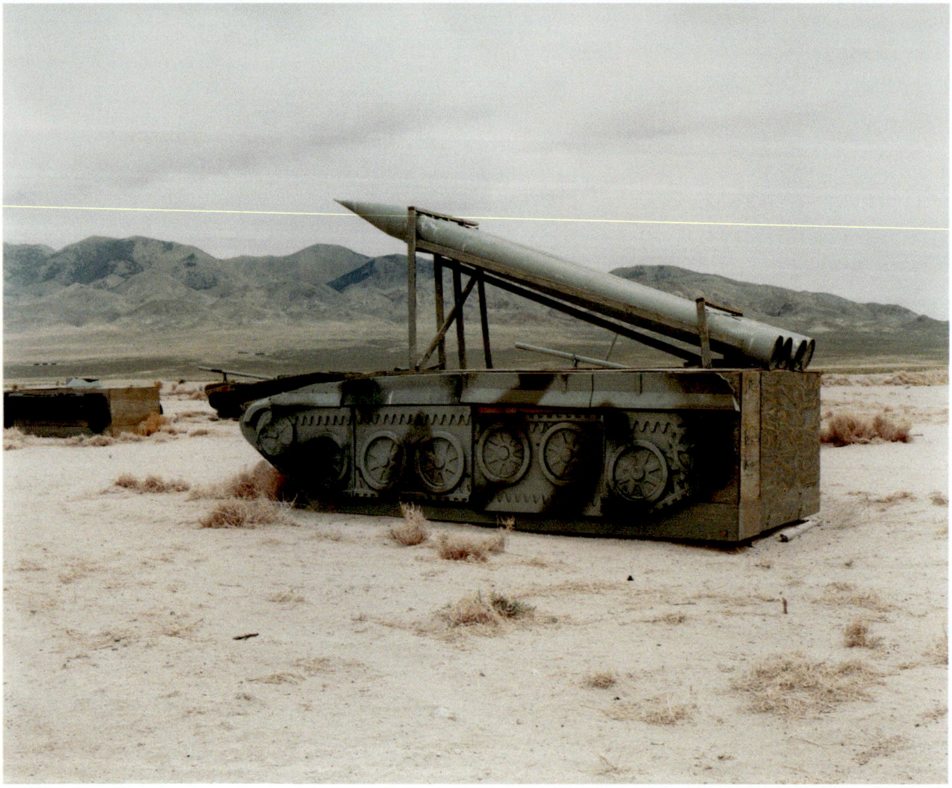

9.6 AND 9.7 *Bravo 17 bombing range, Nevada*, 1997.

These camouflaged tanks and missile launchers are plywood and metal models for bombing targets. Although from the air these models could plausibly pass for the real thing, in this close-up photograph from a vantage point on the ground they look rather ludicrous, a subtle way of suggesting that military uses of the land are a bit absurd. Note how the point of one of the fake missiles mimics the shape of the nose of the actual missile poking up from the playa in the Bravo 20 photograph above (refer to Figure 9.3). In a militarized culture, certain shapes, such as a smooth cylinder convexly curving to a sharp point, are almost instantly associated with weaponry. Here, as elsewhere, Peter plays off of the horizon. He explains:

> The view and representation of the line is a very important element in all of my work, particularly in Nevada, where the horizon is how we define our landscape. People on the East Coast don't have the same horizons we have. You have to go to the shore to see a horizon like we do. That crisp edge on the mountain ranges is so much a part of us.

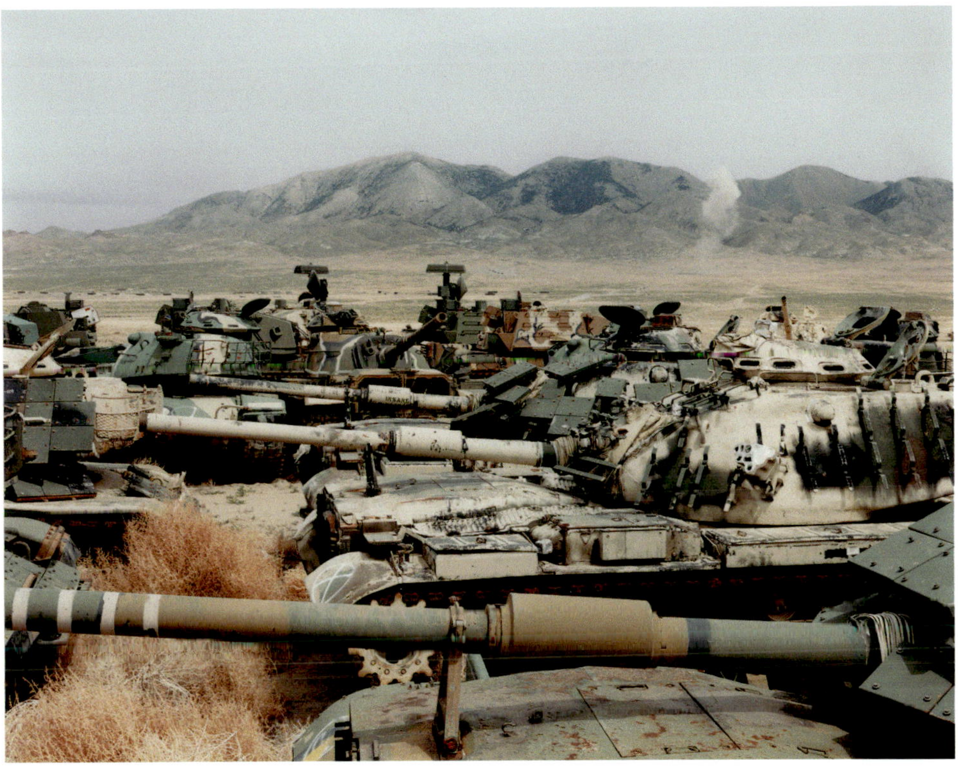

9.8 *Tank field, Bravo 17, Nevada*, 1997.

These camouflaged tanks are not models and are not yet bombing targets. Instead, what we have here is a parking lot for retired tanks. The tanks are no longer mobile, but when desired, they can be loaded onto a semi and dropped off at designated target locations in Dixie Valley. Peter, who has not served in the military, found it curious and interesting to crawl around on these tanks as if they were a playground. "These complexities of a tank," he recalls, "they're impressive war machines, dinosaurs of the arid desert, silent sentinels to the unspoken conquest." As he was setting up to make this photograph, an explosion occurred and a puff of smoke appeared in the distance as if on cue. Serendipitously, the military was exploding live ordnance recovered at the site. Workers haul out-of-date ordnance to a big pit and set explosives to it so that there won't be dangerous, unexploded ordnance lying around the area.

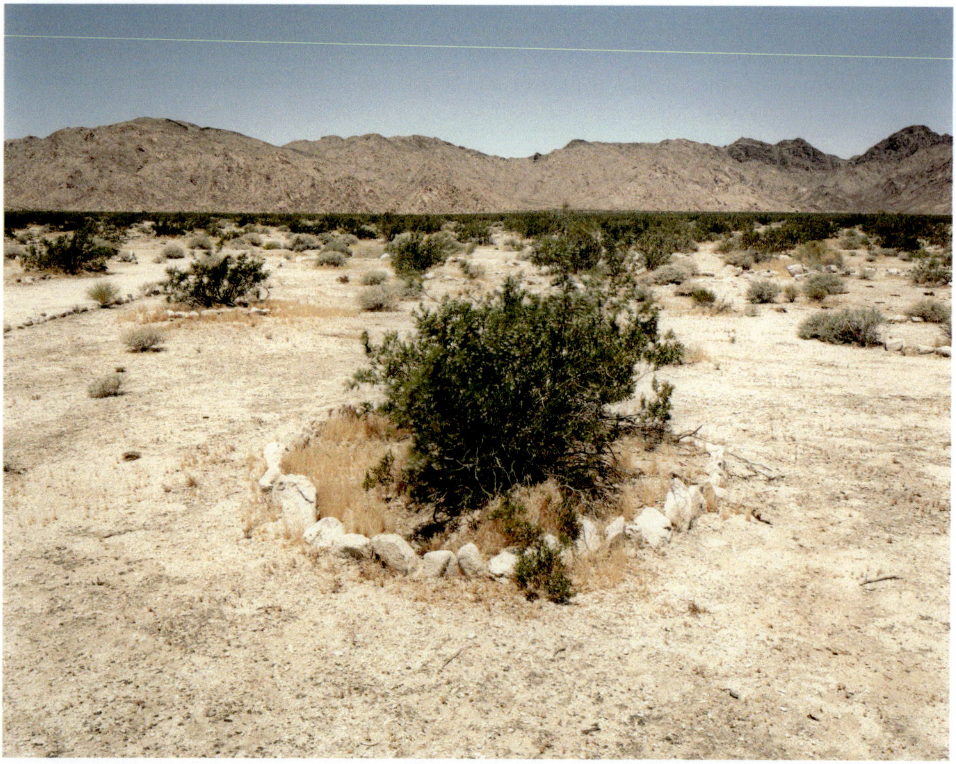

9.9 *Desert Training Center, California*, 1993.

Geometrically this composition mimics the image of the blood-red pond at Bravo 20 discussed above (refer to Figure 9.2): Centered circle in the foreground, straight horizon line above the circle, from which rise a distant row of irregularly cragged mountains, overspread by a narrow band of sky. From top to bottom, an irregular line, a straight line, and a circle—perhaps a visual formula for a militarized desert landscape, with the circle being an artifact. Here we have an arrangement of painted white rocks circling a creosote bush. But why? The viewer becomes an archaeologist, attempting to reconstruct the past from remnant traces on the landscape. Looking more closely, one can discover other rows of rocks, extending beyond the frame of the photograph on both the left and right edges. Peter made this photograph at the DTC, also known as "Patton's Camp," a remote site in the Mojave Desert used during World War II to prepare and toughen up American troops to fight in North Africa. Why the rows of painted white rocks? Let's pursue that question in the next image.

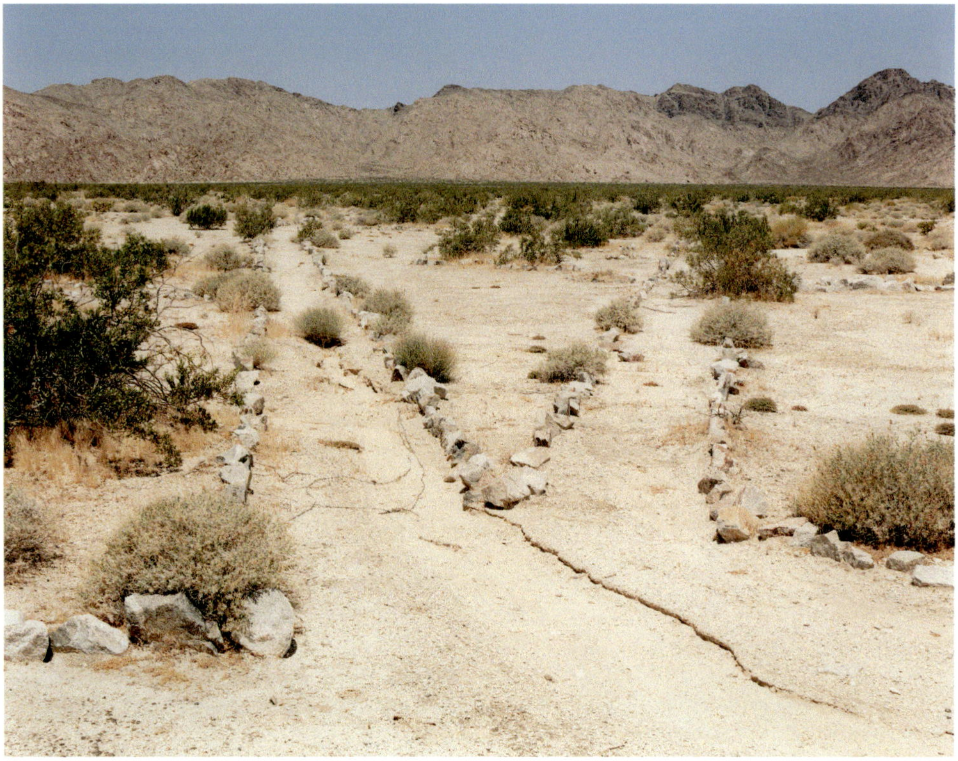

9.10 *Desert Training Center, California*, 1993.

More rows of painted white rocks. Two paths diverge on a sandy plain—a whimsical takeoff of Robert Frost's poem in which two roads diverge in a yellow wood. What do you do with over 100,000 enlisted men stationed in a remote desert on home ground? When they are not marching, shooting, and doing simulated war maneuvers, apparently you tell them to collect rocks, paint them white, and place them in neat rows. In this way, we might speculate, you simultaneously keep idle hands at work and impose military order in the wilderness. Orderly rows of rocks were used to line paths and roads, delimit housing areas, and ornament prominent shrubs as if designing a garden. With the troops and barracks long gone, the visual evidence that this generic western desert had a human history consists of these curious rows of rocks, not all that different from the rock arrangements made by Native Americans that can still be found throughout the American West, perhaps signaling a universal pattern language. The alert reader may detect that these last two images form an "O" and an "M," respectively, spelling out "OM." "OM" subliminally suggests the mystic syllable—considered the most sacred mantra in Hinduism and Tibetan Buddhism—that appears at the beginning and end of most Sanskrit recitations, prayers, and texts.

Works cited

Benti, Wynne. "Military Withdrawals of Public Lands—Public Input Needed." 28 October 2016. *Desert Report: News of the Desert from Sierra Club California/Nevada Desert Committee*. Desert Report, 2018, http://www.desertreport.org/?p=1824. Accessed 6 December 2018.

Bravo 20 Bombing Range. Video by Peter Goin. Edited at the University of Nevada, Reno Studios & Photography Department, 1988. 6:56 min.

"Fallon Range Training Complex Modernization: Expansion of Land Ranges, Airspace Modifications, and Public Land Withdrawal Renewal." Environmental Impact Statement, Public Scoping Phase. August 2016, www.FRTCModernization.com. Accessed 6 December 2018.

"Fallon Range Training Complex Modernization Update." Fallon Range Training Complex Modernization, US Navy, 2021, https://www.frtcmodernization.com/project-updates. Accessed 21 May 2021.

"Final Environmental Assessment for Withdrawal of Bravo-20 Bombing Range, Naval Air Station, Fallon, Churchill County, Nevada." Department of the Navy, Western Division, Naval Facilities Engineering Command; Department of the Interior, Bureau of Land Management District Office, Carson City, Nevada. Contract No.: N62474-79-C-5502. January 1980, https://books.google.com/books?id=I_U3AQAAMAAJ&pg=PA3&lpg=PA3&dq=Final+Environmental+Assessment+for+Withdrawal+of+Bravo-20&source=bl&ots=_t6R_mZSOF&sig=qhY-i9lSsx0t4oUbQQq_HZOcDFcY&hl=en&sa=X&ved=2ahUKEwiPoKuu1ZjfAhUOTKwKHd-WIC7UQ6AEwAHoECAkQAQ#v=onepage&q=Final%20Environmental%20Assessment%20for%20Withdrawal%20of%20Bravo-20&f=false. Accessed 11 December 2018.

"General Patton's World War II Training Ground in the Mojave." *Mojave National Preserve*. National Park Service, 2015, https://www.nps.gov/articles/pattonmojave.htm. Accessed 7 December 2018.

Goin, Peter, "Bravo 17: Militarizing the Great Basin." *High Desert Journal*, vol. 1, no. 1, 2005, pp. 23–27.

Hulse, James W. *Nevada's Environmental Legacy: Progress or Plunder*. University of Nevada Press, 2009.

Martin, Gary. "Nevada's Lawmakers Vow to Protect State's Military Bases." *Las Vegas Review-Journal*, 25 May 2017, https://www.reviewjournal.com/news/military/nevada-lawmakers-vow-to-protect-states-military-bases/. Accessed 29 Nov. 2018.

The Misfits. Dir. John Huston. Screenplay by Arthur Miller. Perf. Clark Gable, Marilyn Monroe, Thelma Ritter, Montgomery Clift, and Eli Wallach. United Artists, 1961.

Misrach, Richard, and Myriam Weisang Misrach. *Bravo 20: The Bombing of the American West*. Johns Hopkins University Press, 1990.

Quinn, Elizabeth. "From the Editor." *High Desert Journal*, vol. 1, no. 1, 2005, p. 3.

Ruhge, Justin M. "Desert Training Center/California-Arizona Maneuver Area (CAMA)." *California Military Department Military History and Museums Program*. California Military Department, 2016, http://www.militarymuseum.org/CAMA.html. Accessed 10 December 2018.

Seekatz, Sarah McCormick. "Desert Deployment: Southern California's World War II Desert Training Center." 15 March 2015, *KCET*. Public Media Group of Southern California, 2018, https://www.kcet.org/shows/artbound/desert-deployment-southern-californias-world-war-ii-desert-training-center. Accessed 7 December 2018.

10

LANDSCAPE IS ARCHITECTURE

Reno, Nevada, is a geographical irony. Just east of "The Biggest Little City" lies the Great Basin Desert, an arid expanse nationally regarded as a wasteland. As we saw in the last chapter, large swaths of this landscape have been sacrificed to the military for bombing ranges and training centers. Yet just west of Reno, nestled high in the Sierra Nevada at approximately 6225 feet above sea level, sits sparkling Lake Tahoe, considered one of the most naturally beautiful places in the United States. Straddling Nevada and California, Lake Tahoe is a popular tourist destination, visited by more than three million people each year. Peter Goin moved to Reno in 1984 and immediately began exploring and photographing the diverse landscapes in his "backyard." To get an idea of the water profile of this region, bear in mind that Nevada is the most arid state in the nation. Storms brewing in the Pacific Ocean typically drop rain and snow on the western slopes of the Sierra Nevada and are mostly wrung out of moisture by the time the weather system reaches Nevada on the east side of the Sierra, the so-called rainshadow. Goin became fascinated by aridity—by how we perceive and treat arid lands and by how we regard and manage water. Much of Goin's work explores arid landscapes of Nevada and the American West, including projects that embrace the oxymoron *Arid Waters*, as one of his edited books is entitled.

Enter Lake Tahoe, whose outflow gives birth to the Truckee River, which provides precious water to Reno and points east. Peter wants his work to engage ideas and encourage public dialogue. In the military projects—including *Nuclear Landscapes* and "Militarizing the American West"—Goin questions the ethics of creating sacrifice zones. Whereas one might expect that a photographer working at Lake Tahoe would assemble an album of colorful scenic vistas targeted for a tourist market—indeed, many such books do exist—Goin's approach is radically different. When he began to become acquainted with Lake Tahoe, what surprised him was how many people he encountered seemed to think that Lake Tahoe is pristine wilderness. They had no idea how heavily managed the Tahoe basin is or that 150 years ago it looked very different than it does today. At Tahoe, Goin became interested in studying the evolution of landscape by documenting changes through time. His chosen methodology already existed under the name *rephotography*, namely making new

DOI: 10.4324/9781003212607-10

photographs—rephotographs—of places recorded earlier, dating as far back as the nineteenth century, and displaying these photographic pairs side by side to facilitate comparison.

Goin's 1992 book *Stopping Time: A Rephotographic Survey of Lake Tahoe*, with essay and captions by historian C. Elizabeth Raymond and archival assistance by Robert E. Blesse, provides fifty-five pairs of comparative photographs that illustrate landscape change. In her essay on changing landscape and image, Raymond observes that:

> The physical alteration of the Tahoe landscape has been extensive, as the photographs demonstrate. The modern terrain is an artifact, created by almost 150 years of active human intervention in the basin. Images have also played a role in the process, however, because the different ways that people have viewed nature at Tahoe have inevitably affected their environmental choices.
>
> *(12)*

Raymond's essay explores both kinds of change, "in the actual Tahoe landscape, and in the historical perception of it" (12). *Stopping Time* documents Tahoe's evolution from the train- and steamer-accessed resort period of the mid- to late-nineteenth century; through an intensive logging, mining, and industrial phase that lasted for just over two decades during the second-half of the nineteenth century; to the 1913 concrete dam at Tahoe City, which effectively made Lake Tahoe a managed reservoir, generating power and supplying water to Reno and other downstream users; to the birth of the gambling era in 1931, spawning a period of highway construction and automobile tourism; to regeneration of the forests and the modern-day incarnation of Lake Tahoe as a world-famous outdoor recreation destination, saddled with all the problems that attend urban growth.

Today, Tahoe's environmental quality is closely monitored and managed to strike a tenuous balance between commercial development and unspoiled "natural" splendor. Raymond nicely encapsulates the shift in Tahoe's economic potential, noting that by the twentieth century, "tourism, rather than timber, became the principal industry, so that protection, rather than extraction of natural resources, became the watchword" (12). And Goin contends that "over time, and in various ways, the Lake Tahoe basin has become a prototype of the managed landscape" (2). In conversation, Peter expands on the environmental significance of Lake Tahoe: "Every issue you need to deal with in environmental conservation exists here. Everything—air, water, economic survivability, fragility, species diversity—it's all here. It's a perfect case study. As Lake Tahoe goes, so goes the rest of us."

Before plunging into the details of the *Stopping Time* project, it will be helpful to brush up on rephotography. Simply put, rephotography consists of pairs of then-and-now images of the same subject. As art critic Lucy Lippard explains in her essay "Time and Time Again: Photography / Rephotography," this method dates back at least as far as the French Revolution of 1848 when barricades were photographed at different stages of the war (167). Before-and-after images have been used as a scientific tool to document change and historically to illustrate local histories. In the art world, rephotography gained attention through the three-year Rephotographic Survey Project (RSP) from 1977 to 1979, in which Mark Klett as chief photographer teamed up with Ellen Manchester, whose background was in photographic history, and JoAnn Verburg, who specialized in photography and museum work, along with photographers Gordon Bushaw and Rick Dingus to rephotograph nineteenth-century government survey photographs of the American West. Major government-funded geological and geographical surveys of the West, such as those led by Ferdinand V. Hayden, Clarence King, and Lt. George Wheeler, hired skilled photographers to accompany the expeditions and document the sites. As members of a survey party, landscape photographers, including William Henry Jackson, Timothy O'Sullivan, John Hillers, Andrew J. Russell, and Alexander Gardner, performed a scientific function. However, today their work is studied by art historians, and even in their own day they produced limited-edition albums.

Klett's RSP, which culminated in the 1984 book *Second View*, paid homage to that select group of nineteenth-century survey photographers by repeating more than 120 of their photographs. The RSP team developed precise techniques of triangulation in order to pinpoint the exact camera location (vantage point) of the historical images, and they sought to replicate the camera lens, angle of view, time of day and year, and weather conditions of the originals to produce a modern-day identical twin that could be used to study landscape change. In the process, they discovered that the survey photographers had made aesthetic choices in vantage point, framing, and image composition in order to create a "Best General View," namely the best possible representation of a given landscape. Discovering their choices in order to replicate them yielded valuable information about visual sensibilities in the nineteenth century, revealing codified ways of seeing and representation. Goin was involved in publishing some of Klett's rephotographic pairs in works that he edited or coedited, including *Arid Waters* from the 1991 Water in the West project and the first issue of *Nevada Public Affairs Review* in 1988, whose special focus was "Nevada's Evolving Landscape."

The first step in rephotography is to locate and select historical images. In the case of Lake Tahoe, a surprising dearth of historical photographs existed, very few of which were made by professional photographers. Goin and his team combed numerous repositories as well as private collections. A hindrance was that many photographers at Lake Tahoe point their cameras toward the lake itself, conveying minimal information about land use along the shore. Similarly, most tourists compose generic, scenic shots rather than views rich in topographical information. Nevertheless, diligent research discovered many hundreds of photographs preserved in repositories such as the Seaver Center for Western History, the Nevada Historical Society, the Special Collections Department at the University of Nevada, Reno, the California Department of Transportation, North Lake Tahoe Historical Society, and The Bancroft Library. Goin and crew utilized both well-known and less celebrated images by professional photographer Carleton E. Watkins, who received a commission in 1876 to photograph the Carson and Tahoe Lumber and Fluming Company's operations at Glenbrook, Nevada. Another valuable archive was albums made by the Los Angeles firm Putnam and Valentine, a company that specialized in picturesque photographs of the West.

Since Goin's interest is in the evolving landscape, he and his collaborators sifted through the images seeking representative views of human-impacted landscapes at different locations around Lake Tahoe and along the Truckee River from Tahoe City to Truckee. The historical images date from the 1870s to the 1950s and document a range of human alteration of the landscape.[1] Strikingly, as a set, the modern rephotographs generally show the locations looking *more* pristine—more forested, less polluted—than they did some one hundred years ago.

The second step in rephotography is determining the vantage point and camera/lens/film type of each historical image. Goin soon came to realize that the nature of his project dictated a more flexible interpretation of rephotography than Klett's project used. Klett's RSP rephotographed the work of a small and select group of professional survey photographers, which meant that camera, lens, and film types could be readily determined and held constant for the modern image remake. However, the Lake Tahoe photographs derived from a miscellany of sources, including commissioned surveys, commercial scenic landscape photographers, road construction documentation, and vernacular views such as tourists make to remember their trip. There was no common camera, lens, or film, nor season, time of day, or weather in the collection of historical images he chose to rephotograph. Using black-and-white film as in the historical views, Goin then opted to employ standardized materials of superior quality to most of the original equipment—"80mm lens (645 format), 150mm and 210mm lens (4x5 format), panorama lens (Widelux), 125 ASA 4x5 film or 120 film, or Polaroid type 55 film" (*Stopping Time* 8). Goin's work thereby establishes a systematized Lake Tahoe survey database that future photographers will be able to rephotograph to study change and continuity.[2]

For several reasons, vantage point selection required further adjustments from the literal practice of precise rephotography. When Peter located an historical vantage point and camera direction, it would often be the case that the modern view was completely obstructed by regrowth of vegetation—almost always dense second- and third-growth forests. A precise rephotograph might well consist of a frame filled with close-range pine needles and tree trunks or a longer-range screen of trees, blocking the horizon and more distant landscape features. While such a precise rephotograph would certainly yield some information about landscape change, more comprehensive topographical information would be shown by shifting the vantage point to where Goin's camera had a clear view of the landscape subject of the historical image. In effect, the historical photographer and Goin each chose the "best general view" available at the time their respective images were made. Had Goin insisted on precise rephotography, he would have faced still other problems. In some cases, there was simply not enough information in the historical photograph, such as horizon and prominent landscape features, to determine an exact vantage point. In other cases, the original vantage point no longer physically existed. For example, in the Cave Rock photograph pair, the historical image was made on a hillside that was subsequently dynamited away to construct a tunnel for Highway 50. Goin therefore adapted to circumstances, reinterpreting the original image "by providing a contemporary example of a *similar* view" (9).

There is also a conceptual difference between Klett's RSP and Goin's Tahoe rephotography project. Klett wanted to replicate the professional photographs of specific official surveys that spanned the American West. Klett estimates that his crew logged between 20,000 and 30,000 miles of driving over the three-year project (Klett, *Second View* 2). In contrast, Goin was interested in studying a particular place, the Lake Tahoe basin, which necessitated that he draw from the work of many photographers. In essence, the RSP followed a few photographers to many places, while Goin tracked many photographers to one place. Over the course of his project, Peter got to know Lake Tahoe very well and was able to develop an intuitive methodology that included a feel for the contour of the land and knowledge of how a given lens—wide-angle or telephoto—expands or compresses the sense of space in the historical image. The project became personally significant, in that so much time spent in the field changed Peter's relationship to the place. As he reflects:

> I've been over that landscape in many, many different ways and hiked to a number of the peaks and done rephotography and experienced the lake on water, on land, and from the air. I've hiked up and down and all over. What happens is that you develop this relationship with the landscape, this knowledge of the landscape that's intimate. It's like being in a relationship. You start to learn secrets and you learn patterns of behaviors and moods. I think that those experiences identify who a person is. Being an individual who walks through the landscape as much as I have at some point has meaning in the sense that I am less a visitor and more *of* the place, more of Lake Tahoe.

One might assume that rephotography is a purely technical operation, leaving little room for artistic or conceptual expression. However, when attentively examined, every one of Goin's photographs becomes a portal for the exploration of ideas. In fact, for him, the very process of making the rephotographs spawned new insights, which, however subtly, became woven into the images as a subtext. In case the visual clues are *too* subtle for the average reader to notice, this chapter aims to elucidate the undergirding ideas in order to enrich your understanding and enjoyment of the images.

As noted in earlier chapters, Peter's work shares an affinity with the New Topographics movement of the 1970s, which abjures landscapes of natural beauty rendered in a romantic style in favor of unpicturesque, human-altered landscapes, documented dispassionately in stark detail. Typical subjects of the New Topographics artists include new subdivisions, trailer parks, streetscapes, and industrial buildings. In the work of Lewis Baltz, for example, which depicts the transformation of southern California by apparently soulless urban sprawl, landscape is real estate. While the "landscape as real estate" formula certainly could be applied to Lake Tahoe, fifty-seven percent of whose shore is privately owned and dotted with mansions, Peter began to think more broadly than that, arriving at the insight that at Lake Tahoe landscape is *architecture*.

What does it mean to say that landscape is architecture, and how do Goin's images express this idea? If you know the history of Lake Tahoe, you begin to realize that the landscape surrounding the lake is human-altered and heavily managed. Consider the pervasive clear-cutting of Lake Tahoe's forests in the nineteenth century and the milling and processing businesses at places like Glenbrook, which the historical images picture. Now consider today's forested slopes, pine-scented air, white sandy beaches, and famously clear water of the lake. As Peter observes in conversation, "This is an industrialized, extractive landscape that's been essentially restored and built as a gateway to the mega-wealthy to experience a truly spectacular large-scale lake." Think about it. The whole Aloha chain of lakes in Desolation Wilderness is dammed. "We just say it's a mountain lake and we embrace the simulated environment," Peter notes:

> but it's not 'virgin' wilderness. I mean, it goes back to the Mesoamerican ruins when you think that it's architected landscape. It's constructed. It's managed. You are building topography. So you look up here, and you say, OK, well, what exactly is natural? Well, it's a highly altered environment.

The pairing of historical images with rephotographs highlights landscape change and introduces history and human alteration into the conception of a place that many people think of as untouched, pure, and timeless.

The "landscape is architecture" idea acknowledges that houses and other structures have been built. More radically, it suggests that even the natural-seeming landscape is an artifact of human architecting, in the form of forest management, fire-suppression policies, dams controlling lake levels, erosion control, flood mitigation, sewage disposal, air pollution control, invasive species eradication, building codes, and strict regulations concerning the landscaping of private property. In 1969 the US Congress ratified a bistate compact between California and Nevada, creating the Tahoe Regional Planning Agency (TRPA) to oversee development at Lake Tahoe, the first bistate regional environmental planning agency in the country. The compact was revised in 1980 to give TRPA authority to adopt and enforce environmental quality standards via ordinances. Raymond wryly notes that "the landscape is now vigorously managed so as to appear, paradoxically, pristine" (21).

Far from being a timeless wilderness, the Lake Tahoe landscape has changed drastically over the years, and *landscape change* is another important underlying idea of *Stopping Time*. As Goin states, "The evolution of this modified 'landscape as architecture' is the subject of this rephotographic survey" (*Stopping Time* 9). A key word in this formulation is "evolution." "Evolving Landscape" is a phrase that Peter has used in essay and publication titles and denotes a way of thinking that transforms our associations of *landscape* from a static picture to a dynamic process in which humans play a part. As we've already seen, after locating the historical vantage point, Peter then sometimes strategically adjusted his own vantage point to convey the best contemporary view of that place. A very astute reader may even pause to wonder what happened in the landscape that created the need for a different vantage point. Transcend two-dimensional thinking and put yourself procedurally into the process of rephotographing, and now you can imagine the landscape three-dimensionally.

One interesting challenge in creating a comprehensive visual survey of landscape change in the Tahoe basin is that the modern rephotographer necessarily relies on historical photographs. But what if no one took a picture of a given place? That question prompts another—why do people make photographs? In the later part of the nineteenth century, most of Lake Tahoe's visitors were viewing the lake for the first time and documenting their experience with photographs. Peter explains that:

> The photograph became a visual 'memory' of the landscape. These photographs were not intended to record the topographical characteristics of the landscape, since the study of how landscapes evolve was not a prevalent concern during the nineteenth and early-twentieth centuries.
>
> *(Stopping Time 4)*

What we begin to realize is that *time* plays a role not only in what a given landscape looks like but in the character and extent of the photographic record. Peter Goin had been mulling over the role of time vis-à-vis photography for a long time, even penning an essay in the 1970s entitled "Time and the Photograph." That essay opens with some broad generalizations, such as the following:

> Photographs are slices of time ranging from a millisecond to hours, even days. Time is not a constant, but a variable that is consciously used to convey expression, meaning and irony. It is a formal element in the construct of the photograph. It measures the passage of the ages while containing the countless fragments called moments. Time is contained, explained, restrained, defined and even ignored in photographs. Although it has a critical relationship to the subject, as a variable, it can also be neutral, held constant to allow some other variable to function in the expression of meaning.
>
> The image of time is complex and often illusive. The photograph can record the passage of time as a clock measures the seconds. The photograph can also represent time through the subject, or use time to convey the essence of the object. It can inspire contemplation, or create apprehension. Time is structure, motion and sentiment. Time is light and it is history. A photograph can be timeless, or put into the context of the times. Infinitely variable, how photographers use time and how they interpret it can lead to a better understanding of the photograph.
>
> *(2–3)*

Goin's essay then applies these overarching principles to art history, tracing changes in the representation of time in photography from the nineteenth century to the present and tying innovations in camera and film technology to the evolution of photographic traditions. Engaging in rephotography for the Lake Tahoe project heightened Peter's awareness of time, so much so that he entitled his book *Stopping Time*.

What might "stopping time" mean? To begin with the obvious, each photograph isolates and preserves in a frozen image a single instant in the flow of time. In the 1950s, Henri Cartier-Bresson, a noble descendent of photojournalism, spoke of his desire to record "the decisive moment" in a photograph that conveys the essence of a situation, capturing that fleeting moment in which action, gesture, and psychology coalesce to express the significance of a transitory event. A famous example is Bresson's photograph of a man stepping off a plank just before his foot will land in a flooded lot. This "slice of life" approach to time in photography seeks out evanescent moments—often in human lives—that may never occur again. At the other end of the spectrum, nineteenth-century government survey photographers generally omitted humans from the scene, making landscape photographs that lent support to the prevailing cataclysmic theory of geology, the idea that landforms are the result of sudden, major catastrophes such as an earthquake, volcano, or meteor strike rather than gradual evolutionary processes. Somewhere in between these two extremes lies the Lake Tahoe rephotographic survey, which records changes in the human-altered landscape. In both the historical images and the modern rephotographs it is often possible to assign an approximate date to the images based on visual evidence, such as signs, buildings, roads, trains, and automobiles. Whereas the original image may have been made for one purpose, its meaning changes when paired with a modern rephotograph that encourages comparison between then and now, two photographs in their respective "time" that together illustrate an evolving landscape.

While all of these meanings are implied by "stopping time," there is another, more subliminal message available to the thoughtful reader, and that is the experience of contemplative time. In contrast to photographs that freeze a passing moment are those that evoke a sense of timelessness, of time standing still. In this sense, the photograph becomes a portal to a form of meditative relationship. Peter explains in conversation:

> But the other side is that moment when we're not encumbered by the structure of time. It's that moment when you are taking a long hike and you sit down on a granite boulder and you overlook the landscape and you just kind of feel it rather than consume it. I think there is a difference between feeling a place and consuming a place. Visually we consume landscapes. But if we are not consuming it but experiencing it or feeling it, it does require a shift in our own expectations, because usually we go to a landscape the way a drunken sailor looks for a pretty woman. There's one! Instead of consuming, really try to embrace the greater complexity of the place you are in and *listen* as you see. Let the sounds dictate where you look or let the feeling of the breeze or let the coldness of the granite soak through your pants. Be present. Whatever it might be, there is something about being enriched by that feeling rather than the consuming of it.

Considering the juxtaposition of historical photographs and contemporary rephotographs invites contemplation and opens up a different way to think about and inhabit a place.

Like the photographs in *Stopping Time*, the image pairs in this chapter are arranged geographically, proceeding clockwise around Lake Tahoe from Glenbrook Bay, Nevada, on the east shore. Curiously, in this series, roads are a recurring motif, and we might pause to ask why. Well, have you ever noticed the viewpoint pullouts along highways? At those parking areas, have you ever seen dozens of people all making the same snapshot? Some road signs in national parks even feature an icon of a camera, essentially directing drivers to stop and take a picture. The fact is that the age of the automobile has affected the photographic record insofar as roads provide easy access to views. Recalling now the importance of the *line* in Goin's work, especially in *Tracing the Line*—the US-Mexico border survey—we can now trace the *line* to Lake Tahoe, where it becomes a road or rather a network of roads around the lake, sometimes entering and even more often enabling the historical photograph.

As you peruse these photographic pairs, we invite you to enjoy them in the spirit of *Thirty-Six Views of Mount Fuji*, a book with which Peter was familiar from his days of managing a bookstore in San Francisco and which was on his mind during the Lake Tahoe survey. During the Edo Period in Japan, the artist Katsushika Hokusai (1760–1849) created thirty-six exquisite block prints of Mount Fuji, depicting the iconic mountain from every imaginable angle and context. There are views of Mount Fuji as seen through the rain and mist, as framed by trees, windows, and bridges, upside-down as reflected in a pond, in morning, noon, and night, from the outside and from indoors, of people climbing the mountain and working in the rice fields below, of birds, ocean waves, and mythological creatures with Mount Fuji as a distant backdrop. "What we're talking about," Peter explains, "is understanding spiritually the complexity and multi-faceted emotional relationship to a landscape that you can't really know from only one linear interpretation." In Hokusai's series, there is even a portrait of a spider's web with a leaf caught in it. "Where is Mount Fuji?" we ask. In that case, Mount Fuji is *implied* by the light gray sky that surrounds and conceals the top of the mountain. In this series of Lake Tahoe views, if you don't see the lake, think about where it might be and how the photograph implies its presence.

■

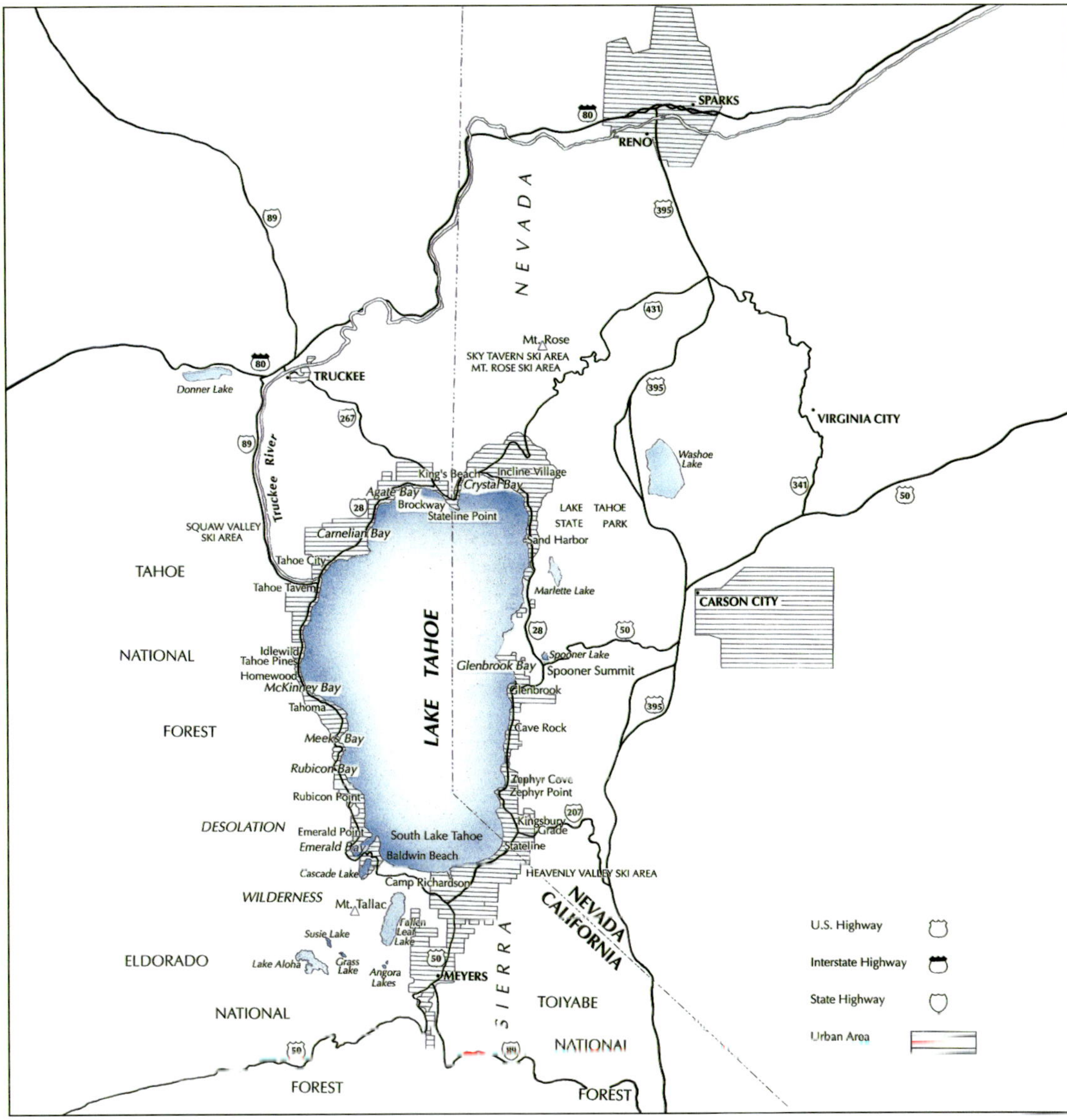

10.1 *Map of Lake Tahoe*, by Kathi Rick, 1991, adapted from *Stopping Time*.

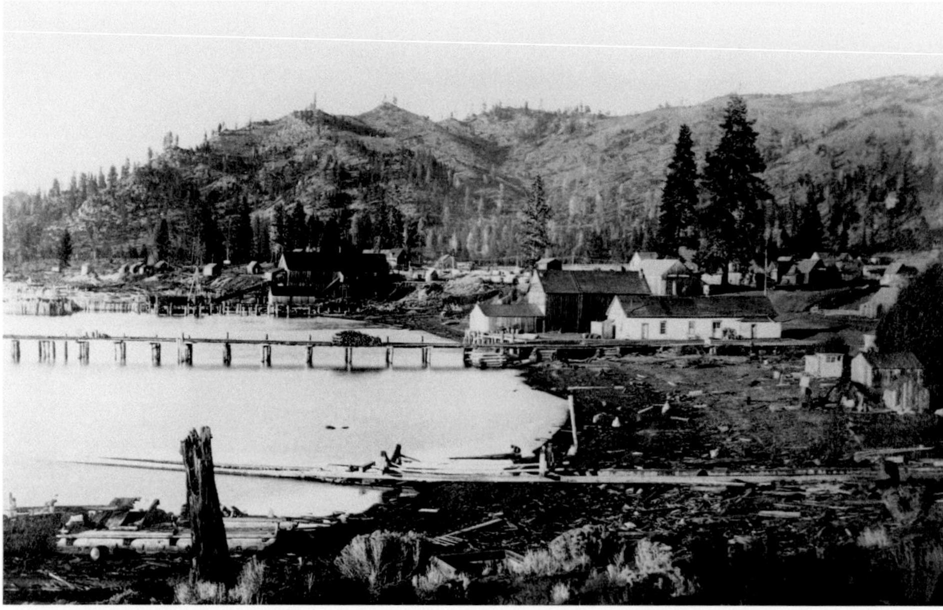

10.2 *Glenbrook Bay, Nevada*, c. 1880s, courtesy of Special Collections, University of Nevada, Reno.

The contour of the ridgeline in this pair of photographs on the east shore of Lake Tahoe is clearly recognizable, but forest regrowth in the modern image obscures the sharp outline visible in the historical image. The really dramatic topographical difference in these images is the location of the shoreline, which encroaches farther out into Lake Tahoe in the historical image, almost as if the lake has grown larger in the intervening century. However, what appears to be shore in the 1880s image is, in fact, floating debris, a dense mass of sawdust and logs from the enormous logging and milling operation at Glenbrook. Furthermore,

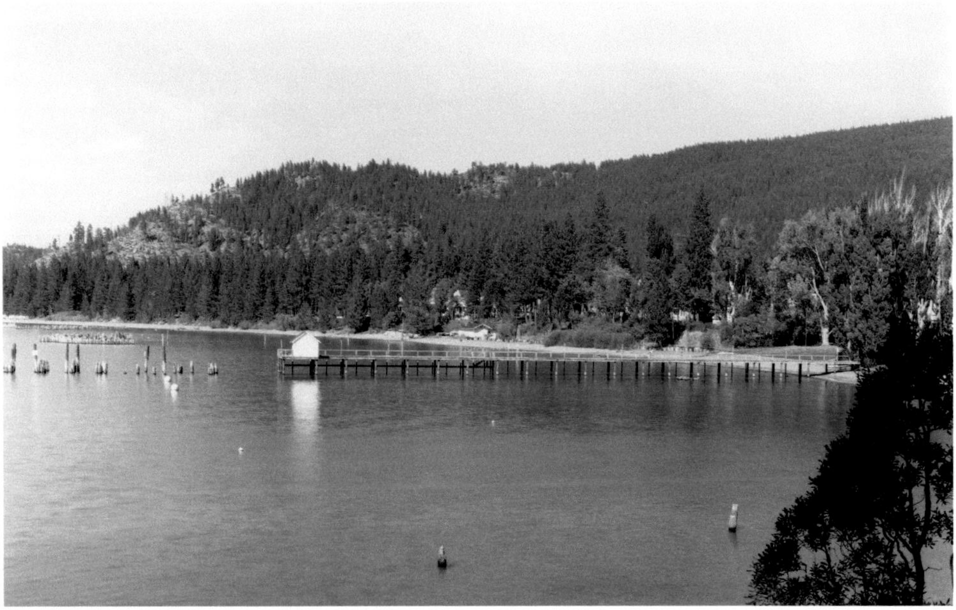

10.3 *Glenbrook Bay, Nevada*, c. 1988–1989.

construction of the concrete dam at Tahoe City in 1913 created the potential for the lake to rise slightly more than six feet above its undammed level, in effect expanding the size of Lake Tahoe during wet years. Interestingly, the 1990s image—with forested hills, open water, and mere glimpses of human habitation—looks far more "natural" than the historical image. A person viewing only the modern rephotograph of Glenbrook may never suspect that the region passed through a heavily industrial phase.

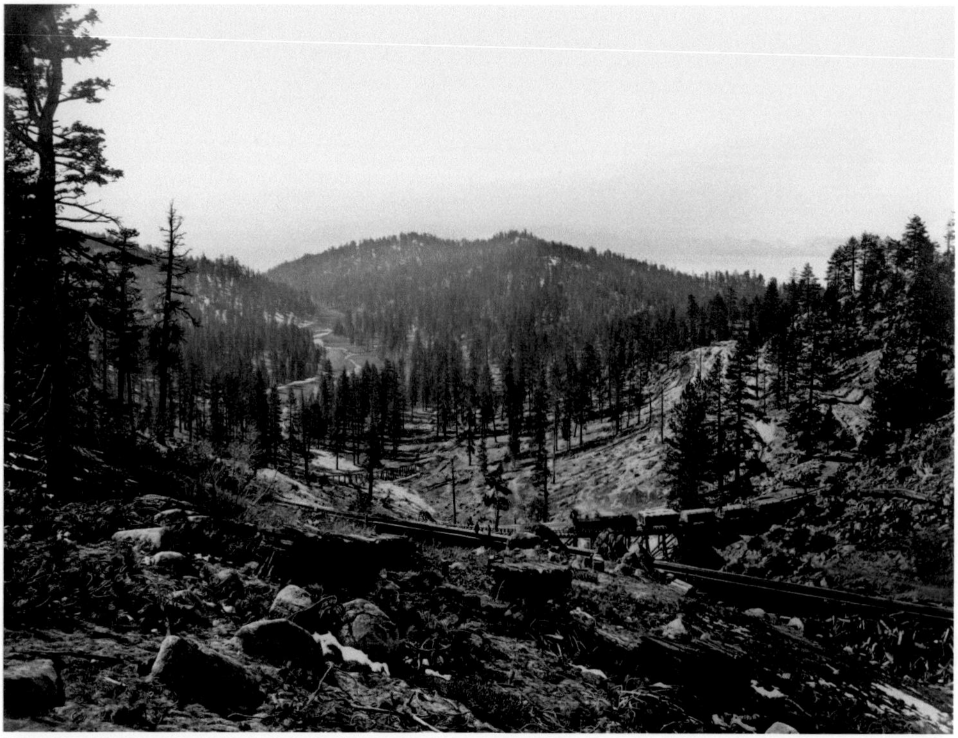

10.4 *Slaughterhouse Canyon, above Glenbrook, Nevada,* 1876, courtesy of James Herz Collection, Special Collections, University of Nevada, Reno.

Goin found this rare Carleton E. Watkins photograph in a private collection. This 1876 image depicts the Carson and Tahoe Lumber and Fluming Co. railroad, which hauled logs and lumber from the Glenbrook mill to an operation at Spooner Summit that supplied wood and timbers to the booming mining town of Virginia City, Nevada. The slow shutter speeds of nineteenth-century cameras, which would have blurred images of men at work, tells us that Watkins asked the train to stop and the men to stand still for this picture—this image was staged. Haze from the Glenbrook milling operations makes it difficult to discern Lake Tahoe in the background at right. So much regrowth has occurred since 1876 that Watkins' vantage point is now hemmed in by trees; therefore, Goin chose a slightly different location

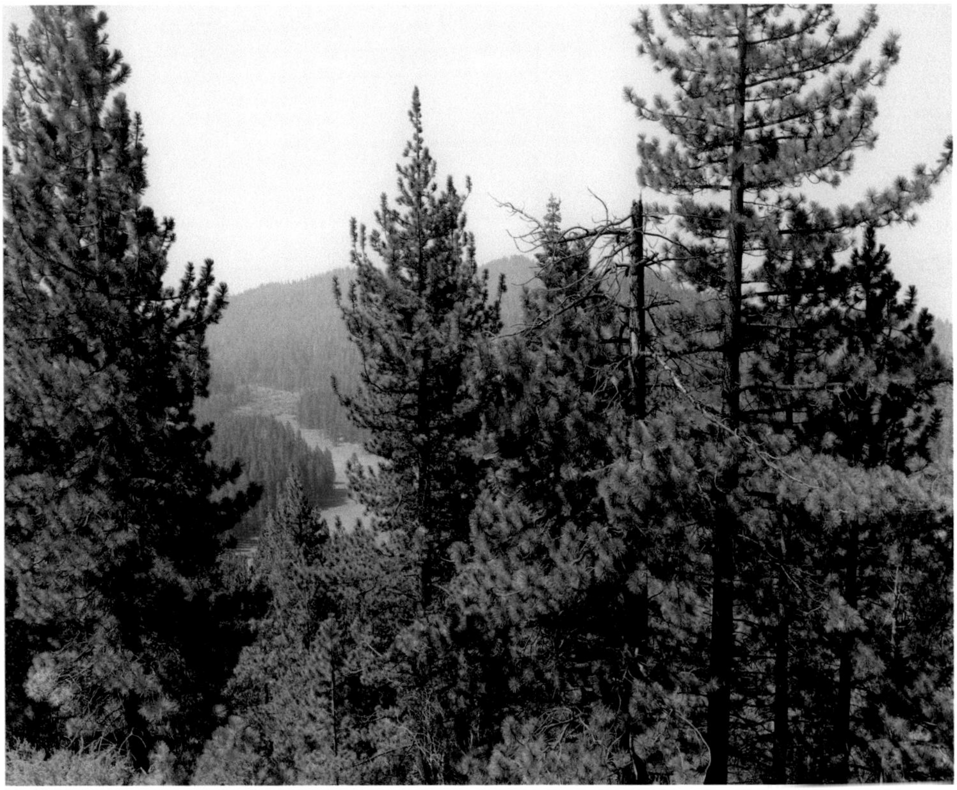

10.5 *Slaughterhouse Canyon, above Glenbrook, Nevada*, c. 1988–1989.

to make the rephotograph so the view could include Slaughterhouse Canyon. This time smoke from forest fires so reduced visibility that Lake Tahoe cannot be seen at all. While one might assume that forest regrowth is a sign of healthy regeneration of the ecosystem, in fact, dense stands of predominantly single-species trees (Jeffrey pines), coupled with decades of fire suppression, create vulnerability to beetle infestation and catastrophic fires, such as plagued the Angora area in South Lake Tahoe in 2007, where 3,100 acres succumbed to the intense fire, and where 242 residences and sixty-seven commercial buildings burnt to the ground, and another thirty-five homes were damaged.

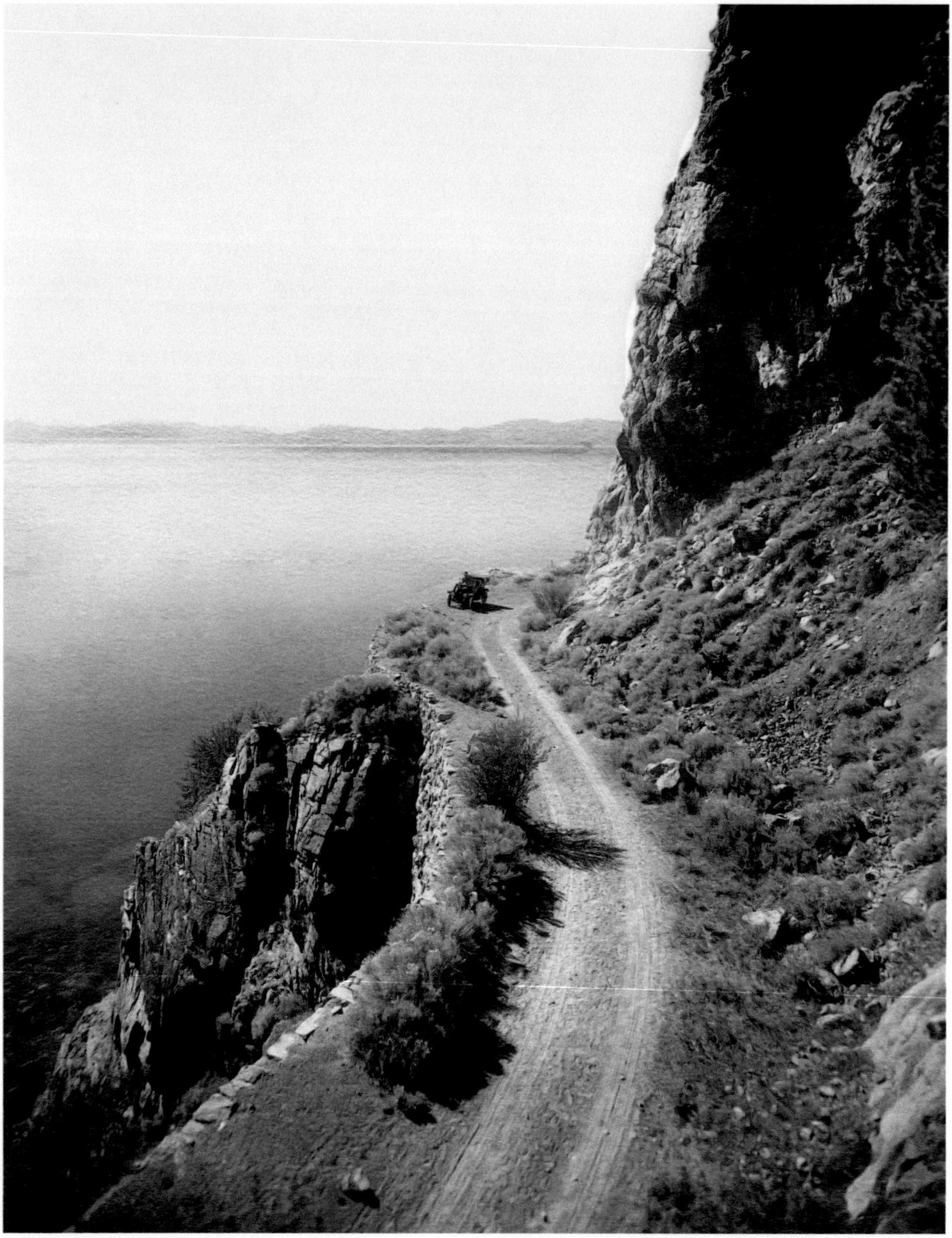

10.6 *Cave Rock*, c. 1918, courtesy of North Lake Tahoe Historical Society.

This historical image of Cave Rock, south of Glenbrook, Nevada, was made by Putnam and Valentine, a Los Angeles-based team (c. 1880s–1930) whose company produced albums of picturesque views of the American West, primarily California. Carlton Valentine often included his wife and their car in the scene for scale. Rephotographing this view presents a puzzle as the historical vantage point no longer physically exists, having been blasted away to construct first one tunnel (1931) and then a second (1957) for the modern Highway 50. In Goin's image, note the fracturing of Cave Rock, caused by dynamiting for the tunnels, re-architecting the landscape. If you were to choose a vantage point

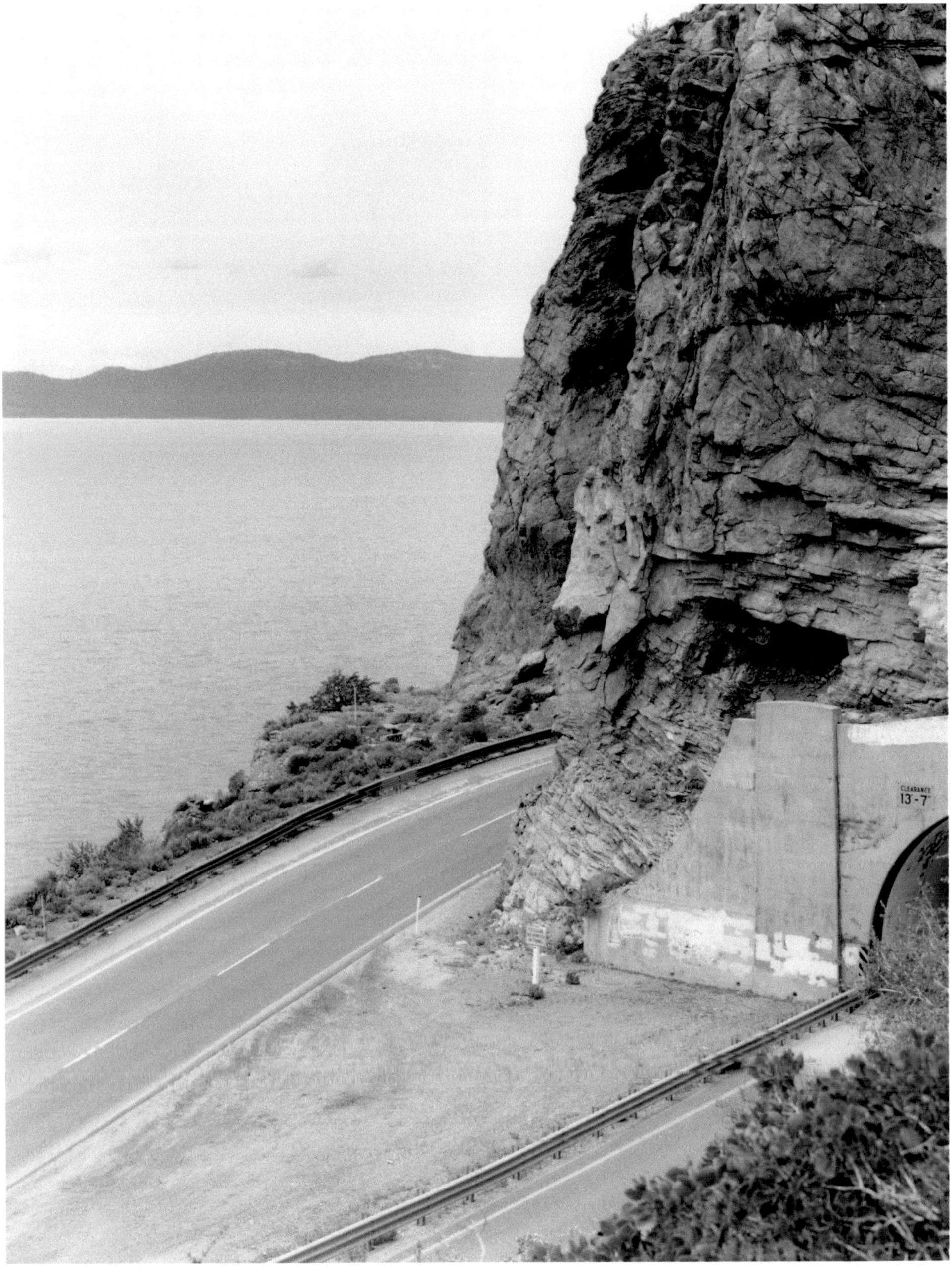

10.7 *Cave Rock*, c. 1988–1989.

on a boat in the lake, you would be able to see the west-facing cave for which Cave Rock is named, a sacred site for the Washoe people. The ethics of blasting the rock and running a highway directly beneath a sacred site were apparently not questioned at the time of construction, the overriding priority being ease of automobile travel to promote tourism and the recreation economy. The sign between the east- and west-bound roads reads, "Pedestrians not permitted in tunnel," another indication that tourism of the mid-twentieth century catered to cars.

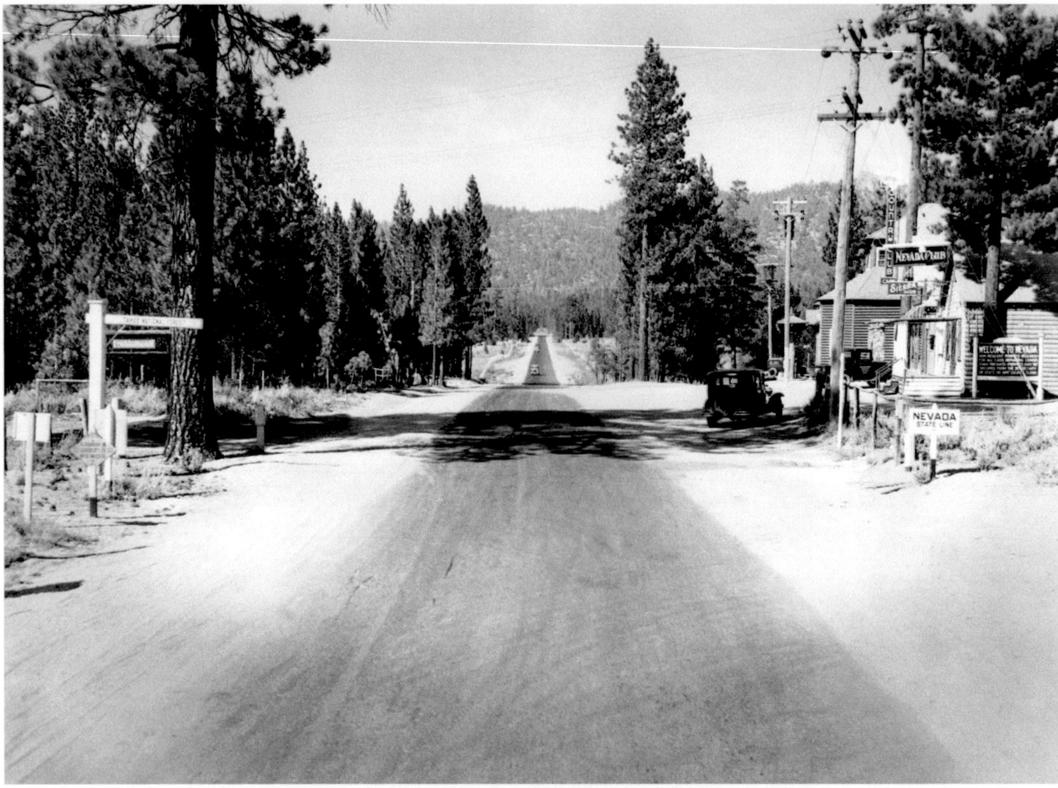

10.8 *Stateline, Nevada*, c. 1930, courtesy of Nevada Historical Society.

In the 1920s, Stateline, at south Lake Tahoe, was a small, seasonal town straddling the border of Nevada and California. The legalization of gambling in Nevada in 1931 and improved access in winter after World War II had a tremendous impact on the South Lake area, which sprung up with high-rise hotel-casinos, becoming the most urban area at Lake Tahoe. Note the "Nevada State Line" sign on the right margin of each photograph. To rephotograph, Goin set up his tripod and 4x5 camera with black cloth in the left turn lane in the middle of Highway 50. He owned a white Jeep Cherokee at the time, which

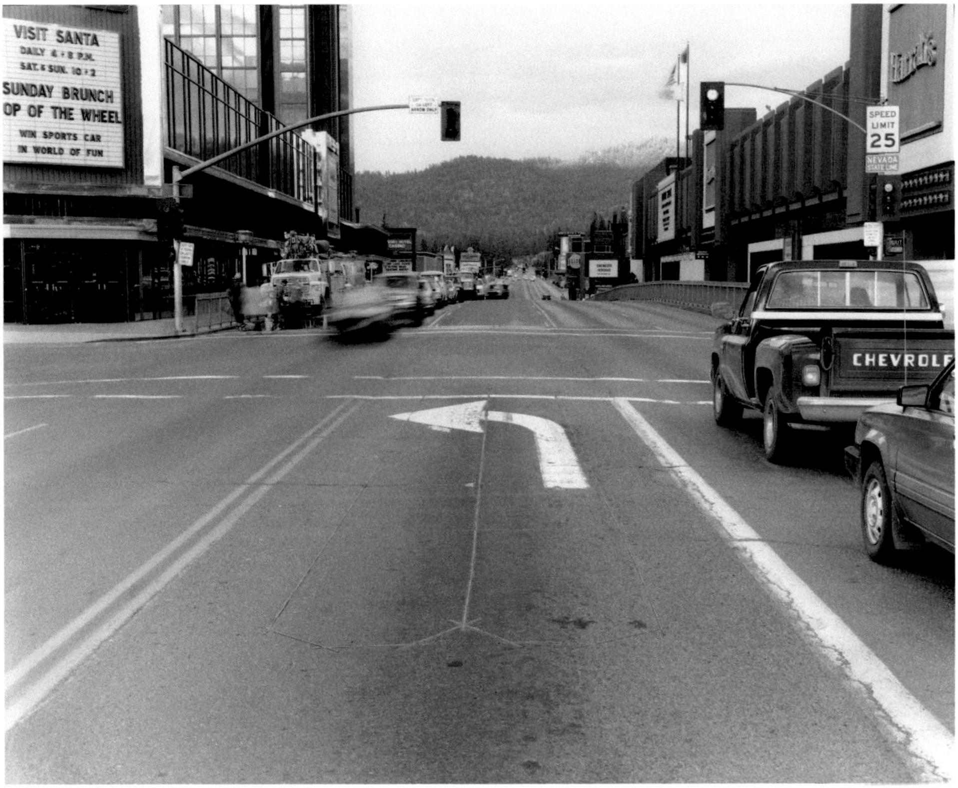

10.9 *Stateline, Nevada*, c. 1988–1989.

resembled an official state vehicle, and he parked his truck in the lane behind him, putting the flashers on. A highway patrol officer arrived on the scene almost immediately. Fortunately, the policeman offered to direct traffic while Goin made the photograph. Goin chose a slow shutter speed to create a blur of motion and the feel of commotion of oncoming traffic and men at work. Traffic is not necessarily an image that comes to mind when picturing scenic Lake Tahoe, but it is a reality, especially on weekends and during the bustling summer months.

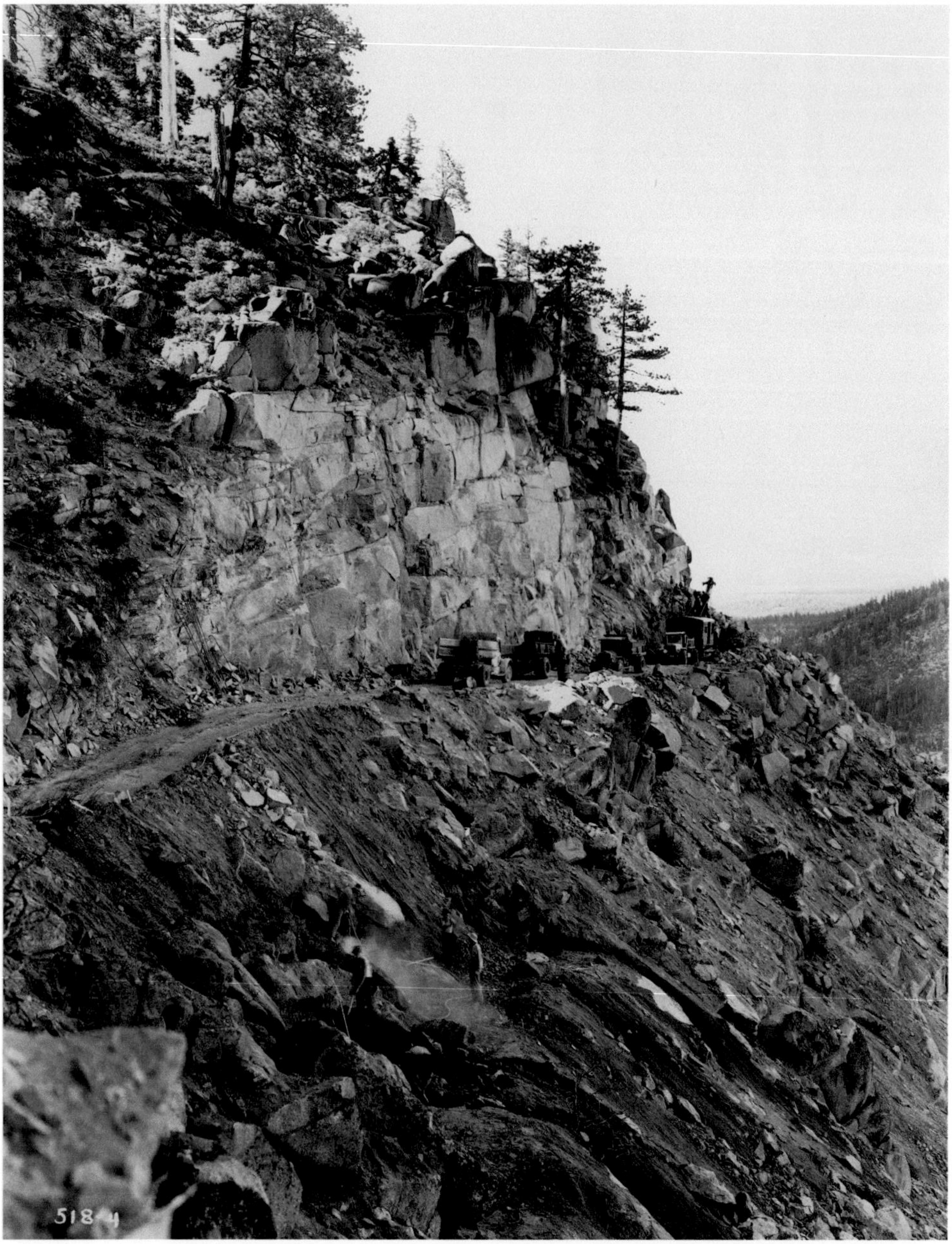

10.10 *Echo Summit*, 1938, courtesy of Caltrans Transportation Library and History Center.

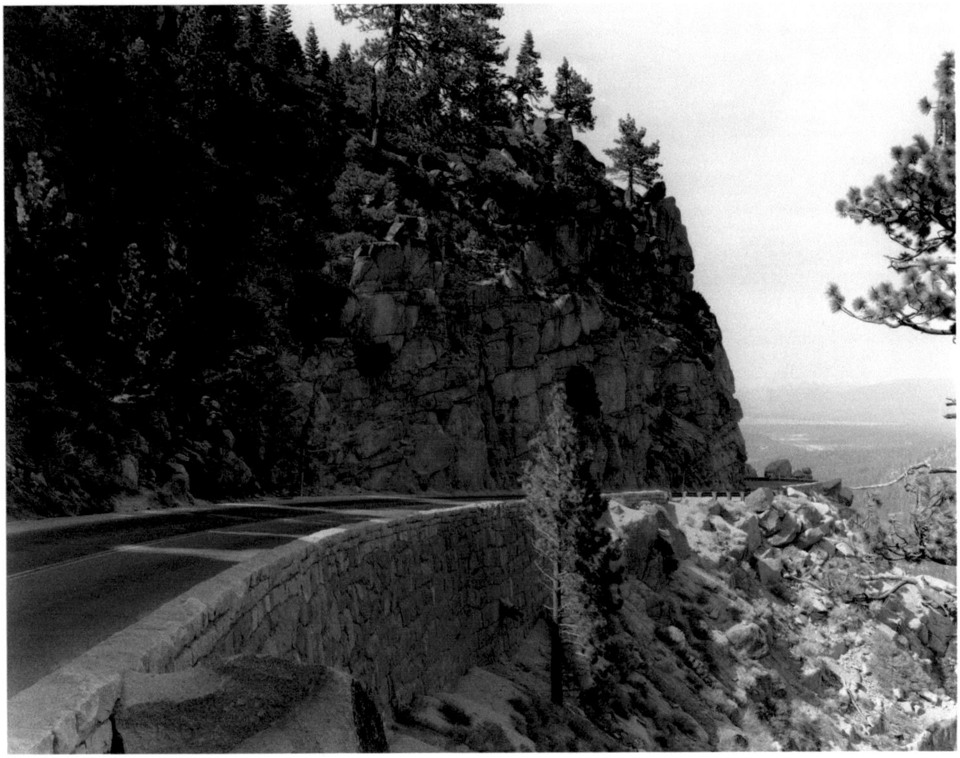

10.11 *Echo Summit*, c. 1988–1989.

The historical photograph from the California Department of Transportation, photographed in 1938, documents the expansion of a prior wagon route over Meyers Grade to make it suitable for automobiles. Highway 50 is a major connector between Sacramento and Lake Tahoe and Carson City. Building this highway was hazardous due to the steep, rocky slope. Men at work in the lower center of the historical image prepared the hillside for the beautifully constructed rock wall in the modern image. In this pair of images, note the difference in effect between the vertical-format original, which highlights work trucks, and the horizontal-format rephotograph of the completed, momentarily vacant road. Both images feature the same rocky cliff (one can identify individual rocks) and are artfully framed by tree branches in the upper right foreground. In Goin's image, however, the cliff lies in the shade, its relative darkness diverting attention to the lighter area to its right, Lake Tahoe. The curve of the rock wall whose top just catches the sun creates an arc that likewise sweeps the viewer's attention toward the distant lake, clearly visible in the rephotograph while almost unnoticeable in the historical image.

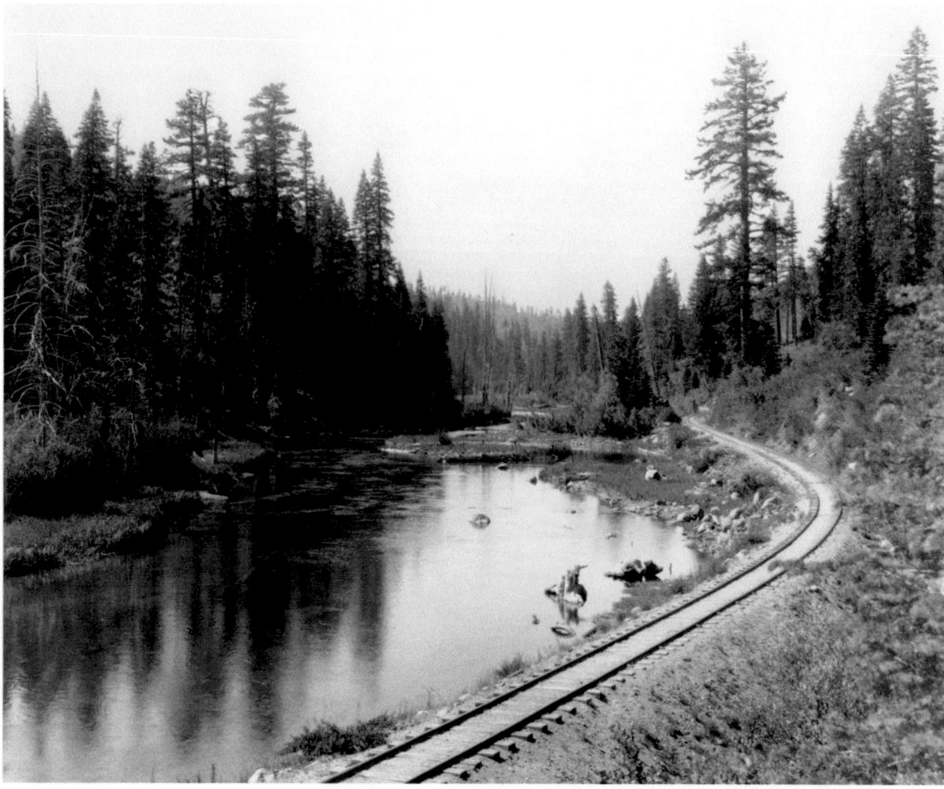

10.12 *Truckee River Canyon*, c. 1918, courtesy of North Lake Tahoe Historical Society.

The railroad in the historical image was built in 1900 by the lumber baron Bliss family to transport tourists from Truckee, California (accessible by the Transcontinental Railroad), to their new resort at Tahoe City, emphasizing the economic shift in the history of Lake Tahoe from logging to tourism. From Tahoe City, tourists could board steamers to reach other lakeside resorts. The railroad was scrapped during World War II and eventually converted to a bike path that serves today's outdoor-recreation-seeking tourists. Determining the vantage point of the historical photograph was difficult due to the absence of landmarks and

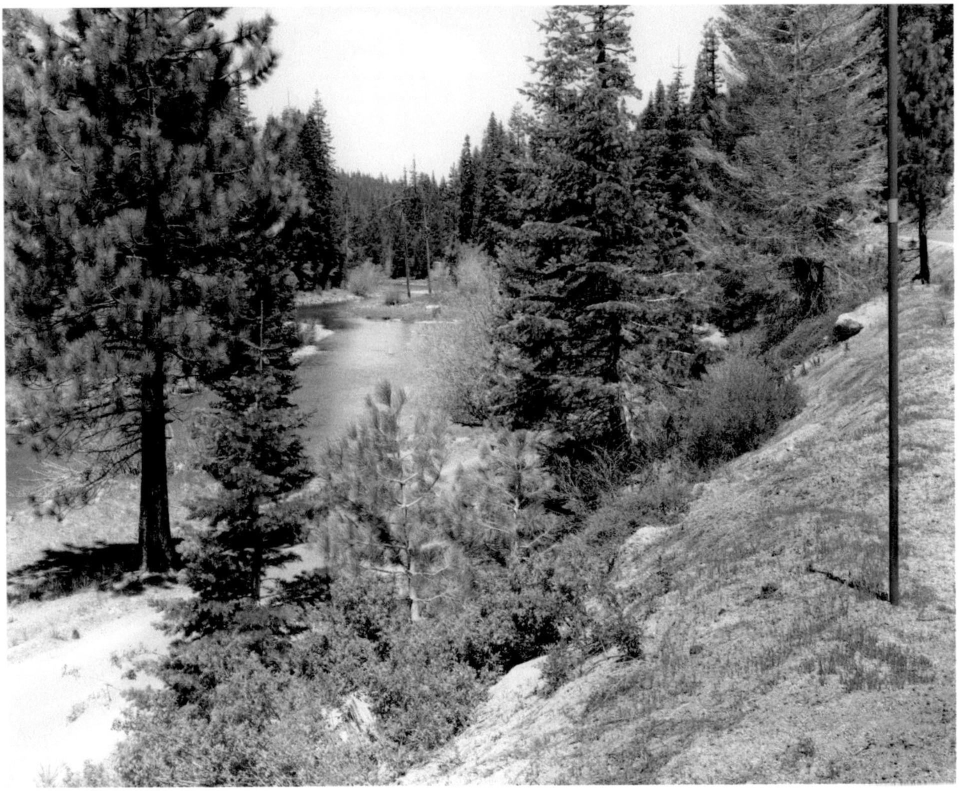

10.13 *Truckee River Canyon*, c. 1988–1989.

the new growth of dense thickets of trees and shrubs that completely obscures the horizon. For several weeks, Goin traipsed along the river's edge, integrating himself into the landscape to finally find this location. Thus, for him, this rephotograph triggers an embodied memory unavailable to the casual viewer. A dam at Tahoe City regulates release of Lake Tahoe's outflow into the Truckee River. In summer, in years of adequate flow, this stretch of the river is often choked with an armada of rafts, whose paddlers enjoy yet another use of Lake Tahoe's waters.

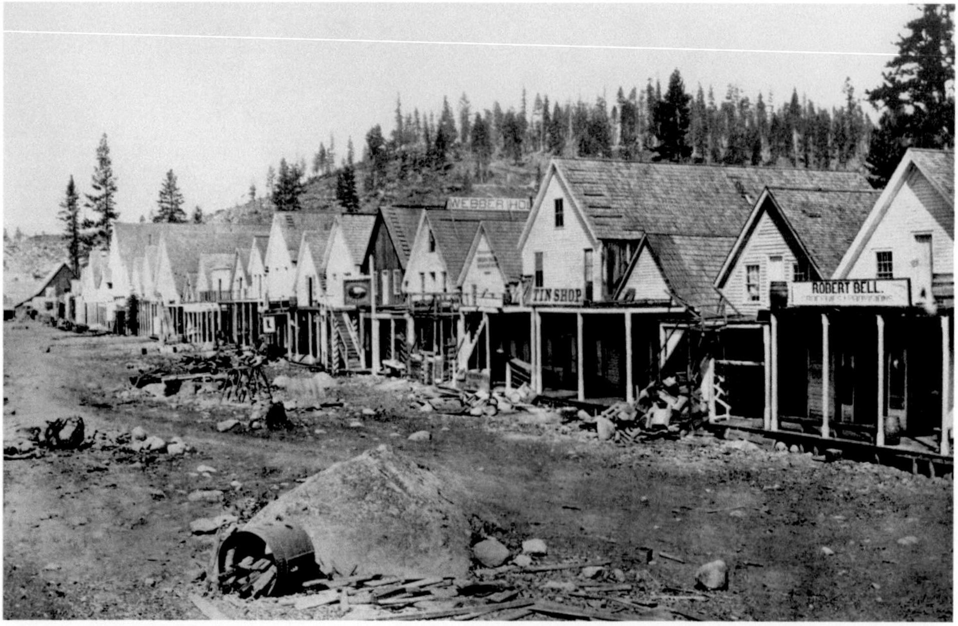

10.14 *Truckee, California*, c. 1870, courtesy of California Section, California State Library.

This pair of photographs documents the transformation of the built environment. For many people in the nineteenth century, Truckee, California, was the gateway to Lake Tahoe. Originally a stage stop, Truckee in 1868 was served by the Central Pacific Railroad, which increased access. The historical photograph was made around 1870, when Truckee was a lumber-milling town of just over one hundred people and somewhat of an eyesore, with piles of rubble heaped along a dirt street and a row of utilitarian buildings, including a good proportion of saloons. Curious is the absence of people in this image, creating a look of abandonment. The slow shutter speed of early cameras renders moving people transparent,

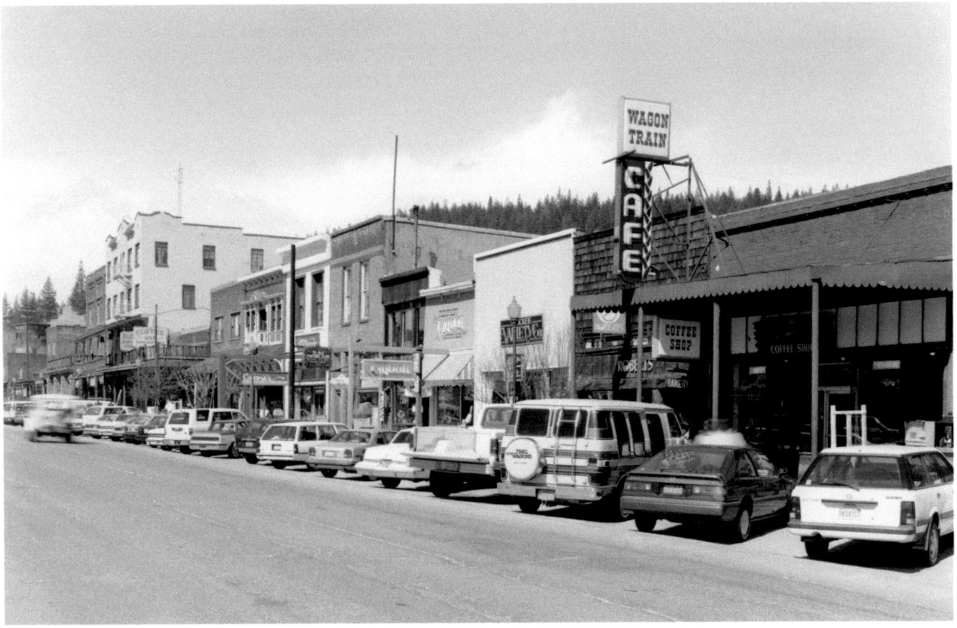

10.15 *Truckee, California*, c. 1988–1989.

which probably accounts for their absence in this image, an interesting metaphor in the historical narrative. Today the mountain town of Truckee, easily accessible via exits from Interstate 80, caters to tourists arriving by automobile, who ski in winter and hike, bike, and enjoy water sports in summer. The row of parked vehicles underscores the fact that ours is a car culture. The tidy row of storefronts includes upscale boutiques, coffee shops, bakeries, jewelry stores, souvenir shops, and restaurants. As we noted in the photograph of Glenbrook Bay, the hills behind Truckee, which were virtually denuded in 1870 by logging, are now fully forested. In both cases, rephotography registers regrowth.

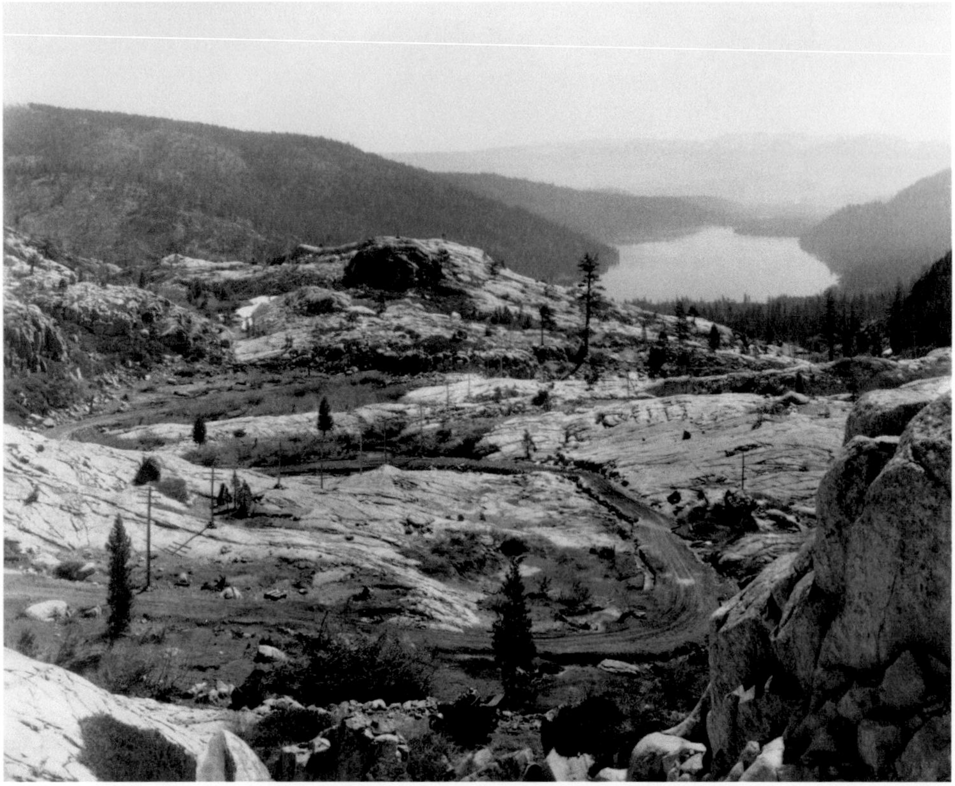

10.16 *Donner Lake, California*, c. 1918, courtesy of North Lake Tahoe Historical Society.

The scenic, historical photograph of Donner Lake made by Carlton O. Valentine characteristically includes his car and wife, here centered at the lower edge of the image. In this romantic rendering, the woman, wearing a hat for sun protection, relaxes in the open air of a convertible, a far cry from the desperate fight for survival faced in the winter of 1846–1847 by the infamous Donner Party, after whom Donner Lake is named. The serpentine curves of the Lincoln Highway, pictured in the Valentine image, are still visible in the rephotograph, even though the road was realigned in 1923 and eventually paved. However, the new

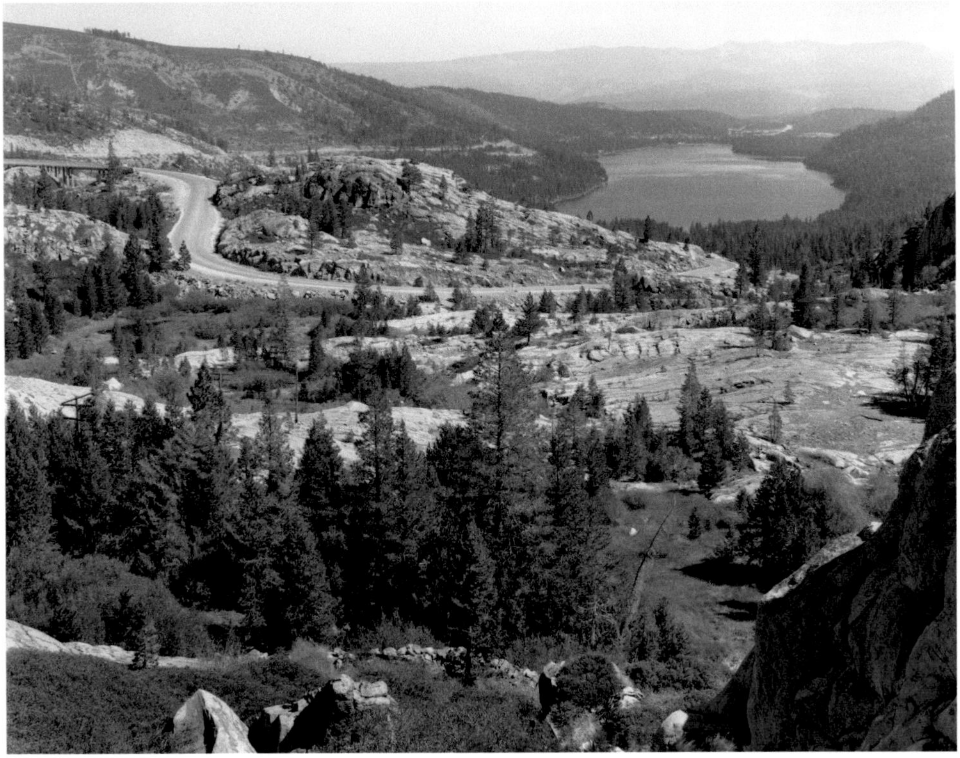

10.17 *Donner Lake, California*, c. 1988–1989.

road is now known as the "old" Donner Pass road, a scenic byway that has been superseded by Interstate 80, the four-lane freeway that slashes like an arrow from the upper left edge of the image to the town of Truckee below, at the eastern edge of Donner Lake. In researching images of Lake Tahoe, Goin discovered photographs of Donner Lake that were erroneously labeled "Lake Tahoe," causing him to speculate that an unscrupulous Truckee tour guide, paid to take customers to Lake Tahoe, ushered his unsuspecting clients to nearby Donner Lake instead.

Notes

1 In ongoing research, Peter Goin and his collaborators have discovered many more historical images of Lake Tahoe than they could accommodate in *Stopping Time*. Furthermore, after *Stopping Time* was published, people began contacting Goin and giving him access to rare photographs in their private collections. Goin has published many Lake Tahoe images in Arcadia's local history book series with selected rephotographs by Goin and his associate, Scott Hinton, Coordinator, Digital Labs and Faculty Research in the Department of Art, University of Nevada, Reno. In order of publication, these titles include *Lake Tahoe* (2005), *South Lake Tahoe: Then & Now* (2010), *A Maritime History of Lake Tahoe* (2012), and *Emerald Bay and Desolation Wilderness* (2018). In addition, Goin is working on two new books: *The Nature of Lake Tahoe: A Photographic History, 1860–1960* (2021) and *Re-Visiting Lake Tahoe,* both with the University of New Mexico Press. Providing contemporary photographs for these projects required that Peter raise funds to buy a "photographic research vessel"—a boat—so he could make photographs of the shore from vantage points on the lake.

2 Rephotographing rephotographs has precedent. In 2004, Mark Klett et al. published *Third Views, Second Sights: A Rephotographic Survey of the American West*, rephotographing 110 sites from Klett's 1970s RSP.

Works cited

Blesse, Robert E., and Peter Goin, eds. *Nevada Public Affairs Review*, no. 1, 1988. [Issue title: "Nevada's Evolving Landscape"]

Goin, Peter, ed. *Arid Waters: Photographs from the Water in the West Project*. Text by Ellen Manchester. University of Nevada Press, 1992.

———. *Emerald Bay and Desolation Wilderness*. Arcadia Publishers, 2018.

———. *Lake Tahoe*. Arcadia Publishers, 2005.

———. *A Maritime History of Lake Tahoe*. Arcadia Publishers, 2012.

———. *The Nature of Lake Tahoe: A Photographic History 1860–1960*. University of New Mexico Press, 2021.

———. *Re-Visiting Lake Tahoe*. University of New Mexico Press, forthcoming.

———. *South Lake Tahoe: Then & Now*. Arcadia Publishers, 2010.

———. *Stopping Time: A Rephotographic Survey of Lake Tahoe*. Essays by C. Elizabeth Raymond and Robert E. Blesse. University of New Mexico Press, 1992.

———. "Time and the Photograph." Unpublished essay, ca. 1979.

———, and Lucy Lippard. *Time and Time Again: History, Rephotography, and Preservation in the Chaco World*. Museum of New Mexico Press, 2013.

Klett, Mark, Kyle Bajakian, William L. Fox, Michael Marshall, Toshi Ueshina, and Byron Wolfe. *Third Views, Second Sights: A Rephotographic Survey of the American West*. Museum of New Mexico Press, 2004.

Klett, Mark, Ellen Manchester, JoAnn Verburg, Gordon Bushaw, Rick Dingus, and Paul Berger. *Second View: The Rephotographic Survey Project*. University of New Mexico Press, 1984.

Lippard, Lucy. "Time and Time Again: Photography / Rephotography." Peter Goin and Lucy Lippard. *Time and Time Again: History, Rephotography, and Preservation in the Chaco World*. Museum of New Mexico Press, 2013, pp. 167–170.

11
HUMANATURE

Where do ideas come from? How do early projects plant the seeds for later ones? How does an artist's body of work cohere? These questions, which propel this study, become somewhat vexing when accounting for Peter Goin's 1996 book *Humanature*, an apparent outlier in his oeuvre. While most of Goin's mid- and late-career work is set in the American West, *Humanature* ranges throughout the United States, particularly the American South. *Humanature* includes more close-ups of people and animals than Goin's other work, in which landscapes predominate. And *Humanature* seems less interested in documenting a particular place or topic than a philosophical idea. (And, by the way, how do you photograph a philosophical idea?) So where did this project come from and how does it fit with the rest of his work? I hypothesize that the projects immediately preceding *Humanature*—specifically *Nuclear Landscapes* and *Stopping Time*—put pressure on Peter personally to develop a philosophy, a way of understanding the relationship of humans to nature, which would not only explain what he had witnessed but also offer hope. This philosophy, encapsulated in the portmanteau "humanature," is articulated in Goin's manifesto *Humanature* and is implicit in virtually all of Goin's work. This chapter, in which the "humanature" philosophy is explained and illustrated, is therefore crucial to understanding Peter Goin's art. It will be helpful to begin by drawing on conversations with him that make some of these connections explicit.

Researching and photographing *Nuclear Landscapes* was deeply troubling. To be personally exposed to dangerous levels of radiation, to witness landscapes made lethal by radioactive contamination, and to realize the global reach of increased levels of radioactivity caused by the nuclear testing era—well, it was profoundly depressing. On a personal level, Peter struggled to find a source of hope. Paradoxically, hope would emerge from the very realization that human actions had changed the planet for the worse. What that means is that nature is not "out there," separate from and immune to human action. Humans and nature are interconnected. Once you accept that nature has been and continues to be *shaped by* humans, then it becomes possible to imagine people becoming a force for positive shaping. In essence, we must accept responsibility for playing a managerial role vis-à-vis nature.

DOI: 10.4324/9781003212607-11

As Aldo Leopold so memorably argued in "The Land Ethic," like it or not, humans alter nature. Rather than denying our effects or wishing ourselves out of creation, why not learn to be intelligent citizens of the biotic community? For example, we could design wildlife corridors through a city! We could, as Ian McHarg's path-breaking book is entitled, *Design with Nature*.

Peter's work on *Stopping Time* provided a different trail toward this same fundamental insight. What intrigued Peter about photographing landscapes around Lake Tahoe is that so many people he encountered believed that the Tahoe Basin was wilderness. Whereas Peter was aware of the intense human management that has created the *look* of pristine beauty of Lake Tahoe and its environs, many people experienced the area as "untouched," "unspoiled," "pure," and "natural," a place where they could get away from civilization. In the best virtual reality experiences, you feel like that digital world is *real*. You forget that you are experiencing a technologically created simulacrum. In some senses, Peter realized, Tahoe is a simulacrum. It is a *constructed* landscape designed and managed to simulate our conception of wilderness. In other words, wilderness is not a real place on Earth. As Roderick Nash posits in *Wilderness and the American Mind*, wilderness is an idea. For Peter, it is perfectly fine to enjoy the smell of the pines, the feel of a fresh mountain breeze, the sounds of birds—he likes to do it himself. But he thinks that people should understand the degree to which places have changed and have been altered by humans. Why? Because rather than fancying that management does not exist or is not necessary, we can be mindful, more informed managers, managing not just for scenery or for timber harvest, but for such values as biodiversity, wildlife habitat, and a functioning ecosystem. There is still so much to learn.

When you talk with Peter, this inferno of ideas about humans and nature can be set off by a single spark—the word *virgin*. He recalls a time shortly after college when he was driving through the American South and saw a sign that read "Virgin Forest." Behind the sign was a plantation of young pine trees laid out in neat rows. That sign stuck with him and ignited a chain reaction of thoughts, such that "virgin" became a key to understanding culture. As he will tell you:

> I always love the word "virgin." In other words, there's no sexual consummation; there's no "intercourse," which historically used to mean conversation not sexual coupling. And so really, "virgin?" "Virgin" landscapes? So, we take a forest planted in rows and we say, "this is a virgin forest." There is a little bit of the introduction of the concepts of Christian purity in the way that we relate to some of these topics, and the role of religion in the environmental movement is also of interest. You can see how this portal, that *one single word,* opens a landscape of literary and poetic interpretations that involve cultural critique on a profound level.
>
> And, you understand, wood and paper companies use "virgin" to indicate that it's a forest that is due to be harvested. We also speak about "old growth." The grand dames of the forest survived. The widows of the confrontation that we call logging. But the virgin forest—this is new, nubile, young female trees. And then, I mean, do we harvest women? And then you look at our society, and you look at paternalism and how it treats women, and you talk about exploitation of resources and attitudes, and you start going, "wow there's a lot of associative possibilities." It's really one of those words that offers an entry to a whole culture.

Goin's introductory essay to *Humanature* reprints a copper engraving of Adam and Eve, made by German Renaissance artist Albrecht Dürer in 1504. Dürer's engraving depicts Adam and Eve naked in a forest, standing on opposite sides of a central tree, from which a snake coiled around a branch places a fruit in Eve's hand. Drawing from our knowledge of the Bible, we can infer that Adam and Eve will eat the fruit of knowledge and thereby be expelled from paradise. Goin opens his essay with a litany of examples of environmental destruction in the United States—of water, oceans, forests, land, air, and wildlife—concluding, "Although the Garden of Eden has been an enduring image of earthly paradise, our planet is now a contaminated garden" (1). But lest we think that environmental despoliation is strictly an American or modern problem, Goin quickly adds that concerns about deforestation and soil erosion, for example, date back at least as far as classical Greece, Imperial Rome, and Mauryan India. In other words, "The tradition of managing, altering, and spoiling the environment is not so much a human problem as it is an indication of the human condition" (2).

As a register of Christian-influenced culture, however, the biblical story of Adam and Eve is symptomatic of a pathological worldview. In the biblical story, when God expels Adam and Eve from Eden, humans are in a stroke forever banished from nature, and humans and nature also become *ideologically* separated. Our culture tends to think of humans as distinct from and in opposition to nature rather than as a part of nature or *of* it. Goin's view is that although "we have abused our garden," we still live in it (2). Essentially, his view enjoins us to acknowledge our condition and become more enlightened gardeners. Rather than idealizing nature as pure, virgin, untouched wilderness, we should accept that Earth is a garden shaped by human choices. Let's learn to make better choices.

When Peter set out to investigate human-made nature, he gravitated to the American South, where there had been four hundred years of Euro-American settlement history. Peter was awarded a year-long sabbatical from the university for 1991–1992, and coincidentally received an Artist Fellowship award with accompanying funds from the National Endowment for the Arts. In concert with the Center for Documentary Studies at Duke University, Peter became a National Endowment for the Arts Fellow at Duke University. He was provided an office, telephone, university identification and library card, and he was expected to give a series of public talks.

He remembers this year as a special time for him and his family. Peter's wife, Chelsea, studied folklore with the excellent faculty at the University of North Carolina while working toward her PhD in anthropology at the University of Nevada, Reno. Their daughters were young; Kari was four and Dana not quite two. Money was tight. Peter drove their belongings across the country in a small U-Haul trailer and rented an unfurnished apartment. He built a dining room table and coffee tables out of plywood and two-by-fours and purchased other furnishings from thrift stores. When the girls began to mark on the apartment walls with crayons, Peter deflected them and had them use the crayons to color the unpainted wood of the tables. From a large packing box, he made the girls a playhouse and a cardboard palm tree (see Figure 2.7). He fondly recalls their living quarters as "funkadelic." While Dana was in daycare, Kari would often come along with Peter and play while he photographed.

One of Peter's public talks at the Center for Documentary Studies was closely related to the *Humanature* project: "The Evolving Landscape: A Photographic Exploration of the Relationship Between Humans and their Environment." When people learned of his work, they suggested places to visit. Peter also read voraciously, gleaning ideas for topics and places to photograph. He roamed up and down the Atlantic seaboard, making photographs in North and South Carolina, Tennessee, Virginia, and Florida. After his sabbatical year, Peter spent another couple of years photographing sites in the American West—California, Nevada, and Arizona—to make the book national in scope.

The hidden-in-plain-sight key to interpreting the images in *Humanature* is the "Suggested Readings" list that Goin includes at the end of the book. Aha! one realizes, these are the ideas that undergird this body of work. Early works include Aldo Leopold's *A Sand County Almanac* (1949), a keystone work of environmental thought that proposes the radical idea that "We abuse land because we regard it as a commodity belonging to us. When we see land as a community to which we belong, we may begin to use it with love and respect" (viii). And:

> All ethics so far evolved rest upon a single premise: That the individual is a member of a community of interdependent parts. . . . The land ethic simply enlarges the boundaries of the community to include soils, waters, plants and animals, or collectively: The land.
>
> *(203–204)*

Leo Marx's *The Machine in the Garden* (1964) introduces the metaphor of nature as a garden, into which American industry intrudes and disrupts. Other book titles signal their own challenge to the way we conventionally think about nature: *The End of Nature* by Bill McKibben, *The Control of Nature* by John McPhee, *The Social Creation of Nature* by Neil Evernden, and *Artificial Nature* edited by Jeffrey Deitch and Dan Briedman. Other titles unsettle our cherished notions about the balance of nature, for example, Daniel Botkin's *Discordant Harmonies: A New Ecology for the Twenty-First Century* and James Gleick's *Chaos: Making a New Science.* And several works draw parallels between human dominance of nature and men's dominance of women under patriarchy, such as Susan Griffin's *Woman and Nature*, Carolyn Merchant's *The Death of Nature: Women, Ecology, and the Scientific Revolution*, and Irene Diamond and Gloria Orenstein's anthology *Reweaving the World: The Emergence of Ecofeminism*. These are the conversations and debates that Peter is joining.[1] But, whereas the works on the "Suggested Readings" list are prose, Peter's preferred language is visual. While the books he references seek to *tell*, he aims to *show*. Nevertheless, it is important to understand that Goin's photographs often reference larger intellectual debates as well as allude to other artworks, such as Dürer's, and photographic movements, such as the post-New Topographics.

This chapter reprints fifteen photographs selected from *Humanature* plus one previously unpublished image. To create a context for understanding the limited number of images reproduced here, it will help to get a sense of the larger range of photographs included in *Humanature*. In that book, the photographs are arranged into eight thematic galleries: Trees, The Zoo, Beaches, The Mine, Reclaimed Land, The River, Dams, and Wildlife. Each gallery is a visual exploration of the idea that nature is an illusion. As in the scene in *The Wizard of Oz* when Dorothy's dog, Toto, pulls back the curtain and exposes the old man whose machinations have created the specter of a grand wizard, Goin's photographs and captions pull back the veil to reveal the human manipulation that has shaped what a naive viewer might have assumed was untouched nature. The following review of the photographs in *Humanature* suggests the range of visual evidence that Goin marshaled to lend photographic support to the radical notion that nature is not natural. Everywhere you look—if you know how to look—you will find signs of human tampering. Better than "nature," the neologism "humanature" more accurately communicates the reality of the omnipresent imprint of the human hand on our environment.

In the Trees gallery, for example, which is mostly set in North Carolina, we see a big, round *drain* in a natural-looking swamp in a forest, betraying that the swamp is, in fact, human-made. A similarly disarming image features the Chattooga River, site of the film *Deliverance*. While the right side of the image looks like new vegetation is rewilding a formerly logged streambank, the left side shows a measuring gauge for flood control as well as a culvert for high-water diversion. A photograph of the Duke Forest in Durham features images of tree stumps—an obvious sign of human intervention—and a photograph of a length of moss-colored green cloth draped down the trunk of a tree, looking like artificial bark and perhaps being used as part of a scientific study. Another image of the Duke Forest educates visitors via a prominent sign at the forest edge, which reads, "Naturally seeded 1926, Thinned, 1939, '44, '49, '54, '59, '84, Prescribed Burn 1981, '83, '85, '87." A four-photograph fire series documents a Nature Conservancy prescribed burn to reduce undergrowth. The upshot? Every forest in the Trees gallery carries a human signature.

The Zoo gallery consists of a series of photographs made at the North Carolina Zoological Park near Asheboro. The zoo has a "Creating the North American Landscape" program that, quite literally, constructs different physiographical regions of North America, where representative species are displayed. Goin's photographs feature irregular structures of wire mesh, which will be sprayed with cement to become artificial rocks for a black bear exhibit; a similar faux rock outcropping construction project for the "wolf landscape"; construction workers building a Sonoran Desert exhibit; workers planting a tree near a human-made swamp for a waterfowl exhibit; and plastic orange fencing threading through a scrappy stand of trees to designate "nature areas" that are to remain undisturbed. A final pair of images shows a large construction crane arm poised to lower workers down the rocky cliff of an abandoned stone quarry in South Carolina. The workers removed rock and reshaped the cliff face to create a nesting ledge for peregrine falcons, which disappeared from the eastern United States by 1970 due to DDT and are being reintroduced in a peregrine restoration effort. To make these photographs, Peter and his 4x5 camera were lowered in a narrow cage, as if he were an exhibit in a zoo. One expects a zoo to be a place built by people for displaying live wild animals. But how wild are animals when they are relocated from their natural habitat to a small, fabricated replica? Can humans reverse the damage they have done and actually restore wild animal populations by reintroducing remnant survivors to a human-constructed "natural" habitat? Is there such a thing as *wild* or *natural* anymore? Are those even useful concepts? The Zoo gallery asks us to *see* and invites us to question why humans are now designing natural landscapes for wild animals.

Photographs in the Beaches gallery are concentrated on the Atlantic seaboard, specifically several shores in North Carolina as well as Virginia Beach, Virginia. Here, as in previous galleries, the intent is to look behind the scenes, revealing evidence of human engineering in settings that most people unthinkingly experience as natural. As in The Zoo gallery, the Beaches chapter includes several photographs of construction-in-progress.

For example, a view of Surf City, North Carolina is dominated by wooden beams, boards, and survey markers, depicting the construction of an access structure, while another view shows four burly workers building a parking lot. Another image shows an uncluttered, natural-looking beach, perforated with a line of small, orange plastic flags that warn, "Buried Telephone Cable." Several work-in-progress photographs of Virginia Beach catch bulldozers and dump trucks in the act of hauling, moving, and sculpting sand in an effort to build the beach back up from naturally occurring wave-action erosion. A number of photographs document various strategies for preventing erosion—for example, a slope stabilized with sand-colored sandbags—or for encouraging sand deposition, such as a stone wall extending perpendicular to the shoreline.

The theme of human-caused geomorphology carries forward into the gallery The Mine, which reveals the profound earth-shaping and landform-creating consequences of two enormous copper mines in Arizona. At Miami, Arizona Goin depicts a series of plateaus and basins formed by mine tailings, the waste material left behind after extraction of ore via chemical leaching processes. If the captions did not inform you of the history of these sites, you might be tempted to think they are naturally occurring plains and ponds. However, there is no mistaking the hand of man in the terraced slopes and snaking haul roads of the Clifton-Morenci Pit, one of the largest open-pit copper mines in North America. This is earth-moving and landscape-shaping on a vast scale. The work in the Mines gallery prefigures Peter's interest in mining and post-mining landscapes, which forms the subject of an entire book, *Changing Mines in America*, about ten years later.

Revegetation begins to mask the effects of a drastically human-altered landscape in the next gallery, Reclaimed Land, set in Tennessee and North Carolina, a theme reminiscent of Goin's earlier Erie Canal work on defunct stretches of the canal system being overgrown and sometimes hidden by plants. Copperhill, Tennessee, has been severely impacted by the Tennessee Chemical Company's strip mining operation. The area was extensively logged and, since 1889, was blanketed in poisonous fumes from the copper smelting operation. The fumes prevented reforestation for nearly one hundred years. Only when smelting halted could the region begin to be reforested; but rather than attempting to restore the original biodiversity of the area, the Chemical Company planted vast tracts of the fume-resistant and commercially valuable loblolly pine and the quick-rooting pitch pine, a tree crop that will one day be harvested to feed the region's many lumber and paper mills. Ironically, some of fastest-growing and lushest vegetation to reclaim damaged landscapes in the American South is kudzu, a non-native vine introduced from Japan and promoted, planted, and distributed by US Government agencies in the first half of the twentieth century as a cover plant to prevent soil erosion. Kudzu adapted so well that it can engulf entire structures and has become known as "the vine that ate the South." The final set of images in the Reclaimed Land gallery star the kudzu vine in both its leafless dormant and profusely green stages of landscape takeover.

The River gallery takes us first to South Florida where a new generation of landscape engineers is working to solve problems caused by the work of the 1960s and the 1970s engineers. The Kissimmee River, connecting a chain of lakes south of Orlando with Lake Okeechobee, creates the headwaters for the entire Everglades system. Historically a floodplain incredibly rich in biodiversity, 103 miles of this meandering river were transformed by the US Army Corps of Engineers into a straight, 56-mile ditch named C-38, built for flood control. The straight ditch eliminated the filtering role of the floodplain wetlands, and the ecological consequences were devastating—ninety percent of the wildfowl inhabitants were lost, along with six species of fish. Since 1984, the Kissimmee River has been the focus of the first nationwide attempt to restore a river of this size and ecological importance. Goin's photographs tell a visual story of the restoration effort, including images of a scaled model, diversion dams, and a marshland boardwalk and rain gauge. The allegory illustrated by The River gallery is that it requires human intervention to repair ecological damage caused by human intervention. Additional images in The River gallery underscore this paradigm, for example, Los Angeles Conservation Corps workers removing graffiti from rocks along a tributary of the San Gabriel River; a stretch of the "wild" Nantahala River in North Carolina popular for white-water kayaking, whose release is in reality controlled by the Nantahala Dam; and helicopters being used to restock rainbow trout in the Chattooga River of South Carolina.

It is perhaps self-evident that the gallery entitled "Dams" will focus on reservoirs, namely human-made lakes in natural settings. Indeed, this is the case as Goin visits a number of reservoirs in the arid American West, making images set in California, Idaho, and Nevada. More surprising—and discomfiting—in this gallery is the preponderance of photographs that show reservoirs during *low* water years. Examples include the Arrowrock and Anderson Ranch Reservoirs in Idaho and the Rye Patch Reservoir in Nevada. These bleak scenes of parched landscapes reveal the exposed striated "bathtub rings" formed by a succession of receding water levels. Viewing these images, one realizes that although lakes are not naturally occurring in these places, it looks disturbingly *unnatural* or somehow not right when they dry up. It not only looks dystopic, but there is a massive die-off of fish, a sad reminder of the vulnerability of the biotic community. A pair of ironic photographs in this gallery implicitly critiques the extravagant use of water in the desert West, where a water-guzzling way of life creates a "need" for reservoirs in the first place. One such image shows an emerald green golf course at the Marriott Hotel's Desert Springs Resort in Palm Desert, complete with rows of planted palm trees, most of which, as the caption notes, are exotic to the United States. Water in the arid West is a topic that Peter is passionate about and that will propel future projects, such as *A Doubtful River* and *A New Form of Beauty: Glen Canyon Beyond Climate Change*, both of which we will explore in future chapters.

The final gallery of *Humanature* represents something of a departure from Peter Goin's specialty of landscape photography. In the Wildlife gallery Goin investigates animals that have been managed and manipulated by humans to the point that the reader wonders how well "wild" applies. This is exactly Goin's point—that all of nature, including wildlife, has been affected by human dominion. A taxidermy-themed set of photographs shows people's skill in "bringing wildlife back to life," as the McKinsey Wildlife Taxidermy shop in South Carolina advertises. Goin's photograph displays their craftsmanship in making caught-in-action mounts of a turkey, duck, fish, and deer head. Another pair of photographs profiles the work of Sergeant Tony Robinson in North Carolina, who makes lifelike, radio-controlled decoy deer, which he places in the forest to catch poachers. A group of photographs tells the story of efforts to reintroduce or breed endangered species. One image shows a wildlife biologist in South Carolina stuffing a wild turkey into a cardboard box for transport to another state in ongoing restorations efforts by the National Wild Turkey Federation. The most heart-wrenching photographs were made at the Carnivore Preservation

Trust in Pittsboro, North Carolina. Goin's images show workers checking on caged animals, including a jaguar that the center rescued from abuse, the almost extinct tayra or tree otter, and some young binturongs.

Fear and promise. Regret and hope. Revulsion and resolve. These are some of the conflicting emotions that the galleries in *Humanature* provoke. If seeing is believing, Goin offers undeniable visual evidence that nature is an illusion and that, as he says, "the forest primeval is now a human product" (22). Once we discard the fallacious notion that humans are separate from nature and "only when we comprehend and respect *humanature* will we evolve constructively as a species and begin to live again in harmony with the planet we so rightly call *home*" (22). That call is the take-home message of this revolutionary book. The photographs displayed below from *Humanature* may prod you to ponder this brave new world, the age of the Anthropocene. Thoughtfully considered, these photographs become portals to much larger debates.

Finally, be aware that Goin—like the Wizard of Oz behind the curtain—may have had a hand in doctoring the scenes that he photographs. Whereas in other projects Peter chose a hands-off approach and did not alter or arrange elements of the landscape prior to making an image, in *Humanature* he sometimes first interacted with the landscape. He might have moved this or removed that or added something new. The photographs are not simply documentary. He explains, "There's an artificiality to the photographs reflecting the artificiality of the interpretive construction of the concept of nature." In other words, for this project Goin developed a meddling methodology consistent with the *humanature* philosophy. However, just as we are so often oblivious to traces of human management of nature, the casual viewer will fail to notice the subtle signs of Goin's tricksterish tampering. Be alert.

■

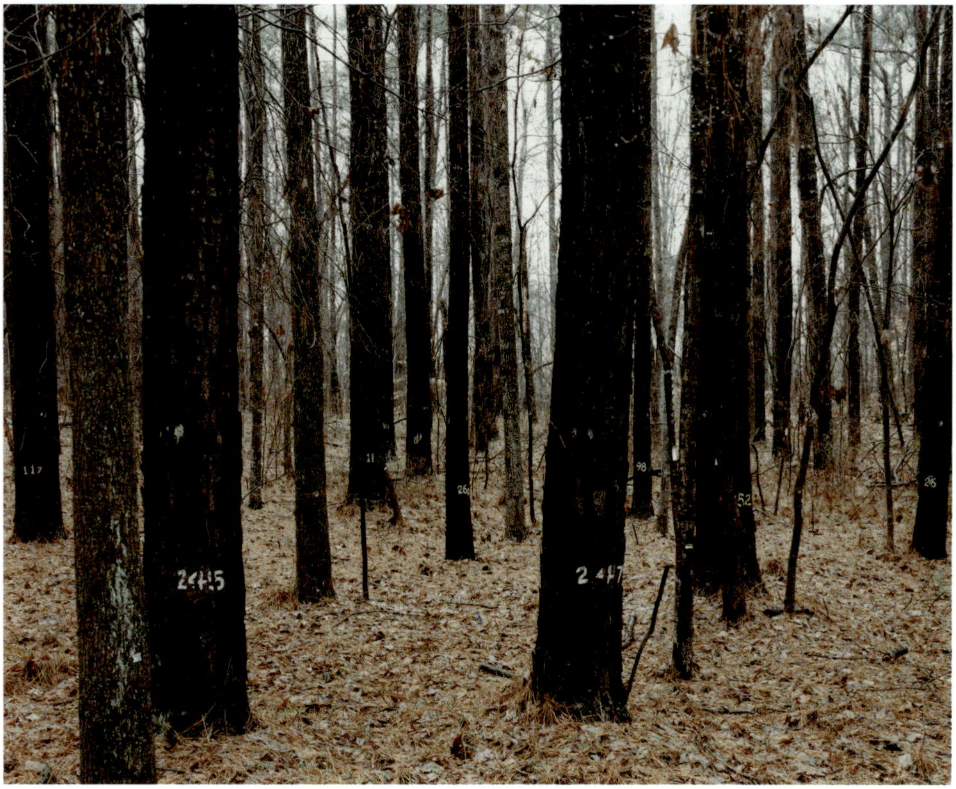

11.1 *Numbered trees in Duke Forest, Durham, North Carolina*, 1991.

The intrusion of white numbers converts this natural-looking forest into a plot of trees managed and monitored by researchers. As Goin notes, "there's something that's deconstructivist about the process of numbering things. It goes from the lyrical to the mundane." The trees in this university research forest are numbered because scientists researching transgenics have introduced genetic material from other organisms into these trees to see how they might affect the growth patterns, durability, and assorted other qualities. Genetically modifying (GM) trees offers the prospect of engineering trees to maximize economically desirable traits, such as fast growth, reduced lignin content to facilitate pulping, frost resistance, and disease resistance. At the same time, GM trees raise ethical questions. This image hauntingly references genetic engineering during the Holocaust, during which Nazi scientists branded prisoners with numbers on their arm. To make this photograph, Goin waited out a storm under a tarp until after a rain had dampened the tree trunks, heightening the contrast between darkened bark and white numbers.

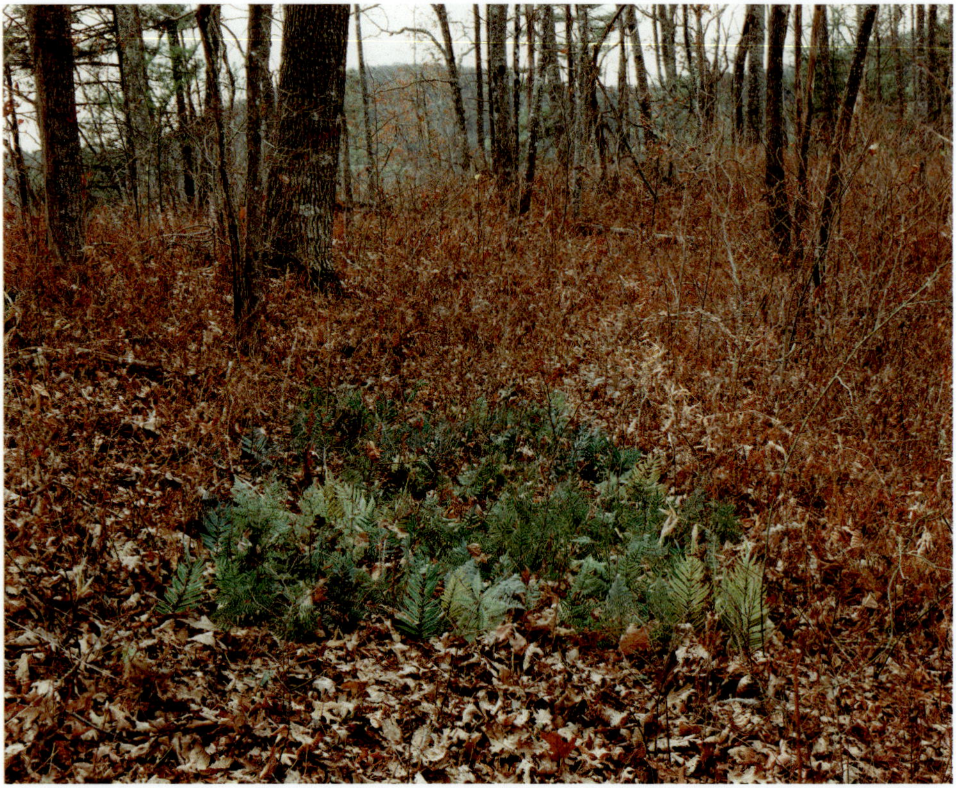

11.2 *Faux ferns*, 1991.

This little patch of artificial ferns explores the phenomenon of plastic plants, a synthetic thread in the *Humanature* tapestry. In search of visual evidence, Goin began visiting cemeteries, looking for plastic flowers. In a cemetery in North Carolina he found a discarded pile of plastic ferns. He returned to the site in the fall to ensure a color contrast between fallen brown leaves and bright green ferns. He then, à la Andy Goldsworthy, planted the ferns in an arrangement that plays off the dynamic compositional element of the triangle, an optical illusion that makes the fern patch appear larger than it is. It's interesting to think about plastic plants, plastic being made from oil and oil made from decomposed plants. Our culture values the color green. We plant lawns in the desert. We buy artificial plants and trees to simulate nature in our houses and offices. We place plastic plants in the park-like environment of the cemetery, evergreen notes of permanence in a space that reminds us of mortality. On a psychological level, it works. We can be tricked. While traditionalists may prefer real flowers, Peter Goin finds something hopeful about our degree of comfort with the illusory environment.

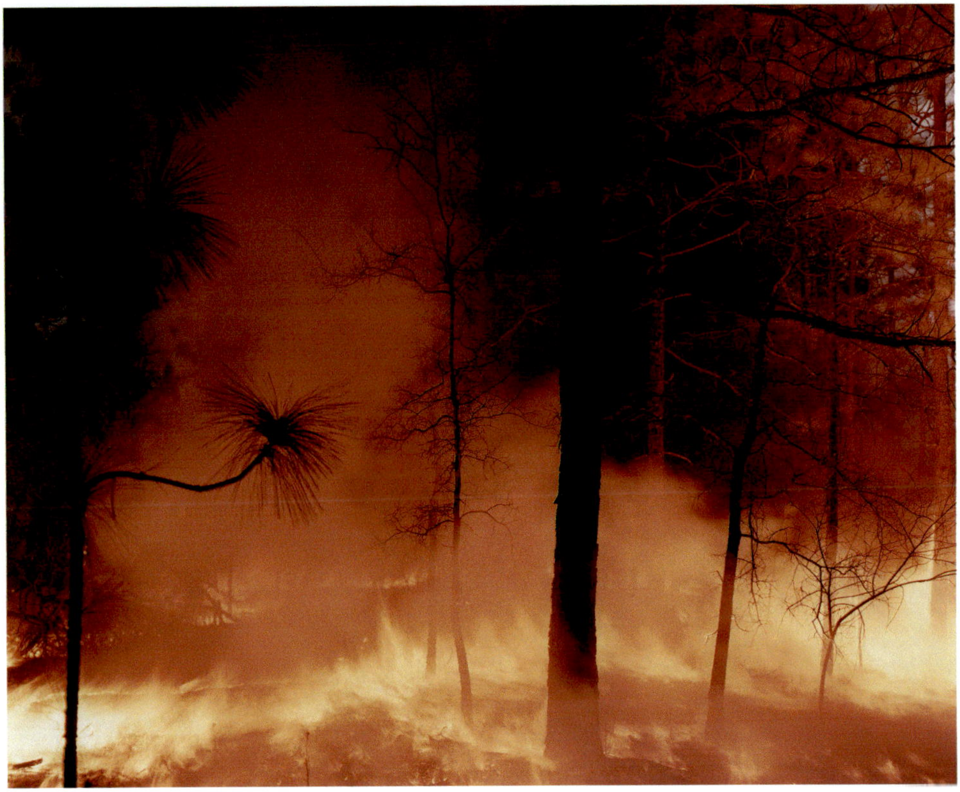

11.3 *Active burn, Peachtree Rock Preserve, South Carolina*, 1992.

As the caption in *Humanature* instructs, "Since 1968 government policy and practice has been to set controlled fires in an attempt to mimic natural processes, such as lightning strikes." This photograph records "an active burn conducted by Nature Conservancy personnel at its Peachtree Rock Preserve in South Carolina, a three-hundred-acre tract it purchased in 1980." Fire ecology is a relatively recent science, following years of fire prevention and suppression, symbolized by Smokey the Bear. As we learned the hard way, fire is part of a natural cycle that rebuilds soil and renews forests, whereas fire suppression permits the growth of a shrubby understory that fuels a much hotter, more destructive burn when fire does strike, as it inevitably will. Today periodic, human-controlled burns maintain forest health; however, "control" is never assured, as unpredictable wind conditions and quirky fire behavior can wreak havoc with the best laid plans. In fact, capricious fire played an active role during the making of this image when Goin's photographic cloth ignited.

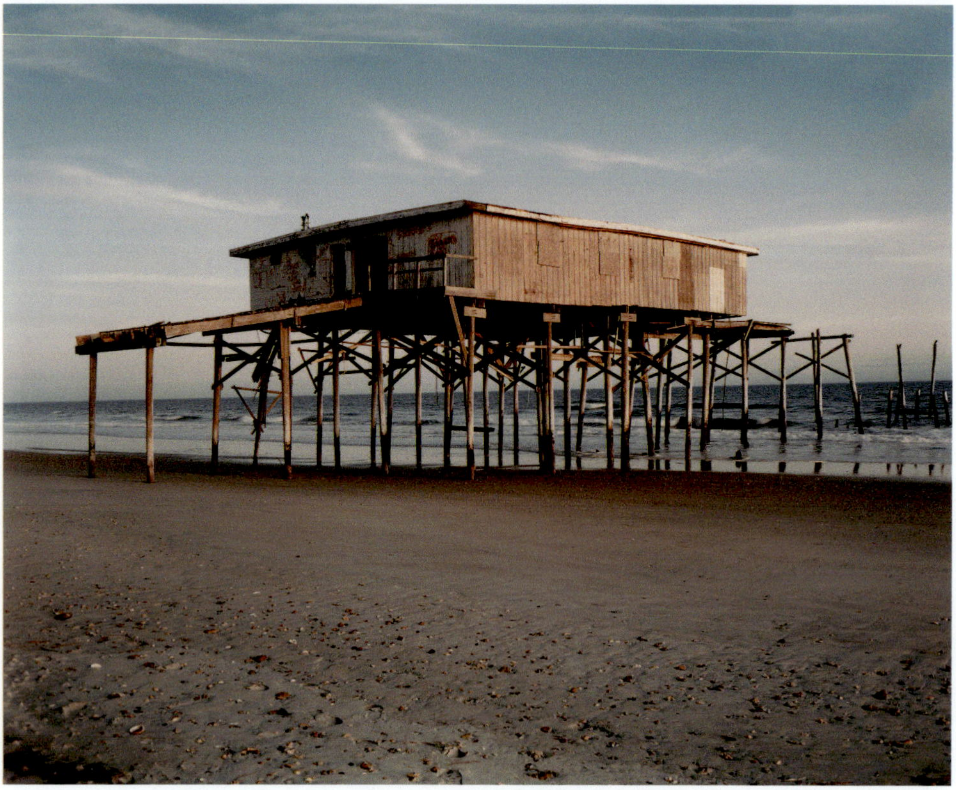

11.4 *Surf City, North Carolina*, 1991.

Looking a bit like a mechanical crab or a prehistoric behemoth skeleton, this abandoned pier building north of Surf City, North Carolina, has been rendered inaccessible by natural damage caused by the Atlantic Ocean. Here human ingenuity failed to account for the dynamism of nature in the form of powerful storms and mammoth waves. As Goin opined in an interview:

> We're fundamentally naïve about the nature of the processes of an evolving biotic community. We assume that this is the way it has always been, and we embrace the idea of a predictable balance or permanence to nature. But the only constant is change. Our inability to grasp change as a constant is profound and is the root of so much failure to integrate into biotic communities.

This elevated building visually epitomizes the way we have separated ourselves from the natural world, subscribing to a belief system that casts humans and nature asunder. Goin framed this image such that the horizon line cuts across the bottom of the pier, creating the illusion that the building is standing on water. The stick-like support structure bespeaks the fragility of a propped-up belief system. Lesson: Adapting and integrating is better than trying to resist or control the uncontrollable.

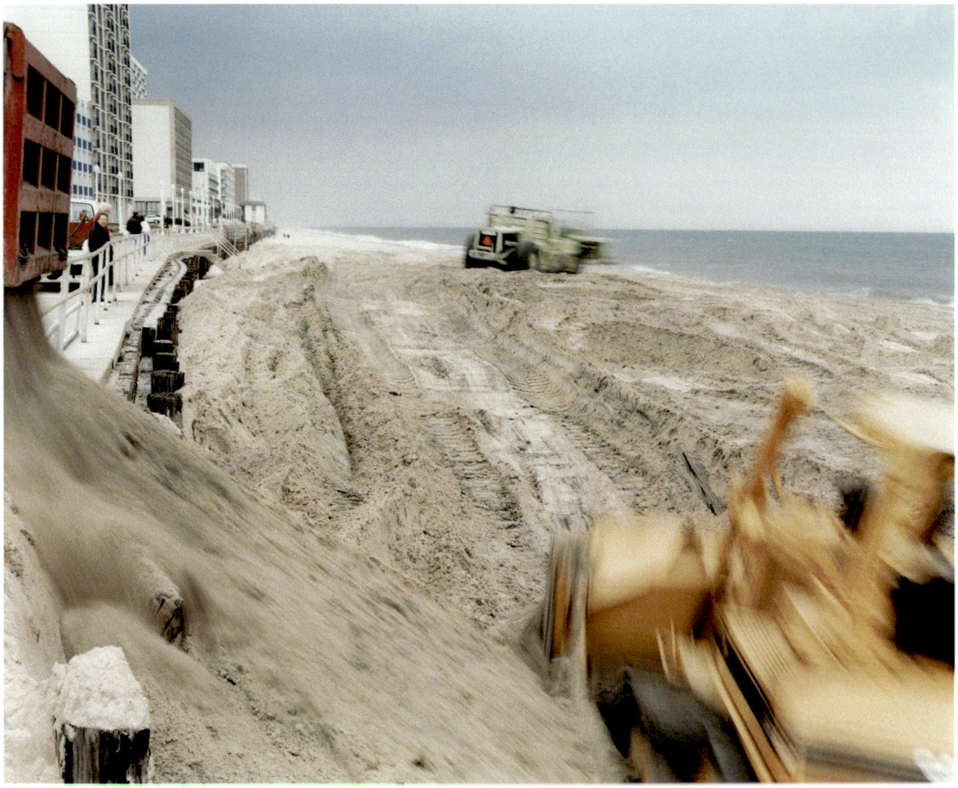

11.5 *Virginia Beach, Virginia*, 1992.

When tourists visit Virginia Beach, most probably don't see bulldozers, because these heavy machines do their work before the tourist season begins. Up and down the Atlantic seaboard, vast amounts of money are spent each year to relocate thousands of cubic yards of sand to rebuild popular beaches from natural erosion caused by currents, wind, and storms. Sand is hydraulically dredged from nearby inlets, stored in stockpiles, loaded onto dump trucks, and hauled to the oceanfront. Bulldozers then grade the sand and establish desired elevations, in this case following architectural drawings made by the Engineering Division of the Virginia Beach Department of Public Works. When we go to the beach, we often don't realize that we are enjoying a constructed environment that has been shaped to conform to our expectations of what an ocean beach should be like. Human-made beaches are a hopeful metaphor for *humanature* in that they exist at the interface of the uncontrollable ocean and human efforts to integrate and adapt to it. In this photograph, the slow shutter speed creates a flow of motion of diagonally cascading sand and industrial-age dozers, counterpointed by the static horizon line, creating a perspective of eternity reminiscent of a Renaissance painting.

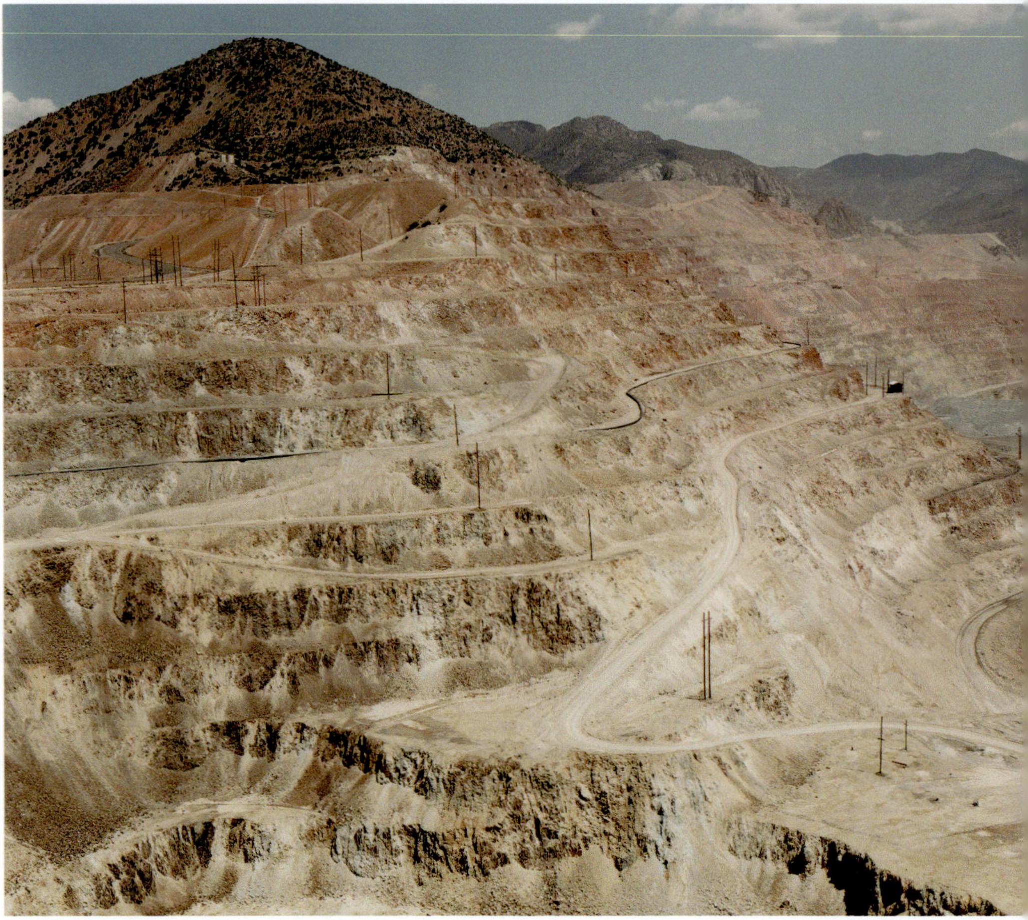

11.6 *Clifton-Morenci Pit, Arizona*, 1993.

Behold the *humanature* landscape produced by large-scale extractive industry. In this diptych of the Clifton-Morenci Pit in Arizona, one of the largest open-pit copper mines in North America, mountainsides have been dynamited away and terraced in order to access and transport the ore. A less visible intervention at this site occurred in 1906 when Chase Creek—one of the tributaries of the San Francisco River that carved this canyon system—was channelized with slag to control floods, making the area easier to develop. The enormous *scale* of modern open-pit mining is staggering. The water fountains used to aerate the

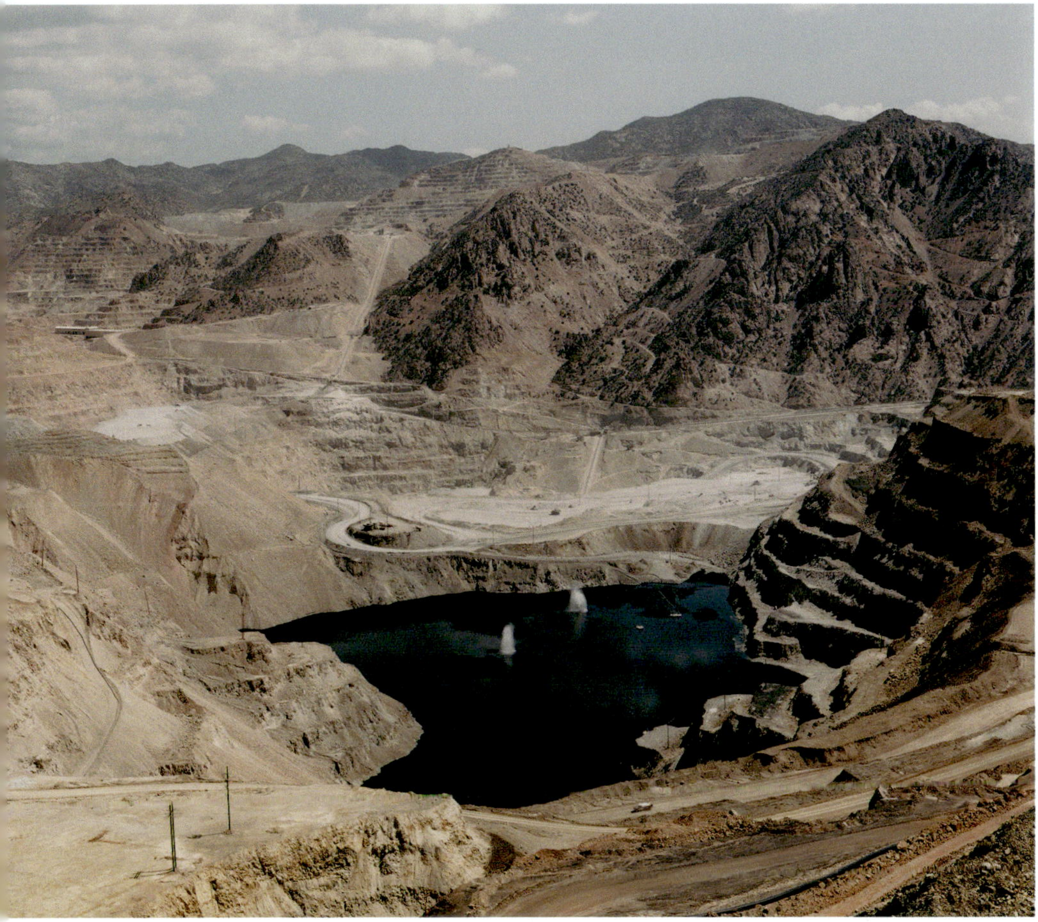

retention basin in this pit are approximately fifty feet high, and the tiny dots near the edge of the pond are actually huge haul trucks about three stories tall. An adult would be shorter than a truck tire. In operations such as Clifton-Morenci, humans are literally reshaping the Earth and constructing the global habitat. However, these giant mine sites are generally out of sight, out of mind. Goin's coauthored book *Changing Mines in America* (2004) draws public attention to mining and provocatively recontextualizes extractive industries.

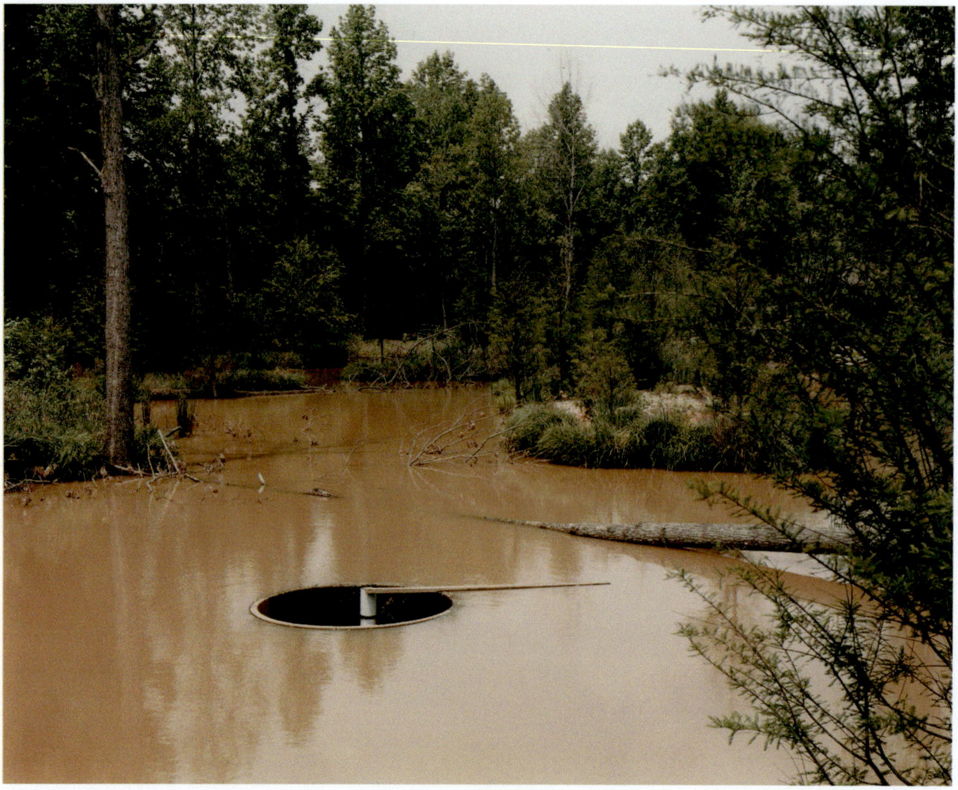

11.7 *North Carolina Zoological Park, Asheboro*, 1992.

Southern swamps hold a unique place in American mythology. In our national consciousness, swamps figure as foreboding wilderness, a place infested with alligators, snakes, leeches, spiders, and rodents. A place where runaway slaves hid and where escaped convicts lurk. Swamps are so undesirably represented that we do not think of them as human-made. Surely visitors to this southern swamp would do a double take to see this *drain* in the middle of the water. What? Are swamps humanly constructed?! In many cases, the answer is yes. Human water management in America in the form of dams, ditches, diversions, and levies has not only drained swamps but delineated their boundaries. However, swamps *look* natural, and it can be difficult to make a photograph that shows the human intervention. Fortunately, Goin timed his visit to the North Carolina Zoological Park in Asheboro to coincide with the annual cleaning of the swamp habitat for one of the exhibits. To clean the swamp requires draining the water, which suddenly reveals the human engineering behind this visual spectacle of a swamp. While this photograph is a quintessential image of a simulated environment, Goin contends that aren't *all* landscapes to some degree simulated?

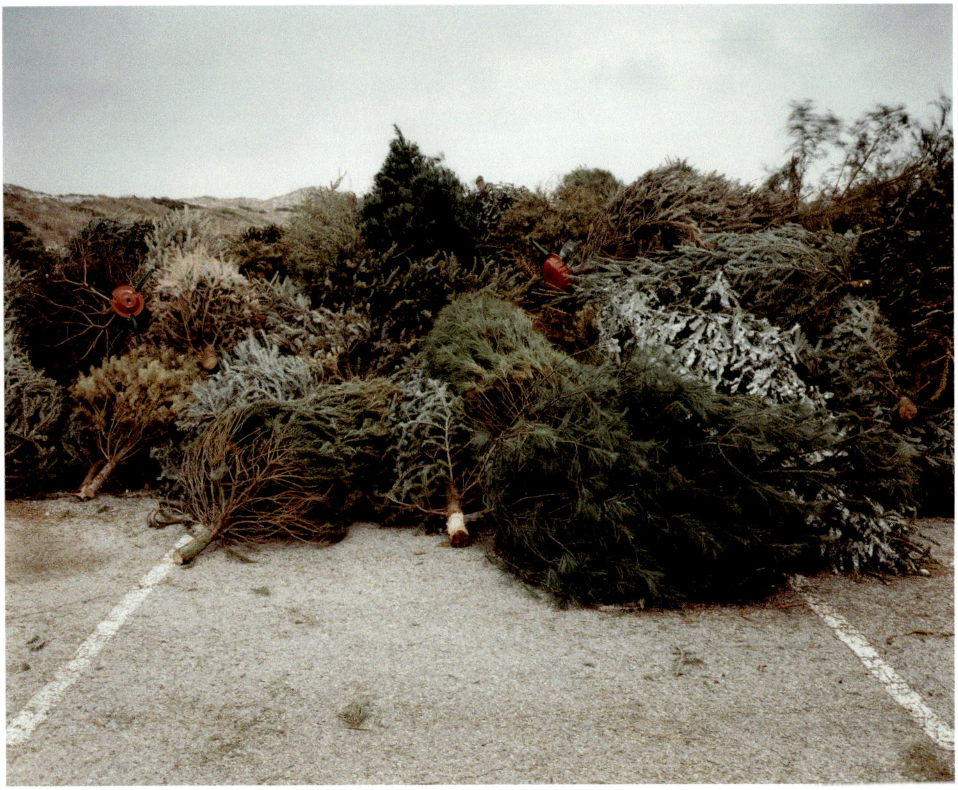

11.8 *Fort Macon, North Carolina*, 1992.

In this parking lot in Fort Macon, North Carolina, sits a pile of discarded Christmas trees of assorted species, some sporting fake snow flocking, others with red metal tree stands still attached. In the background, what looks like a snowy ridgeline is actually a sand dune, as Fort Macon is on the Atlantic coast. The Fort Macon State Park website touts the "unspoiled" shoreline and "undisturbed" natural beauty of salt marshes, estuaries, and dune fields. What we learn in *Humanature*, however, is that these recycled Christmas trees will be strategically buried to catch blowing sand in order to establish new dunes. Fort Macon, then, is actually a human-managed shoreline with "humanatural" beauty. The trees themselves are also humanatural, harvested from tree farms. A dye fed to the tree through the roots keeps the needles green throughout the Christmas season. Christmas trees are a prime example of bringing nature into the house. A green tree—like a new sand dune—is a sign of renewal. This image contains a design element of an equilateral triangle, the white lines of the parking space slanting inward to direct our gaze to the tip of the top tree.

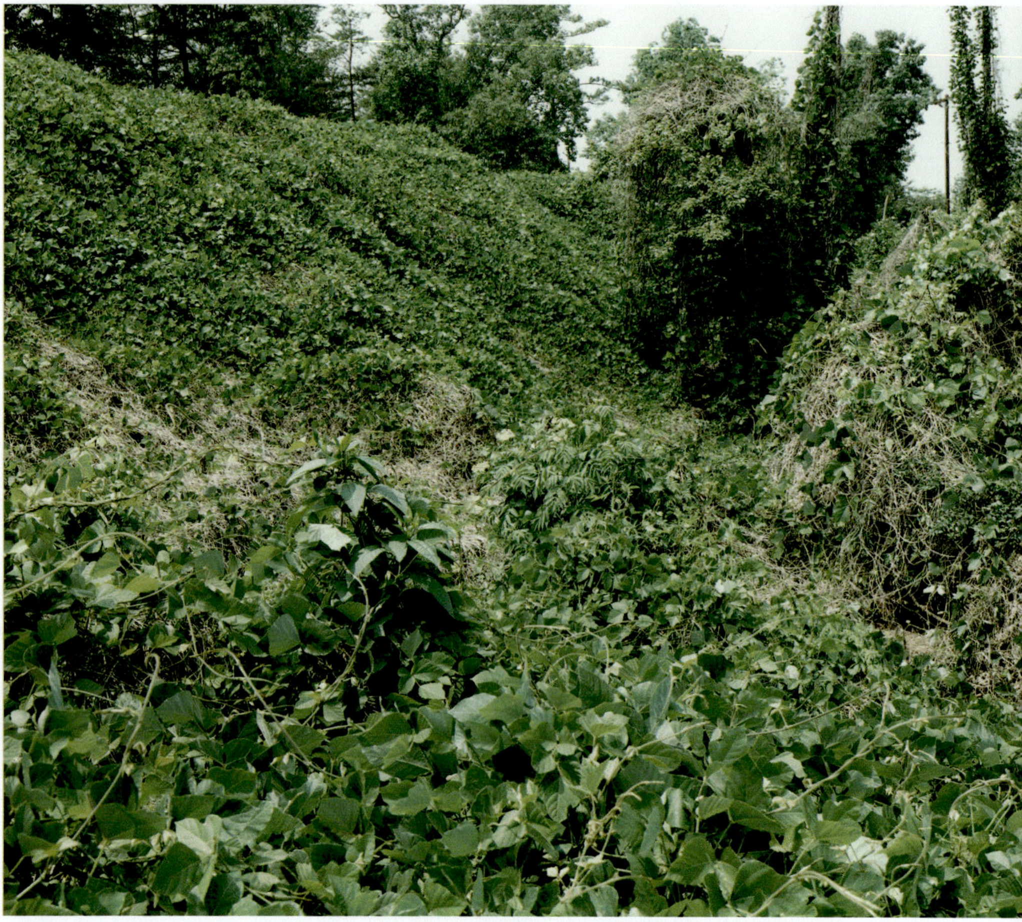

11.9 *Kudzu along Highway 421, North Carolina*, 1992.

Kudzu (*Pueraria lobata*), a pervasive vine indigenous to Asia, has become a symbol of the American South, just as the tumbleweed, introduced from Russia, is an icon of the American West, and the eucalyptus tree, from Australia, helps define a California sense of place. This photograph, made along Highway 421 in North Carolina, is a portal to debates over invasive species and natives versus non-natives. During the 1930s, in an attempt to prevent soil erosion, the US government paid farmers to plant kudzu. By 1970, kudzu was classified as a weed, and federal money was appropriated to eradicate it. Easier said than done.

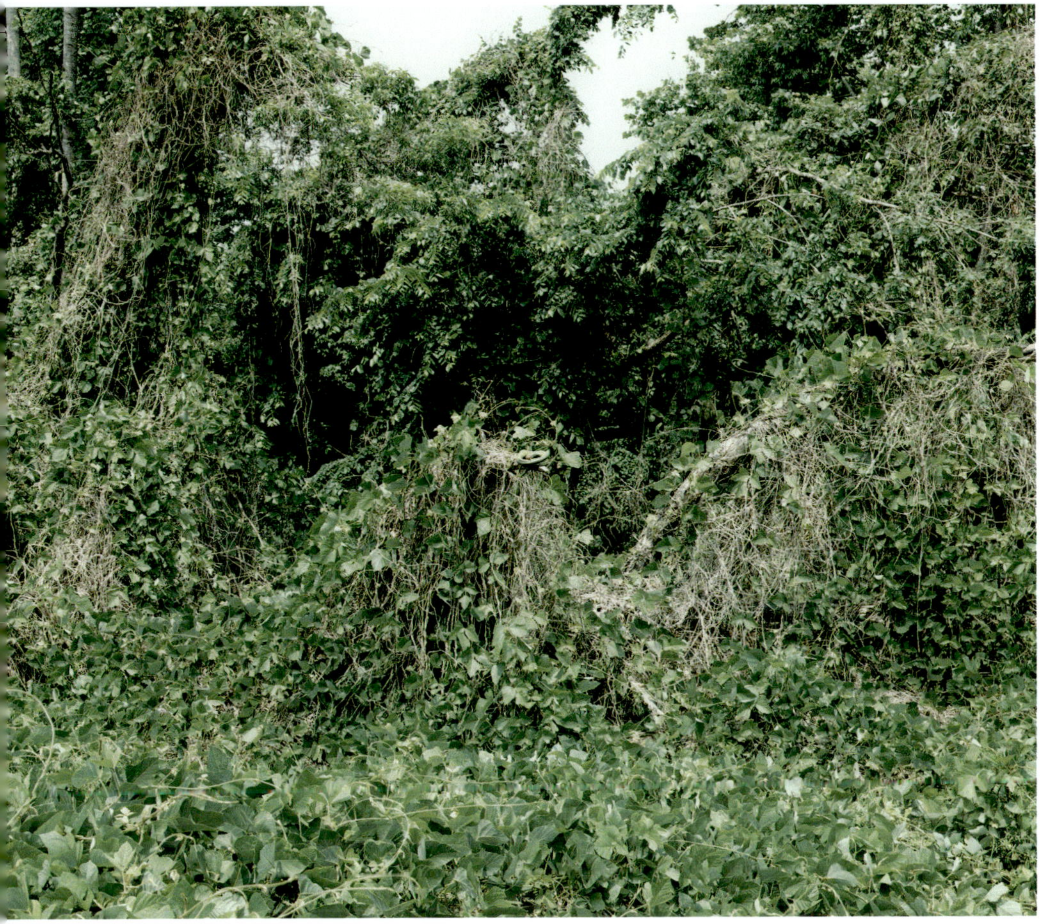

By 1992, according to the US Department of Agriculture (USDA), kudzu covered seven million acres. Goin's decision to make a diptych references art history. If you look carefully, you'll find another subtle reference to art history: A green snake in the center of the right-hand image. Ah, a snake in a garden of kudzu, descendent of the snake in Dürer's *Garden of Eden*. Goin placed a rubber snake into the scene to make a point: "In photography people expect the native, unobstructed, objective view, but I'm giving you a clue that there's always a lyrical component."

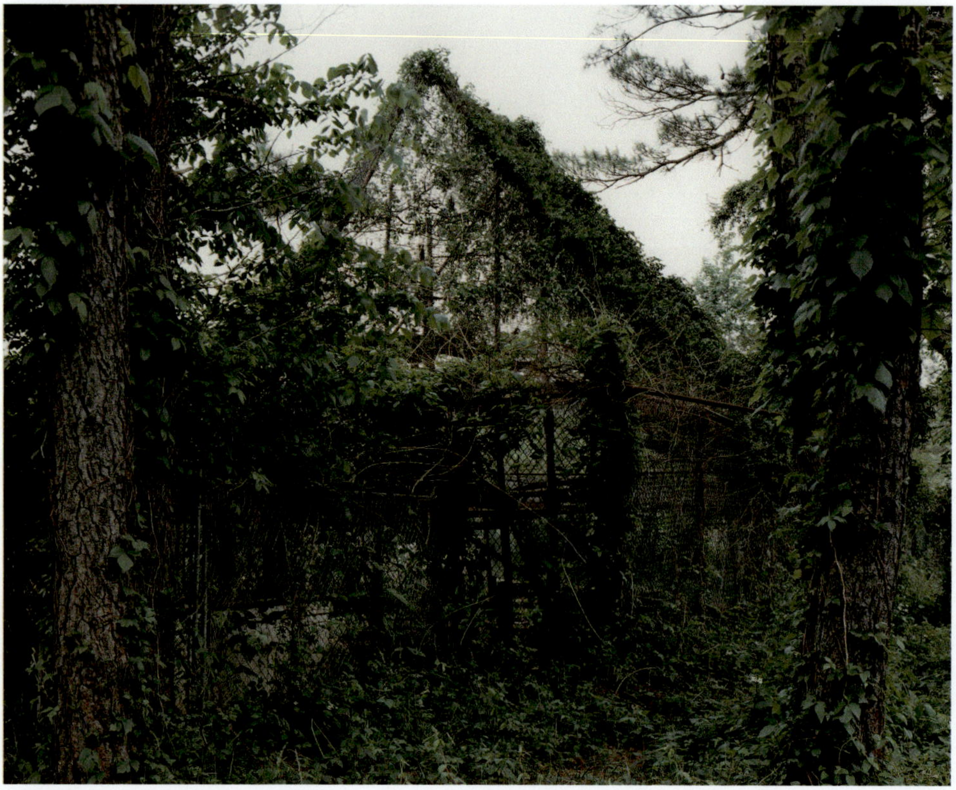

11.10 *Carnivore Preservation Trust, Pittsboro, North Carolina*, 1992.

Multiple species of vines engulf the black leopard cage at the Carnivore Preservation Trust in Pittsboro, North Carolina. Graphically, vertical brown tree trunks frame the green-encapsulated cage, and the cage itself contains a triangle, a prevalent design element in Goin's work. Historically, one can imagine workers clearing out underbrush to build this metal cage, a human structure asserting itself in a natural setting. But over time, nature is reclaiming the site, vines taking advantage of the cage to climb toward the light. This image, then, epitomizes the dynamism of the nature-culture relationship. Then, of course, there is the humanature phenomenon of caged wild animals, which in the case of the Carnivore Preservation Trust is done to develop a genetic sperm bank of rare and endangered species to guard against total extinction. Metaphorically, we might think of Noah's Ark, an enclosure that saved animals from the oncoming flood so they could replenish the Earth after the deluge. In this spirit, zoos contain the hope that we will one day restore the damaged Earth, creating natural habitats of human design, where caged wild animals can be set free and learn a feral style of surviving.

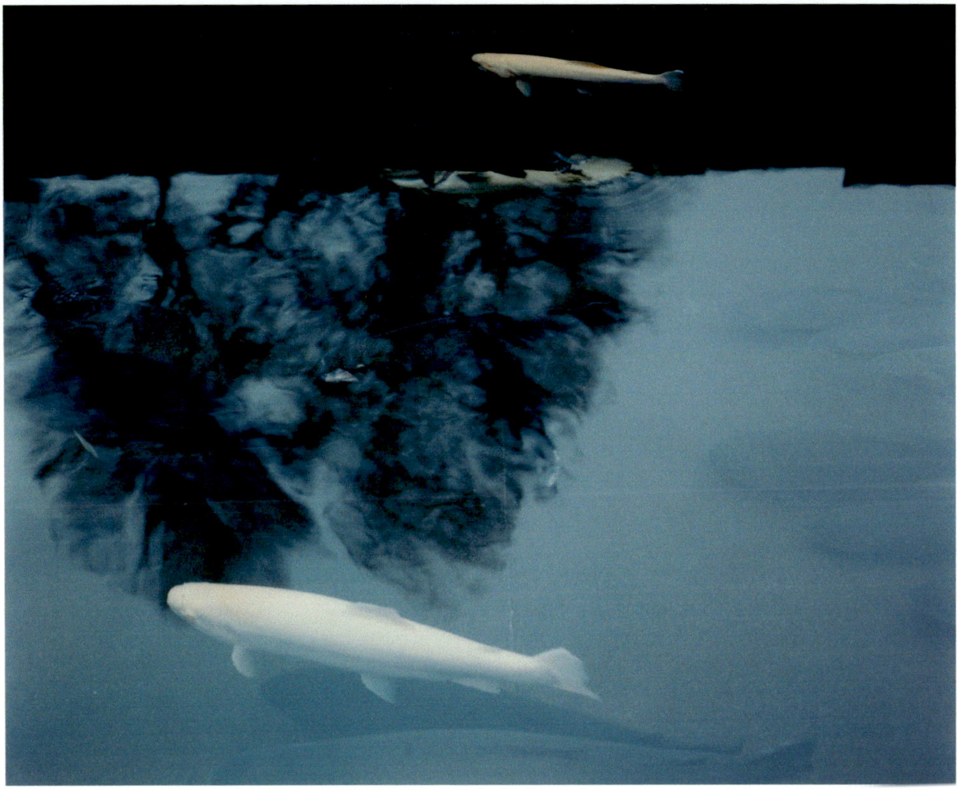

11.11 *Albino rainbow trout in the Walhalla National Fish Hatchery, US Fish and Wildlife Service, near the border of North and South Carolina*, 1991.

Albino rainbow trout? That is an oxymoron along the lines of jumbo shrimp or genuine imitation leather or . . . human nature. As the nearly imperceptible gray forms in this photograph hint, rainbow trout have evolved their coloration to *avoid* being seen; however, recreational fishermen have become a vector in evolution by exerting pressure on fisheries to breed trout that are easier to see. Stocking fish in rivers and lakes is a standard practice, and it is hard to say whether the fish you find in nature are "natural" or hatchery-raised. That line has been blurred by *(aqua)culture*. This image was difficult to make because fish do not stop to be photographed. Goin wanted to include at least two albino fish, because while one is an anomaly, two constitute a pattern. Serendipitously, there are three white fish in this photograph, and it's a nice optical illusion that the top fish is floating in air. The inverted tree reflection in the water integrates into the concept of our distorted view of environments and alludes to *artificial nature*, another oxymoron.

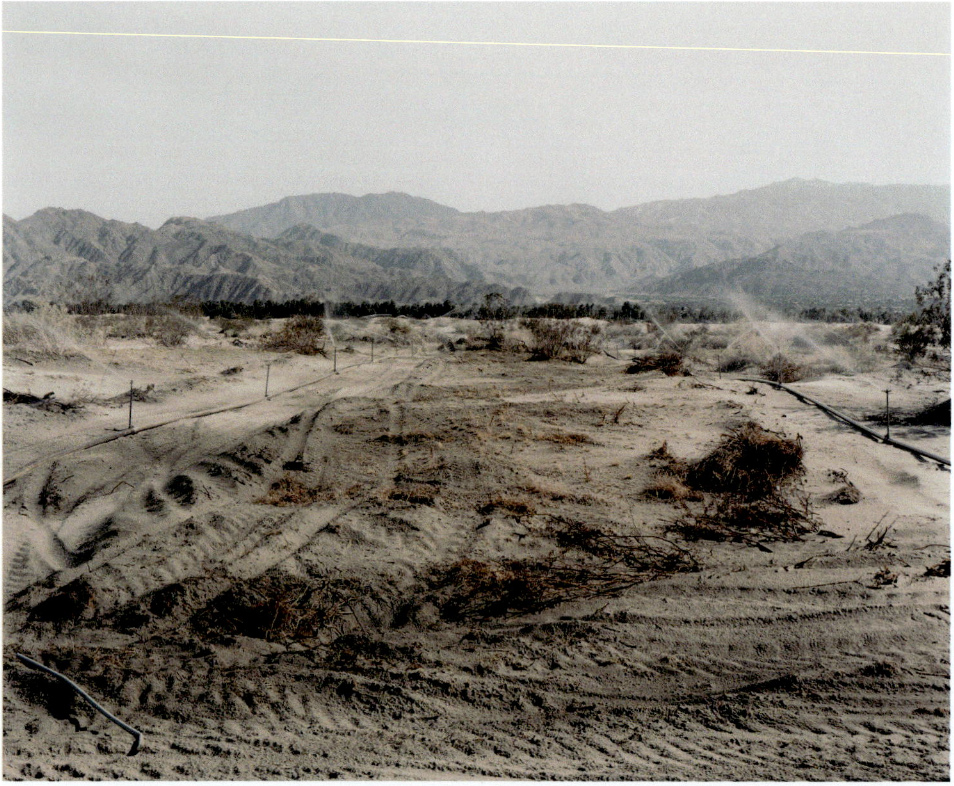

11.12 *Palm Desert, California*, 1987.

Does anything strike you as absurd that sand is being watered at Palm Desert, California? Much of Goin's *Humanature* project is kin to Jonathan Swift's *Gulliver's Travels*. As a satirist, Swift employed the absurd to critique elements of society. The problem with satire is that some people miss it. It is possible for someone to look at this image of a future golf course and completely miss the irony that we create verdant expanses of green in the desert so adults can hit little white balls and chase them around this vast landscape, which is legally defined as agricultural space. To Goin, golf courses in arid environments are a theatrical performance in the tradition of French farce. There are other humorous references in this image. Peter had been reading a book about the calculation of pi (π). Now observe the shape made by the two sprinkler lines intersecting with the horizontal fringe of vegetation. Goin delights in inserting veiled references that few will ever discover: "I read and I have many influences. I find it fascinating to weave these in however subliminally. I think what I'm doing is saying, 'Look, look, there is more here than meets the eye.'"

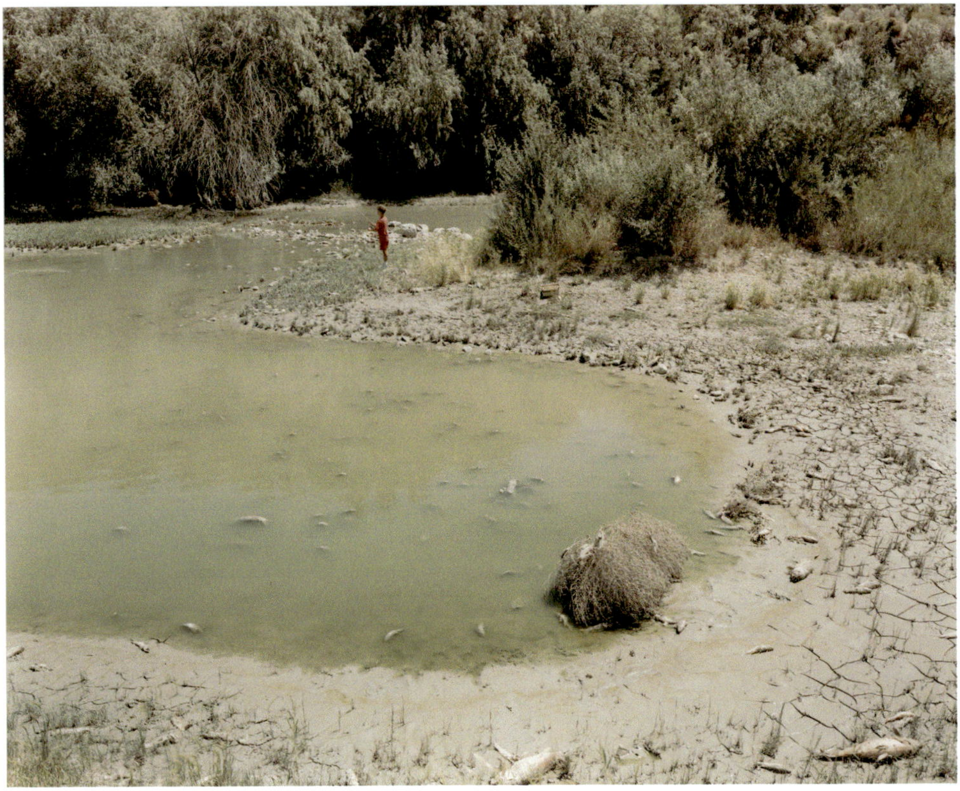

11.13 *Below Rye Patch Dam, Nevada*, 1992.

This is not a staged image. There really was a little boy dressed in red, fishing in the stagnant water below Rye Patch Dam in Nevada after the final eight hundred acre-feet of water were drained in 1992 for irrigation. The dead fish littering the shore and in the shallows succumbed to oxygen depletion in low water, angering anglers, who complained that priority in water allocation favored farmers over fish. Rye Patch Dam was conceived by the US Bureau of Reclamation in 1933 and built in 1935–1936 to impound water from the highly variable Humboldt River to irrigate agricultural fields near Lovelock, Nevada. In drought years the Humboldt River dries up, and in periods of extended drought Rye Patch Reservoir itself has gone dry. Hence, this photograph references climate change and the humanature issues of water management and reservoir construction in the arid West. It also evokes the innocence of childhood in the spirit of Mark Twain's *Huckleberry Finn*, children being unaware of the conflicts and calamities that rock the adult world. This boy is apparently also oblivious to the stench of rotting fish. Goin sees this as a sad image about human idealism and presumed innocence in the face of environmental decline.

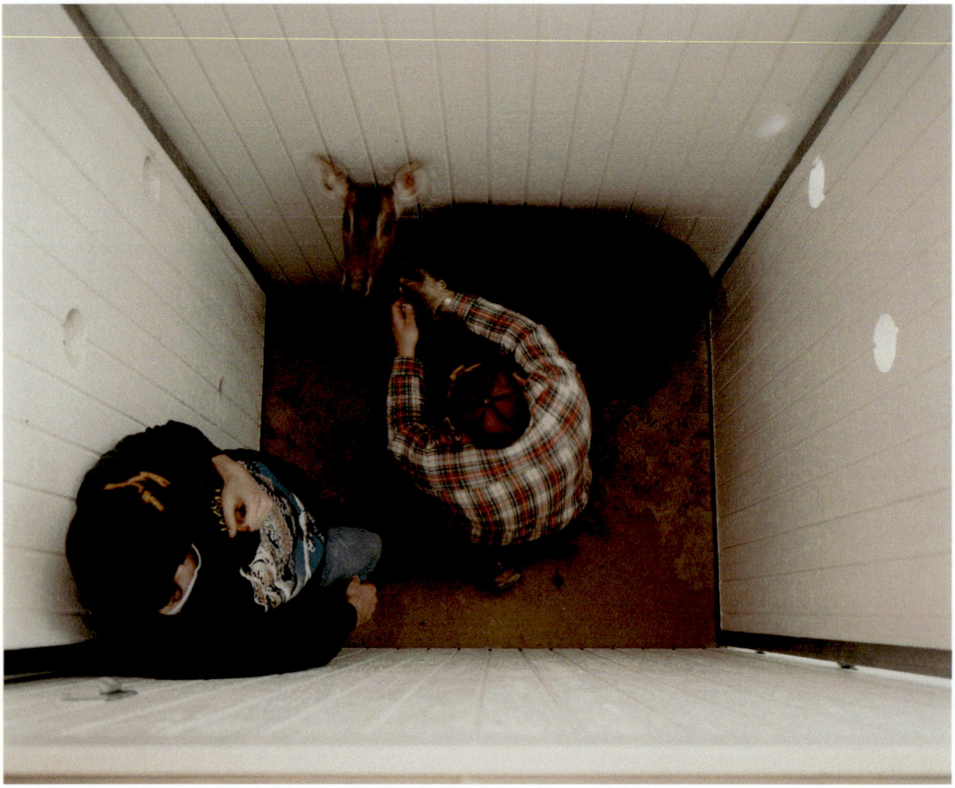

11.14 *National Zoo Conservation and Research Center, Fort Royal, Virginia*, 1992.

Human intervention in nature becomes uncomfortably hands-on in this image of workers taking a blood sample from a deer at the National Zoo Conservation and Research Center in Fort Royal, Virginia. The deer is an endangered species called the Eld's deer, native to Southeast Asia. In 1991 the National Zoo's reproductive physiologists used preserved, frozen-thawed Eld's deer sperm to artificially inseminate seven females, including this doe. While in the abstract it may be a noble goal and necessary measure to use artificial insemination to preserve a species, this image, which Goin made while precariously balancing on beams above the room, evokes a feeling of claustrophobia and empathy for the female deer, who is cornered and overpowered by two men, however well meaning.

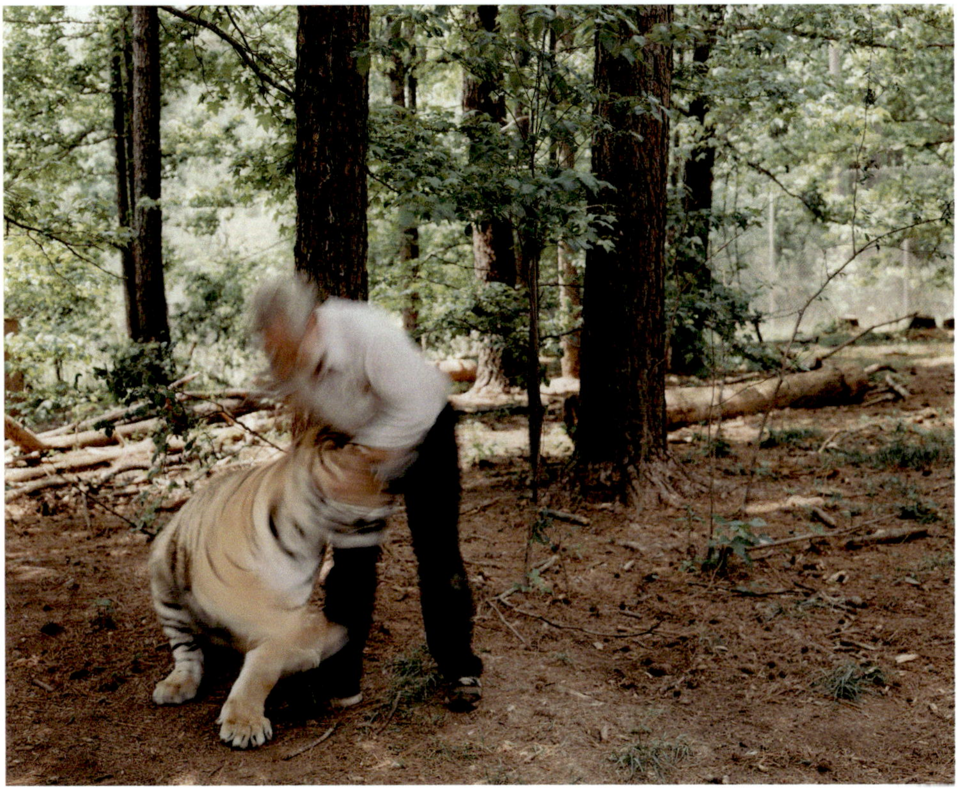

11.15 *Man and tiger, Carnivore Preservation Trust, Pittsboro, North Carolina*, 1992.

Here, at the Carnivore Preservation Trust in Pittsboro, North Carolina, founder Dr. Michael Bleyman plays with Romeo, a rescued tiger. Romeo's great-great-great-great grandparents were the last of his ancestors to live free in the wild. Romeo himself was bred in a Missouri Zoo from illegally transported parents and then ended up in a pet store in Wake Forest, North Carolina, where he was sold to a customer who did not properly care for him. Wake County humane society officials confiscated Romeo, and he was brought to the Carnivore Trust, which protects rare and endangered rainforest carnivores. On the right, in the background of this sun-dappled, park-like setting, one can faintly discern a tall, chain-link fence. Goin wryly notes that the word *paradise* traces its roots back to a Proto-Iranian word meaning "walled enclosure," which the Greeks borrowed to mean "park for animals." Our fascination with zoos dates back thousands of years, reflecting the antiquity of the *humanature* relationship.

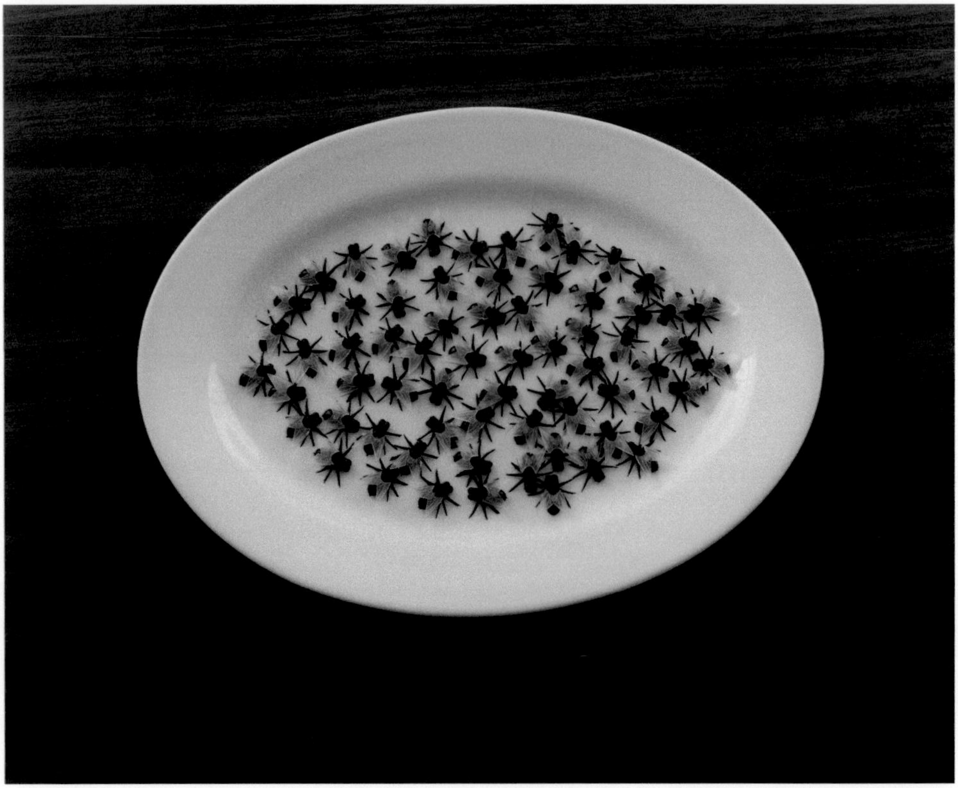

11.16 *Plastic flies*, 1991.

Humanature concludes with this futuristic image of plastic flies on a porcelain platter. In terms of art history, this picture references a photograph of fish heads from the German Bauhaus movement. The straight-down camera angle creates a vague sense of disorientation in the viewer, who has no way to tell which way is up. The platter of artificial flies is served on a faux wood formica table, made from a photograph of wood, laminated onto the table-top. Plastic flies and faux wood are an entrée into the concept of the simulacra or simulated environment. Goin concocted this arrangement of human-made flies to reference genetic engineering, presaging the future of humanature coevolution. At the time, California agriculture was threatened by fruit flies, and newspapers were reporting a USDA-sponsored initiative known as Mediterranean Fruit Fly Preventative Release Program, a birth-control process, whereby millions of scientifically sterilized male medflies are released into infested areas. Female flies who breed with these human-altered male flies produce infertile eggs. If you are sitting at an outdoor café, you can legitimately wonder whether the fly that lands on your food is a natural or a faux fly.

Note

1 In the mid 1990s, when Peter was formulating his concept of *humanature*, other thinkers were arriving at similar conclusions. Bruno Latour in *We Have Never Been Modern* argues that modernity is the result of an artificial separation of three realms that are in actuality intermeshed: Science, politics, and discourse, or nature, culture, and language (1993). Donna Haraway's concept of "natureculture" posits the inseparability of the technological and the organic and interrogates entrenched binaries of human/animal and nature/culture (Haraway 1997, 2003). More recently, Anna Tsing proposes the term "blasted landscapes" to refer to radically disturbed landscapes, such as clear-cut forests. Tsing's observation echoes Goin's notion of *humanature*, as she writes, "Watching landscapes in formation shows humans joining other living beings in shaping worlds" (152).

Works cited

Goin, Peter. *Humanature*. University of Texas Press, 1996.

Haraway, Donna J. *The Companion Species Manifesto: Dogs, People, and Significant Otherness*. Prickly Paradigm Press, 2003.

———. *Modest_Witness@Second_Millennium. FemaleMan©_Meets_OncoMouse™: Feminism and Technoscience*. Routledge, 1997.

Latour, Bruno. *We Have Never Been Modern*. Harvard University Press, 1993.

Leopold, Aldo. *A Sand County Almanac and Sketches Here and There*. Oxford University Press, 1949.

Tsing, Anna Lowenhaupt. *The Mushroom at the End of the World: On the Possibility of Life in Capitalist Ruins*. Princeton University Press, 2015.

12

THE POSTMODERN WEST

There is rarely a lull in Peter Goin's work flow. He likens his work to an eight-burner stove. While one project is on the front burner, several more are simmering on the back burners. Even as Peter was working through the edits for *Humanature*, he was ramping up activity in other areas. A wellspring of multiple new projects originated in the photography conference "The Political Landscape" that Peter attended in 1989 in Aspen, Colorado. As conference participant Ellen Manchester writes, "The Aspen conference brought photographers together with western historians, environmental writers, and filmmakers to discuss common concerns about the state of the landscape and environment of the American West" (18–19). Peter gave a presentation on his *Nuclear Landscapes* work and participated in a planning meeting spearheaded by Manchester, a photography researcher and project director who, with photographer Robert Dawson, was in the process of organizing a group of photographers under the auspices of a new Water in the West project. Peter's involvement in the Water in the West project initiated a broad range of activities and publications, discussed in Chapter 14 "Arid Waters."

The project discussed in this chapter can be traced to a casual conversation Goin had at the Aspen conference with western historian Patricia Nelson Limerick, author of the influential *Legacy of Conquest*, a revisionist history that ushered in what became known as the New Western History. New Western History poses a direct challenge to Frederick Jackson Turner's "frontier thesis" that American character was shaped by a steadily advancing westward frontier, where American traits of democracy, individualism, pragmatism, optimism, and restlessness were forged. Limerick and other new western historians, such as Richard White, Donald Worster, and William Cronon, challenged Turner's model of hardy pioneers moving west and being shaped by their encounter with empty land. They argued that the so-called "frontier" was not so much a line separating settled land from empty land as it was a "contact zone" between diverse groups of people, including Euro-American settlers of various religious and ethnic affiliations, Hispanic settlers pushing north from Mexico, and many different groups of indigenous Native Americans. Whereas Turner envisioned the frontier settlement process as innocent, new western historians viewed it as a violent clash

DOI: 10.4324/9781003212607-12

whose aim was conquest. And, whereas the old western history starred the pioneers, new western historians highlighted the minority groups that previous histories had relegated to the background, including the environment itself, which became elevated from a mere stage to an actor in the unfolding drama of the making of America.

After the Aspen conference, Limerick stayed in touch with Peter and invited him to join a team of scholars who were working on a new atlas of the West or, as the book was ultimately entitled, *Atlas of the New West*. The project was headquartered at the University of Colorado's (CU) Center of the American West, an institution founded in 1986 by Patricia Limerick and Charles Wilkinson "to survey this distinctive region, bring together people from every Western cultural tradition, and foster dialogue ranging from public lands to community development" (back cover flap, *Atlas*). For the purposes of the Center and the *Atlas*, "this distinctive region" refers to the US Intermountain West, from Canada to Mexico and from the Sierra Nevada and Cascades mountain ranges to the Rocky Mountains. The Center's website states:

> The Center of the American West takes as its mission the creation of forums for the respectful exchange of ideas and perspectives in the pursuit of solutions to the region's difficulties. We at the Center believe that an understanding of the historical origins of the West's problems, an emphasis on the common interests of all parties, and a dose of good humor are essential to constructive public discussion.

A "dose of good humor" is reflected in the Center's very name—Center *of* rather than Center *for* the American West—and, as we shall see, humor spices up the flavor of *Atlas of the New West*, published in 1997.

What exactly is the "New West?" Admittedly, as Patricia Limerick notes in her concluding essay, the "New West" is "a phrase with a history," dating at least as far back as 1869 when Samuel Bowles published *Our New West*, heralding an American West just made more accessible by the completion of the Transcontinental Railroad (151). Fast forward 125 years. According to *Atlas of the New West*, the New West of the 1990s is a place where "Ski resorts, fur shops, and espresso bars have edged out feed stores, cowboy bars, and greasy spoons. Walled estates, golf courses, and factory outlet malls weave through ranching and mining towns" (12). Above all, it is a place where "a 'cowboy economy' based on extracting natural resources" (logging, mining, drilling, farming, and ranching) has transformed into a "postindustrial service sector" (services, information, ostentatious wealth, and tourism) (12). Fittingly, the *Atlas*'s textual ornament, gracing the title page and heading every chapter, is a designer silhouette of a bull sniffing a laptop computer. "The American West," the authors argue, "is the archetypal case of an American region yanked from its historical and myth-based sense of place into hyperdevelopment and plugged-in modernity. Such a transformation forces startling contrasts between image and reality, between old and new" (12).

Atlas of the New West was spearheaded by CU geography professor William E. Riebsame, whose research focuses on natural resources of the American West. The project's Director of Cartography was CU geography professor James J. Robb, known for having produced some of the first 3D computer-generated terrain maps. Charles Wilkinson, a law professor at the CU specializing in natural resource management in the American West, wrote the introductory essay, and CU history professor Patricia Limerick contributed the concluding essay. Peter Goin, the only collaborator outside the CU, was invited by Limerick to join the project as photographer. Goin's participation created controversy, which we'll discuss. But first, a look at the *Atlas* itself.

As one might reasonably assume, the main content of *Atlas of the New West* is maps, forty-six of them to be exact. In a sense, as cartographer Robb remarks, most maps are "one-page books," representing great effort from a number of people (9). Maps that a reader might expect to find include topographical base maps; a map of the region's towns and cities; a road map; a population density map; a map of county populations; several different demographic maps showing migration patterns, race and ethnicity data, and political trends; a land ownership map; and a map of annual precipitation. Less expected maps include a map of microbreweries; a consumer map of gourmet coffee shops/*New York Times*/Land Rover/Orvis/and Patagonia Outfitter shops; a map of Hispanic Radio and Indian Radio stations and Gay, Hispanic, and Indian newspapers; maps of Superfund sites and Nuclear testing, mining, storage, and production sites; maps of ski areas and gold medal trout streams; a "places rated" map derived from *Money* and *Outside* magazine features and almanacs; a map of light pollution from the International Dark Sky Association; a map of "Ghost Houses," that is, second homes; a Retirement Hot Spots map; a map of New Age centers; a "Wired West" map that shows internet, fiber optic cable, and cellular phone coverage; and a "Wild Future" map that compiles various large-scale ecosystem protection visions. Sidebars, brief essays, and illustrations augment and help readers interpret the data displayed in the maps. Taken together, the chapters offer a fresh portrait of the New West's geography, infrastructure, water, demographics, lifestyles, environmental problems, and future trends.

As the whimsy of some the map topics demonstrates, the *Atlas* is laced with witticism, puns, and playfulness. Although the *Atlas* is a physically big book—12 ½ inches tall × 9 ½ inches wide—it is far from ponderous to read. On the contrary, it is light and enjoyable, striking a new and unusually engaging tone for a reference work. For example, a map of annual precipitation is entitled "Arid, Extra Dry" like the deodorant. The road map section is entitled "A Road Runs Through It," a playful allusion to Norman Maclean's acclaimed Montana memoir, *A River Runs Through It*, made into a film directed by Robert Redford. An essay on a proposed water project that would divert one-half the flow of the Animas River in southwest Colorado is entitled "Jurassic Pork." A sidebar entitled "Heaven and Hell on Earth" consists of a list of western place names that allude to heaven—Mountain of the Holy Cross, Angel Peak, Jacob's Ladder—or Hell—Dante's View, Devil's Slide, Hell's Half Acre. A sidebar that identifies two permanent depositories for the nation's most dangerous radioactive waste is entitled "Rest in Nuclear Peace." And a section on contemporary writers is entitled "Home on the Page."

That you can literally laugh your way through the *Atlas* raises questions about humor in the context of the American West. Certainly, humor has legs that stretch back to the tall tale tradition and to Old West humorists Bret Harte, Mark Twain, Dan DeQuille, Artemis Ward, and their fraternity. In *Atlas of the New West*, humor is used to make a potentially dull topic delightfully compelling, to lampoon political chicanery, to delicately broach serious problems, and to expose the folly of our ways. Rhetorically, humor is a tactic capable of conveying the incongruities of "a postmodern West, where old and new combine to create something different" (12).

Postmodernism as a literary and artistic style likewise combines old and new to make something different. It mixes high art and pop culture and splices together seriousness and comedy. It musters and disarms history and myth. It embraces kitsch and exhibits awareness of capitalism's infiltration into every corner of our lives. *Atlas of the New West* is a maverick reference work that flirts with a postmodern style in order to characterize a postmodern West. As the *Atlas*'s photographer, Goin embraced the project's postmodern sensibility. But it took some wrangling before the *Atlas*'s editorial team would embrace Peter.

When Patricia Limerick invited Peter Goin to join the *Atlas* project, the other team members first wanted to vet his work, and he was asked to submit sample photographs of the American West. When the team saw his images, they balked. The main point of contention was Peter's skies. They were not that cobalt blue that the team felt was a hallmark of the West. "They don't have Kodachrome skies" was the dismissive comment that Peter received. He countered that:

> Kodachrome makes a very saturated film base for slides because the quality of light of traditional slides gets washed out when being projected, and Kodachrome film compensates for that loss. But you'd be hard-pressed to find a Kodachrome-colored sky in the American West, because it's not that kind of light.

The team continued to dig in their heels, some members arguing that Peter's photographs had too much of a color range, others that the images looked too muted, too Mediterranean. Another problem, according to the team, is that Peter's photographs required the viewer to scrutinize the image and think rather than featuring the obvious scenics that they were used to, such as the spectacular Arizona skies. Over a couple of the team's objections, Limerick insisted that Goin would be a valuable contributor, and he was brought onboard.

But the battles were not over. The next disagreement was about the role of photography in the *Atlas*. The team assumed that Peter would illustrate their text, providing decorative images on the topics they wrote about, similar to the function of photographers in *National Geographic* magazine. For example, one mini-essay in the *Atlas* is about Missoula, Montana's civic revitalization, epitomized by the new hand-carved wooden carousel in the center of town. Peter was asked to take a picture of that merry-go-round. He resisted. In his view and that of Limerick, Goin deserves the equal status of an author, not a servant. Goin insists that each of his photographs "is a vessel for scholarly intention and complexity. It can and should stand equally with any writing as long as the content and commitment are relevant and parallel and coexisting." If a map can be considered a book, a photograph can be considered an essay. Each of Goin's images carries intellectual content, available to the viewer who takes the time to read it closely.

For this battle, Limerick brokered a compromise, permitting Peter to submit twenty, full-page photographs representing *his* interpretation of the New West, while allowing other contributors, such as some of the geographers, to submit their own illustrative photographs, which would be run as rows of thumbnail images. Peter himself compromised by ponying up and photographing the Missoula carousel, albeit in his own style, namely a close-up of a single, wooden horsehead in the process of being painted. The caption reads, "Painting a red apple on a new, hand-carved carousel horse, Missoula, Montana, January 1997. Carousel volunteers and adoptive families tuck memorabilia—family photographs, clippings, mementos—into the hollow basswood horses" (41). Peter is happy with the image because it lyrically recontextualizes the role of the horse in the American West. At the end of the day, Peter and Patty won the showdown, and their victory is more than personal. Since the entire point of the *Atlas* is to display evidence that characterizes a New West, it is fitting that the photographs revise and not simply reprise stereotypical views of the mythic West. As Goin explains in conversation, his photographs "are all about recontextualizing the West as this place that's been heroically defined. Now we're asking: Who are we, what are we, where do we live, what is this place, and what are the emerging paradigms."

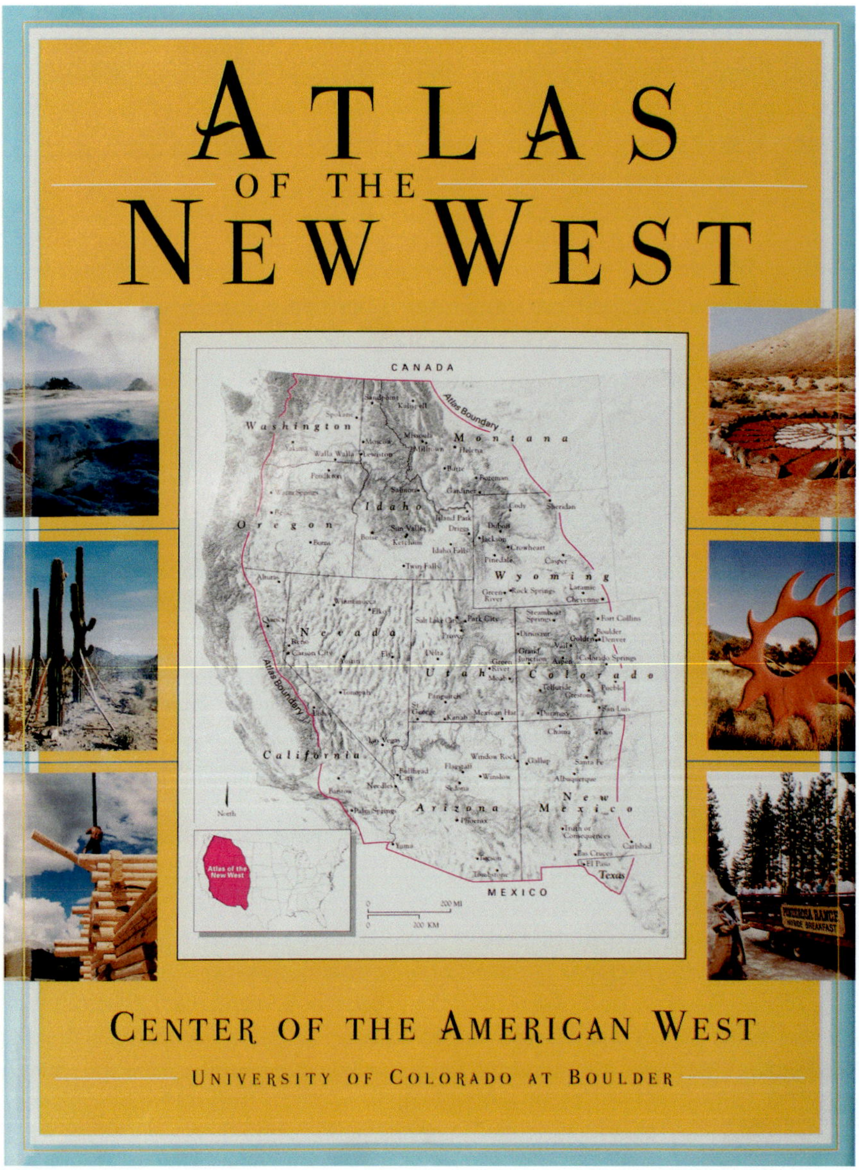

12.1 Atlas of the New West.

The front cover of *Atlas of the New West* is a good place to get a preliminary sense of Peter's perspective on this topic. Perhaps the first thing to note about the book cover design is the background colors of sky blue and ochre. The absence of the color green subliminally communicates the aridity of the Interior West, the region outlined on the black-and-white cover map, whose topographical shading imparts the mountainous character of the region. A vertical strip of three of Goin's photographs on the left side of the cover predominates in blue tones, while the matching photo strip on the right margin favors brown tones, the combination of blue and brown underscoring the cover's ambience of aridity. The images themselves proceed from top to bottom and left to right in a sequence that begins with an unpeopled vista and ends with a crowd of tourists at a theme park. Each image uniquely summons the mythic West, but with a New West twist. The first photograph, for example, looks like an iconic Yellowstone geothermal landscape. However, the caption informs us that the image is of The Needles at Pyramid Lake, Nevada, a little known natural area off the radar of most tourists. The unstated subtext of this image is that The Needles is located on the Pyramid Lake Paiute Tribe's Reservation, and the tribe closed public access to The Needles in the mid 1990s to protect it from trashing and vandalism. Goin obtained special permission from the Tribal Council to photograph there. Encoded in this seemingly innocent landscape image, then, is a complex history of Native versus Euro-American conflicts, along with thorny issues surrounding the "loving it to death" desecration of western lands.

The second image in the strip may elicit a chuckle due to the incongruity of tall saguaro cacti in the Arizona desert absurdly being held upright by wooden supports as if they are props on a stage set. The caption informs the viewer that these are "Saguaro at Eagle Mountain residential and golf development, east of Phoenix, Arizona." As Peter explains, the saguaro have been uprooted from somewhere and transplanted here:

> so we're recreating the landscape in a memory of the mythic West for a new housing development because we now have romanticized the saguaro as the identifier of 'cool, we now live in the desert.' It's a simulacrum of the mythic West. We're recreating this kind of historical faux-narrative of who we are and what we were.

The next image of a log home under construction near Buena Vista, Colorado—likewise nostalgically recreates the look of the frontier past but in a New West, affluent incarnation. This custom log home will be a multistoried, multimillion dollar mansion. The High Country Log Homes website constructs a have-it-all fantasy of upscale, downhome, rustic, luxurious western living. "*A log cabin mansion?*" Peter shakes his head incredulously, "Really? It's just over the top."

Moving to the brown-toned strip of photographs on the right edge of the cover, our eye is caught by two circular designs. The top image is an environmental artwork made from different colors and sizes of rocks laid out on the desert floor in the shape of a wagon wheel from the Old West. Only by reading the caption and zooming in for closer inspection do we learn that this "outsider art" piece in Nevada's Black Rock Desert is entitled "Ground Zero." The inscription on the gateway rock reads, "THE ALTERNATIVE TO NEGOTIATION." Suddenly we are catapulted into the Atomic Age and reminded of the New West use of Nevada as a nuclear test site. Juxtaposed below this violent image stands a clay-colored upright sculpture of a stylized sun, rising from a pastoral, sun-dappled field of dry grasses. The caption identifies the location of the sun sculpture as "in the Crestone community of spiritual followers, Colorado." Shazam!, now we are in in the New Age. An *Atlas* map entitled "New Age in the New West" explains that "Dramatic scenery, rich Native American heritage, and isolation attract New Age practitioners to the West. Channelers, healers, healing centers, and communes populate the region's nodes of crystal power and psychic vortexes, like Crestone, Colorado, or Sedona, Arizona" (114).

While New Age practitioners adopt ancient spiritual traditions, tourists enjoy watching—and sometimes playing a part in—reenactments of the Old West. The final image of the front cover's photograph gallery shows a trailer-load of tourists on the Ponderosa Ranch's Hayride Breakfast being held up at gunpoint by an outlaw clad in cloak, chaps, and cowboy hat. Goin's caption: "Lori Johnson, a college student from North Dakota, robs the Ponderosa Ranch breakfast haywagons every fifteen minutes" (157). Ponderosa Ranch, Western Studio and Theme Park, is nestled in the pines upslope from Lake Tahoe and was one of the stage sets for the popular 1960s television series *Bonanza,* which chronicled Ben Cartwright and his three sons on their thousand-acre spread near Virginia City, Nevada. *Bonanza* was so popular that the real Virginia City, Nevada, a cosmopolitan and opulent mining town back in the day, restyled itself as a tourist attraction, facing its elegant brick buildings with wooden false fronts to conform to tourists' expectations of the Old West and staging hourly faux gunfights on Main Street. In 1967 a *Bonanza* theme park opened at the Ponderosa Ranch site, becoming a lucrative "bonanza" of profit for its owners until the park closed its doors in 2004, some thirty years after the *Bonanza* TV show ran its final episode. So, this single image speaks to the way that film and television westerns played a role in creating the mythic West and how the tourist economy of the New West perpetuates and capitalizes on that fictitious history by selling tickets to participate in the theater of the myth. Ironically, people on the Ponderosa Ranch hayride have paid to be robbed.

While subtle and not-so-subtle touches of humor enliven much of Goin's oeuvre, his sense of humor is most on display in the *Atlas* images. The editorial team of the *Atlas* struggled with "how to blend the strange and humorous with the familiar and serious" (12). And they admit that their "droll response" to the West's "newfound, upscale trappings" coexists with a sense of loss for what has been edged out by the "fast-paced, place-uprooting, and disquieting transformation of region" (12). They "anguished" whether they "had embarked upon an enterprise fundamentally optimistic or dour" (12). The unsettling ambiguity of tone smoldering under the mirthful surface of text and image is one of the most provocative aspects of *Atlas of the New West.* How do we respond to change?

■

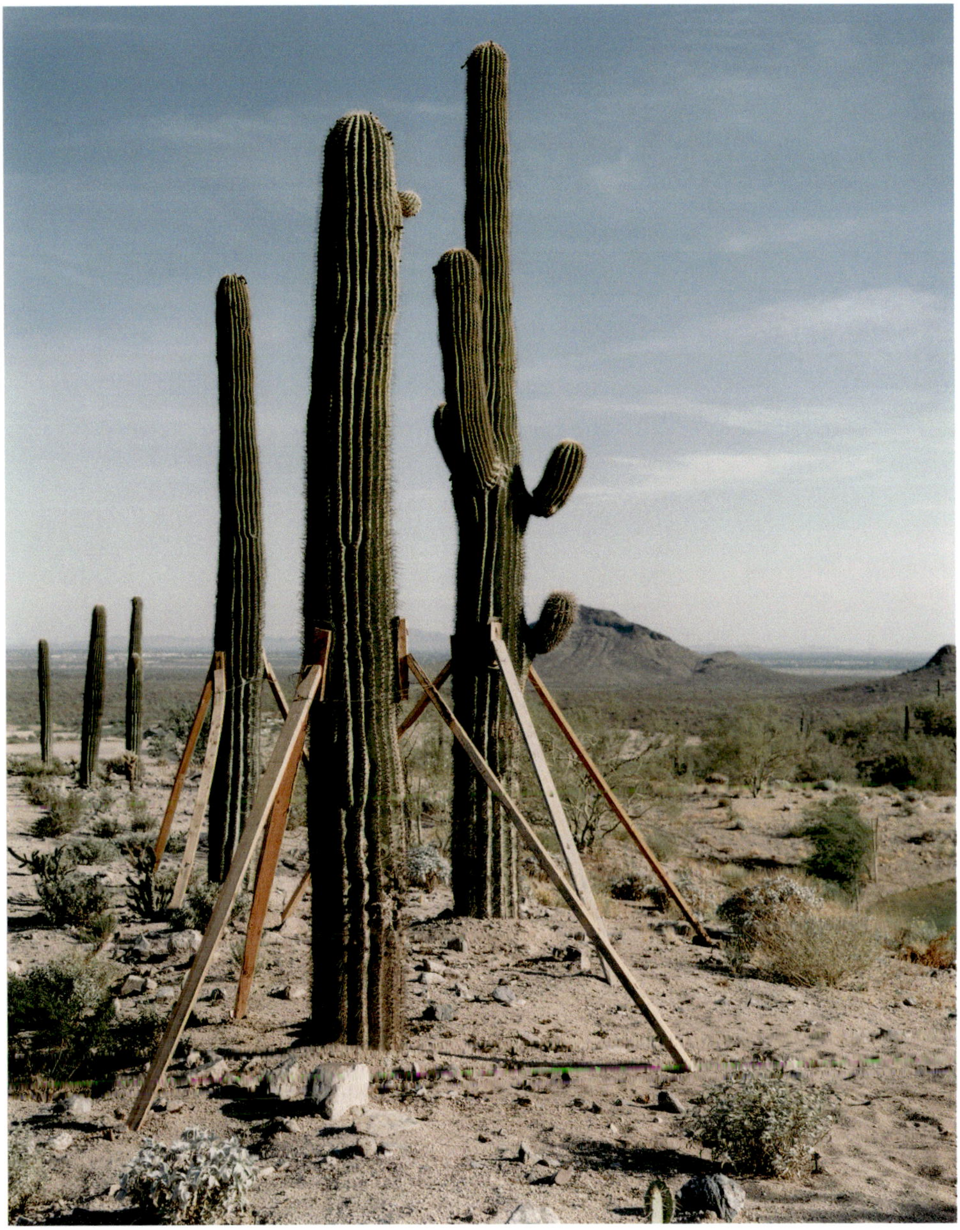

12.2 *Saguaro at Eagle Mountain residential and golf development, east of Phoenix, Arizona*, 1996.

As discussed earlier in this chapter, these sizable saguaros are propped up by wooden supports, because they had been recently transplanted to landscape a new planned subdivision on the outskirts of Phoenix. The sprawling city of one million people at the time—1.7 million people in 2021—is faintly visible in the background. In the early settlement period saguaros were razed to make way for urban growth, but in the New West the saguaro has become a prized symbol of desert living, and this iconic cactus is now imported and sometimes poached to be replanted in clusters or rows, chic sentinels to new housing developments, municipal buildings, and shopping malls. This set of saguaros has been put in place before the streets have even been laid out, probably to establish the mythic western setting for marketing materials. In terms of the composition of this image, note how the wooden supports form triangles—a favorite geometrical motif of Goin—and how the horizon teeters on the apex of the triangles, signaling perhaps the precarious balance between the desert ecosystem and urban encroachment. The horizon is the future, and this image is all about the constructed future.

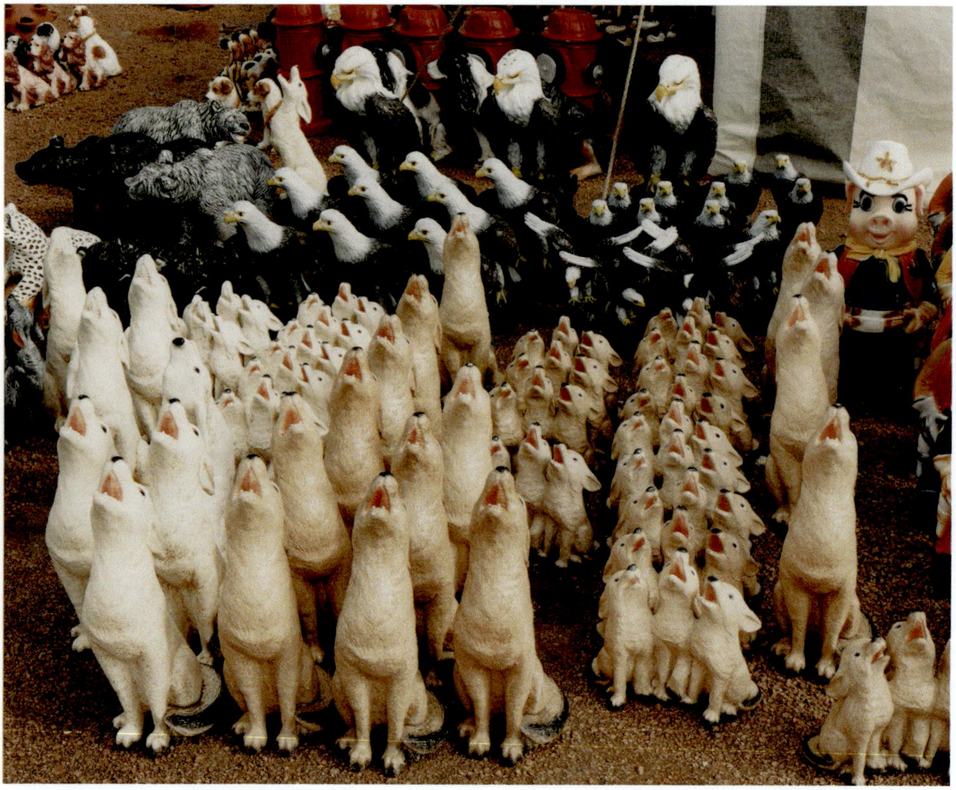

12.3 *Coyote howls at Old Divide Trading Post, Colorado*, 1996.

In *Atlas of the New West*, these plastic, for-sale howling coyotes of assorted sizes and hues echo the book's notes of humor and irony. The postmodern West commodifies the mythic past and packages wildlife into kitschy souvenirs, even as human impact threatens the survival of the actual wildlife whose manufactured replicas decorate our yards. In similar fashion, new development destroys wildlife habitat, and we then name the offending subdivisions after departed species—Quail Ridge, Wolf Hollow, Bear Valley Estates, Bighorn Park, Golden Eagle Court. We in the New West tend to pacify landscapes and calcify romance. Lawn ornaments are a further irony in the arid West, where lawns themselves are not native and require irrigation, all to simulate the genteel, green landscape of England. Private and public landscapes reveal a tremendous amount about our cultural values, just as decoration establishes the stage by which we narrate our identity. Peter used this image for a family Christmas card, with the wish, "We hope you have a howlin' holiday! ~ Dana, Kari, Peter Goin."

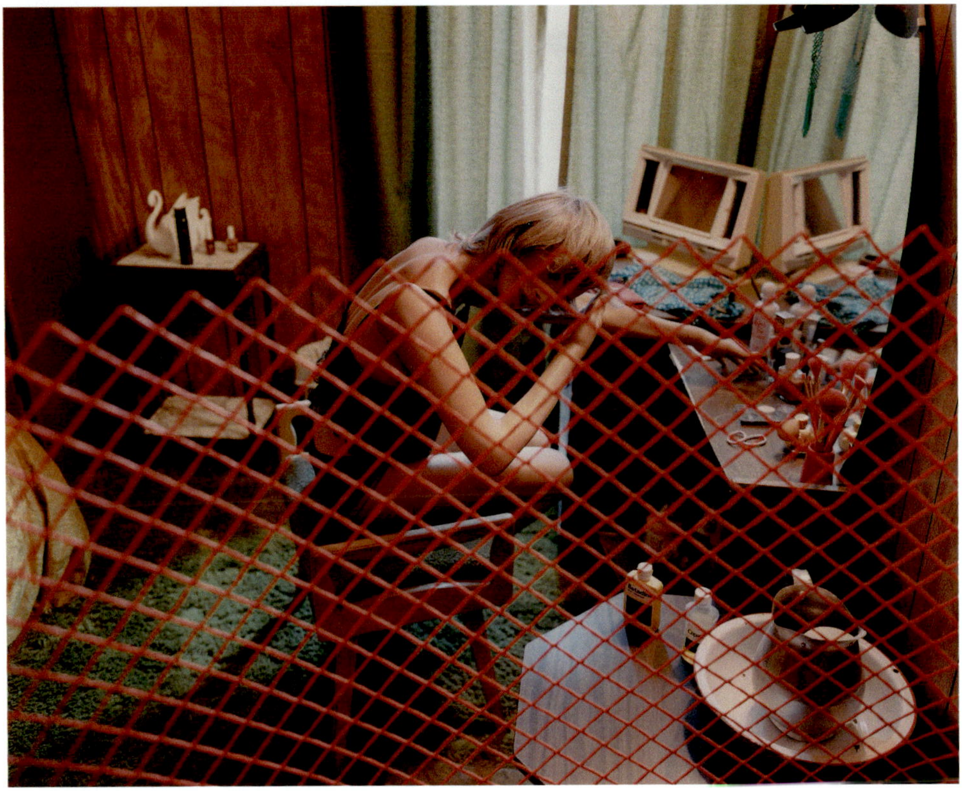

12.4 *Oasis Bordello Museum, Wallace, Idaho,* 1996.

Passing through northern Idaho? You might want to stop at the offbeat Oasis Bordello Museum in Wallace. As the story goes, this bordello had been doing a banging business in the heart of the Silver Valley mining district since the 1800s and was well thought of by the community for sponsoring the high school football team and contributing to local charities. The Oasis prospered for over a century until the FBI shut it down in 1988. The bordello responded by freezing everything in time, leaving the rooms as is, replacing the prostitutes with mannequins, and opening the doors as a museum. This image illustrates how we turn history into myth, fossilizing metaphors and glorifying Old West activities such as prostitution, gun fights, and stage robbery. What sets Goin's image apart from Oasis photographs on the internet is the screen of orange fencing Goin includes in front of the tableau. To Goin, the fence becomes a sanitizing tool that separates visitors from the forbidden, signaling that it is not to be experienced, it is to be viewed.

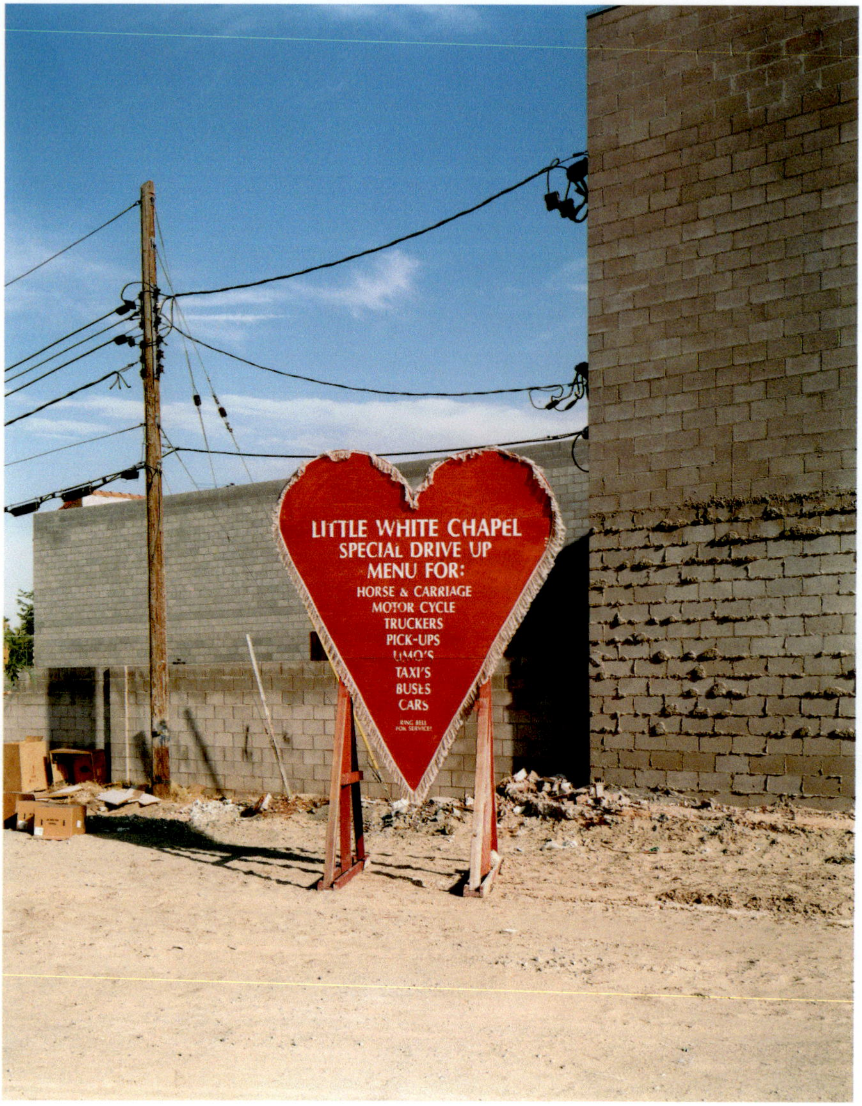

12.5 *Little White Chapel, Las Vegas, Nevada*, 1996.

"Special Drive-Up menu for horse & carriage, motor cycle, truckers, pick-ups, limo's, taxi's, buses, cars. Ring bell for service." This bright red, heart-shaped sign stands out against a decidedly unromantic backdrop to advertise a wedding business down the road. The "world famous" Little White Wedding Chapel and drive-thru Tunnel of Love offers a menu of wedding packages, including Pink Cadillac, Stretch Hummer, and Elvis Tribute ceremonies. "We're a drive-by culture," Peter observes:

> We have drive-by weddings, drive-by shootings, drive-by restaurants. This little box, this kind of institutionalized conceptualized container, or vessel, of privacy allows us to have all of these kinds of experiences without ever leaving the car. It reflects how intrinsically important our own sense of private mobility is.

"Cars are very important to romance in our life," Goin continues, "I mean limos, people go out to dinner, they have the car, they borrow the car, they make love in the car, they travel together in the car. It's a passage technology in people's lives." This New West image introduces that whole cultural aspect of how we incorporate the life of the car into our society and culture.

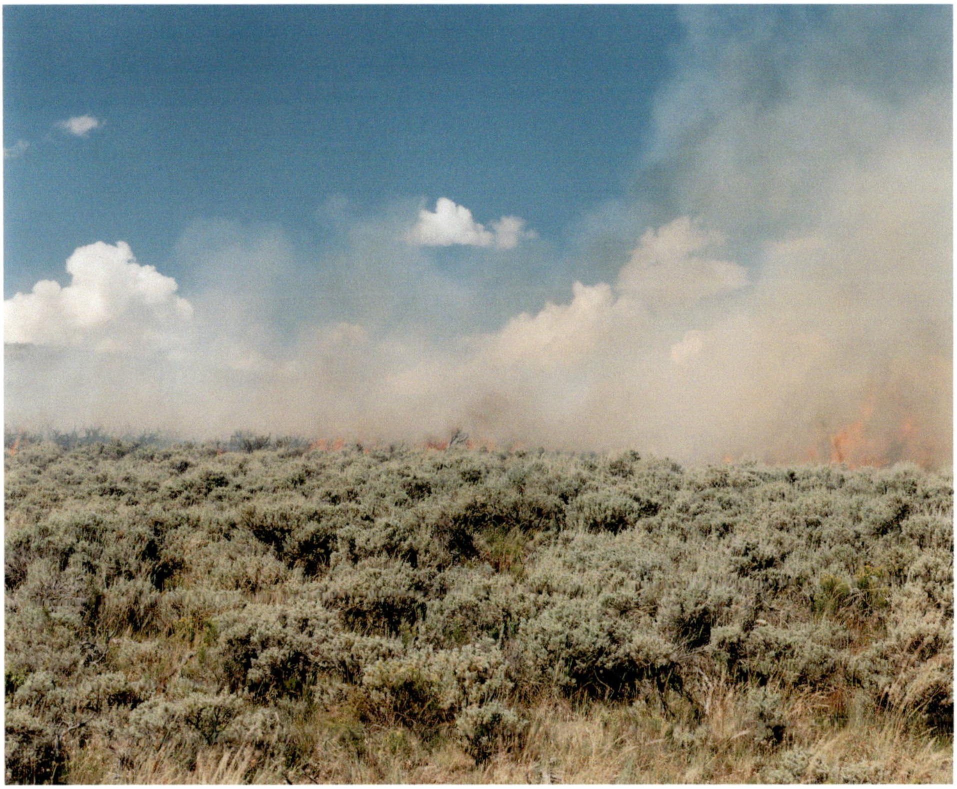

12.6 *Sagebrush fire, Colorado,* 1996.

If sagebrush is a symbol of the mythic West, wildfires that burn millions of acres annually pose a major challenge in the New West. In 2018, for example, according to a report from the Congressional Research Service, nearly 22,000 wildfires burned more than 7.0 million acres in the West, where most of the wildfires occur on federal lands. Wildfires not only threaten homes and businesses, but they destroy habitat and leave the landscape vulnerable to invasive, fire-adapted non-natives such as cheatgrass, which dramatically changes the ecology of the region. This image asks us to contemplate the scale and importance of fire ecology, which is transforming the environment of the New West. Goin happened to notice smoke from this emergent wildfire in Colorado when he was in the area working on *Atlas* photographs. He immediately pulled out a map and navigated a maze of dirt backroads to a good vantage point to set up a tripod for a 4x5 photograph. When firefighters arrived on the scene, they couldn't figure out how Goin beat them to the scene, and a sheriff, suspecting arson, held Goin for questioning. Hours later, the fire chief drove up and said that the origin point of the fire had been identified several ravines over, so Goin was released.

Works cited

Center of the American West. Regents of the University of Colorado, 2019, https://www.centerwest.org/. Accessed 7 March 2019.

High Country Log Homes. High Country Log Homes, 2007, http://highcountryloghomes.net/. Accessed 7 March 2019.

Hoover, Katie, and Laura A. Hanson. "Wildfire Statistics." *Congressional Research Service*, 31 October 2019, https://crsreports.congress.gov/product/pdf/IF/IF10244. Accessed 12 April 2020.

Manchester, Ellen. *Arid Waters*, edited by Peter Goin. University of Nevada Press, 1992.

Riebsame, William E., James J. Robb, Patricia Nelson Limerick, Charles Wilkinson, and Peter Goin. *Atlas of the New West*. W. W. Norton, 1997.

Thompson, Jerry L. Essay. *Walker Evans at Work: 745 Photographs Together with Documents Selected from Letters, Memoranda, Interviews, Notes*. By Walker Evans and Jerry L. Thompson. Harper & Row, 1982.

13

SENSUAL ANTHROPOMORPHS

Throughout Goin's photographic career, most of his images were made to serve a book project. As such, the majority of Peter's fieldwork has been project-driven, and his succession of books forms the basis for many of the chapters in this retrospective. However, there are exceptions, and Peter's Joshua Tree National Park work is one of them. The Joshua Tree photographs explore the concept of sensuality, a vein implicit in much of Goin's oeuvre and explicitly developed in the Joshua Tree series. While Goin typically gravitates to human-altered landscapes, in this series he is drawn to anthropomorphic landscapes, namely natural features that look like parts of the human body.

The opportunity to photograph at Joshua Tree National Park arose through the park's Artist-in-Residence program. Peter had enjoyed two other artist residencies to date—the Light Work residency in Syracuse, New York, in 1981 and a residency at the Appalachian Environmental Arts Center in Highlands, North Carolina, in 1991. The Joshua Tree National Park residency appealed to him for several reasons. First, there are a number of abandoned mines within and near the park, which gave Peter a chance to pursue his growing interest in post-mining landscapes. Second, a National Park offered a special opportunity to study a place that most visitors experience as wilderness but that is, in reality, highly managed *humanature*. And third, Joshua Tree fed Peter's abiding interest in arid landscapes of the American West. In 1996, he applied to the Artist-in-Residence program at Joshua Tree National Park, which allows artists to live in the park for a four-week stay to pursue their own creative practice. Shortly after Goin was notified that he had been awarded a residency, his wife died, rendering him a single parent and making a four-week stay impracticable. Program administrators graciously allowed him to space out his residency to four or more one-week stays over the course of several years, roughly 1997–2002.

Artists-in-residence at Joshua Tree today are housed in a furnished one-bedroom apartment at Black Rock in the northwest corner of the park. During the years Peter had the residency, however, artists were given access to an employee campground in Hidden Valley. Public campgrounds at Joshua Tree are typically fully booked, and campsites are packed closely together. If you camp there, make reservations well in advance, and expect

DOI: 10.4324/9781003212607-13

to hear talking, radios, clinking rock-climbing gear, and recreational vehicle generators in a transitory outdoor community that will be bustling and feel oddly urban. Peter's situation at the employee campground was entirely different as he was the only one there. It was very still and quiet. The solitude and isolation from car headlights and lanterns facilitated long-exposure nighttime photography. Peter invited his former college roommate Dennis to join him and pitch in as a fieldwork assistant. The two made quite a pair. As Peter recalls:

> Always with Dennis there's laughter. He has a great spirit, and a wonderful sense of living life to the fullest. He was great at helping set up the camp and assisting with the tasks and support required for eight to nine days of concentrated photographic explorations. He would often drive, allowing me to scan the horizon and navigate. He's larger than life—6'5", 275 pounds, more or less—and would do the heavy lifting.

As we shall see, Dennis also served as a model in Peter's performance pieces.

It may help to set the stage. Joshua Tree, established as a National Monument in 1936 and designated a National Park in 1994, is located in southern California, about twenty-five miles east of Palm Springs. It straddles the border of the Mojave and Colorado deserts and is exceedingly dry, with annual precipitation averaging only four inches, which often comes in the form of brief torrential summer thunderstorms. Temperatures can vary hugely from more than one hundred degrees Fahrenheit in the summer daytime to below freezing on winter nights. Day-night temperatures can fluctuate by forty degrees (Fletcher). While the eastern section of the park, in the Colorado desert, contains a mix of low-lying creosote, cholla cactus, and ocotillo, the western section, in the Mojave, supports a comfortably spaced congregation of Joshua trees (*Yucca brevifolia*), rising up from the light-colored sandy ground and standing out against the sky, displaying their Dr. Seuss-like bodies of multiple "arms" ending in bunches of rapier-like "fingers." According to Jane Rodgers, Vegetation Specialist for Joshua Tree National Park:

> By the mid-nineteenth century, Mormon immigrants had made their way across the Colorado River. Legend has it that these pioneers named the tree after the biblical figure, Joshua, seeing the limbs of the tree as outstretched in supplication, guiding the travelers westward. Concurrent with Mormon settlers, ranchers and miners arrived in the high desert with high hopes of raising cattle and digging for gold. These homesteaders used the Joshua tree's limbs and trunks for fencing and corrals. Miners found a source of fuel for the steam engines used in processing ore.
>
> Today we enjoy this yucca for its grotesque appearance, a surprising sight in the landscape of biological interest.

Forming a backdrop for this forest of uncannily humanoid trees, humplike conglomerations of eroded granite boulders dot the landscape, beckoning as a natural playground. As you approach any one of the many sensuously rounded rock formations, you are likely to see people climbing and clambering over them, resembling a party of spiders on holiday.

On Peter's first stay, he was diligently project-driven, becoming more open to spontaneity in subsequent years. That first year, he hiked to all the historic mine sites and made a series of photographs of Eagle Mountain Mine—a private inholding surrounded by Joshua Tree National Park—to illustrate a chapter of his coauthored book *Changing Mines in America*. In later visits, he explored the possibilities of nighttime photography and performance pieces, styles he had been experimenting with for many years in annual trips to Lake Powell, which we will explore in Chapter 19 "Artifacts of the Future." Nighttime photography requires long exposure times, the use of a tripod, and a cable-release to control the shutter. The black silhouettes of Joshua trees against a night-blue sky might look like dancers in a prehistoric drama. Light from a campfire might cast a golden glow over the primitive scene. There is a sense of the extraordinary or otherworldly in the nighttime images of Joshua Tree, an escape from mundane daily life into a realm of adventure and sharpened senses.

Performative pieces, such as the one featured in this chapter (see Figure 13.3), presented another way for Peter to creatively engage with place, as he orchestrated scenes that depict a spectral human figure appearing in or melting into the desert's evening shadows. Performance pieces required multiple exposures, the first made at last light—that transitory moment when the foreground is in the shade and the mountains in the background are still in the sun. One can almost feel the evening chill coming on. The second exposure is made after dark. For this exposure, Peter hung a blue plastic tarp between two Joshua trees and placed a lantern behind the tarp. His model, Dennis, then posed between the tarp and the lantern such that his shadow was cast against the tarp. This exposure required Dennis to remain stationary from thirty seconds to two minutes. Any movement would create a blur or even result in a missing limb. The effect is of a translucent, ghostlike human figure emitting a bluish aura and standing weightlessly in a natural setting of dry grasses, gray boulders, and tall Joshua trees.

For Peter, these lyrical photographs evoke a rich history of associations, both personal and historical. When Peter was a boy, his family lived in Indonesia and he enjoyed watching Indonesian puppet shows, in which characters appeared as animated shadows, acting out the story on a screen backlit by fire. That childhood memory inspired the dancing figure photograms that Peter made when he lived in San Francisco (reproduced on the last page of this book) and again in performance pieces staged at night in a variety of places throughout his career. In the context of Joshua Tree, Goin's performance pieces reference the film genre of the western. He explains:

> Joshua Tree to me was always a place of ghosts. It was always a place where history was evident from the most recent nineteenth-century ranchers and thieves who lived there who would rustle cattle. It's a perfect den—in fact, there have been numerous bands of desperados who hid out here. In the classic western movie tradition, they actually did set scenes in places that were really hard to find. There's all these rocks. Outlaws could set up a camp, and they were hidden, and they could protect themselves. You're not going to find people that easily in there.

Beyond westerns, the wilderness look of Joshua Tree belies a broader history of generations of people who came before. Goin continues:

> We know that there are these antecedents of our own existence—the aboriginal, sports recreationists, ranchers, nomadic tribes who came through. We know that there are tribes that have identities there and have their own origin stories. The landscape is

> imbued with the absence and therefore the presence of these antecedents. The performative process was really a way of conjuring them up. When you look at this kind of an image, it's about looking at a landscape, and emerging from the landscape are these ghost-like apparitions of those historical antecedents.

Recalling the making of the Joshua Tree performance pieces, Goin explains that:

> The performative really alters the engagement with sense of place. You look at the sites, and you design them during the day. You set it up and then wait for darkness to fall. So, you have the stage, and doing this little performance creates a social cohesion with whomever you're with. Peter and Dennis had a lot of fun making nighttime performance images. Because Peter was using 4x5 color film, there was no telling at the time whether the images came out well or not.

By day Joshua Tree was a good time as well. Peter and Dennis had a system worked out. Dennis would make breakfast in the morning while Peter drank coffee, read, prepared the film and cameras, and enjoyed sitting in the morning sun. The two would hike and explore during the day. Peter would make dinner and clean up. They laughed a lot and enjoyed being alive, together, outside, and on assignment. In life's journey, this was a difficult time for Peter due to his wife Chelsea's recent death. He found the trips to Joshua Tree to be a helpful diversion. As he puts it:

> The process of using the landscape to further the solace of life's contempt is a very active and dynamic process. The good news is that in the practice of Zen and in understanding how to deal with the forces of life's energy, that being in the landscape does provide some solace. You experience it and redirect so that when you're boulder scrambling you're not really thinking about what's going on. You're just making sure you don't fall and break your leg and that you're having that experience. And you're finding a route and you don't want to get lost. And it's a pretty interesting landscape. You know, it's a young boy's dream to climb on these boulders and go through these washes, and the vegetation and the diversity, the biotic diversity in Joshua Tree is phenomenal, and it's always interesting.

Although the two were playful, they did have a purpose, and that was, in Peter's words:

> to try to come to terms with that landscape and a way of representing it, to approach it with an open mind and see what ultimately might come of things, to interpret it creatively from my own perspective through my own lens.

For Peter to be an artist-in-residence entailed making photographs. A camera, padded with towels, was stowed in his knapsack, a relatively small instrument that made a large impact on his experience of the park.

Author Susan Sontag in her 1977 classic essay collection *On Photography* takes up the topic of the role of photography in the modern world of proliferating images. She argues that photography mediates not just our view of the world but our *experience* of it. Sontag asserts in the influential essay "In Plato's Cave" that in carrying a camera, one adopts a "mentality which looks at the world as a set of potential photographs" (7). "Photographs really are experience captured," she writes, "and the camera is the ideal arm of consciousness in its acquisitive mood" (3–4). However, she continues, a camera conditions the type of experience that one is likely to have: "A way of certifying experience, taking photographs is also a way of refusing it—by limiting experience to a search for the photogenic, by converting experience into an image, a souvenir" (9). Sontag speculates that in contrast to actively participating in events, "taking photographs has set up a chronic voyeuristic relation to the world" (11). We take pictures rather than taking action.

Peter is certainly aware of Sontag's theories of a camera as a mediating technology, but his own experience at Joshua Tree was that rather than "limiting experience," as Sontag phrased it, his camera *facilitated engagement* with the landscape. The camera became a complement to the process of being conscious in the expression and the experience of that landscape. "For example," he explains,

> if you're walking on a path, your mind might be thinking about struggles at work or at home; whereas, a camera really is a physical reminder that there is this external reality that in many ways invites interpretation. I don't always presume the parameters of a project when I enter into a landscape or an idea. Occasionally, and this is an example, I'm interested in just synthesizing a more organic experience.

One such experience, which Goin expressed in a series of black-and-white images, was of first noticing and then being unable *not* to see sensuality in Joshua Tree's rock formations. "All of a sudden it becomes funny," he recalls, "because it's like this whole landscape has now been transformed into this deconstructed assortment of bodies, most, but not all, of them female." The most vulgar intrusion of sensuality into his consciousness was male and arose in the camp. As he recalls:

> We were camping by a place where when we would eat breakfast early in the morning, when the sun would rise, it would cast a shadow of a stump of a Joshua tree that was ridiculously clinically phallic. It's just like really in an erect position. It's over the top, so much so that I was there for a week and refused to photograph it because, first, I didn't think that anyone would actually believe that this existed without my creating it, and second, it just looked too vulgar. Couldn't we just have a hint or suggestion of the sensuality of a landscape, or must we just take this and poke it in your face?
>
> Every morning we saw this. Here's two guys going, "well, this is just over the top, what is going on here?" And we didn't do anything except sit there and drink our coffee. And then you start laughing. After a while, you just think it's hilarious. And one day we get up in the morning and it's overcast, and so there's no shadow. And you're sitting over breakfast and you start saying, "Ah, I feel good. I feel ready to take on the landscape." And then the sun comes out and you go, "Oh my!" It's almost like this kind of interactive response. I finally made the photograph on the last day because I said, okay, you can't tell people this story without having the proof of what that shadow is.

As a matter of fact, there is a sensuality to many landscapes. There's the appearance of velvet on distant hills in California as the sun sets or the waving wheat in an agricultural landscape.

Beauty, in part, is defined by this sensuality. At Joshua Tree, the eroded, rounded boulders with their cracks and crevices, their bumps and protrusions, can look almost pornographic. And once you start seeing the landscape anthropomorphically—as human body parts—there will be no end of examples. As Peter observes:

> It's hard not to see it because it's actually everywhere because of the way that granite erodes. You've got shadows and shapes and you have bushes in crevices, and after a while you say, 'Well, that's really vulgar.' And you start laughing after a while. The obviousness of it became so incredibly vulgar that after a while I just started photographing it whimsically as an interesting sideline.

Peter's sensual images of granite boulders at Joshua Tree reflect an artistic interpretation of the natural landscape that has precedent in art history. In the history of art, beauty has often been represented in the figure of a nude female. We project our desires onto the landscape, and because we are sexual beings, we perceive the world sexually. For example, we name a mountain range in Wyoming the Grand Tetons, French for "large breasts." Many people assume that Georgia O'Keeffe's large paintings of flowers depict female genitalia despite O'Keeffe's vehement denial of that reading. The American photographer Henry Callahan overlaid nude photographs of his wife, Eleanor, onto the landscape, merging the identity of a nude female onto Mother Earth. Photographer Edward Weston made a series of erotic photographs of female nudes posed in natural settings, such as at Oceana Dunes, California. German artist and surrealist photographer Hans Bellmer made photographic collages, wherein he took images of miscellaneous body parts and rearranged them into grotesque, headless assemblages of random appendages—arms, legs, breasts, butts. His images abstracted human body parts from the human, isolating erotic shape and form.

Susan Sontag uses the word "voyeuristic" to describe the relationship of the photographer to the world, and she further notes that "Being a professional photographer can be thought of as naughty, to use [Diane] Arbus's pop word, if the photographer seeks out subjects considered to be disreputable, taboo, marginal" (13). Whether Goin's photograph series of sensual boulders at Joshua Tree strikes you as naughty or not—as vulgar or as beautiful—will depend on how you view sensuality. The Joshua Tree series is published here for the first time; however, Goin would continue to explore sensuality in his work, explicitly so in the *California Agriculture* project, in which he made images of what he calls "agriporn."

When we arrive at a new landscape, such as Joshua Tree National Park, pictured here, the initial view is typically from our means of transportation. In America, the view is often out of a car window, and the windshield is our frame. This image is about how the landscape is contextualized within the view of our cars. As Goin explains:

> We travel through these landscapes and say we experience their physical beauty, but it can only be appreciated when we're removed from it because we need the air-conditioning. When you're out there, you're overwhelmed by 110-degree heat and wind, or whatever. The landscape tends to consume us.

■

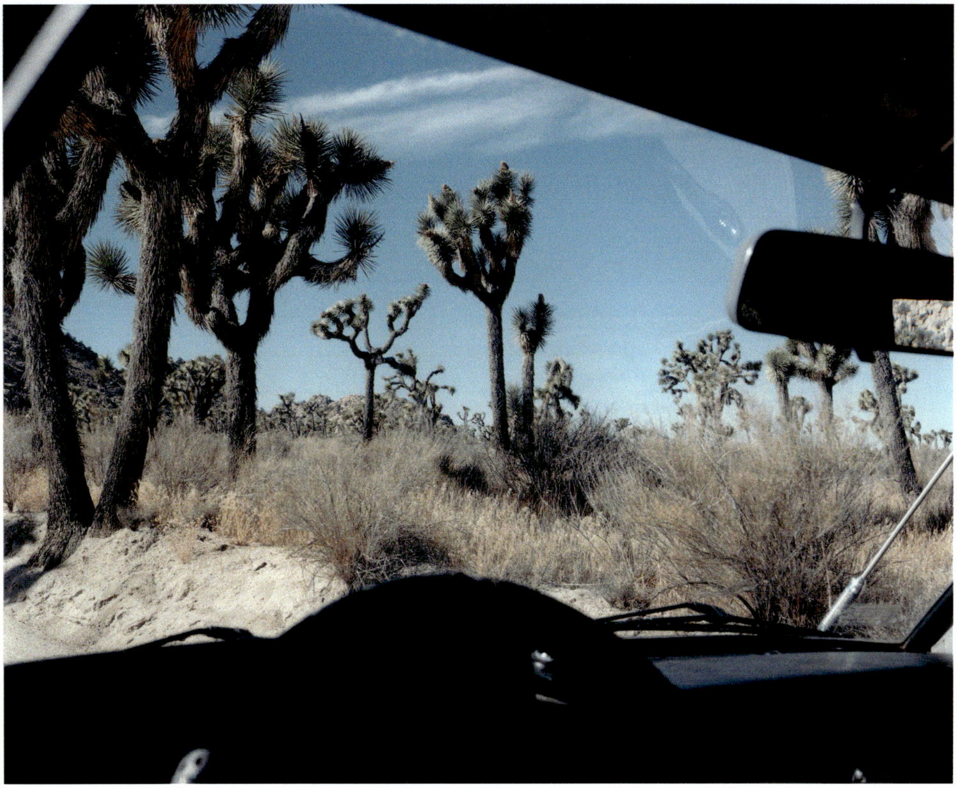

13.1 *Joshua trees through car window*, 1998.

There is a tease in this image. Where is the road? And how can the photograph be so sharply focused if the car is in motion? Goin actually set up his 4x5 camera on a tripod inside his parked car and made an angular view off to the right. These details signal the craft and artistic intent of a photograph that one might otherwise mistake for a simple snapshot.

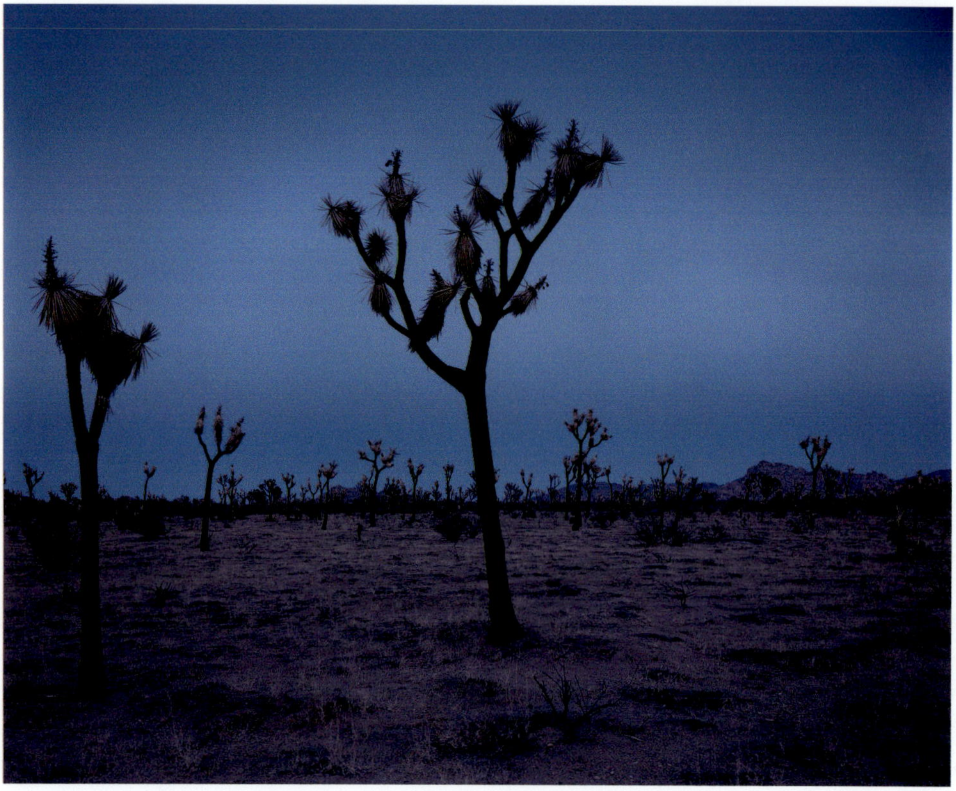

13.2 *Burned Joshua tree forest*, 1999.

When we imagine a forest fire, most of us don't tend to picture a forest of Joshua trees. And when we view the landscape in a national park, we tend to assume that it has looked that way since time immemorial. This image challenges both of those paradigms. At Joshua Tree National Park, most of the trees are likely to be second and third growth, since the original trees were cut down by nineteenth-century pioneers and miners for fence posts and firewood. Human-caused forest fires have also had a significant impact on the landscape, which is becoming more vulnerable than it used to be to invasive, non-native vegetation that fuels an increasingly destructive fire cycle. The somber tone of this post-apocalyptic image, which Goin made at dusk, is about the nature of the place that has been so affected by human action and alteration. The impending darkness of night provokes contemplation.

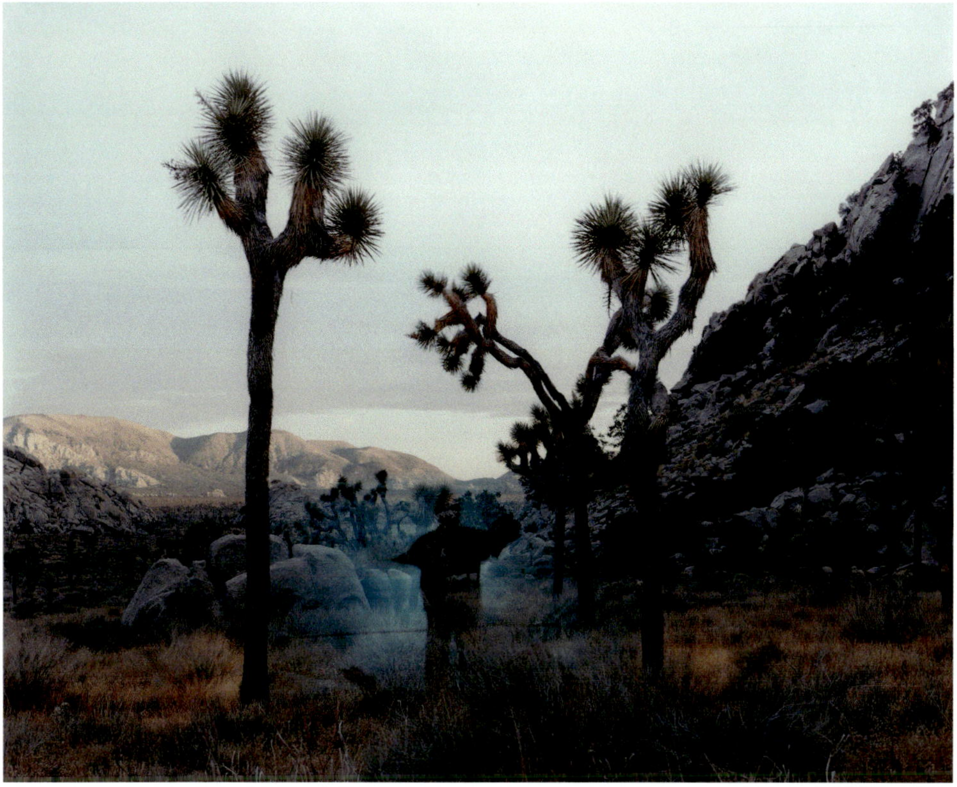

13.3 *Performance piece at Joshua Tree National Park*, 1998.

Most of Goin's photography adopts a documentary style, but a series of nighttime performance pieces made over many years indulges a more overtly lyrical sensibility. Reminiscent of a modern-day, ephemeral pictograph, this image made at the Joshua Tree Artist-in-Residence campground in Hidden Valley conjures up the park's human history. While all photographs reference time, this one deconstructs it via a double exposure, the first made at last light and the second made at dark with a lantern placed behind a man who stands near a suspended blue tarp such that he casts a shadow onto it. In this process, multiple perspectives of time are contained within the same image. In the Hidden Valley campground at night, unfamiliar noises coming from random locations, along with the flickering light of flashlights roving who knows where, created the sense of an animated landscape, one that might be inhabited by ghosts. Referencing both Indonesian shadow puppetry and film westerns, this image offers a fresh interpretation of Joshua Tree National Park, an imaginative departure from the more typical scenic shots featured in glossy calendars. This photograph also opens up a different way of engaging with place—theatrically, by creating performance art.

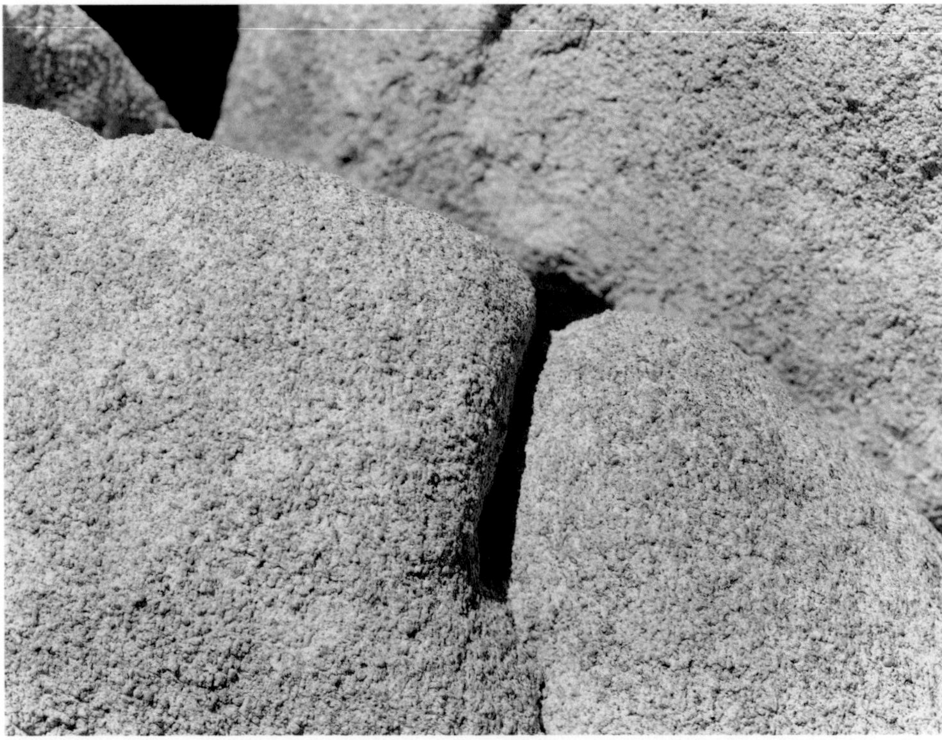

13.4–13.11 *Eroded granite boulders, Joshua Tree National Park*, c. 2002.

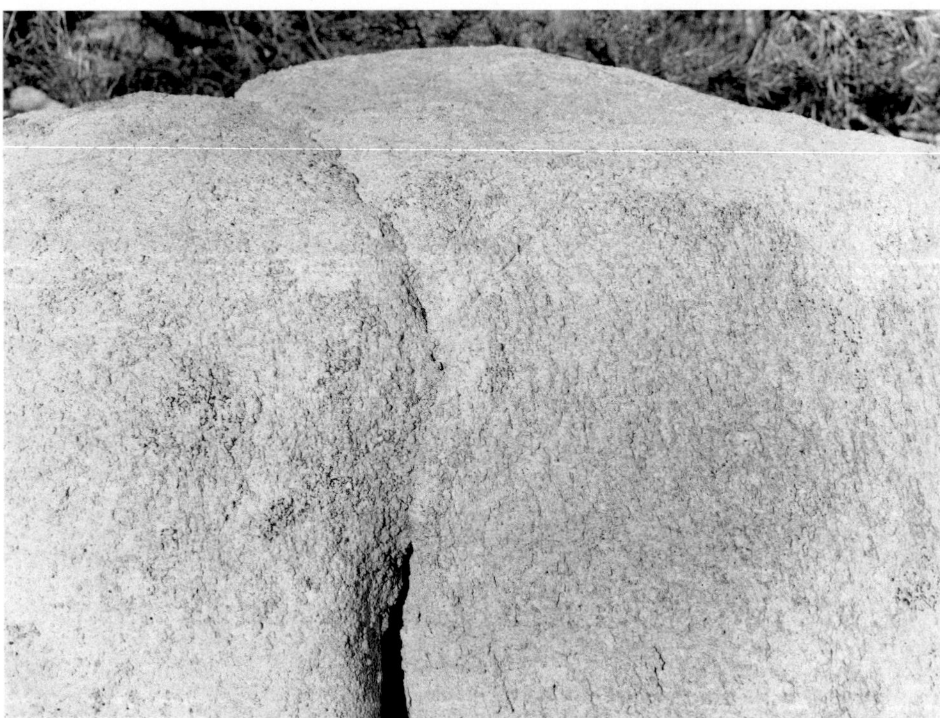

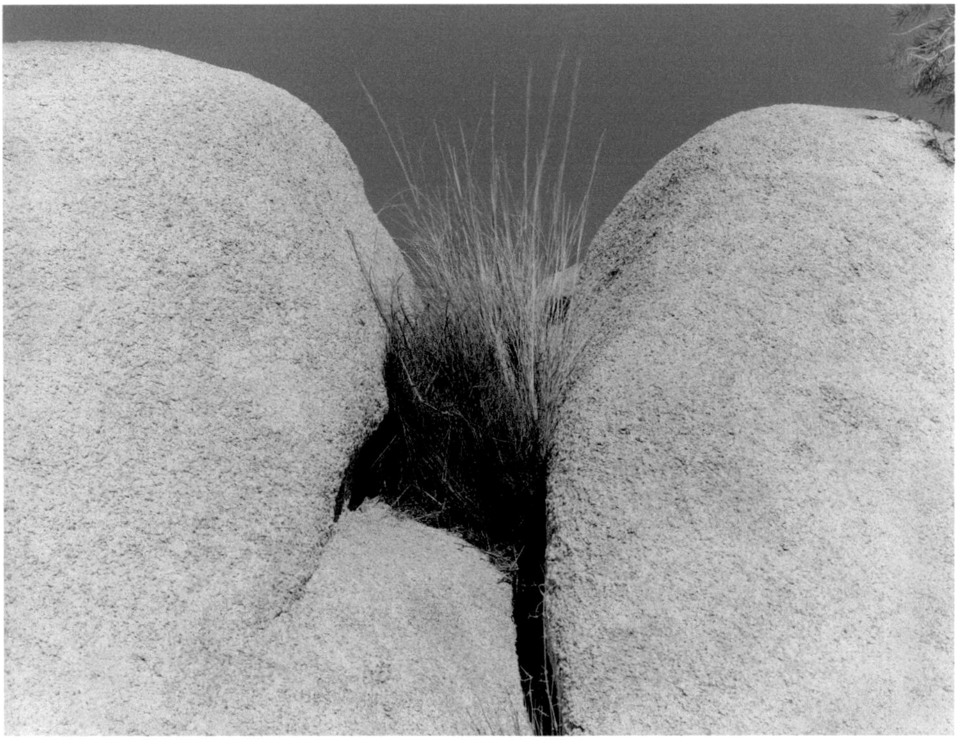

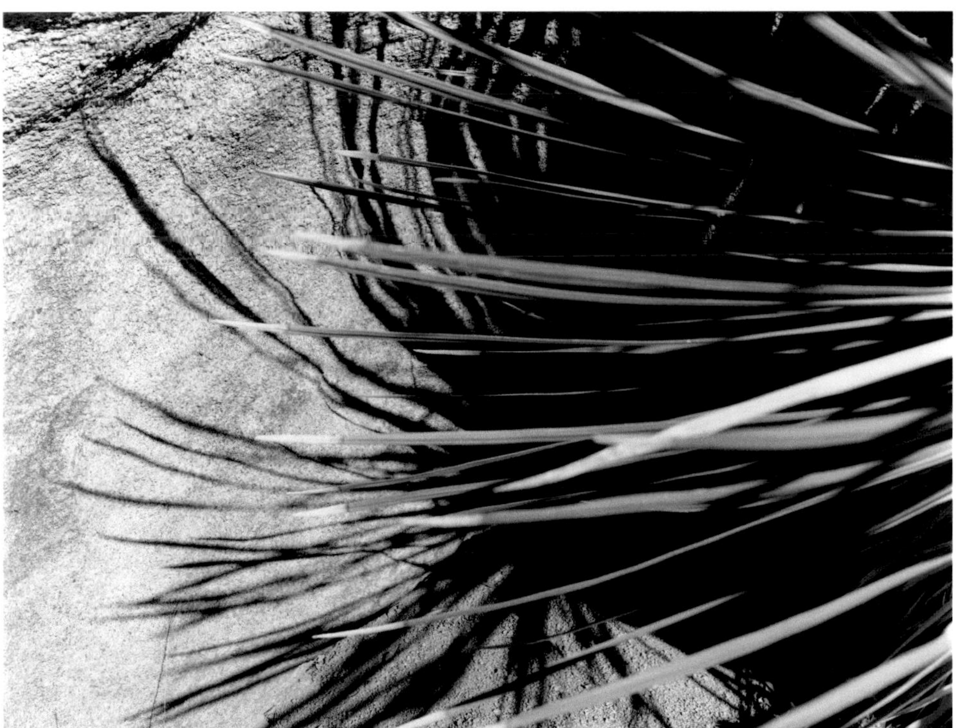

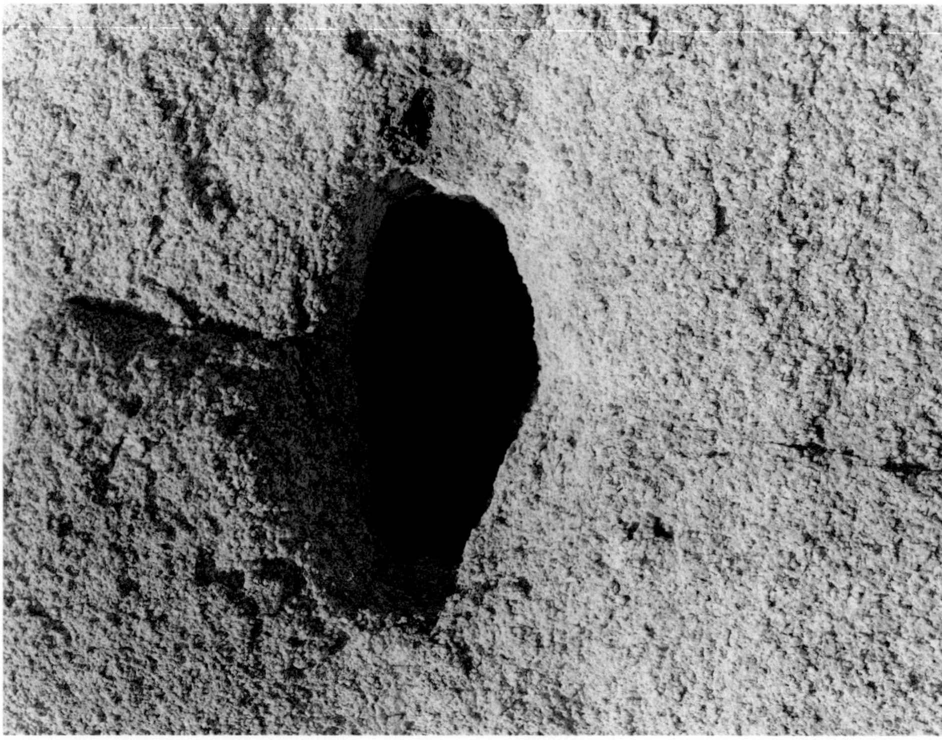

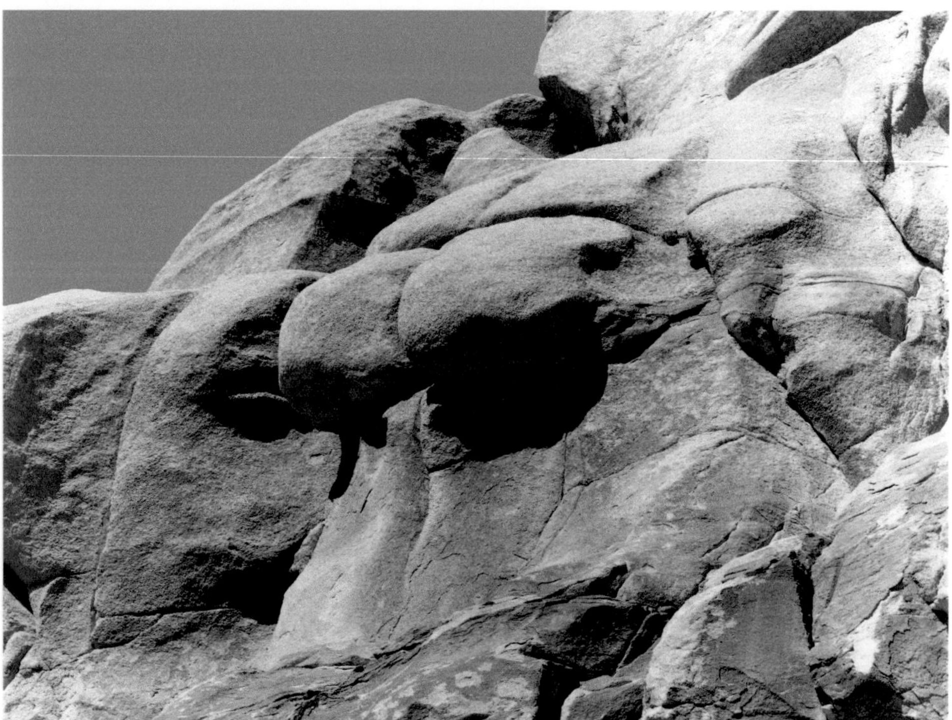

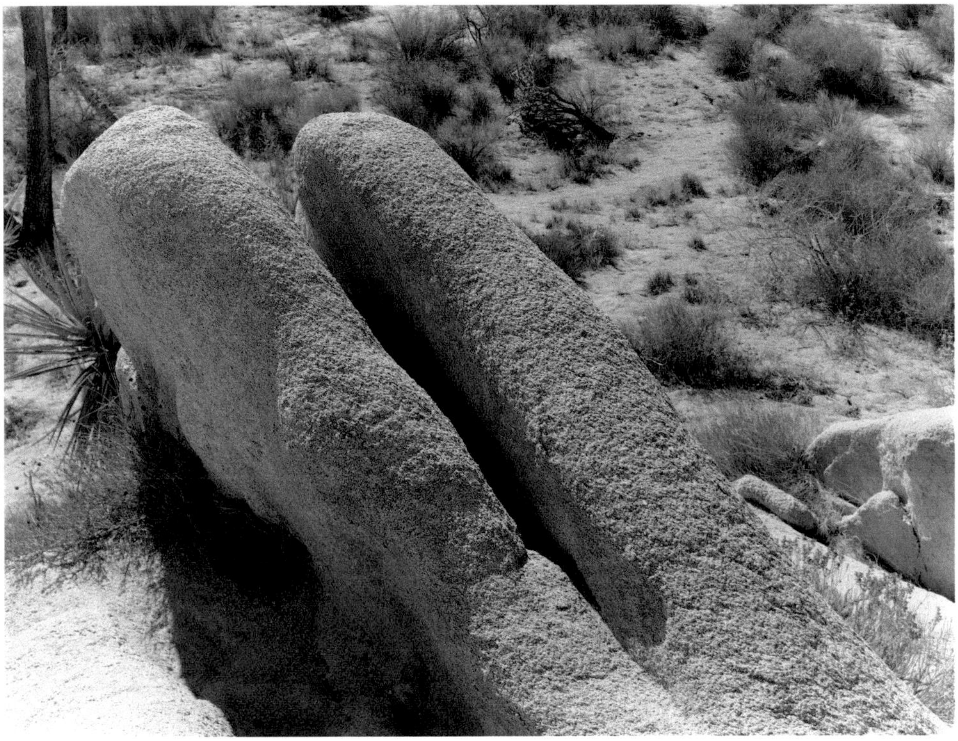

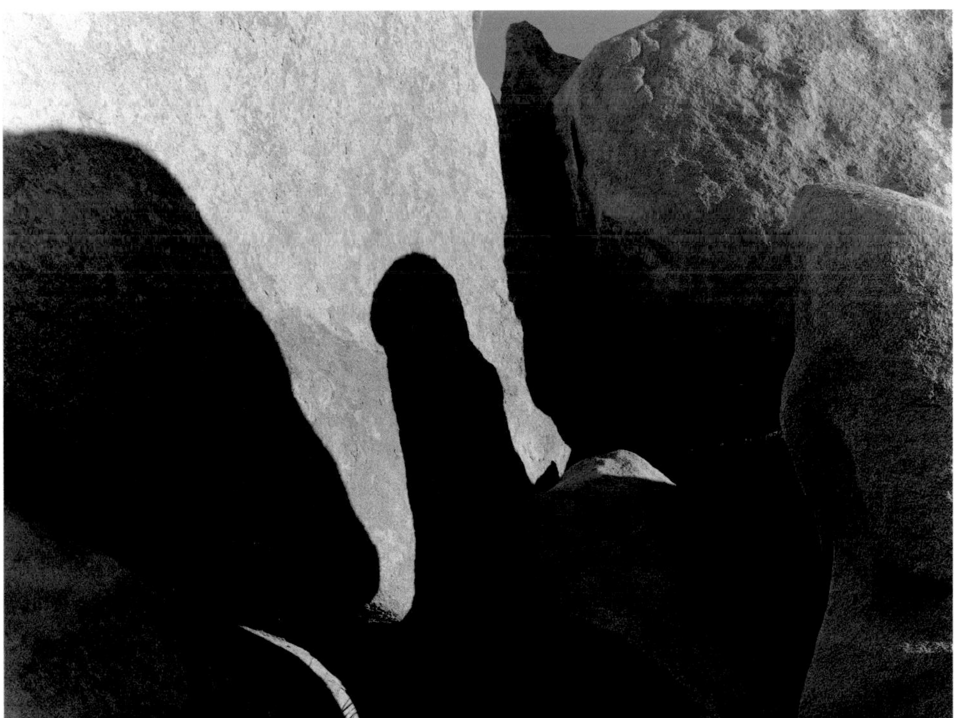

What you see when you look at this series of black-and-white nature photographs will depend on how your mind works. We carry our point of view with us. The sensuality and anthropomorphic shapes of granite boulders at Joshua Tree took Goin by surprise. He and his friend Dennis hiked, clambered, and camped amongst the beautifully rounded granite formations, relishing the physicality and tactile dimension of their explorations. Once they started noticing the anthropomorphic nature of the granite, it became so obviously referential that they started seeing male and female genitalia in abundance. At first refusing to photograph such suggestive images, Peter eventually relented and began documenting the sensual shapes that they saw. Black-and-white film highlights shape by removing color. There is a long history of sensuality, nudity, and eroticism in the art of painters and sculptors. Photography is a newer medium that initially mimicked painterly compositions. In some cases, early photographers would sketch out a tableau that they would then construct and photograph. As photography evolved, it developed its own language and means of expressing sensuality, such as the strategic framing and close-up view employed in this series, reminiscent of photographer Linda Connor's image of a "pubic mound" in White Sands, New Mexico.

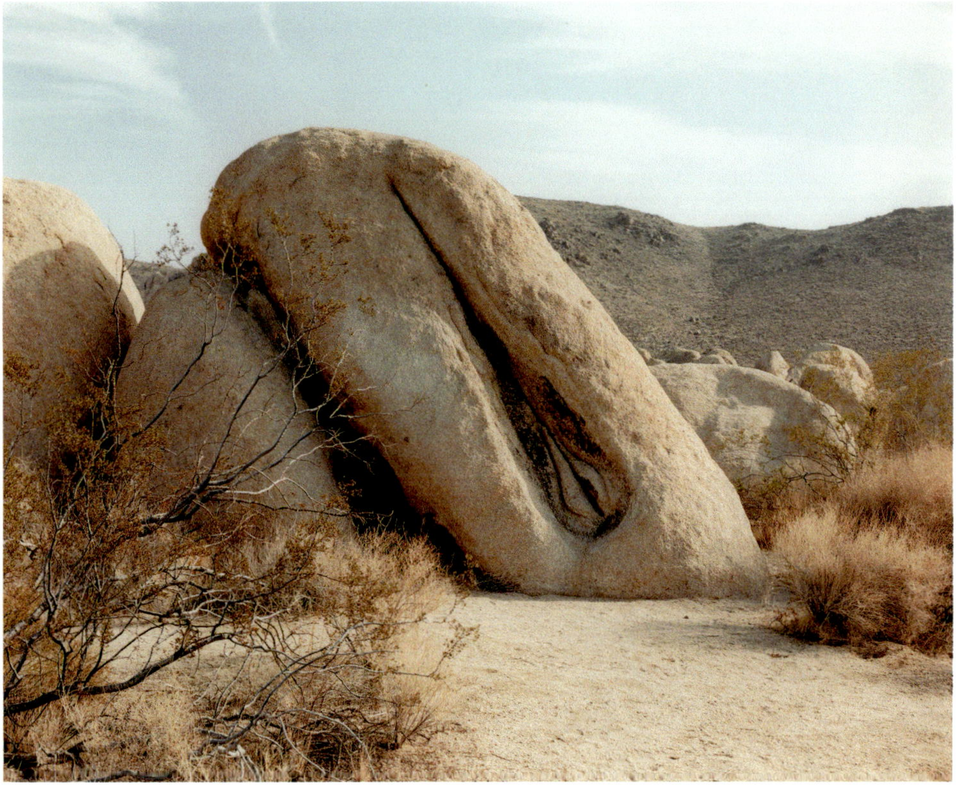

13.12 *Granite boulder, Joshua Tree National Park*, 1996.

Many photographers have taken pictures of the curves of a nude woman to make her body resemble a landscape, but here that relationship is reversed, the landscape looking like a female body. Sensuality is an element throughout Goin's work, just as sensuality is an element of our own identities. To exclude it seems unnecessary. To include it is certainly not required, but it does appear in a variety of different places. At Joshua Tree, sensuality was so omnipresent that Goin decided to intentionally engage it. A park ranger told Goin that locals call this rock the "Desert Queen." Others call it "Yoni," the Sanskrit word for the vagina that is loosely translated as sacred space or sacred temple. Goin was told that this site was and is used by local aboriginal people to train their young girls in the ways of menstruation and the world. As such, this image references the many ways that people interact with the landscape, including intimate and erotic experiences.

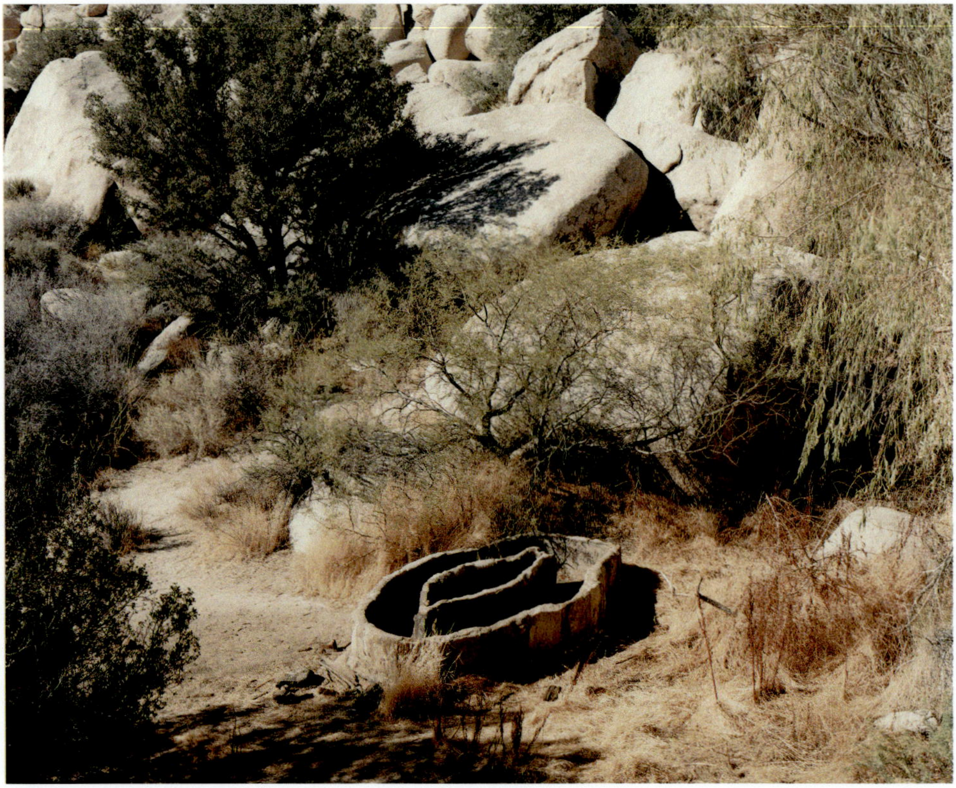

13.13 *Watering trough, Joshua Tree National Park*, 1994.

In the context of Goin's sensuality series, this concrete structure near Barker Dam appears to be a sculpture of a vagina. It was built by ranchers in the early 1900s and used as a cattle trough. The area is called Ivanpah Tank, "pah" being the Paiute word for water. Peter originally made this photograph for the *Water in the West* project, in which a consortium of photographers contributed images documenting the presence, absence, and uses of water in the arid American West. In that context, this trough of a dry water tank, like Goin's photographs of dry reservoirs during a drought, represents the scarcity of water and the structures we build to contain water. Only later, when Goin came across this photograph after making the sensuality series at Joshua Tree, did he notice the vagina motif. It is an interpretation that others have shared—one Instagram image hashtagged "#itsjustapussy" is captioned, "some dirty #cowboys must have built this watering hole."

Works cited

Fletcher, Kenneth. "Joshua Tree National Park." *Encyclopaedia Britannica*, 2013, https://www.britannica.com/place/Joshua-Tree-National-Park. Accessed 30 March 2019.

Rodgers, Jane. "Joshua Trees." *Joshua Tree National Park*, 2016, https://www.nps.gov/jotr/learn/nature/jtrees.htm. Accessed 30 March 2019.

Sontag, Susan. *On Photography*. Picador, 1977.

14

ARID WATERS

If *Atlas of the New West* encouraged Goin to think about the contemporary American West from the perspective of tourism, and the Joshua Tree National Park artist-in-residency afforded the opportunity to explore a place's visual structure and identity, his next project would seriously investigate what it means to *reside* in the arid West. How can we as a diverse culture live sustainably in an environment of limited water? And how can photographs contribute to public understanding and debate on this issue? These pressing questions undergird sustained work that Peter carried on for more than a decade, some of that time under the umbrella of the collaborative Water in the West Project, culminating in several publications, including the centerpiece of this chapter, Goin's coauthored book *A Doubtful River*.

"Water, water, every where,/Nor any drop to drink," complains the narrator of Samuel Taylor Coleridge's "The Rime of the Ancient Mariner," as he recalls futilely gazing upon an ocean of saltwater. That irony—the *illusion* of abundant water—takes on a different twist in the American West, as Water in the West Project co-coordinator Ellen Manchester explains in her introduction to *Arid Waters*. When one beholds the verdant suburban lawns and sprawling golf courses, the public swimming pools and water parks, and the cascading waterfalls fronting shopping malls and resorts, and when one is cognizant of the rapid population growth of western cities such as Las Vegas, Phoenix, Tucson, Los Angeles, and Denver, one might assume that there is water in abundance. People in the American West certainly seem to live *as if* water is plentiful and even unlimited. And that's the problem. With the exception of the Pacific Northwest, the land west of the one-hundredth meridian receives an average of less than twenty inches of rainfall per year, the amount below which irrigation is required to grow crops. In the arid West, as author Wallace Stegner contends in his essay "Thoughts in a Dry Land," "You have to get over the color green; you have to quit associating beauty with gardens and lawns" (45). Furthermore, as Goin and his colleagues in the Water in the West Project might add, you have to become *aware* of water—its scarcity, how we manage it, and how it is used and misused.

The name "Water in the West Project" derives from the name that photographer Robert Dawson gave an ambitious photographic project that he began in 1983 to document the effects of large-scale agriculture on California's Central Valley and, more broadly, to address "environmental and social issues of the arid American West" (Dawson and Goin,

DOI: 10.4324/9781003212607-14

"Pyramid Lake Project," 3). Dawson wanted to extend that work geographically, conceptually, and in terms of public impact by collaborating with other photographers, historians, and writers. Toward that end, Dawson and Ellen Manchester, who is a photography editor and researcher, convened an informal planning meeting with photographers Peter Goin and Mark Klett in 1989 in Aspen, Colorado, followed by an official first meeting of the Water in the West Project in March 1990 in Sausalito, California, attended by nine founding members—Laurie Brown, Gregory Conniff, Robert Dawson, Terry Evans, Peter Goin, Wanda Hammerbeck, Mark Klett, Ellen Manchester, and Martin Stupich. A group of advisors from other disciplines challenged the group to think about "the photograph's role as a record of cultural attitudes, how photography can function as a tool for social and political change, and how artists might provide new and unique perspectives on critical environmental issues" (Manchester 20). At Goin's urging, the group drafted a "preamble" that gave direction and definition to their work:

> The Water in the West Project is a collaboration of a group of photographers whose art shares strong ties to the landscape of the western United States.
>
> The Project's primary goal is to create a body of work which will contribute to the increasingly urgent dialogue about the future and quality of life on earth as sustained by increasingly limited natural resources.
>
> As a collaborative project, Water in the West represents a wide range of interests and concerns from agricultural practices in western Kansas to water rights on the Paiute Reservation at Pyramid Lake in Nevada. At the core of the project is a concern for how water use and perceived needs have shaped our natural and social landscape.
>
> Water in the West intends to serve as a clearinghouse for ideas and as a structure to encourage collaborative and interdisciplinary work through the establishment of a comprehensive photographic archive, public lectures and symposia, and a continuing series of publications and exhibitions.
>
> *(22–23)*

Peter Goin was an extraordinarily hard-working member of the Water in the West Project, hosting the group's second meeting at the University of Nevada, Reno, in fall 1991. That meeting featured a public symposium whose invited speakers included Pyramid Lake tribal member Joe Ely and prominent environmental historians Roderick Nash, Donald Pisani, and Donald Worster. Goin organized a photography exhibition at the Sheppard Fine Arts Gallery at the University of Nevada, Reno, which displayed both museum-quality prints and work-in-progress materials from each member of the project. In addition, he curated a second exhibit in an adjoining gallery that involved students, who researched and selected historical photographs from the university's Special Collections and University Archives Department, contrasting documentary photographs of the 1906 Newlands Irrigation Project with promotional images from Las Vegas casinos, sporting a "water as plenty" theme. The Reno conference spawned two books—an anthology of reprinted essays from political scientists, historians, and journalists, entitled *A River Too Far: The Past and Future of the Arid West*, edited by Joseph Finkhouse and Mark Crawford, and the photography collection *Arid Waters: Photographs from the Water in the West Project*, edited by Peter Goin. The Water in the West Project met regularly throughout the West for nearly a decade, suspending formal meetings in 1997, after which group members donated their work to the Center for Creative Photography at the University of Arizona, which has preserved the work and related materials in an archive.

Peter Goin and Robert Dawson first met in 1982 and began collaborating on a Pyramid Lake Project in 1988, a collaboration that subsequently became part of the Water in the West Project (Dawson and Goin, "Pyramid Lake Project," 3). The Pyramid Lake Project is described in the 1992 special issue of *Nevada Public Affairs Review* that Peter edited on "Dividing Desert Waters" and is more fully developed in their 2000 coauthored book *A Doubtful River*, with text by Mary Webb. When they met in 1982, both Dawson and Goin had been politically active for a long time, both having been student demonstrators during the Vietnam War and Goin having worked as a political aide in Washington, DC, during college. Both men eventually pursued a career in art and "discovered the power of photography to help understand our society, and its potential as a tool for social change" (3).

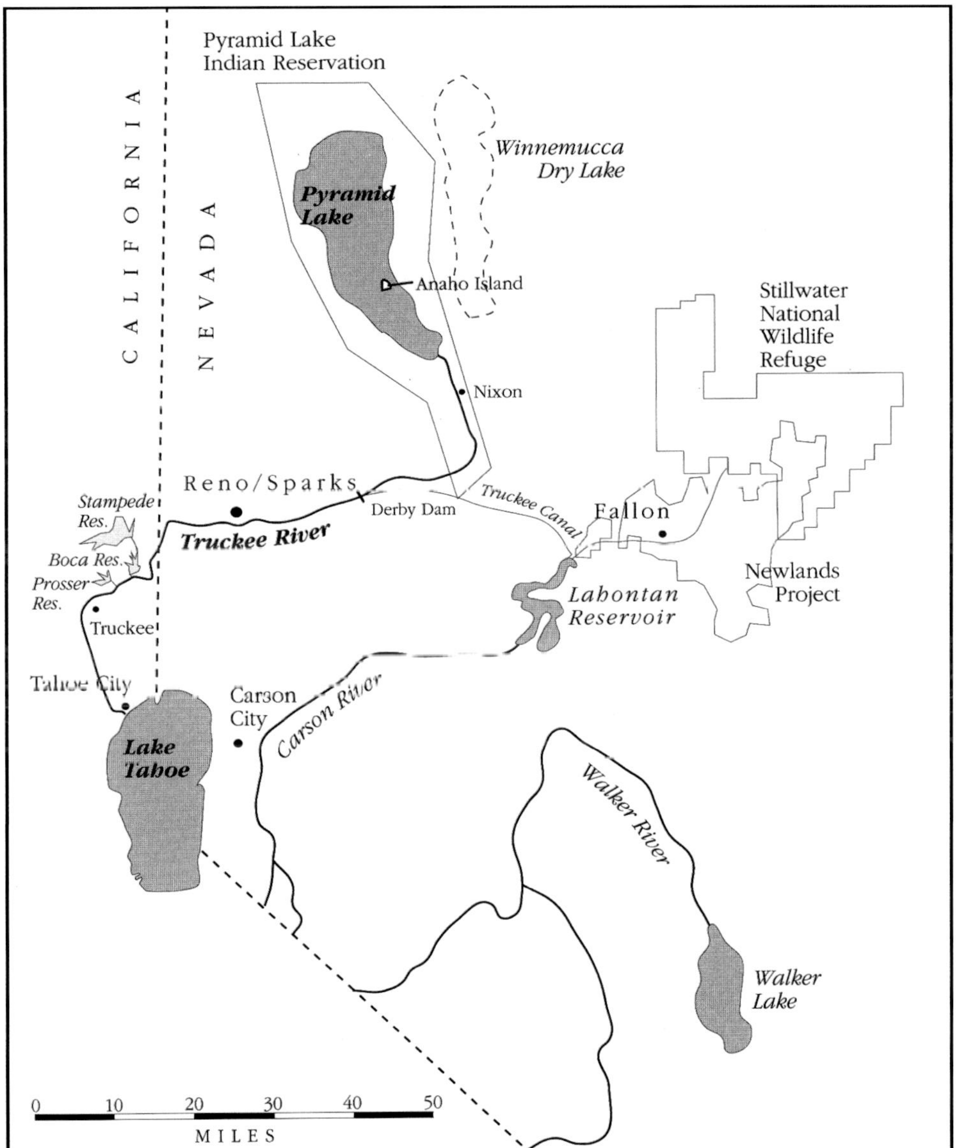

14.1 *Map of the Truckee River,* reprinted from and courtesy of *Nevada Public Affairs Review.*

In the Pyramid Lake Project, Goin and Dawson sought to photographically survey the entire 121-mile length of the Truckee River watershed, from the river's headwaters at Lake Tahoe, through the Reno-Sparks metropolis, to its natural terminus in the landlocked Pyramid Lake on the Pyramid Lake Paiute Tribe Reservation, and to its diverted branch that irrigates croplands in Lahontan Valley. What makes the modestly sized Truckee River highly significant is that it is the site of the first federally funded irrigation project in the West, the Newlands Project, named after US senator from Nevada Francis G. Newlands. Senator Newlands drafted the 1902 Newlands Reclamation Act, whose intent was to open areas in the arid American West to settlement and farming through federally funded dams and irrigation canals. Newlands' vision was that diverted water made available to agriculture would make the desert bloom and support yeoman farmers in the Jeffersonian ideal of a democratic citizenry comprised of independent landowners. Sales of western public lands were to fund the initial construction of dams and canals, and subsequent sales of irrigated property and water rights would, in turn, fund future dams in a self-renewing process overseen by a new federal agency, eventually named the US Bureau of Reclamation. Ultimately, the Newlands Reclamation Act led to the damming of most major western rivers.

And it all started on the Truckee River with the construction of Derby Dam in 1905–1906, diverting a portion of the Truckee River via the Truckee Canal to grow successful harvests of "Hearts-O-Gold" cantaloupes, turkeys, and alfalfa in Fallon. As Dawson and Goin explain in their *Nevada Public Affairs Review* lead essay:

> The Pyramid Lake Project is a photographic survey of the entire Truckee River and Pyramid Lake watershed. This water system is a good example of the critical water issues facing all arid regions of the West. By focusing on specific regional issues, we hope to illustrate broader concerns affecting our society as a whole. These include such issues as expanding cities versus preserving agricultural land; unchecked development versus vanishing wetlands and open space; and the history of water development in this region.
>
> *(3)*

Despite the historic significance of Derby Dam as the first arid lands reclamation project in the West and as a prototype of federal intervention in western water law and appropriation, Derby Dam and the Truckee River had been all but neglected in scholarship on water in the West. Marc Reisner's *Cadillac Desert*, for example, which *Newsweek* hails as "the definitive work on the West's water crisis," mentions the Truckee River only one time, Derby Dam not at all. Donald Worster's magisterial history *Rivers of Empire: Water, Aridity, and the Growth of the American West* gives the Truckee River one sentence and never mentions Derby Dam by name. Goin and Dawson sought to put the overlooked Truckee River and Derby Dam on the map. As Goin explains in conversation:

> It's about the landscape and about how this place becomes emblematic of a water-use debate that is common everywhere in the American West, and the Truckee River is the core that hadn't been addressed. The Truckee River is a perfect case study. We have all of the relevant constituencies, from incoming wealthy, non-participatory citizens to people moving here wanting a Midwestern landscape to the Paiutes to the

> ranchers to the farmers to the truckers. Everything is here. Yet, unlike the Mississippi or the Colorado River, the Truckee is small enough to make a tellable story.

Photographs illustrating the "Dividing Desert Waters" special issue begin upstream at Truckee, California, and conclude downstream at Winnemucca Lake, Nevada, "lake" in this case meaning "dry lake," as the water diversion at Derby Dam caused Pyramid Lake's adjacent, overflow lake to the east to dry up. Political, social, and environmental issues exposed for public consideration in *Nevada Public Affairs Review*'s stream of photographs include water shortage, water management, restricted access, recreation, new home construction, urban growth, air quality, homeless encampments, trash, water contamination and treatment, industry, infrastructure, water damming and diversion, Indian Reservation history and culture, wildlife, and natural beauty. Dawson and Goin challenge themselves and their readers to forge personal answers to important questions:

> How can each of us engage in the public debates of our time? What can we do to make a difference in our society? What can we do to avoid slipping into the nihilism and disillusionment so prevalent in our culture?
>
> (3)

Their own answers, delivered in the language of photography, are to *engage* with rather than evade problems, bringing important issues into sight, into mind.

Another possible answer to the question of how to make a difference is *collaborate*. Work together. Indeed, as we have seen, the rationale for expanding Dawson's original, individual Water in the West Project into a group project was the sense that a collaborative effort would be more effective than individuals working independently. Up until this project, Peter had mostly photographed on his own. He had worked with field assistants, yes, but he had not embarked on a formal collaboration with another photographer. Working on the Pyramid Lake Project with Robert Dawson entailed physically following the river, frequently together, traveling by car, on foot, and in an airplane, exchanging ideas and each making photographs on the same trip—two different perspectives on the same subject, influenced by the meeting of minds. To then invite author Mary Webb to join the team as the Pyramid Lake Project evolved into the *A Doubtful River* book project presented yet another challenge and opportunity for collaboration. While Goin and Dawson's photographs are "useful in understanding this place," Webb's text tells stories of "the people who depend on and adjudicate the waters of the Truckee River" (*Doubtful River* 7, 6). The book's title phrase, "a doubtful river," alludes to John W. Audubon's 1849 characterization of the Arizona desert as "a doubtful country, where one is uncertain as to the friendliness of the Indians" (160). In contrast, what is doubtful about the Truckee River is its fickle flow and its questionable ability to sustain the many users who rely on it. The Authors' Introduction to *A Doubtful River* explains the book's purpose:

> The landscape of northern Nevada is divided yet woven together by the waters of the Truckee. Connecting a tourist recreation and gaming economy, a rural ranching oasis, and an aboriginal belief system, the Truckee River offers evidence of the layered complexity inherent in creating a sense of community defined by shared needs. We hope that this book can contribute to an understanding of the debate over this

> watershed's environmental and political future. We invite the reader to consider how the culture of this arid land conceives of water as a commodity, an abstract legal right rather than the most basic physical source of life.
>
> *(5)*

Because *A Doubtful River* focuses on one river in one place, is published by the "home" press—the University of Nevada Press—and appeals to a local audience, it could be characterized as a bioregional project. A bioregional perspective envisions place less by political jurisdictions—cities, counties, electoral zones, and states—and more by ecological realities, such as mountains, valleys, watersheds, climate, and geology. A bioregional address for Reno, Nevada, USA, for example, might be the Truckee Meadows, along the Truckee River, at 4500 feet elevation, on the east side of the northern Sierra Nevada.[1] A bioregional way of thinking about place knits together people who may be divided by political boundaries, uniting them and the non-human biotic community around a shared resource in a specific place. The book enjoins residents of northern Nevada to learn about, understand, and identify with the place where they live. Identification with a place can be jump-started with information and nourished by stories, both of which *A Doubtful River* supplies.

Identification with a place means that *who* one is will be intimately tied with *where* one lives. Peter argues that:

> however you interpret the arid West, these are issues that matter to us—our climate, our relationship to the land, the water that flows through the soil. These things are of fundamental value. We can't exist separate and apart from the place in which we live.

True identification entails harmonizing one's lifestyle to fit in with the natural limits of the local environment. "But," Peter laments, "we don't live in concert, we live in opposition to the landscape." Having grown up in tropical Indonesia, Brazil, and Mexico, and having lived in the relatively humid US Midwest and temperate San Francisco Bay Area, Peter Goin, who transplanted himself to Nevada in adulthood, has learned to identify with and *live-in-place* in the Great Basin Desert.

Goin has no lawn. Instead, his house is encircled by shade-creating, house-cooling trees, efficiently watered by drip lines. The backyard deck looks down on Last Chance ditch, a nineteenth-century construction that diverted a portion of the Truckee River for use by ranchers and farmers. Although human-made, Last Chance ditch has become a wildlife corridor in the city, exactly the kind of integration of urban and wild that Peter believes is a good way to plan cities. His backyard is a habitat for birds and wild animals such as chipmunks, squirrels, lizards, snakes, raccoons, coyotes, deer, and even the occasional bear and mountain lion. Sitting on his deck, one can watch birds flitting about in the trees and shrubs, hawks circling overhead, scrub jays claiming the fruit trees, and crows exhibiting their corvid antics. Depending on the season and time of day, you might hear songbirds, woodpeckers, owls, raptors, scrub jays, coyotes, and frogs. Goin's is an example of a low-water-use dwelling integrated into the landscape, a yard shared with a host of fellow creatures. Far from feeling like he is making a sacrifice in being frugal with water, Peter *enjoys* living in tune with the natural environment, living *with* nature.

Other residents of the Truckee Meadows, however, seem to flout the idea of xeriscaping, or low-water landscaping. *A Doubtful River* stages a conversation between Mary Webb, who shares Peter's water conservation ethics, and her neighbor, whose sprinklers hiss as she insists on keeping her lawn green even in an extended drought. As the neighbor sees it, restricting one's personal water use backfires by freeing up more water for urban growth and development. This water-guzzling old-timer has a point. The photographs in *A Doubtful River* raise similar issues of rapid development seemingly blind to a severely limited water supply. While one spread of four photographs reveals the alarmingly low water levels—verging on drying up—of Prosser, Boca, and Stampede reservoirs above Reno, a triptych of photographs in the next chapter shows a smoggy winter panorama of pollution and sprawling new development in the hills above Reno.

As in the special issue of *Nevada Public Affairs Review*, photographs and text in *A Doubtful River* follow the Truckee River's flow, from its headwaters at the outlet of Lake Tahoe in the Sierra Nevada, through the former logging-turned-recreational tourist town of Truckee, California, past a series of reservoirs and the Farad gaging station, across the border into Nevada, through the Reno-Sparks metropolitan area and under highway 395, past the town of Lockwood—built on a floodplain—to Derby Dam. Here the team encounters a book-design dilemma: Derby Dam splits the Truckee River into two branches, so how can a linear narrative follow its flow? The northern, natural course of the river flows under Interstate-80 and onto the Pyramid Lake Paiute Tribe Reservation, past the reservation towns of Wadsworth and Nixon, ultimately emptying into Pyramid Lake, which technically is a mesotrophic, monomictic, hyposaline, terminal lake. In plain English, Pyramid Lake has medium levels of submerged aquatic plants and nutrients, undergoes one annual mixing of the colder water layer and the warmer water layer, has abnormally low salinity, and is landlocked with no outlet to the sea. The southern, diverted flow is channeled into a canal that sends a portion of the Truckee's waters to meet the Carson River at Lahontan Reservoir, from which a main canal directs the controlled outflow to a network of smaller irrigation canals in the farming community of Fallon, Nevada. Any leftover water and agricultural runoff replenishes the Stillwater National Wildlife Refuge, listed as a "Globally Important Bird Area" by the American Bird Conservancy and an "Important Bird Area" by the National Audubon Society.

The design solution that the coauthors arrived at was to have Webb's text first treat Fallon-Stillwater and then end at Pyramid Lake, while the photographs reverse that order and end at the Stillwater Wildlife Refuge. What this organizational choice underscores is that text and photographs are not strictly linked, and neither is subservient to the other. Essentially, Webb's text, Dawson's photographs, Goin's photographs, and nearly twenty black-and-white historical photographs represent four different perspectives of equal value, mimicking the layered complexity of the Truckee River itself. Together, the essays and photographs examine "the human manipulation of arid lands" (6).

Historical photographs depict the construction of the Lake Tahoe outlet gate at Tahoe City, California, the wooden irrigation flume system that parallels the Truckee River, promotional images of Reno, including the "Wedding Ring Bridge," from which new divorcees were reputed to throw their wedding rings into the river, and a 1950, knee-high flood in downtown Reno. Further downstream, a 1920 historical photograph of agricultural fields at a bend in the river at Lockwood, east of Reno, is paired with Dawson's 1989 rephotograph of the same location, scraped down to bare dirt and sprouting with newly built houses. Five historical photographs dating from 1905 document the construction and completion of the Newlands Project canals and Derby Dam, the small dam at the heart of the book. Another set of seven historical photographs shows Paiute men and women catching and drying fish caught in Pyramid Lake, an aboriginal shelter, an Indian cattle herd on the Pyramid Lake Reservation, and the faces and feathered headdresses of five Paiute chiefs superimposed above the iconic pyramid-shaped tufa formation near shore. A final historical photograph shows the Stillwater Marsh in healthy condition in 1949. This photograph is juxtaposed against a series of contemporary photographs of the marsh at low water during the 1987–1994 seven-year drought. All the historical photographs tell a visual story of diverse human relationships with the Truckee River. Even though the river may look wild and pristine in some places, it is now and historically has been, from source to terminus, intricately tied to the cast of human characters who live in the watershed.

Goin and Dawson photographed in color and in black-and-white, from the ground and from the air, resulting in a valuably comprehensive photographic survey of the Truckee River, the first of its kind for this river. Dawson's and Goin's photographs are of approximately equal number, and they alternate and are paired regularly throughout the book, such that the untrained eye would have a difficult time distinguishing between the two were it not for each photographer's discretely lowercase initials identifying his photographs. (The images in this chapter are all by Peter Goin.) Some photographs are lyrically beautiful, while others appear starkly documentary, and still others are enigmatic. A significant number of photographs turn trash into treasure, so to speak, giving a conventionally non-aesthetic subject an aesthetic treatment. For example, there are exquisitely composed images of homeless camps along the river in Reno-Sparks, of the boarded up Old Natchez School, of a spray-painted swastika on an announcement board in Wadsworth, of a wrecked car in the snow in Sutcliffe, and of literal debris—broken glass and a window frame on a dry lake bed and a tattered book lying open on the desert in a Paiute junkyard near Nixon. Another motif in the collection of photographs is signs—neon signs, road signs, KEEP OUT and NO TRESPASSING signs, kitschy signs, vintage signs, interpretive signs, motel signs, graffiti, historical markers, movie marquees, and billboards. We *homo sapiens* are logocentric, decorating our habitat with text.

Text is also helpful in understanding the images, and some photograph captions tell a hidden story. For example, there is a deceivingly lovely image of sparkling water in a marsh at dusk, whose caption reads, "Dead fish floating along the shore. Increasing salinity, combined with decreased fresh water to recharge the wetlands, resulted in fish dying at the Stillwater National Wildlife Refuge during the 1987 drought" (143). Many of the photographs feature people doing things related to water. In this category, one may notice images of hydrologists inspecting a water gauge, technicians working at a cloud seeding generator site, crowds enjoying the annual Great Truckee River Duck Race (a fund-raising release of rubber ducks), a golfer at the Aquarange at Hilton Bay, a Paiute fish barbecue and dance, and more. The people pictured represent the many different constituencies ineluctably connected by the Truckee, groups that Mary Webb's text personalizes in stories.

Counterbalancing the obviously utilitarian view of water is a spiritual undercurrent that acknowledges that human beings are water creatures, having evolved from water and whose bodies are approximately sixty percent water. Water is not only a source of physical life, but it nourishes the spirit, especially the presence of water in an arid environment. Some of the most scenic views in *A Doubtful River* evoke a spiritual response, such as cloud reflection on the water at a natural bend in the river, a rainstorm over the desert, a hot spring in winter, or the final image of a waterbird's nest floating on spiral grass.

Regarding that concluding image of the waterbird's nest—reprinted and analyzed at the end of this chapter—there were differences of opinion among the authors about how *A Doubtful River* should end. One view held that the book ought to take the approach of the jeremiad and hammer home stern lessons about the need to conserve water . . . or else. Goin's view, however, eventually prevailed. As he sees it, audiences have become jaded to finger-wagging rhetoric, and a final ultimatum would be depressing. He argued that the book had already made its point and that the ending should offer hope: "You can always find something to complain about, it's easy. The real challenge is not to say what's wrong. The real challenge is to say, 'But there's hope.'" It was a lesson that he learned the hard way during work on *Nuclear Landscapes*. Goin has trained himself to adopt a Buddhist mindset, a Zen attitude: "Do you think that we could just, for the sake of it, admit that life is tough and challenging and difficult? Can we agree with that for a second? OK, now let's go plant a tree." Granted, conflicts exist with competing demands for the severely limited waters of the modest Truckee River. *A Doubtful River* responsibly explores those problems. But, as Peter says, "Let's all just calm down for a minute, and let's understand that water is the spiritual vein of life." Water is not a commodity. Water is life.

While there may have been differences of opinion during the project, as is often the case when there are several cooks in the kitchen, the resulting product is a better, more nuanced view than any single author could have created alone. As such, their collaboration is a good model for shared policy-making over the arid waters of the West. Goin envisions the book as:

> opening a portal to the discussion over a collective responsibility in arid environments, calling for a more sustainable, charitable, collaborative understanding of water, how it could be shared and how we could live more sustainably. That's not critical. That's hopeful.

Not everyone agrees. The reception of *A Doubtful River* has been mixed. The book has triggered some local resistance. Some readers charge that the book is too critical, while others bristle at being told to be more conscious of their water use. One man in Fallon, who had been gifted the book at Christmas, returned it in person to Peter with the remark, "It is my God-given right to have this water." But, in the case of the Truckee River, perhaps stirring up controversy contributes more to the public good than enabling complacency. Although it has been criticized, *A Doubtful River* has also been favorably recognized and respected. The book received the Wilbur S. Shepperson Book Award from Nevada Humanities. In 1994, the Library of Congress in Washington, DC, collected an entire set of 530 images from the Truckee River/Pyramid Lake Project for their permanent collection, and selected images were included in their publication *Eyes of the Nation*. Readers have sent appreciative letters and emails. Furthermore, the book has been assigned in college classes at the University of Nevada, Reno, where it stimulates exactly the kinds of complex debates about water that society needs to be having. Peter's attitude toward his work is philosophical:

> I'm a seed-maker. You throw out seeds, and who knows which one will ultimately become the oak that is memorialized for centuries, and which others become consumed by a passing coyote and tossed out in scat that fertilizes new seeds. Who knows where your role is? And who really cares, because you don't control it anymore at that point. So, I follow the evolutionary strategy of the cottonwood and put out as many seeds as I can.

■

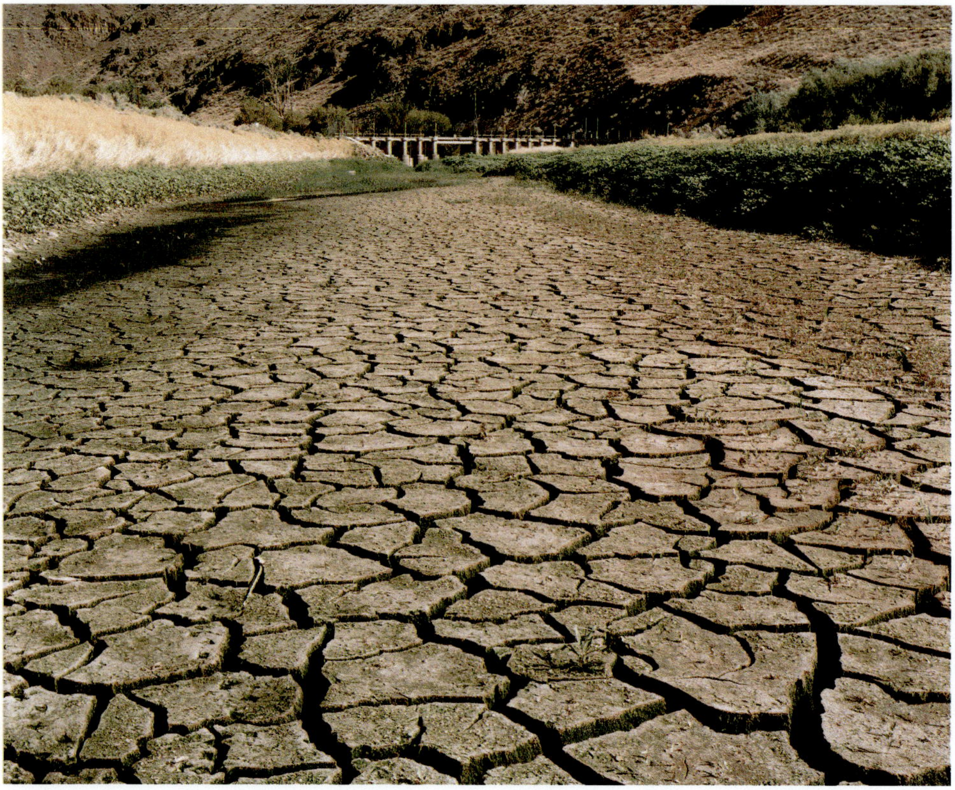

14.2 *Cracked mud and Derby Dam*, 1989.

This modest dam is where it all began—the first federally funded dam in the West, the inaugural project of the Newlands Reclamation Act, passed by the US Congress in 1902 to "reclaim" land going to waste due to lack of irrigation. Derby Dam diverts a portion of the Truckee River to Fallon, Nevada, an agricultural community made possible by the Newlands Act. The irony of this photograph, of course, is that there is no water in the irrigation canal due to an extended drought. What is a dam without water? All the engineering in the world is not going to irrigate crops if there is no water. The camera angle and close focus highlight the cracks in the dried mud, emphasizing aridity, while the diagonal lines formed by the canal edges point to the modest dam. The subdued color palette underscores the desert environment, hardly a verdant landscape despite the government's dreams of making the desert bloom. The fundamental issue raised by this photograph and by *A Doubtful River* is how do we live sustainably in an environment where water is scarce and not guaranteed?

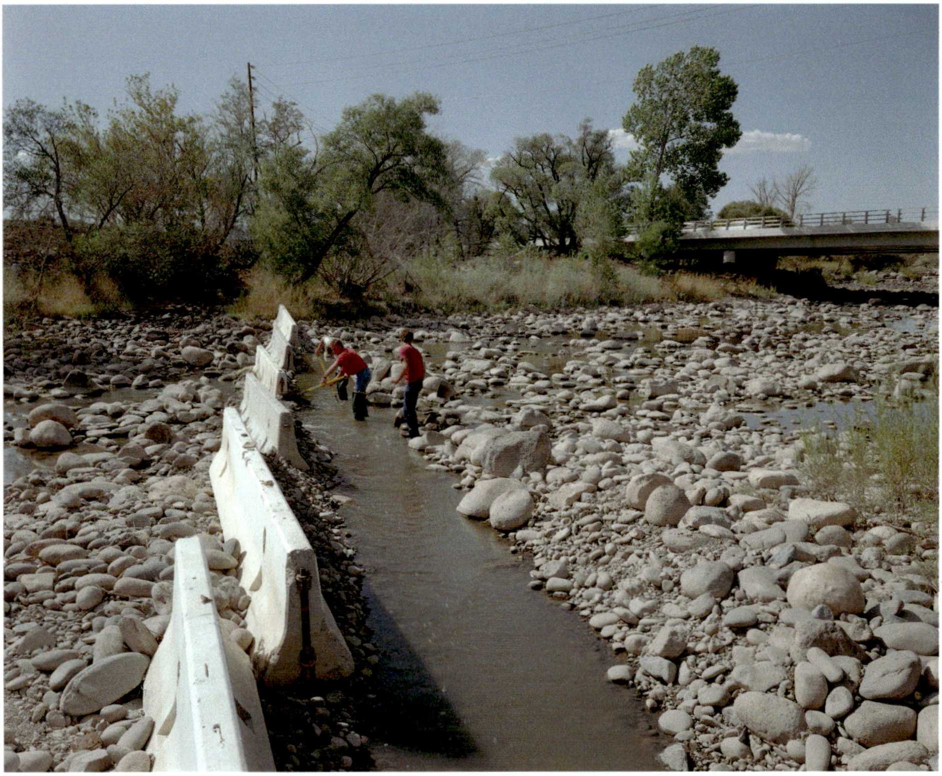

14.3 *In August of the drought summer of 1992, crews divert water from Truckee River to the Glendale Water Treatment Plant, Sparks*, 1992.

This photograph does not require much explanation. It is an image of aridity, bordering on crisis. Drivers whiz by on the McCarran bridge above at right without even noticing that the river is almost dry or that road-divider concrete barriers are being repurposed to capture as much water as possible for treatment and use by the cities of Reno and Sparks. The photograph was not staged. Peter Goin and Mary Webb were out doing fieldwork, saw the workers, clambered down to a good vantage point, and made the image. In *A Doubtful River*, this picture appears in black-and-white to create visual balance and equitable sequencing with Dawson's images. Here, we reprint the color version as an instance of a subliminal element that is true of many of Goin's photographs—the colors red, white, and blue. The workers just happened to be wearing red shirts and blue jeans; coincidence notwithstanding, the symbolic message of this photograph is that this scene, this situation, is part of American history and experience.

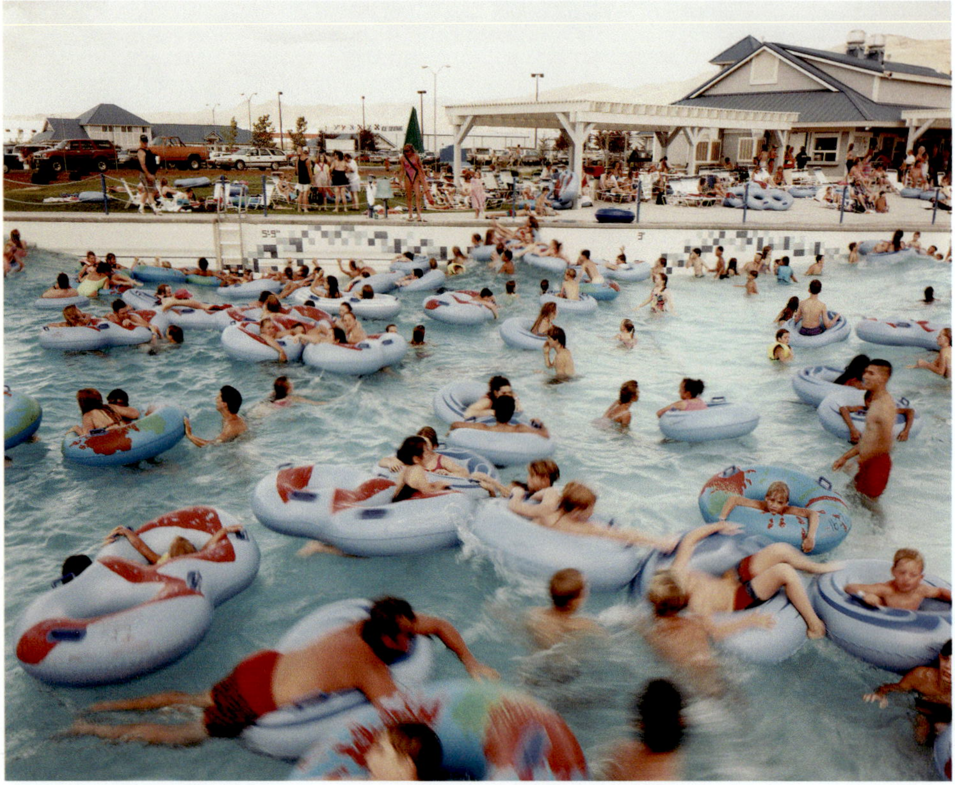

14.4 *Wild Island Family Adventure Park along the Truckee River, Sparks*, 1993.

In *A Doubtful River*, this photograph of the Wild Island Family Adventure Park in Sparks is printed on the page facing the previous image of crews diverting water, an ironic pairing of contrasting images. While one might reasonably assume that the heavy-handed juxtaposition of these images is intended to question whether water parks should be permitted during times of severe drought or to scold people for their cavalier disregard of the water shortage, Goin in conversation offers another angle. Peter's point is that we need water not just to survive but to refresh our spiritual sense of balance and health. Our relationship to water is more than just utilitarian, and water is more than just a commodity. The human spirit thirsts for water. We are water creatures. Even in times of drastically limited supply, we need water for recreation and renewal. Given that the water in this park recirculates and is recycled, it could even be argued that water parks are eco-friendly, allowing people to recreate close to home rather than driving to distant lakes.

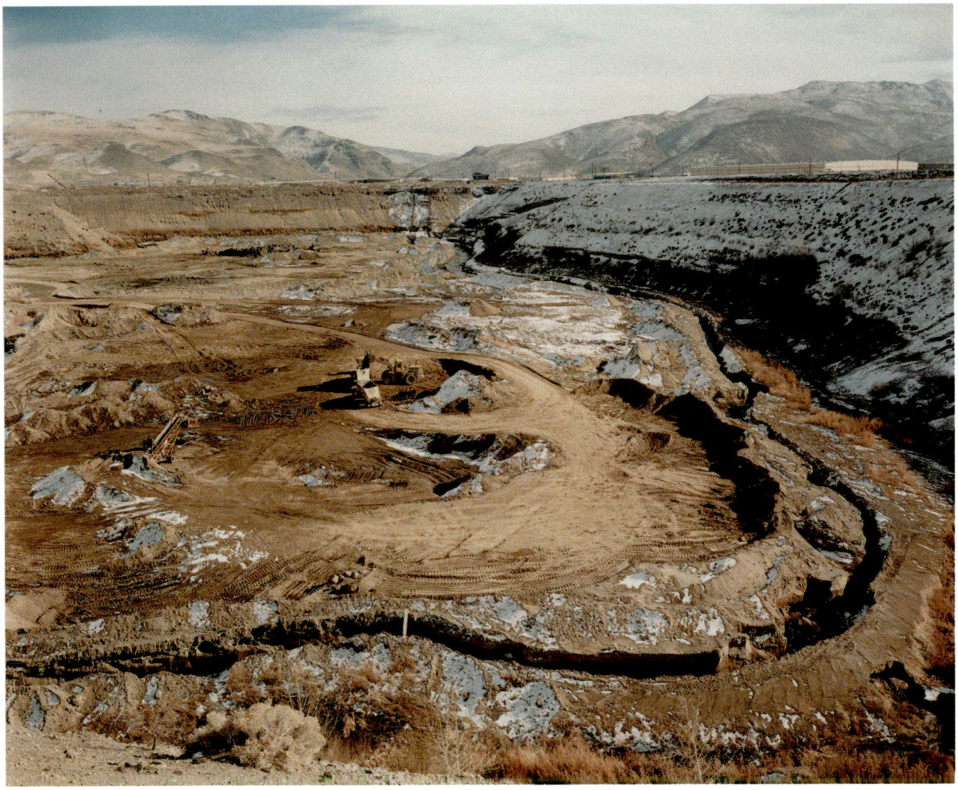

14.5 *Helms Gravel Pit, Sparks, just north of Interstate 80 and the Truckee River,* 1990.

If you visit this site today, you'll be pleasantly surprised. The former Helms Gravel Pit, which operated from 1968 to 1995, has been transformed into the Sparks Marina Park, where people fish, kayak, standup paddleboard, picnic, play volleyball, sunbathe on the beach, play Frisbee on the grass, and walk, jog, skate, and bicycle on a paved two-mile trail around a lake. Looking at the Sparks Marina in 2021, one may never guess that it was once a Superfund Site, where millions of gallons of leaked fuel from a nearby tank farm were pumped out and cleaned up. The site was declared free of petrochemicals in 1996, and millions of federal dollars were made available to the City of Sparks, which planned to create a park and a thirty-feet-deep pond. Fortuitously, the flood of January 1997 caused the Truckee River to overflow its banks and completely fill the Helms Pit to its current depth of one hundred feet. When Peter made this photograph, he had Ansel Adams's concept of previsualization in mind, thinking, "I'm going to photograph this landscape because I *know* it's going to change."

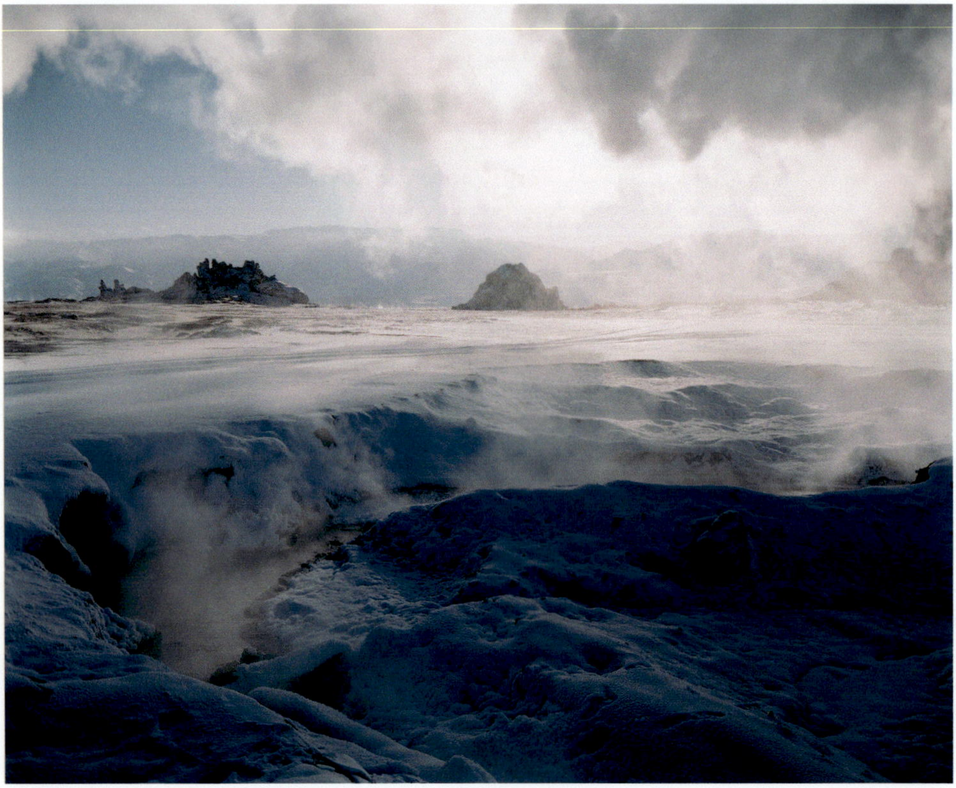

14.6 *Hot springs at the Needles, Pyramid Lake*, 1990.

Warmth in winter. Water in the desert. Snow on a normally brown landscape. Sunbright features on a cloudy day. This photograph embodies contrasts and speaks to the way that scarcity heightens value. Whereas many photographs in *A Doubtful River* depict the human-altered landscape or human presence, the place in this photograph seems almost primordial, untouched by time, forgotten by modernity and development. Goin does not often make scenic nature photographs, an aesthetic vein that in his view has been over-tapped. The charismatic landscapes that are the bread and butter of nature calendars have been photographed ad nauseam—Yosemite, the Rocky Mountains, Utah's Red Rock Country, the California coast, redwood forests. What distinguishes this scenic view is its unexpected location, an Indian reservation in northern Nevada. When people think of Nevada, scenes like this one rarely come to mind. This image asserts that Nevada is wild and beautiful, too. Perhaps more importantly to Goin, this winter view:

> is really about understanding that aridity is much more complicated than simply the absence of water. There are spiritual components and a sense of traditional aesthetics contained within arid landscapes. The Needles is a spiritual site for the Paiute people.

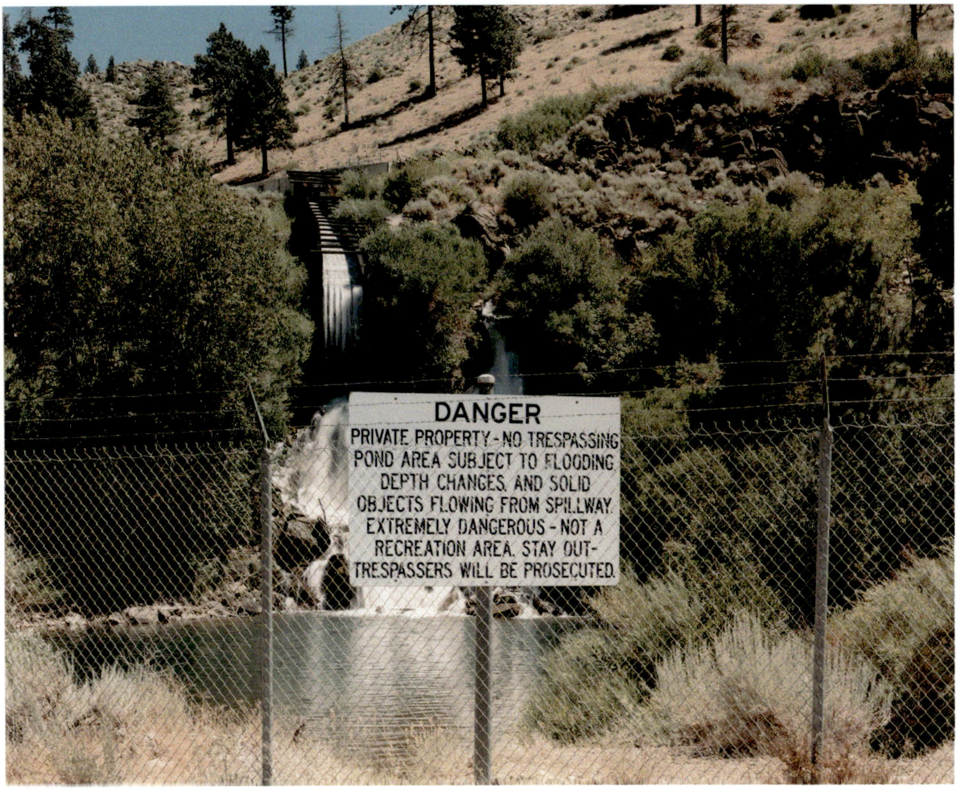

14.7 *"No Trespassing" danger sign along Truckee River at spillway west of Reno,* 1989.

Think of other ways that this waterfall and pool could have been photographed. One obvious alternative would have been to choose a vantage point—either looking through a gap in the fence or standing above or even beyond it—such that the fence does not show. In that case, the view would be of a beautiful pool along the Truckee River, albeit provocative in that the waterfall is artificial, resulting from the water release from a power generating station and flume system. As it is, Goin framed this image so that the fence and the aggressively centered sign deny visual access to the lyrical scene in the same way that the barbed wire-topped fence and "NO TRESPASSING" sign deny physical access. While *A Doubtful River* encourages residents of the Truckee Meadows to better know their river, because of the way the Truckee is managed, there will be barriers to experiencing it fully. As Goin's *Nuclear Landscapes* project attests and this image reinforces, Goin has a long-standing interest in the dialectic of the aesthetic and the anti-aesthetic, the sacred and the profane.

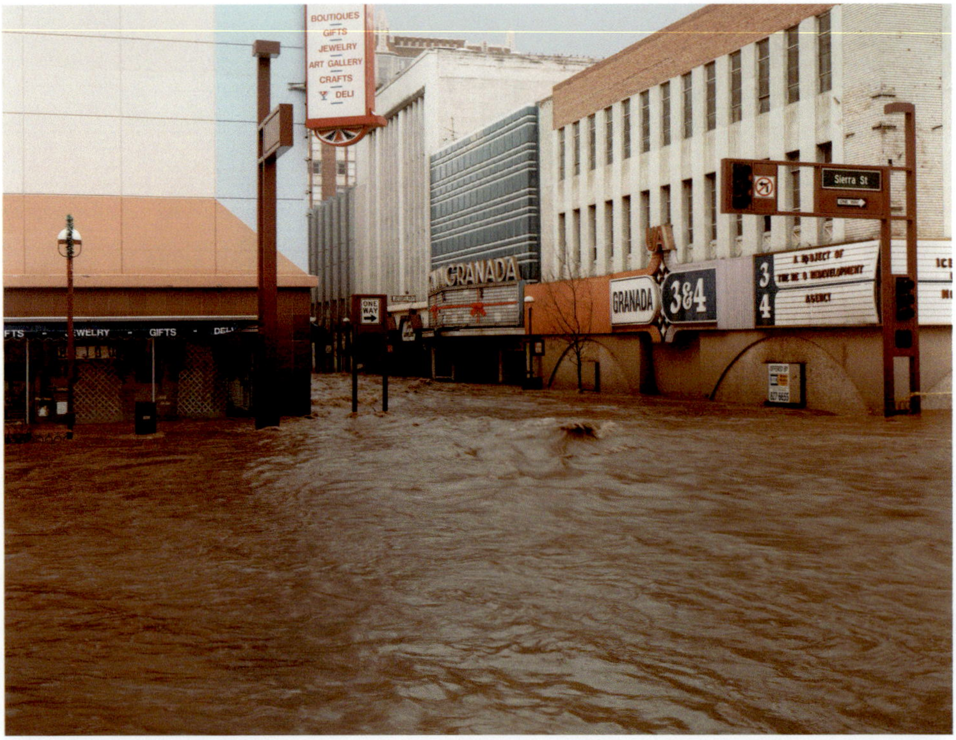

14.8 *January 1997 flood. View looking down First Street at Sierra Street, Reno,* 1997.

A "doubtful" river includes the notion of unpredictability. There can be a seven-year drought followed by an historic deluge. According to a study commissioned by the Nevada Department of Transportation, the New Year's flood of 1997—when warm rain hit a deep snowpack—put more water into the Upper Truckee River than would a one hundred-year flood. *A Doubtful River* was conceived and drafted in the context of drought and was in 1996 being prepared for publication. The authors delayed production in order to include images and narrative about the flood. In this image, note Goin's sense of humor as the "ONE WAY" sign appears impotent to direct the flow of the floodwaters, and the movie marquee positioned above the natural disaster spectacle reads, "A PROJECT OF THE RENO REDEVELOPMENT AGENCY," subtly signaling a cinematic reading of this historic event, when the show was outside rather than inside. To make this photograph, Goin, standing knee-deep in flood waters, framed the image such that there are no mountains or natural landmarks to orient the viewer, making it disconcertingly unclear where the water is coming from.

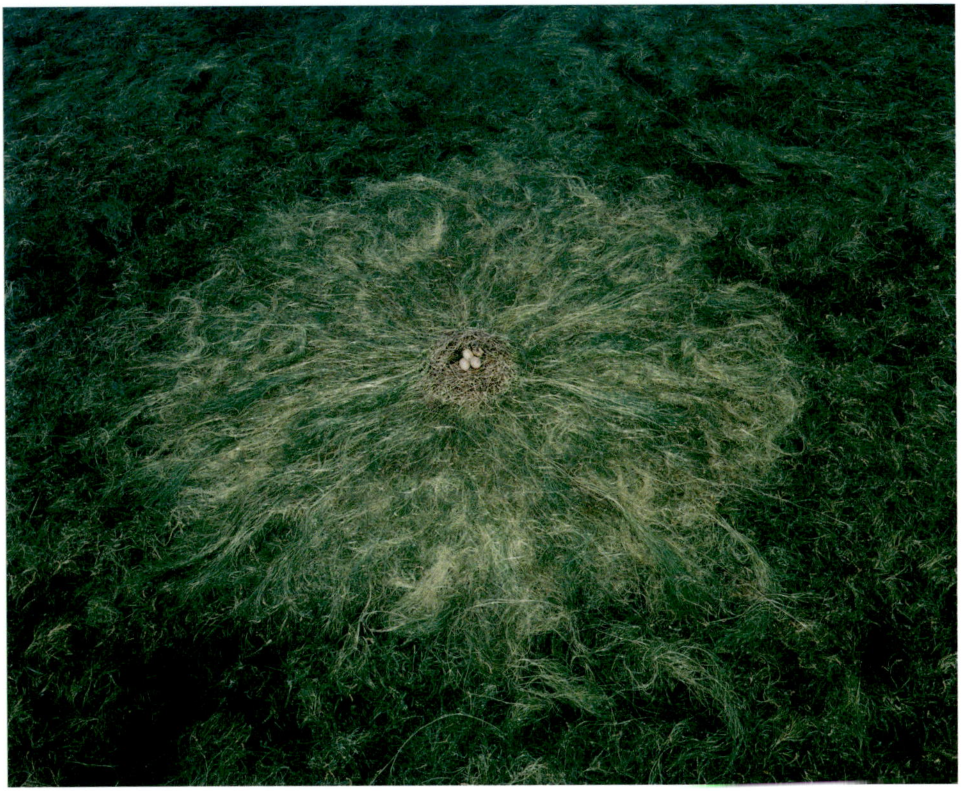

14.9 *Eared Grebe nest and spiral grass, Goose Lake, Stillwater National Wildlife Refuge*, 1995.

Eared grebes are water birds that build their nests upon floating vegetation. In this remarkable image, reminiscent of a terrestrial supernova, made from a swamp boat, Goin managed to photograph an eared grebe nest with eggs. While much of *A Doubtful River* is appropriately alarming, this concluding image is hopeful. "In wildness is the preservation of the world," wrote Henry David Thoreau. The Stillwater National Wildlife Refuge is one of *the* most important stopover points for migratory birds on the Pacific Flyway, attracting more than 290 species of birds. The original Stillwater marshes are a remnant of ancient Lake Lahontan, which formed twelve thousand years ago by melting glaciers. Agricultural water usage in the Lahontan Valley reduced the amount of water reaching the Stillwater marshes, restricting recharge. In 1948 the US Fish and Wildlife Service and the Nevada Fish and Game Commission took action to prevent the complete loss of the marshes, creating the Stillwater Wildlife Management Area. The US Fish and Wildlife Service purchases water to flood the refuge marshes. This story and this image offer hope and evidence that habitat restoration is possible, not just for humans but for birds and other creatures.

Note

1 To learn more about bioregionalism, see Berg's *The Biosphere and the Bioregion*, McGinnis's *Bioregionalism*, and Snyder's *The Practice of the Wild*.

Works cited

Audubon, John Woodhouse. *Audubon's Western Journal, 1849–1850: Being the MS. Record of a Trip from New York to Texas, and an Overland Journey through Mexico and Arizona to the Gold-Fields of California*. 1905. University of Arizona Press, 1984.

Berg, Peter. *The Biosphere and the Bioregion: Essential Writings of Peter Berg*. Edited by Cheryll Glotfelty and Eve Quesnel. Routledge, 2015.

Dawson, Robert, and Peter Goin. "Pyramid Lake Project." *Nevada Public Affairs Review*, vol. 1, no. 1, 1992, p. 3.

Dawson, Robert, Peter Goin, and Mary Webb. *A Doubtful River*. University of Nevada Press, 2000.

"Dividing Desert Waters." Special Issue Editors, Peter Goin, Robert Dawson, and Jill M. Winter. *Nevada Public Affairs Review*, no. 1, 1992, pp. 1–79.

Finkhouse, Joseph, and Mark Crawford, eds. *A River Too Far: The Past and Future of the Arid West*. Nevada Humanities Committee and University of Nevada Press, 1991.

Goin, Peter, ed. *Arid Waters: Photographs from the Water in the West Project*. Text by Ellen Manchester. University of Nevada Press, 1992.

Manchester, Ellen. "Arid Waters." *Arid Waters: Photographs from the Water in the West Project*, edited by Peter Goin. University of Nevada Press, 1992, pp. 1–34.

McGinnis, Michael Vincent, ed. *Bioregionalism*. Routledge, 1999.

Snyder, Gary. *The Practice of the Wild*. North Point Press, 1990.

Stegner, Wallace. "Thoughts in a Dry Land." *Where the Bluebird Sings to the Lemonade Springs: Living and Writing in the West*. 1992. Modern Library Classics, 2002, pp. 45–56.

15

CHANGING MINES

In the last chapter, "Arid Waters," we described how the landscaping around Peter Goin's house in Reno is a great example of *designing with nature* and *living in place* in an arid environment. His yard is attractive, water-conserving, food-producing, and wildlife-friendly. Let us begin this chapter by imagining that Peter Goin has invited us into his house for a dinner party with friends. When we arrive at the front door of *Casa Goin*, we are welcomed by a decorative sign that says, "Please remove your shoes and don't take a better pair when you leave." Inside, the earth-toned tile floor is tastefully arrayed with hand-woven rugs from around the world. Art pieces decorate the walls—an international mask collection, colorful tapestries and textiles, and framed photographs by Goin and other artists. There are bookshelves in most rooms, filled with art books, environmental history and philosophy, cultural criticism, and classic poetry, fiction, and nonfiction. In the living room, an enormous houseplant that reaches the ceiling thrives in a gigantic terra cotta pot as big as a two-person hot tub, next to a row of plate glass windows, creating a sense of nature coming indoors.

The kitchen and dining room of Peter's 1965 home have been tastefully remodeled to have an open floor plan and attractive granite countertops. Everything is clean and tidy. Guests are seated around the dining room table, sipping wine, enjoying a home-cooked meal, and engaging in random conversations. Suddenly Peter gets everyone's attention and says, "Okay, I have a question for you. What in this room, or in your life, does not involve, on some level, mining? Can you name *one* thing?" Somewhat startled, people play along and suggest things—walking, talking, having sex—but all of the suggestions can be quickly shot down. *Every* aspect of modern-day life is tied to the products of extractive industry. We couldn't live without it. Consider our clothes, computers, cellphones, houses and workplaces, recreation, transportation, infrastructure, and food. Mining or drilling will have been implicated at some stage in production or shipping. Even Peter's giant houseplant lives in a ceramic pot made from the earth, is watered—plumbing and faucets containing metal—from a pitcher made of metal or petroleum-based plastic, grows in potting soil procured with metal machines, mixed with metal mixers, sold in plastic bags, and transported from factory to market in metal-based trucks on asphalt roads, all made from the products of mining.

DOI: 10.4324/9781003212607-15

"If mining is so fundamental to our civilization, why do we dislike it so much?" Peter wonders aloud. Why is it so ostracized in academic debate? Why is it so pejoratively defined? These are the kinds of questions that trigger projects and that motivated *Changing Mines in America* by Peter Goin and C. Elizabeth Raymond. According to Raymond, in America, per capita annual consumption of newly mined materials is more than 47,000 pounds, not including precious metals (xvii). In addition to the precious metals gold and silver, the principal products of American mining up until the twentieth century were coal, iron, lead, copper, oil, and zinc, not to mention the more prosaic stone, limestone, sand, and gravel. With the growth of new technologies, Americans now extract a long list of additional materials from the earth's crust—borax, tungsten, titanium, vanadium, magnesium, perlite, bentonite, bauxite, chrysolite (asbestos), and uranium, to name a few (xxiv–xxvi). However, as the introduction to *Changing Mines in America* points out, "Although we zealously use the minerals that mines provide, we would rather not ponder the sacrificial landscapes that produce those minerals" (xxvi).

In *Nuclear Landscapes* Goin explored sacrificial landscapes and in *Humanature* he investigated mine sites. Visiting the sites and making the photographs for the *Humanature* chapter "The Mine" convinced him that there was a mother lode of provocative material across America just waiting to be visually mined. In addition, an experience that Peter had in Nevada left him with powerful emotions tied to mining, a memory that he could not bury. He and a friend were out exploring near Winnemucca Lake, east of Pyramid Lake. They came upon an abandoned mine and ventured inside to have a look. What they saw was spectacular. The walls of the tunnel were made of quartz. There were unopened beer bottles from the 1920s and even a lunchbox. They thought it would be a fun adventure to sleep in the mine near the door, but since they had already set up their tent elsewhere on the ridge, with an overview of the dry lake, they decided against it. The next morning, they were awakened to the sound of a bulldozer scraping and dozing earth and rocks, sealing the entrance to the mine to comply with safety regulations to protect the public from the alluring hazards of abandoned mines. Had the two slept in the mine tunnel, they would likely have been blocked inside. This story is vividly told by Goin in the prologue of *Changing Mines in America* as the origin point for the book: "Since that experience, Goin has explored the physical and perceptual realms of American mining, observing and pondering an industry that has long been censured as noxious and dangerous, at best a necessary evil" (ix).

Goin's collaboration with historian C. Elizabeth Raymond began in a coffee shop. Having enjoyed working together on *Stopping Time*, they met for coffee to stay in touch. After pleasantries, Elizabeth looked at Peter and asked, "So, what's our next project?" "Well, since you asked," Peter replied, and he floated the idea of exploring the visual landscapes and cultural meanings of American mining. Such a project would sustain Goin's strong interest in human-altered landscapes and the nature-culture dynamic, interpreted through an aesthetics informed by the New Topographics and its style of documentary realism. To Peter's surprise, Elizabeth found the idea immediately appealing, and they began planning a collaboration.

Goin and Raymond agreed that the book should be national in scope, investigating a limited number of representative sites, each one deeply researched, rather than offering a superficial overview of the industry. They wanted to explore "drastically altered places" that nevertheless "complicated a simple, linear narrative of environmental exploitation" (ix, x). They were interested not only in active mines but also in post-mining landscapes transitioning into new uses. And they wanted to select sites that were both "visually compelling to an artist and culturally intriguing to a historian" (ix–x). Ultimately, they chose eight sites in different parts of the country, from a nearly two hundred-year span of mining eras, representing several different metals and mining processes, and exemplifying different post-mining futures. They agreed that the book would be image-driven, with Goin's photographs providing primary evidence, not mere illustration. The prologue quotes visual sociologist Eric Margolis, who argues that "Visual knowledge cannot be reduced to verbal description—and vice versa. . . . The grammar and syntax of photographic images are not interchangeable with that of prose" (ix). In essence, *Changing Mines in America* explores mine sites from two irreducible but mutually informing perspectives—image and word.

The book's title pun—"changing mines"—signals two of the project's priorities. First, Goin and Raymond posit that mines *change* and that mining landscapes are *dynamic*, not static. Second, they urge Americans to change their *minds* about mining to reconcile our needs and uses and demands with what we expect from the earth. "Out of sight, out of mind" doesn't serve the environmentalist well, in this case. The authors agree that mining is essential, but they contend that mining landscapes are not entirely waste places. Industry has a responsibility, too, and toxic sites are simply not acceptable. It's complicated.

In the United States there is a disconnect between our behavior—voraciously consuming the products of mining—and our disdain for actual mines. There is another disconnect between our assumptions of mine sites as a dead-end and the reality of a dynamic, evolving landscape. There is a related disconnect between our view of mine sites as waste places and the reality of artful reclamation, rehabilitation, and restoration. And there is a disconnect between how we romanticize the historical mining past yet reject present-day mining. As Raymond notes:

> Mines are reviled places. In 1997, the industry ranked even lower than tobacco in American public opinion. . . . Many Americans, it is safe to say, share a negative view of mines as malevolent sites, dismissing these industrial sites as lost and desolate landscapes.
>
> *(181, ix)*

Environmentalists tend to view mining as a threat, routinely opposing plans for new or expanded mines. Granted, mining has created serious environmental problems, including habitat loss, polluted water, toxic waste, heavy metal pollution, radioactive contamination, acid rain from smelting, denuded landscapes, mountaintop removal, loss of topsoil, and erosion. But regulations have tightened since historical days, and as the authors point out, "adaptive recycling and reuse of mine sites is a constant feature of the American landscape" (181). Further, in terms of aesthetics, they concur with historian Richard Francaviglia that "dismissing mining landscapes simply as 'wasted' places is a form of visual elitism that prevents an understanding of what such landscapes mean, both to the people who inhabit them and to the society that benefits from them" (xxvii). For all these reasons, Goin and Raymond contend that "exacting attention to these sites should be given by a generally oblivious public. It is insufficient simply to dismiss them as waste places" (182).

Answering their own call for more sustained examination and analysis of American mining, *Changing Mines in America* takes on this unpopular topic. The eight case studies are arranged into four thematically and visually related pairs.

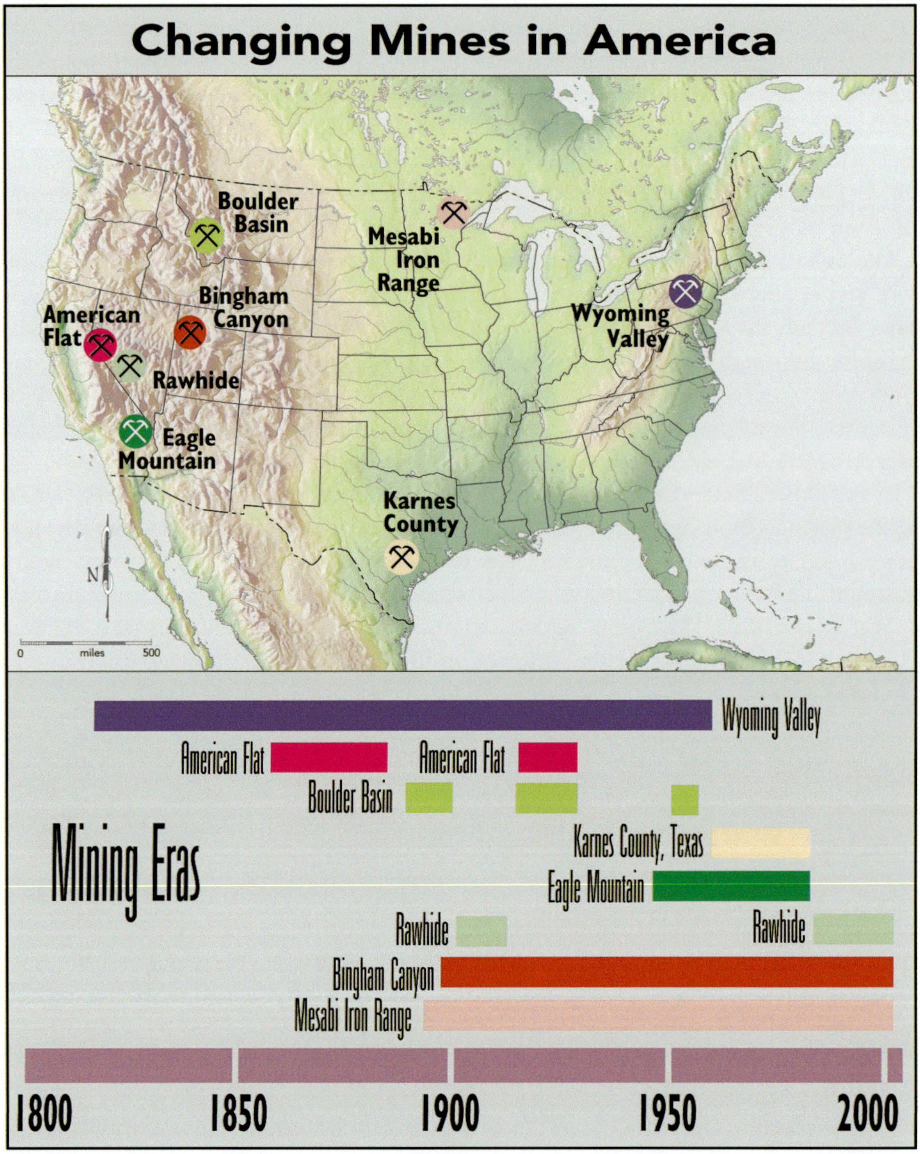

15.1 *"Changing Mines" map,* by Paul F. Starrs.

The first pair examines two historical mining districts, the Mesabi Iron Range in northeastern Minnesota and the anthracite coal field of Wyoming Valley in northeastern Pennsylvania. In contrast to the relatively isolated mines in the American West, these mines, which date back to the 1800s, are situated in densely populated communities that have been inhabited for generations. Closing the mines does not necessarily depopulate the area, but there is a question of what to do with the land in the wake of the mining era. In the Mesabi Range, parts of the post-mining landscape have been replanted, are returning to a post-nature landscape, and are popular recreation areas, while others have been converted to tourist attractions such as the Ironworld Discovery Center, teaching visitors about the mining history of the area. Historical photographs of open-pit mines contrast with Goin's contemporary views of reforested sites; new growth in some cases obscures the human-made terraces and tailings piles, which begin to look like natural features of the landscape. In Pennsylvania, coal mining and processing took place in close proximity to town, such that the long hills of waste rock—bench formations known as culms—are integrated into the townscape and even butt up against some homes. As Goin's photographs document, when these culms are reprocessed to extract additional ore, recontoured to create more usable land, or revegetated to replace scrub with grassland, the look of the area changes so drastically that it can be disorienting to residents, who value the familiarity and heritage of home terrain.

The second pair of mine site studies investigates two former uranium mines with strikingly different post-mining ambience. The first, in Karnes County in south Texas, focuses on a sprinkling of relatively small-scale open-pit mines on private land that were mined for a couple of decades in the mid-twentieth century and then environmentally remediated to such a degree that one can hardly tell from the grassy plains and fishing ponds that there were ever mines there at all. However, as Goin's photographs reveal, barbed wire fences backed up with signs warning, "NO TRESPASSING BY ORDER OF THE U.S. DEPARTMENT OF ENERGY" and "AUTHORIZED PERSONNEL ONLY" as well as odd, off-limits, low mounds of white rock, capping buried radioactive tailings, convey an ominous sense of hidden danger. This KEEP OUT landscape in Texas forms a sharp contrast with the "Come on in!" marketing of six former iron, gold, and uranium mines in the Boulder Basin region of southwest Montana. These old underground mines are yielding new profits in their reincarnation as "health mines," where visitors suffering from a range of ailments, including arthritis and bursitis, come to be healed. Apparently, repeated exposure to the low levels of radon in the mines cures some afflictions, or so sufferers believe. Goin's images show the small business trappings above ground and the electrically lit, hardrock mining tunnels belowground, where hopeful pilgrims sit on repurposed bus benches, taking the cure. This pair of examples underscores the divergent futures that post-mining landscapes can have, futures that are very much tied to perception, one person's poison being another's panacea.

The third pair of site studies is of currently or recently active open-pit mines. Such mines dynamite enormous amounts of earth in order to chemically recover tiny quantities of ore per ton of dislodged dirt. Both the machinery and the pits are supersized and speak to the geologic scale of modern-day mining operations. The Bingham Canyon copper mine in Utah, which has operated continuously for the past two hundred years, is so vast that it can be seen from space. Visual evidence of Bingham Canyon's evolving landscape is jaw-dropping. Black-and-white historical photographs depict a clutter of houses, buildings, streets, railroad tracks, and powerlines in what appears to be a bustling, lived-in mining community of permanent structures. But Goin's color images of the same places, roughly one hundred years later, provide graphic proof of impermanence. An unimaginably enormous terraced pit has seemingly swallowed up and obliterated the canyon and the town. There is no visual evidence other than historical photographs that a town ever existed at this site. Furthermore, the natural contours of the steep canyon that once defined the topography of the Bingham region are today completely gone, replaced by bare dirt, yawning pit with engineered, concentric rings of stair-step ledges diagonally sliced by switchback haul roads. Goin's images catch haul trucks in a blur of motion as well as a geyser-like plume of recently dynamited dirt, evidence that the landscape of the Bingham Canyon copper mine is still in motion. "The Rawhide," Nevada, chapter likewise employs a series of then-and-now images that counterpoint the turn-of-the-century mining boomtown with today's resculpted landscape. The recently closed open-pit goldmine at Rawhide, Nevada, for the last two decades of the twentieth century chewed up, sifted, and leached the mountainous terrain of the nineteenth-century gold mining town of Rawhide, a historical gold district kicked off by marketing hype. In contrast to Bingham Canyon, Rawhide was never much more than a tent city. Time-lapse images made by Goin in 1989, 1994, 1995, and 1998 depict the dismantling of Grutt Hill, which modern-day, open-pit mining methods have transformed from a hill to a pit with the revealingly oxymoronic name of Grutt Hill Pit. Open-pit mining operations such as Bingham Canyon and Rawhide cast humans as agents of the Anthropocene, reshaping the earth's surface in a blink of geological time.

The final pair of mine sites raises the question, "What is the proper post-mining use for a mine?" (xiii). The Eagle Mountain open-pit iron mine in southern California, which closed in 1983, is surrounded on three sides by Joshua Tree National Park, complicating a 1988 proposal to reappropriate the pit as a landfill for domestic trash transported from the Los Angeles metropolis. While using a waste pit as a garbage dump seems like a brilliant post-mining use, the proposal was tied up in court for many years and ultimately dropped in 2013, supplanted by a 2015 proposal for a hydroelectric plant, which would entail converting the former mining pits into two reservoirs. Images of Eagle Mountain include period photographs of thriving community life in the Kaiser Steel Corporation company town, followed by Goin's photographs of tidy rows of company-built, stucco houses—now mostly abandoned—and derelict mining structures, used as a movie set for *Terminator-2:3D*, as well as views of the vacant pit itself, being ever so slowly colonized by pioneering shrubs that can tolerate the Mojave Desert's heat and aridity.

The final site in *Changing Mines in America* is close to home for Raymond and Goin—American Flat, a historic, silver-processing mill site just outside Virginia City, Nevada, and the famed Comstock Lode. When Goin photographed the site in the 1980s and 1990s, American Flat consisted of silent hulks of concrete walls and empty industrial buildings whose windows had long ago been blown out and their remnants cleared away. This once clamorous, working landscape became an unsupervised "outlaw" playground for teenagers and grown adults playing paintball, partying, and spray-painting graffiti. As Goin's photographs bring to life, the once gray concrete walls of the silverworks later exploded in color in a postmodern palimpsest of words, initials, and designs—hearts, nudes, a hand, an eye, portraits, arrows, and blobs. From one angle, the ruins suggest an industrial-era, decorated Parthenon. However, after an ATV rider was killed at the site in 1996 while riding up a set of stairs, in the name of public safety the Bureau of Land Management decided to put an end to decades of vernacular recreation and artworks, closing the site in 1997. The chapter's final image shows a spray-painted fresco and illicitly inviting labyrinth of corners and columns as the backdrop to a government-orange BLM stake planted in front, standing like a visual non sequitur, and warning, "Do Not Enter. American Flat Millsite. Area Closed to all entry. Violations Punishable. No Shooting other than the lawful pursuit of birds and mammals." The post-post-mining future of American Flat is left to the reader's imagination.

Ultimately, *Changing Mines in America* is a radical work, in that it reconceptualizes mining landscapes in positive terms. Rather than dismissing them as dead zones that, as blight, should be forgotten, ignored, or despised, *Changing Mines* recontextualizes post-mining landscapes not for romantic reasons, but for their real possibilities as sites that might still be productive. As Peter puts it in an interview, "I think there is something human about the return of the prodigal idea. Prodigal idealism. I think that I am looking for the return of prodigal idealism." Considered individually, as *Changing Mines* does, mines are places of tremendous complexity and surprising variety, and "mining landscapes can be places of beauty and curiosity, as well as apocalypse" (xiv).

Goin is aware that *Changing Mines*' interpretation of mine sites as evolving landscapes with the potential to be rejuvenated will be met with suspicion by environmentalists. The fact that Barrick Goldstrike Mines, Inc. helped to fund the book's publication will also raise skeptical eyebrows in the environmental community. Peter counters that he is being "positively subversive." In challenging environmentalist dogma, he intends not to tear the movement down but to make it smarter. He laments that the environmental movement has not yet come to terms with extractive industries. The environmentalist response tends to be hypocritical, along the lines of:

> Mining is gross. This shouldn't be happening. They're ruining a landscape. They're the evil doers. But excuse me while I get in my car and drive off to purchase more coffee makers, so I can buy my shovel to dig my garden and plant my organic garden.

As Goin confides in conversation:

> I realize that by publishing *Changing Mines* I become a potential, and I hope superficial, enemy of environmentalism. However, I believe that we have to rethink our collective response to extractive industry. Not just because we're so entirely dependent upon it, but because these landscapes are transformed, and they are transformed in a variety of ways that serve some human purpose and facilitate greater biotic identities.

As Goin has said on more than one occasion, we need to have hope. Too often, environmental rhetoric settles for despair. *Changing Mines*' takeaway manifesto is that:

> mining should be appraised as simply a stage in landscape evolution, not its culmination. . . . At some point mining ends, and the place where the ore was still remains. In a society that is heavily dependent on the products of mining, the intelligent reinvention and reintegration of such sites is imperative.
>
> *(182)*

In *Changing Mines*, Goin's images and Raymond's text run parallel to each other, representing simpatico treatments of mining and selected mine sites. The purpose of the text in the present retrospective of Goin's art is different. Here, text *serves* image. Rather than offering Glotfelty's view of mining, the text in this chapter interprets Goin's photographs and philosophy, revealing the mind behind the lens. While Peter Goin looks at mining, this chapter looks at Peter Goin. The narrative captions in the gallery section at the end of this chapter analyze selected images from *Changing Mines*. In the interest of space, captions in this book are typically limited to one paragraph. However, one paragraph is just the tip of the iceberg of what *could* be said about each image, for Goin's images are multilayered. To tease out those layers in greater detail via a single example, we begin our analysis of images from *Changing Mines* with a fuller discussion of the book cover's image, which depicts two sightseers on a viewing platform overlooking the Old Town Mine in Virginia, Minnesota.

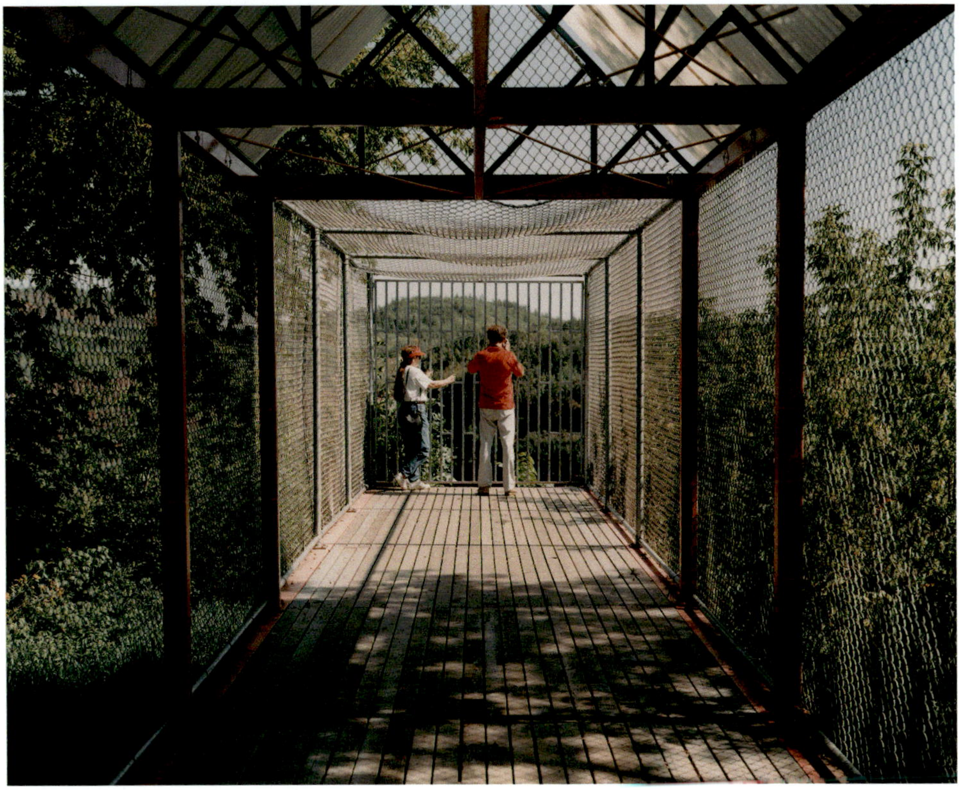

15.2 *The Old Town Mine view in Virginia, Minnesota*, 1996.

In this image, a woman and man stand at the end of an extended cage whose sides and ceiling are made of chain-link fence and whose back wall is constructed of vertical metal bars. The woman grasps two of the bars, while the man takes a photograph through a gap between the bars. Outside the enclosure, greenery is abundant, further enclosing the cage, fancifully suggesting that the cage is protecting the visitors from an encroaching nature that threatens to engulf them. Although the caption in *Changing Mines* informs the reader that the couple is overlooking a massive pit of the retired-from-service Old Town Mine, there is actually no pit visible in the image and the couple appears to be looking at distant forested hills. The color green engulfs the scene, a striking contrast to the sunlit chain-link cage, which stands out like a white square against the background foliage, green also contrasting with the bright red, white, and blue of the couple's clothes.

The first impression of someone picking up this book might be, "I thought this book was about mining. How is this photograph about mining?" Score! That is precisely the question that Goin intends this image to provoke. Just as the couple is viewing a mine site, this image invites us to *view* mining—in a new light. What is striking about the Old Town Mine is that it has been developed into a tourist attraction. One visitor's review on Trip Advisor speaks to the success of the makeover of an underground mine in the Mesabi range:

> Fantastic Experience in a Real Mine. I saw this [mine] as part of a tour group and it was a highlight of the trip. It was a real working mine, then it was dormant for many years, now brought back to life as a tour.

The open-pit landscape of the Old Town Mine is being reclaimed by vegetation as the mine site itself melts back into nature in the ongoing evolution of the landscape. In fact, if the caption and context of this image did not signal that this viewing platform overlooks a mine, one might assume that this overlook was built for visitors to safely view a natural wonder, a Minnesota country cousin to the thirty million dollar glass-floored Skywalk that extends over the Grand Canyon National Park.

The boardwalk and viewing platform pictured in this image, so characteristic of national parks, suggests another subliminal theme encoded in this image, the "landscape zoo," graphically represented here by the iron bars of the cage that structurally separate the couple from the landscape. "Landscape zoo" is a phrase Goin coined to describe the management style of national parks, which he regards as "a controlled magician's act." Whereas most tourists who flock to national parks imagine that they are experiencing pristine nature, Peter understands that these popular attractions are highly managed. National parks "zooify" a wilderness via crowd control, funneling visitors to established scenic drives, designated paths, and fenced lookouts, cutting down trees to open up vistas, and scripting interpretive signs to control meaning. In overt and subtle ways, national parks separate people from the very landscape they come to engage with. Peter explains in conversation that this cover photograph strategically:

> frames the figures and isolates them from the landscape, which of course is functionally happening in the view to begin with. You as the reader don't get to see the view, so the view is divided, which is a subliminal clue of what the cage is doing to the visitors. Their view is controlled. It's managed. There will be things they want to see that they can't from that restricted vantage point.

The nation-state's drive for control is not limited to nature. As Peter continues in conversation, "landscape zoos are very much a reflection of society's desire for order and management of that which is wild, which reflects the desire of these social systems to manage human spirit and human desire." Read psychoanalytically, this image contains quasi erotic references to sexual desire. Historians have documented a pattern in American culture wherein the landscape has been imagined as female—virgin land, "fresh green breast of the New World." In Goin's image, riffing on this historical pattern, the viewer is drawn into a man-made vessel that projects into a green landscape of fertile nature.

While this image thus symbolically taps subterranean currents of desire and control, it references art history on another level as well. Nineteenth-century American landscape painters would often insert distant human figures into their paintings to convey a sense of the vast scale and grandeur of the American land, nationalistically evoking a sense of the sublime. With the growth of American industry, the size and power of technological works and human-altered landscapes likewise arouse a sense of awe and wonder. In his book *American Technological Sublime*, scholar David Nye introduces Perry Miller's notion of the "technological sublime" in his account of how conceptions of the sublime evolve through history, originating in natural splendor and transferring to spectacles of human power, reaching a crescendo in an atomic bomb blast. Goin's similar point here is that modern mining occurs on a scale so vast that it evokes the sublime. The fact that both historical and recently closed mines sometimes become tourist attractions plays into another concept, "technological nostalgia"—or *technostalgia*—explored in Richard V. Francaviglia's *Hard Places: Reading the Landscape of America's Historic Mining Districts*, which investigates the "repelling beauty" of mining landscapes. Goin does extensive research for his projects, and his images reflect a well-read intellect, visually developing concepts that researchers in other disciplines have expressed verbally.

Conceptually, then, as is true of much of Goin's photography, this cover image is complex and multivalenced. There are also visual strategies at work in the composition of the image, such that form conveys meaning, too, and the image succeeds aesthetically. Starting with color, although it was fortuitous that the couple happened to be wearing red, white, and blue, it is no accident that those colors appear in this image. Red, white, and blue appear throughout Goin's oeuvre, signifying his examination of American culture. Beyond color, at least three of Goin's signature design principles are at play in this image. First, note the triangles in the top margin of the image, triangles being a favorite visual element in Goin's work. Second, observe the perspectival lines formed by the boards of the boardwalk and the cage edges, lines of force that pull the viewer's attention into image. Finally, know that Goin is *always* aware of the horizon, and he very deliberately composes each image to place the horizon where he wants it. Here the horizon extends across the page to unite the image, but it is most prominent inside the barred square at the end of the viewing platform, where it creates a framed landscape picture nested within the larger landscape picture. Formally, then, the horizon is caged, suggesting the idea of the landscape zoo. The nested squares in this photograph also suggest visual embeddedness, as readers see the couple seeing and Peter photographs the man photographing. Formally, the receding layers of the view allude to the mirror reflection motif in Alfred Hitchcock films such as *Psycho,* a design signifying psychological layering of character and complexity of motives. Finally, this photograph is an optical illusion. Looked at one way, the square cage recedes away from the viewer, but if you relax your mind, the cage can be made to pop out toward you, creating a kinetic visual experience where the view switches, enacting the process of changing minds.

The point we are making in unpacking the cover image of *Changing Mines in America* so extensively is that Goin's images are portals to much larger conversations. Unfortunately, most of us merely glance at a photograph and move on, taking it to be a snapshot akin to a machine-made photocopy of a view, absent interpretation. We overlook the fact that trained photographers such as Goin are artists and thinkers who use their compositions to convey ideas and provoke contemplation. We invite you to see what you can *see* in the rest of the images selected for this chapter.

■

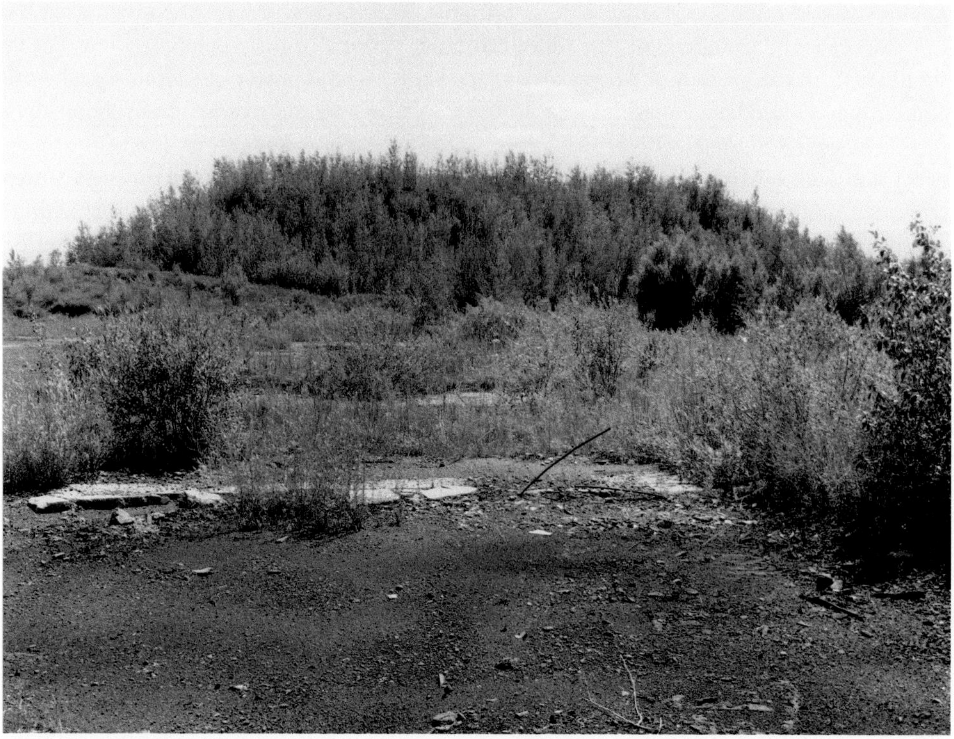

15.3 *Site between Hibbing and Chisholm, Minnesota,* 1978.

These two photographs of a revegetated mine tailings hill in the Mesabi Range were made in 1978 and 1996, respectively. When he made the black-and-white image in 1978, Goin had just returned from an eight-month trip to Central America, where he was struck by how plants were growing over, and in some cases obliterating ancient burial mounds and Mayan ruins. After that adventure, he returned to Iowa to teach, and on an excursion to Minnesota he noticed the reforestation of American mining "ruins" in the Mesabi. Eighteen years later, when he was working on *Changing Mines in America*, Goin returned to the

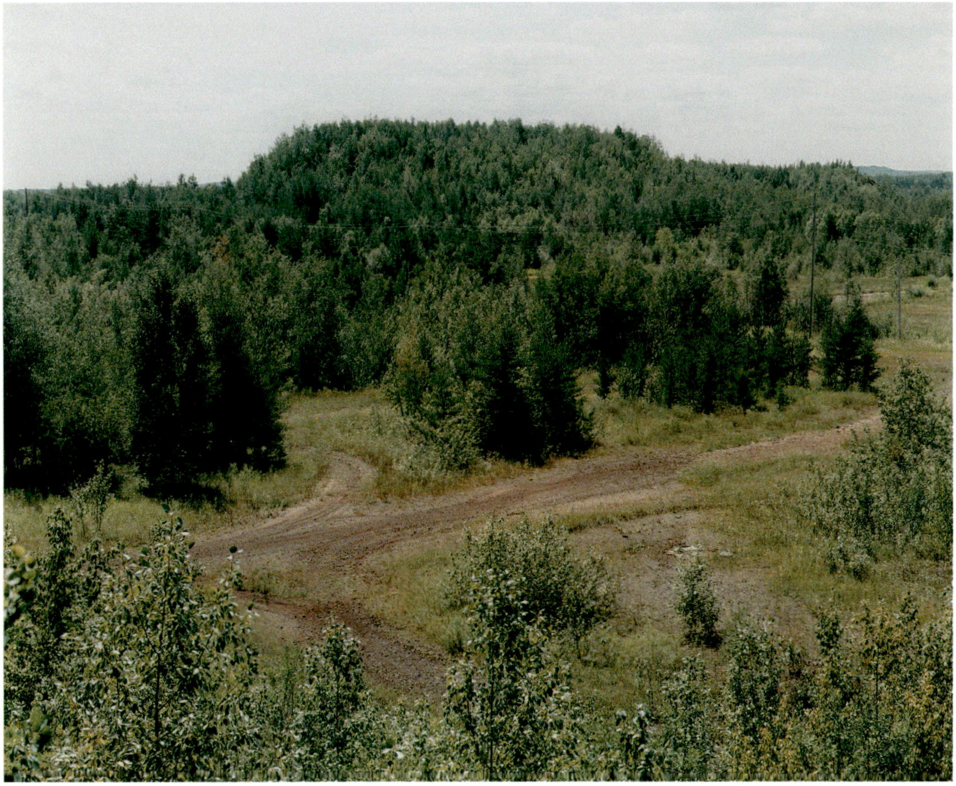

15.4 *Site between Hibbing and Chisholm, Minnesota*, 1996.

Mesabi and made a color photograph at nearly the same site to document further change in the landscape. The multiple dirt roads in the 1996 image are evidence that this recycled landscape has been discovered by recreationalists, who use the roads for snowmobiling, hunting, berry-picking, dirt-biking, and access to power line roads. In recorded history, this Mesabi landscape has evolved from a flat and relatively featureless forest to an industrial landscape—with clear-cut forest, open pits, and manmade hills of waste material—to a re-grown forest of varied topography, laced with roads.

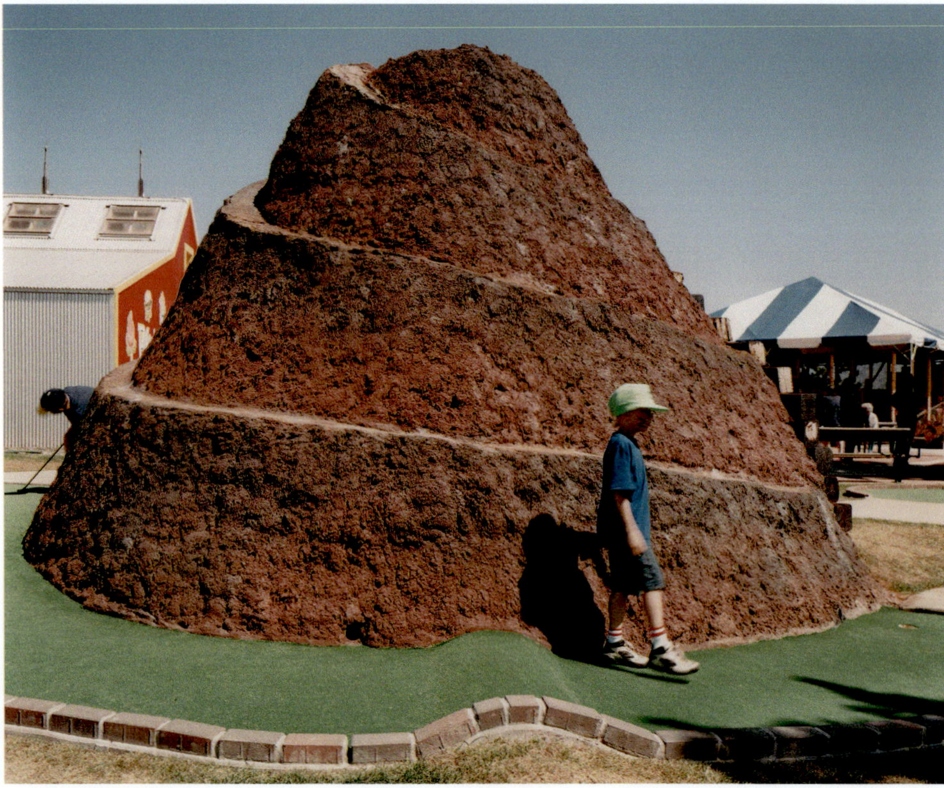

15.5 *Pellet Peter miniature golf course, Ironworld Discovery Center, Minnesota*, 1996.

How interesting that we are now creating simulacra of mining landscapes. Frequently condemned when active, mining can be enjoyed from a safe temporal distance. The Ironworld Discovery Center is a theme park that banks on entertainment as a way to profit from the Mesabi Range's mining heritage even as the actual practice of mining is on the wane there. Ironworld is an example of technology being recontextualized in our imagination, becoming tinged by "technostalgia." Once the threats posed by mining are past, we tend to create a noble story from that history. But we are not content just to look at mining exhibits; we want to interact. Taking a cue from Disneyland, Ironworld boasts mining-themed rides and games, Disney's Matterhorn here appearing as a mountain of mine tailings set in a miniature golf course. This photograph mimics the evolution of the landscape from industrial to postmodern and of human endeavor from blue collar to blue shorts, signifying leisure. Note Goin's signature themes of simulation and humanature, his hallmark design features of the triangle and the colors red, white, and blue, and his characteristic sense of humor.

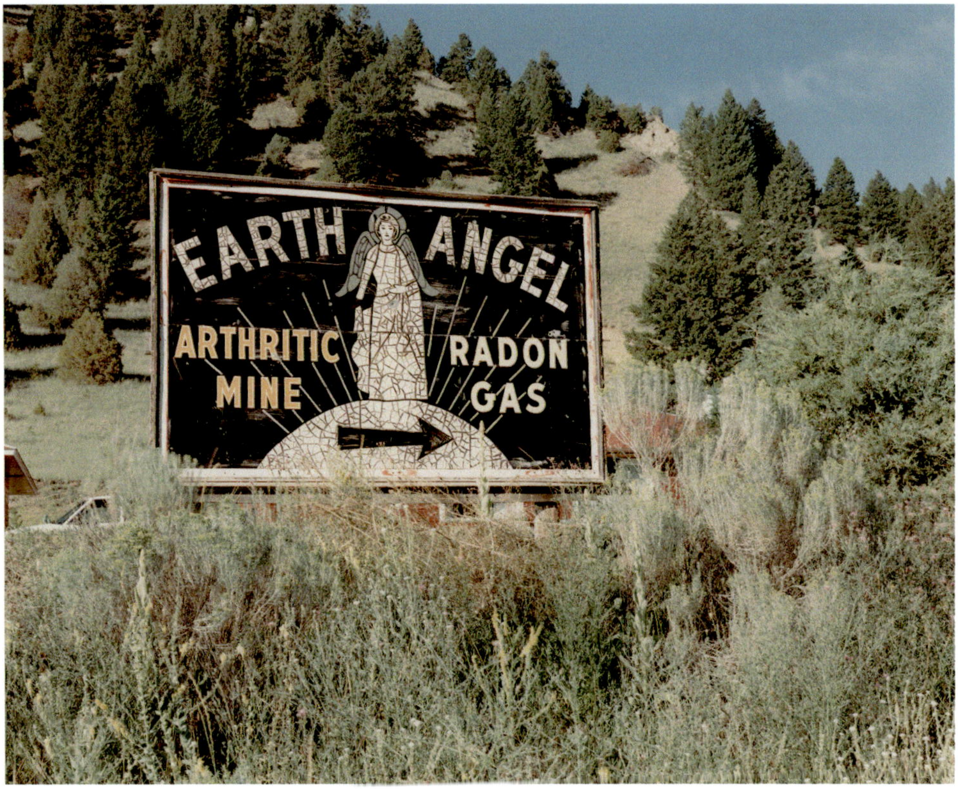

15.6 *Earth Angel Health Mine, Boulder Basin, Montana*, 1998.

Like the Ironworld miniature golf course image, this photograph of an aging highway sign for the Earth Angel Arthritic Mine may elicit a chuckle. An arthritic mine advertising radon gas? Really? This mine began its life in 1929 as a gold and uranium mine. It became a health mine in 1953, touting a faith-healing tenet: "It's All in the Mine." Radon gas, which forms by the natural radioactive decay of uranium, has been identified by the Surgeon General as the second leading cause of lung cancer in the United States. To protect ourselves from this colorless, odorless, tasteless, and chemically inert gas, we install radon detectors in our houses. However, some people believe that in proper doses, radon can cure ailments such as arthritis, bursitis, and more, possibly by stimulating the endocrine system. A half-dozen mines in Boulder Basin, Montana, are now marketing radon instead of mining ore. When Goin went to the area photograph, he brought a friend who suffered from arthritis in her hands. For a week, three times a day, she alternated one hour in a health mine and three hours out. Her arthritis vanished and has never come back. True story.

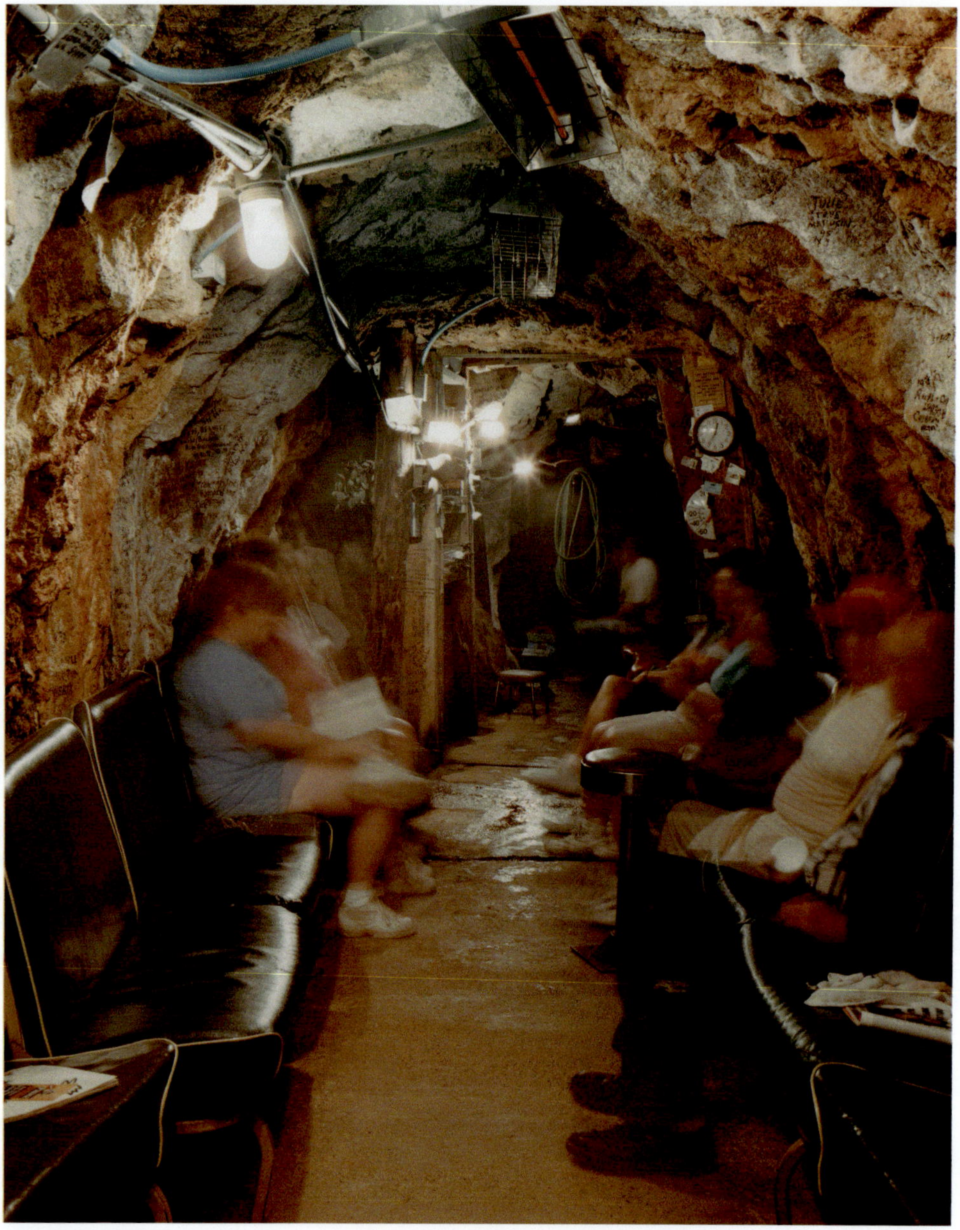

15.7 *Merry Widow Health Mine, Boulder Basin, Montana*, 1998.

The Merry Widow Health Mine's slogan is "Health is Wealth." Here we see what a health mine looks like: A crude tunnel in the rock, lined by cushioned benches, electrically lit, with a damp poured-concrete floor, and inscriptions written on the rock and wooden beams like an underground version of a mountaintop trail register, a record of who has come before. Goin chose a slow shutter speed due to low light conditions and also so that patrons would be blurred and unrecognizable. Some people who visit health mines do not want their pictures taken, particularly the Amish, whose practices forbid photographs in which individuals are recognizable. This photograph conveys the community feel of this mine, wherein patrons make friends and enjoy leisurely conversation, a markedly different demographic and purpose from the original hardrock miners, who labored to carve out these tunnels and extract ore, chasing wealth even as the working conditions may have compromised their health. Look carefully at the figures in this mine. The man in the white shirt seated by himself at the end of the tunnel is Peter Goin, à la Hitchcock, making a cameo appearance in his own film.

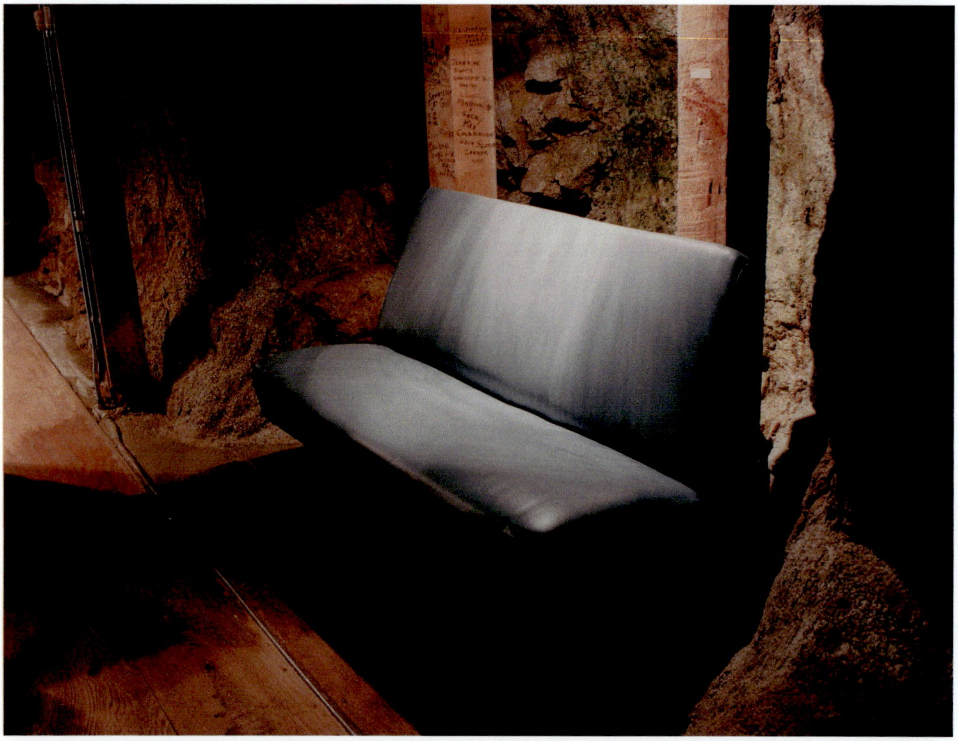

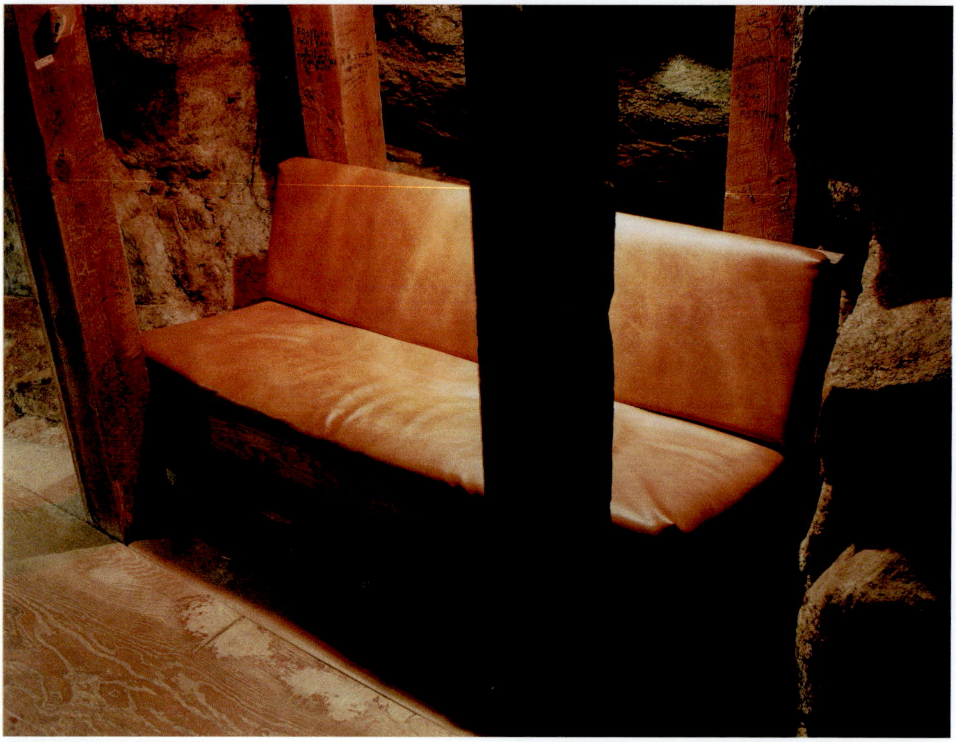

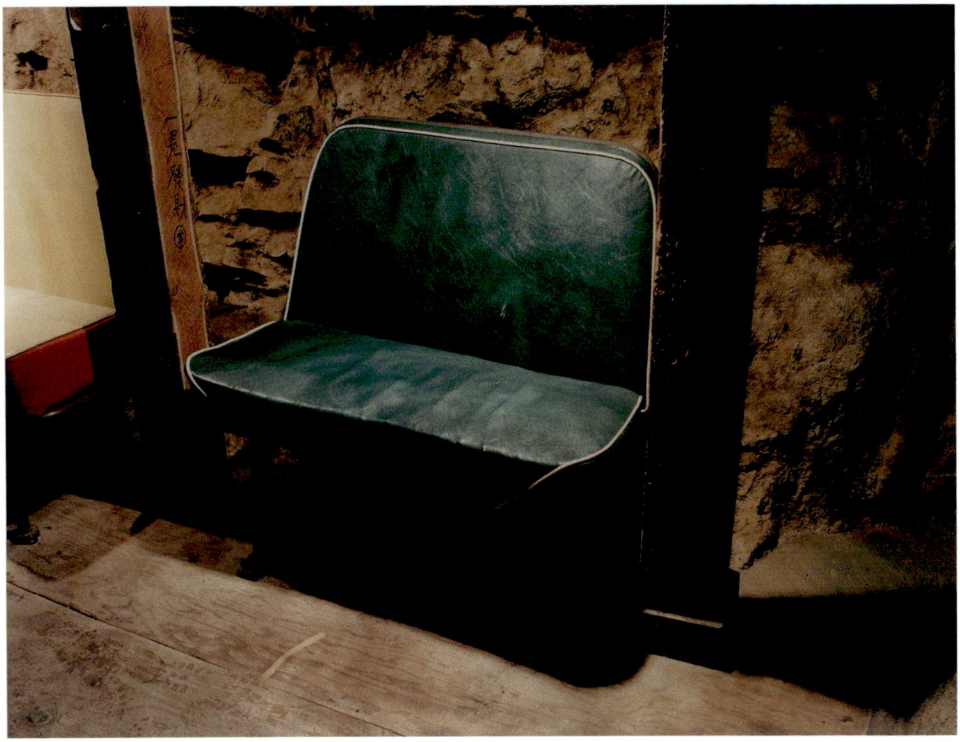

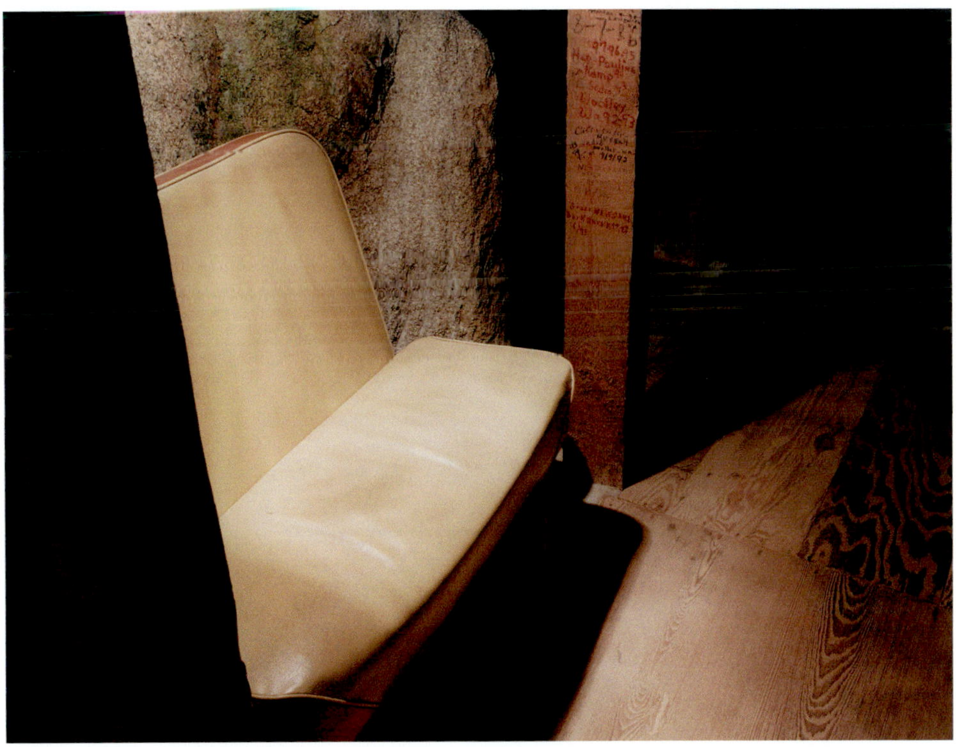

15.8–15.11 *Bus benches at the Free Enterprise Health Mine, Boulder Basin, Montana*, 1998.

Repurposed bus and office benches provide comfortable seating for health-seekers in repurposed mines, striking a *Changing Mines* theme of recycling and reuse. Whereas people in the Merry Widow Mine sat together and chatted, in some mines patrons prefer privacy and contemplation. The Free Enterprise Mine, pictured here, has a warm-looking plywood floor and provides some singleton benches, reading tables, magazine racks, and recliners for solitary sitting and reading. The wooden beams have become texts in themselves in an ongoing vernacular narrative that subsequent visitors may choose to coauthor by leaving their own words. The Chinese characters next to the green bench suggest an international clientele in this unusual, underground destination. In this image grid of four benches, Goin joins a tradition in photography of creating typologies of similar objects to encourage objective comparison and attention to form. Bernd and Hilla Becher, for example, in the 1950s began documenting dilapidated industrial architecture such as blast furnaces, water towers, and foundries, pursuing a type of "industrial archaeology." In the bench typology, Goin may be exploring the idea of *post-industrial* archaeology.

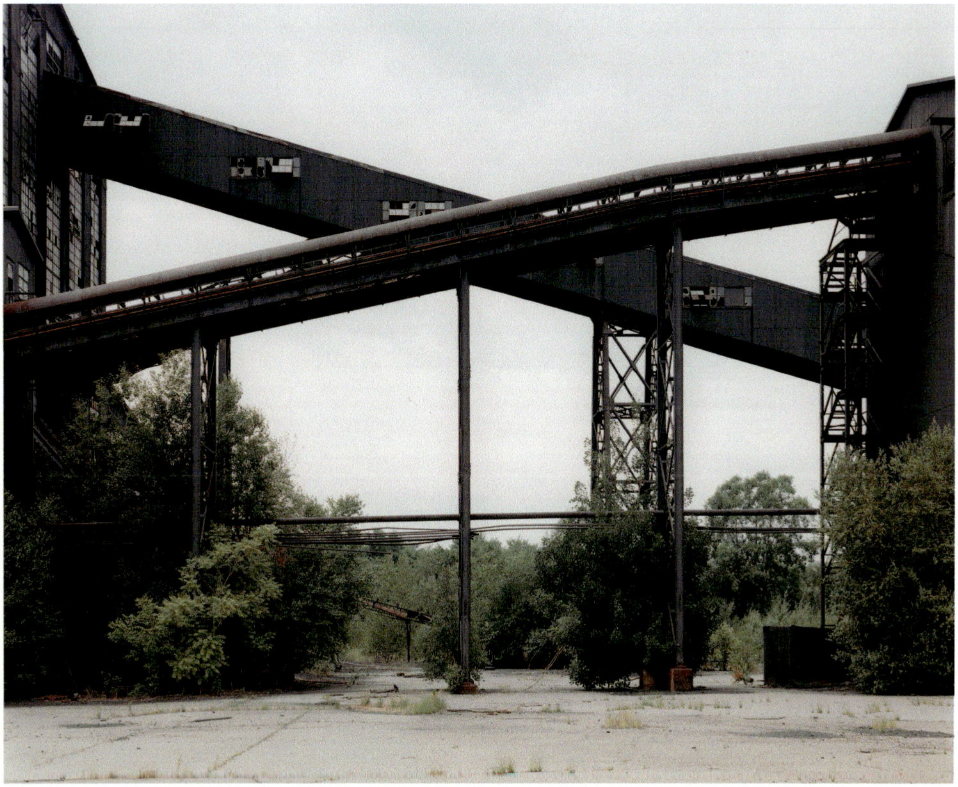

15.12 *Huber or Blue Coal Colliery, Lower Ashley, Pennsylvania*, 1997.

Goin's photograph of this defunct coal breaker references the work of American artist Charles Sheeler (1883–1965), whose abstract paintings of American industrial architecture and skyscrapers celebrate American power, but are also strangely absent of people, mechanism having replaced humanism. In 1927–1928 Sheeler was hired to document Ford Motor Company's River Rouge plant outside Detroit, Michigan, and one of Sheeler's photographs is a prototype for Goin's image of the defunct colliery in Pennsylvania. Goin's image shares Sheeler's ambivalence about America's ideals of power and productivity as well as Sheeler's blend of abstraction and realism. Sheeler documented productive factories, whereas Goin, at the end of the twentieth century, is figuratively rephotographing these same kinds of sites, now in ruins. In the era of film photography, photographers would draw or scratch a big X across negatives that they did not want to be printed, thereby cancelling them out. The black X made by the conveyors in this abandoned factory metaphorically signals that this post-mining landscape has been condemned and cancelled out. However, the green growth beginning to recolonize the area opens the imaginative possibility that this abandoned plot may one day support a new purpose.

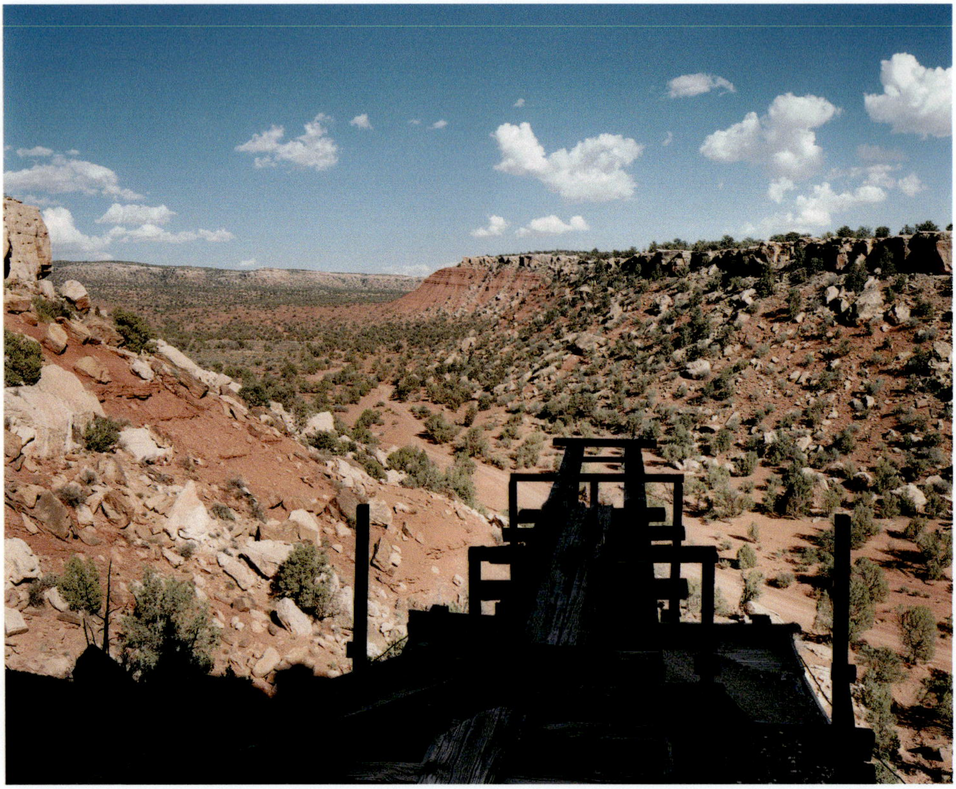

15.13 *Abandoned mine site, Grand Staircase Escalante National Monument, Utah*, 2000.

Had Goin chosen a different vantage point, he could have made an image of the pristine landscape of Utah's canyon country. But he included in the foreground a wooden structure from an abandoned mine site, composing the image so that the old mill is darkened by shadows that sharply contrast with the bright and colorful desert beyond, just as mining itself exists in the shadows of the contemporary imagination. When one explores the outback of the arid American West, it is common to come across old mine sites, which are a kind of structural analog to tumbleweeds, both being invasive yet iconic of the wild West. In this image, the dark mill arm rather sinisterly—and one might say sexually—thrusts into the landscape. Viewed through the rosy lens of technostalgia, abandoned mine sites evoke the Old West and folksy tales of prospectors, burros, and gold strikes. However, this image suggests that mining represents the incursion of humans, history, and change into a land before time. Because we the viewer look down the mill arm, we become the intruder, perhaps sighting down a rifle of industry, aiming to blast out the riches hiding beneath the surface of this innocent land.

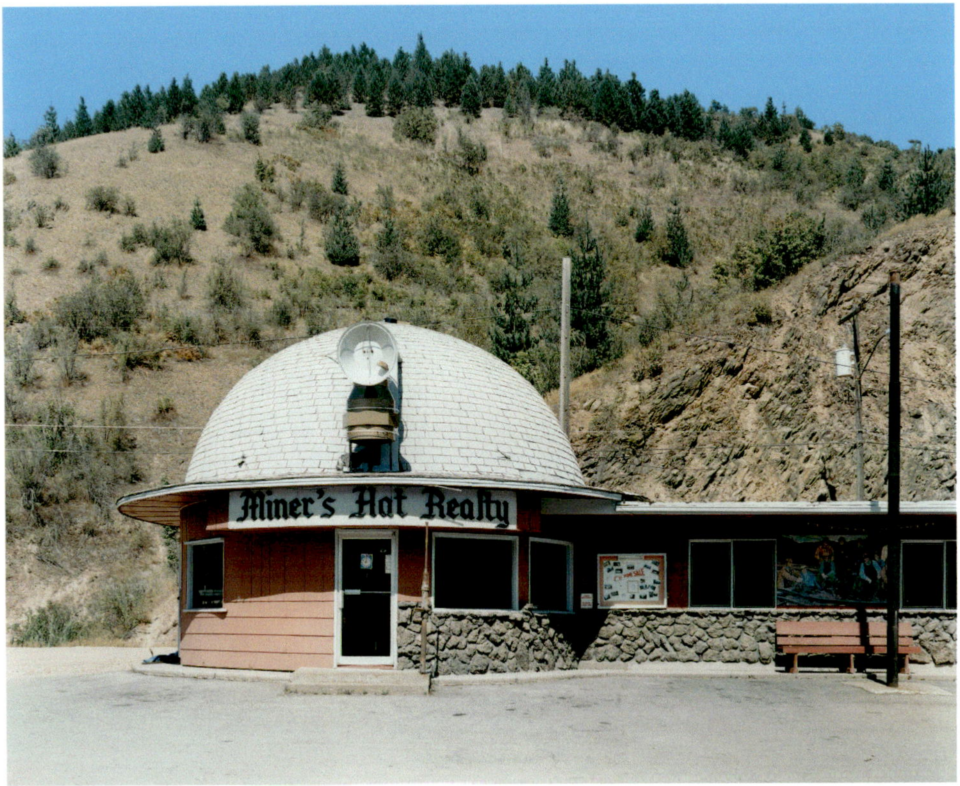

15.14 *Miner's Hat Realty, near Kellogg, Idaho,* 1996.

Goin delights in this bit of kitschy Americana, a realty office in Idaho in the shape of a miner's hat, complete with a headlamp. So many of the issues surrounding mining are serious and intense. This bit of roadside humor offers welcome relief. Environmentalists and people in the general culture tend to perceive mining landscapes as damaged, wounded, abused, and raped—Goin has heard all of those words used. The locale pictured here, for example, is in the Bunker Hill Superfund site, where lead pollution from silver mining and smelting contaminated the soil. Nevertheless, Goin holds out hope that these landscapes can be restored and converted to new uses. This photograph is emblematic of the fact that mining landscapes are being recontextualized as purchasable real estate with its own value, so much so that there are businesses designed to market it. Ideally, we will be able to recycle these landscapes in a way that benefits the local ecology. Then, not only would we have extracted ore that was valuable in human industry, but we would also create an environment that might enable a greater biotic diversity to survive and thrive.

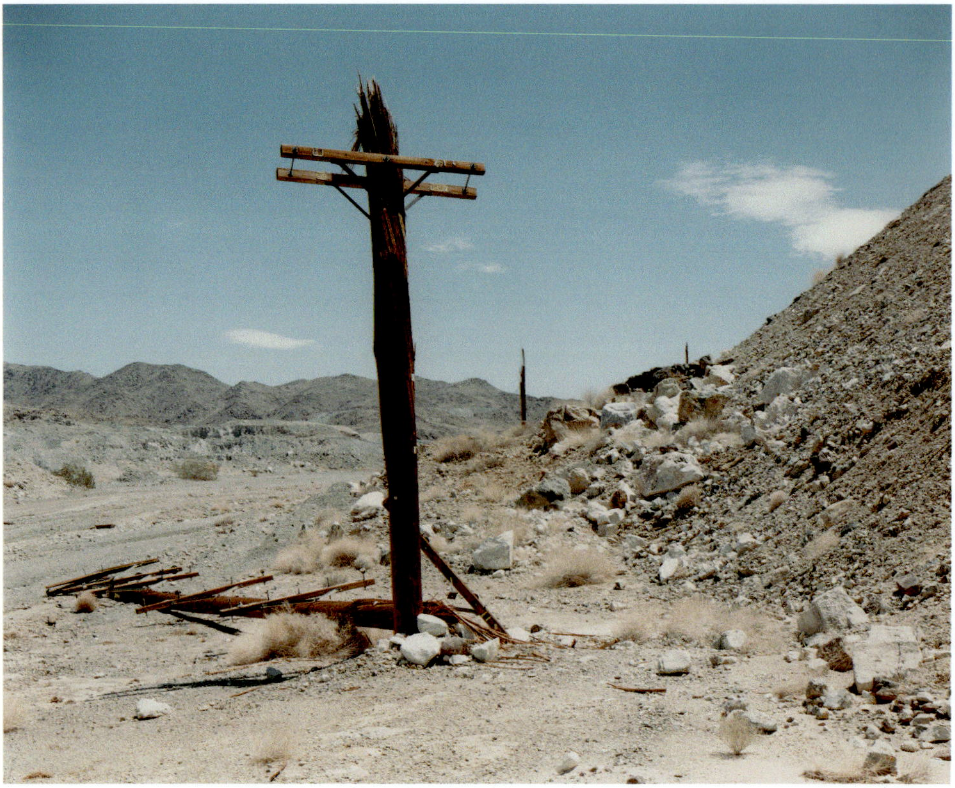

15.15 *Shredded power pole, Eagle Mountain Mine, California*, 1997.

A formalist reading of this photograph would immediately note that the derelict power pole forms a cross, evoking Christian iconography and meanings. What is being sacrificed? Perhaps the land itself, as there is visual evidence of waste piles and terracing from a mining operation now apparently defunct, as inferred by an absence of machinery and the regrowth of desert shrubs. What a formalist reading will not reveal is why this pole has a splintered top. Fully understanding this image requires textual support. The reclamation plan for Eagle Mountain Mine, an open-pit, iron ore mine that ceased operation in 1983, called for removal of the telephone poles. Until removal was completed, workers shredded the tops of the poles to prevent ravens from nesting, ravens being known predators of the endangered desert tortoise. The Eagle Mountain mine site is surrounded on three sides by Joshua Tree National Park, complicating a 1988 plan to convert this abandoned pit into a sanitary landfill for southern California's household waste. A landfill would attract scavenger birds and animals, impacting the ecology of the park, and an operating garbage dump would create light pollution. Reclaiming this post-mining landscape is complex and contentious.

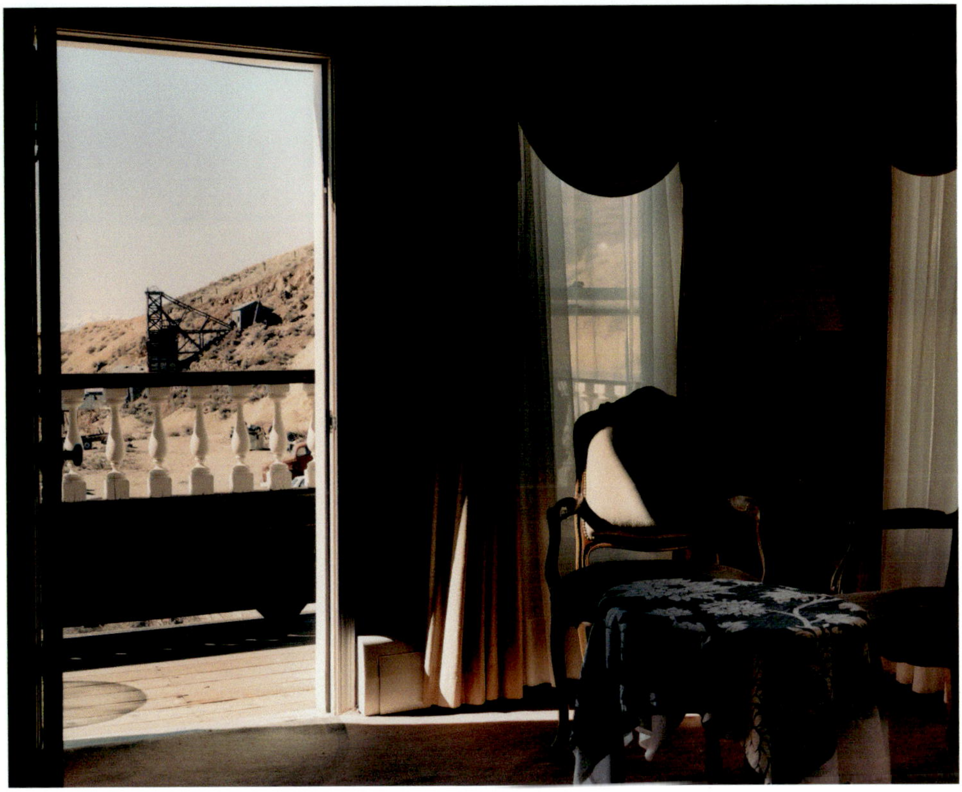

15.16 *View of Yellow Jacket Mine from Room 10 in Gold Hill Hotel, near American Flat, Nevada*, 1998.

Resembling a desert variation of Andrew Wyeth's painting *Her Room*, which pictures a view of the ocean as seen through the windows and open door of a quiet room on the Maine coast, this photograph contrasts the blindingly bright working landscape of the Yellow Jacket silver mine in Nevada with the shadowed gentility of a Victorian-style room, whose gauzy drapes soften the harshness of the clamor and labor that went on outside. Rarely are the site of labor and the refined lifestyle made possible by the profits of that labor so proximate. This vintage hotel, built of stone in 1859, has become a tourist destination, and Room 10, pictured here, is advertised for its view of the historic mine, one of the most notable silver mines on the famous Comstock Lode. Thus, not only does this image contrast inside and outside, but it also conjures a sense of the past in the present. The Gold Hill Hotel has become well known among seekers of the paranormal for being haunted, so perhaps the empty chair in the room is occupied by a ghost.

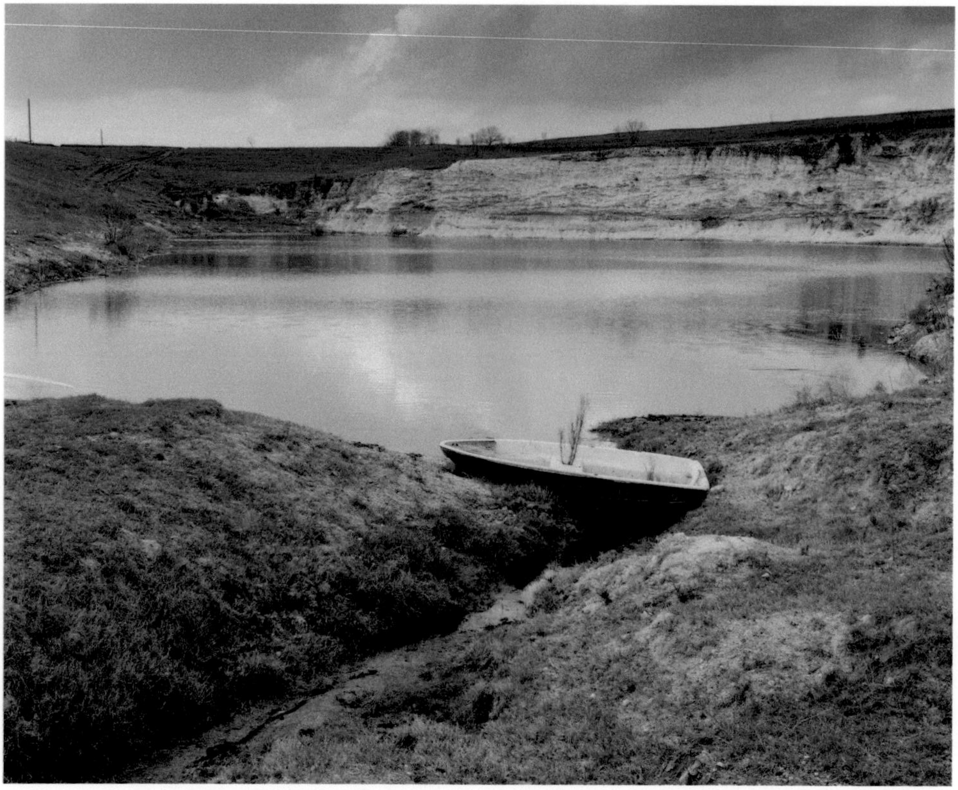

15.17 *Abandoned boat at reclaimed uranium pit, Karnes County, Texas*, 1997.

When localized uranium mining operations ceased in the 1980s, the scattered pits recharged naturally with groundwater and rain. Given the relative lack of water features in the dry brush country of southern Texas, the artificial lakes created from the old pits suggested potential new uses. If radiation levels are within acceptable limits the ponds can be water sources for domestic and wild animals and recreational sites for people. But there may always be that lingering sense of doubt and fear that we associate with radioactivity. How can we *know* for certain that it is safe to swim in or eat fish caught from this pond? Goin's challenge became how to represent the threat of radioactivity that looms over Karnes County, Texas. While some of his photographs incorporate posted warning signs, this image employs light and color—flat light, slime green, dirty white, brooding gray—to convey a sense of unease. Furthermore, the viewer is vaguely disturbed to see plants growing in the abandoned boat in the foreground. That is not normal, and the ominous mood infusing this image—a possible storm brewing on the horizon—correlates with an *emotion* of radioactivity.

Works cited

Francaviglia, Richard V. *Hard Places: Reading the Landscape of America's Historic Mining Districts*. University of Iowa P, 1991.

Goin, Peter, and C. Elizabeth Raymond. *Changing Mines in America*. The Center for American Places, 2004.

Margolis, Eric. "Mining Photographs: Unearthing the Meanings of Historical Photos." *Radical History Review*, vol. 40, 1988, pp. 32–48.

Nye, David. *American Technological Sublime*. MIT Press, 1994.

16

WATER INTO LIGHT

For most of his career, Peter Goin has tended to think and work in terms of book projects. When a book project has been conceived, he then dedicates several years to working in the field, making photographs for that project. Although Goin's Black Rock photographs did culminate in a book by that name published in 2005, the photographs collected in *Black Rock* and discussed in this chapter were made over a span of roughly twenty years (1985–2004), and only toward the end of that period was the idea of a book even on the horizon. What we have in the Black Rock work is a compendium of memories of dozens of trips Goin made solo and with friends, family, and students to the Black Rock Desert in northern Nevada, slightly over one hundred miles north of Reno. Some trips were one-day excursions, while others were overnight camping trips, and still others were multiple day expeditions. On many trips, Goin made no photographs at all. On others, he may have spent several days orchestrating a single image—waiting for the moon to be full, for example, or for a dust devil to sweep across the dry lakebed, called a playa. The Black Rock Desert country is referred to as the "Black Rock" for short, named after a prominent, dark rock formation that emigrants on the Applegate-Lassen Trail used as a landmark. This empty quarter, especially the vast playa itself, drew Peter like a magnet. He wanted to explore it, know it, experience it, and become attuned to the spirit of the place. As he recalls in conversation, "I went out there because the place intrigued me in many ways. It was a place of solace, a place of contemplation, a place of escape, a place of discovery."

Early in the twenty-first century, after about fifteen years of spending time in the Black Rock, Goin began to think about sharing his vision of the place. It is possible that he felt the time was right to offer an alternative perception from the one projected by the increasingly popular annual Burning Man event, which in 1990 moved from a San Francisco beach to the Black Rock Desert and grew exponentially thereafter, topping 25,000 by the year 2000. Goin found a friend and fellow traveler in University of Nevada, Reno, colleague Dr. Paul Starrs, a cultural geographer whom Goin first met while observing the 1997 Reno flood, and who shared Goin's fascination for the Black Rock Desert. The two of them set out "to see what was involved in grappling with the complexities of a piece of the earth"

DOI: 10.4324/9781003212607-16

(*Black Rock*, xii). Together and separately they made trips to the Black Rock, where, "in the field our senses always roam, seeking out the components of place, trying to understand the landscape, eliciting relationships, contrasts, lighting, conversational keys, watching always for essence" (xiv). Their coauthored book *Black Rock* represents a sustained effort to apprehend and communicate the essence of a particular place on earth. This chapter focuses specifically on *Goin's* perceptions and artistic expression of the Black Rock country.

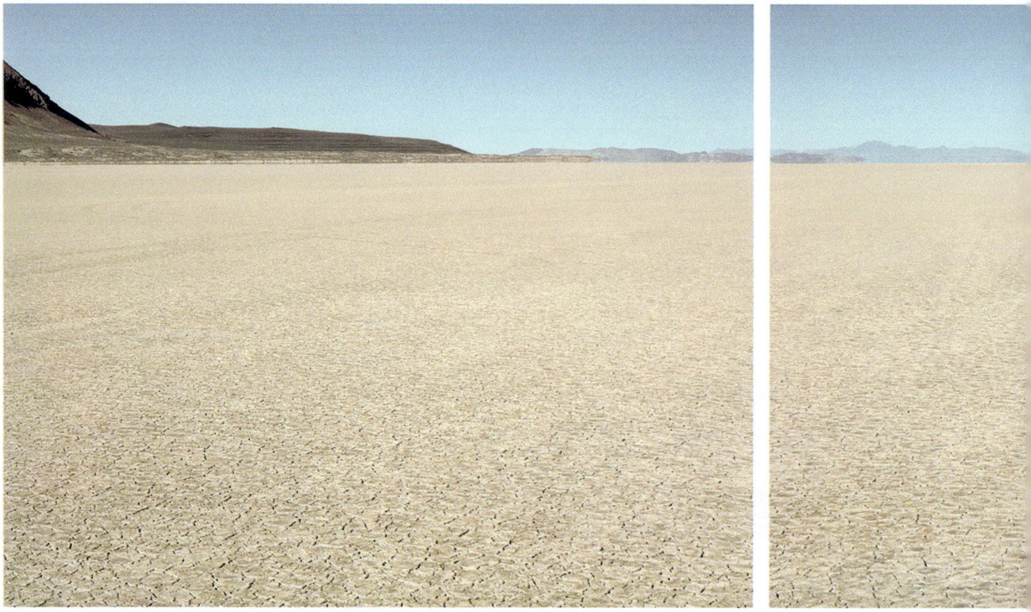

16.1 *Black Rock playa*, 1988.

A picture is the best way to give an immediate sense of the Black Rock Desert, and we offer here a triptych of the playa that Goin made with a 4x5 camera in 1988. For many years, an impressively large, 24 inches × 80 inches, framed print of this triptych was displayed in Peter's living room, where it figuratively transformed a slumpstone brick wall into an opening through which the viewer could enter the expansive space of the Black Rock Desert.

The first thing to note about this photograph is the minimalist color palette—off-white, pale blue, light gray-brown, with a hint of black at the far left. Also, immediately striking is the amount of space this image grants to the lakebed, which fills the lower seventy-five percent of the picture, like an alabaster rectangle anchoring an abstract painting. From afar this white space appears undifferentiated, but up close one begins to discern detail in the foreground—a rumpled, wavy surface and fine network of cracks, indicating the former presence of an immense sheet of water that has completely evaporated. The viewer is confronted with a puzzle or a paradox—a lake without water. At the furthest verge of the playa appear mirages faintly connecting the playa to the sky, forming an optical *illusion* of water. If you search for tracks, you may be able to make out a few disappearing traces of vehicle tracks trending off into the distance, but they are so faint and diffuse that the overall impression is of a land without human traces, pure space. Rising above the horizontal line of the playa is an assortment of low, broad, rugged, and seemingly barren mountains, a geological substrate that apparently precedes or perhaps endures beyond an age of biological life. The dark hill to the left exhibits horizontal striations that geologists have identified as ancient shore levels of the prehistoric Lake Lahontan, an 8,610-square-mile body of water that covered much of western Nevada in the late Pleistocene. A fitting description of this triptych could be extracted from the *Black Rock* chapter that Goin authored: "Space in the Black Rock is not empty, as some writers might assert, but it is filled with light, with the memory of water, with the spirit of the earth, moving, twisting, lurching, resting, active in every moment" (154).

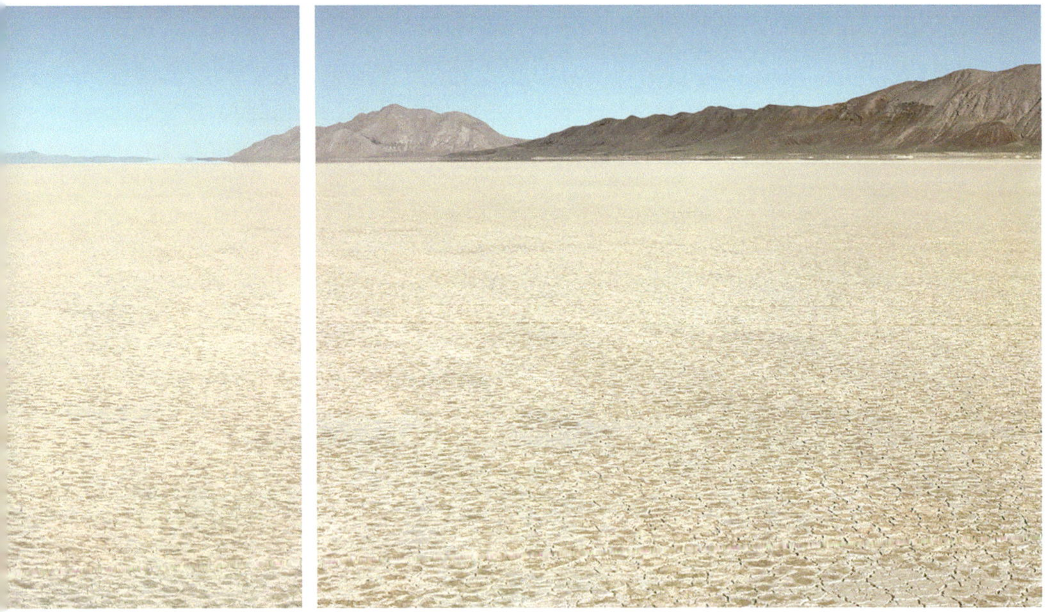

This image of the Black Rock playa contrasts dramatically with the cover image of *Changing Mines in America* discussed in the last chapter, just as *Black Rock* originates from a different sensibility than most of Goin's other work. If we juxtapose the *Changing Mines* image of the couple standing in a caged overlook of a Mesabi Range mine pit with this *Black Rock* triptych, many visual, thematic, and conceptual shifts occur. We move from a view of a proximate skyline partially obscured by foliage to an unobstructed view of a seemingly infinite horizon. We move from Minnesota to Nevada, from a well-watered climate to an arid one, and from the color green to its absence. We move from a mine site to an unexploited part of the earth and from a human-engineered viewing platform above the vista to an undeveloped vantage point on the ground. We move from people in the scene to human absence. We shift from a zoo-like and controlled sightseer's glimpse to a free and capacious gaze. And we move from a single photograph to a triptych. As Peter explains, the triptych format represents the process of viewing: "It's the traditional viewing of things in threes. You look here, then there, then there; side, center, side; one, two, three." This trio of photographs could have been assembled into a seamless panorama, but displaying images in a triptych format creates a conceptual formulation that references the act of viewing. Perhaps the biggest difference between *Black Rock* and Goin's other projects—subtly signaled by this triptych—is that most of his work investigates human-altered landscapes, whereas *Black Rock* seeks to *alter our perception* of an *unaltered* landscape.[1]

To get a sense of how radically Goin's perception of the Black Rock Desert differs from the norm, it will help to review three prevalent ways of viewing that place. The first is a blisteringly negative perception, penned by an emigrant who felt duped into taking the Applegate Cutoff on his way to Oregon. Midwesterner Jessy Quinn Thornton had this to say about his 1848 crossing of the Black Rock:

> The country over which we had passed was dreary beyond description. There were in it no diversities of color or form, to relieve the mind by their variety. The earth was iron, and the heavens brass. Every thing was parched and arid; and all those sources of beauty, which, from their being so generally diffused through nature, are usually regarded as things of course, were here dried up by a hot sun beaming down upon sand and rocks. Here was none of the living luster of a gay and beautiful spring, dressed in robes of the richest green, smiling upon the wooded hills and the grain-covered valleys, or laughing and dancing along the brooks and rivers. Here were none of the rich glories of autumn, laden with delicious fruits. There were neither sounds of melody to charm the ear, nor sights of beauty or grandeur to please the eye, and delight the heart.

In perceiving the Black Rock Desert, Thornton can only "see" what it lacks and fantasize about what is not there. Thornton's is the view of a homesick traveler, who compares the Black Rock to his beloved homeland and finds it wanting. As will be further developed below, Goin's view is entirely different. His is the view of someone who has chosen to make Nevada his home and who loves the Black Rock's uniqueness. Where others see ugliness and absence, Goin finds beauty and presence.

A second perception of the Black Rock is widely shared by the people who organize and attend the annual Burning Man event, which by 2013 had reached seventy thousand people. These revelers or "feral mendicants," as Paul Starrs refers to them, regard the playa as a tabula rasa or empty stage for human performance (xiv). As Goin explains:

> For some people, the Black Rock is a slate of utter blankness. The tabula rasa in Latin is literally a 'clean slate'. . . . Another way of putting it is to say it's a template suitable for the release of human energy, or perhaps it's some gargantuan canvas, suitable for the painting of human activity.
>
> *(155)*

People who purchase tickets to Burning Man are sternly warned on the Burning Man website that they are "no mere 'attendees' but rather active participants. . . YOU are Burning Man. Create your own form of participation! Create an art installation, a Theme Camp, a Mutant Vehicle, or performance." Radical self-expression is one of the ten core principles of Burning Man, and the culture of carnival that has evolved encourages people on the playa to flamboyantly act out alter egos and fantasies that get repressed in the workaday settings of normal life.

As an artist, Peter Goin enjoys going to Burning Man, and the *Black Rock* chapter entitled "Drama" includes a gallery of photographs of Burning Man devotees in costume. In 1993, when Burning Man attendance had reached approximately eight hundred people, Goin made a video entitled *In Search of Ritual: The Burning Man*, which was selected as a finalist for the 1994 Emmy Awards. Rather than employing a single, authoritative narrative voice, the video splices together voices and images from a variety of participants and organizers, recorded live at the event itself, sharing what this polyvalent, neo-pagan art festival means to them. Creatively honoring the Burning Man creed of "no observers, only participants," the video crew actually participated in the event, as shown by the camera angles during the filming of the ceremonial burning of the forty-foot-tall wooden man. As Goin explains, "the resulting montage was not intended to tell a single story, but instead reflects the chaos, energy, and spirit of the ritualized event." Nonetheless, although Goin enjoys the spectacle of Burning Man, for him the Black Rock Desert has its own essence that gets overlooked and drowned out when Black Rock City temporarily takes center stage.

A third common perception of the Black Rock can be named "the glance." American landscape photographers dating all the way back to Timothy O'Sullivan, who in the late nineteenth century crossed the American West in a mule-drawn carriage, have a propensity for what might be called "drive-by shootings." As Peter explains in conversation:

> In the contemporary era, you drive in your car, you see something, you get out of your car, you go make the photograph, come back, get in your car, and drive on. It's almost as if the representation of the American West is through glances of other people, not those who live there.

While such "glancing" photographs may be visually compelling for their formal compositional design, they do not reflect the understanding of a person who lives in that place. Instead of glancing at the Black Rock, Peter says in conversation:

> I want to be of the place and create images that represent the deep-seated integrity of having a connection to place so that the photographs are more contemplative or reflective of understanding what this place may mean in a vernacular, spiritual kind of way. It's not just a glance for me. It's a very serious investigation, and that's why each photograph tends to be expansive, because it's a portal through which I'm viewing, not just the place, but the experience of the place and the transformation of the place over time.

A single photograph or even a collection of photographs cannot adequately convey Goin's gradually accrued understanding and deepening love for the Black Rock. Therefore, even though Paul Starrs wrote most of the text, Goin chose to author one of the chapters of *Black Rock*, entitling it "A Good Place to Make Love." He lifted that phrase from an inscription on a stone tablet that is part of a mile and-a-quarter-long environmental art installation known as "Dooby Lane" or "Guru Road" on the southwest edge of the playa. As Peter interprets that phrase in reference to his own experience in the Black Rock, "The love is not necessarily erotic or sexualized love, although that is certainly possible, but love in the sensuality of observing and the freedom of feeling" (152). In a published interview with French art critic Julie Ramos, Goin elaborates, "I have come to realize that this arid, often inhospitable, and unforgiving landscape possesses a character both spiritual and sublime." After two decades of getting to know it, Goin in conversation describes the Black Rock Desert as "a spectacular landscape that can be mad and threatening and dangerous and alluring and engaging. It is a little bit like defining love." Goin's specific insights and formulations, expressed in "A Good Place to Make Love," in conversation, and in his photographs, are a gift that can guide readers toward re-perceiving the Black Rock Desert not as empty, ugly, and barren, but as replete, spectacular, and spiritual.

One way that Goin creatively transforms our view of the Black Rock is to respin negative impressions of emptiness into positive connotations of space. In emptiness, he perceives not lack, but abundance. Such places are rare. He tells Ramos, "as the global habitat is ever increasingly dense with concrete and the detritus of human self-focus, images of rediscovery and contemplation infuse the sublime paradox with awe of a different nature." In conversation, Goin characterizes the sublimity of expansive environments such as the Black Rock by referring to Mircea Eliade, a historian of religion who wrote *The Sacred and The Profane*: "If you think about combining those two in a landscape it evokes the sublime, the combination of awe and terror that defines these vast spaces in landscapes that are overwhelming in our effort to try to understand them." In "A Good Place to Make Love," Goin shares what Black Rock sublimity feels like to him, "The sheer space of the place invades my psyche. Direction is ambiguous, altered as the spirit within the earth requires a different compass than north, south, east, and west" (154). Dwarfed by that space, one can feel profoundly insignificant. But human insignificance can be a good thing, Goin says:

> You're in the middle of nothingness. The concept of nothingness goes all the way from Sartre's *Being and Nothingness*, to Zen concepts of the meditative space, to mind separated from body—that kind of alternative experience—to the most profound recognition of quiet, to the transformation of space.

Goin reminds his readers that "most of the world's religions originated in arid places. The paradox of aridity nurturing, even provoking, spiritual growth is clearly apparent" (152). Through Goin's words and images, one can learn to value and indeed yearn for the Black Rock playa as a meditative, contemplative space.

Another way that Goin alters readers' perceptions of the Black Rock Desert is through his re-enchantment of light. Whereas the average person will anxiously apply sunscreen and grab their sunglasses to shield themselves from the blindingly bright light, Goin figuratively exposes his imagination to the light and thereby develops poetic insight. As Goin contemplated the nature of light on the playa, an artistic revelation struck him: Metaphorically, conceptually, water evaporates into light. As the sunlight crosses the playa, it's as if a tide emerges and recedes. It is perhaps impossible to apprehend that metaphorical idea in an image alone, but reading Goin's "A Good Place to Make Love" establishes a way of seeing that the reader can then bring to his photographs or to their own experience. Try reading this passage shortly before visiting the Black Rock Desert for yourself and see if it alters your perception of that space:

> Light reflected from every angle erodes the obstacles that create shadows. It is almost as if within the evolutionary cycle water has become light. Light bends around the ranges, through groves, into gullies, held within plants until the last fading glimmer dissolves into dusk. Light molds the landscape, giving it dimension or flattening it so thoroughly that huge mountains appear two-dimensional like line drawings against a slate-blue backdrop. Light massages alluvial plains and velvetizes craggy, volcanic outcroppings from the Miocene era, some fourteen to twenty million years ago. At times, light is movement, like the shadow from some ancient and enormous sentient being stepping between basin and range. At sunset, the light recedes like a tide inching

> rhythmically ever backward across the arid playa. These shadows account for Earth's movement, just as its curvature is visible across the barren playa. Midday shadows are absorbed in the reflective glory of the playa's brightness. They become liquid, evaporating like water in the hot sun.
>
> *(152–53)*

The Black Rock Desert is a place of paradoxes. A bright white landscape is named the *Black* Rock Desert. In the triptych that opens this chapter, a flank of the black rock formation, after which the desert is named, creeps into the extreme left edge of the photograph like a small clue or subtle hint of the color black. Again, Goin imbues this observation with poetic force, helping his readers view that place in a new light:

> That "black" itself is actually the absence of light is but one of many inherent contradictions within this mystical desert. Black absorbs all light, owning the spectrum selfishly as if the rock contains its own gravitational force. And yet the Black Rock is surrounded by light: Mountains, valleys, and rivers of light, cascading over the ridges and horizons like an ephemeral flood. It is as if light itself has shaped the landscape, not the traditional erosive influences of water, temperature, wind, and gravity. (152)

Whereas other guides to the Black Rock Desert, such as Sessions S. Wheeler's 1978 *The Black Rock Desert*, enrich one's experience of that seemingly blank landscape by sharing its interesting human history, Goin's words and images offer the possibility of connecting with the Black Rock aesthetically, sensually, and spiritually. According to Wheeler, the Black Rock Desert was once proposed to become a giant dump for San Francisco's garbage. One of Wheeler's incentives in writing his book was the hope that "more widespread knowledge of the beauty and history of this unusual part of the earth may help to save it from destruction" (7). If we perceive a place to be a wasteland, then that place becomes vulnerable to being trashed. Perception is everything. *Black Rock* by Goin and Starrs joins Wheeler's cause a quarter century later, helping readers learn to respect and even fall in love with a place that, until the rise of Burning Man, was little known and seldom visited, and whose virtues are still difficult for many to grasp. Rather than offering easy mental access to the Black Rock Desert's landscape by dumbing it down, so to speak, Goin opts to highlight the region's contradictions:

> That sensuality is integral to aridity redefines the sublime, that combination of awe and, well, terror. This becomes fertile ground, both topographically and intellectually, as the language of aridity—parched, barren, monotonous—merges with the language of sensuality—voluptuous, aesthetic, indulgent, and luxurious. The contradictions, the paradoxes, are intrinsic; the opposing elements create a form of synergy that may be elusive, but is still definitive.
>
> *(152)*

As Goin told interviewer Ramos, the *Black Rock* images "are artifacts of energy unexplained, beneath and beyond the horizon." As his comment suggests, much of the challenge of photographing the Black Rock is that its essence—energy, air, wind, spirit, silence—is invisible.

As we turn now to a closer analysis of *Black Rock*, the book, it is evident that Goin and Starrs sought a poetic design and impact. The book is dedicated to "the knowledge and spirit of the place, itself," and its Table of Contents resembles a poem:

> where the pavement ends . . .
> playa
> the view of god: overhead
> light
> thus the thunder spoke: water
> drama
> looking to land: profile
> signs
> a good place to make love
> gerlach
> lying below: subterranean
> earth
> immanence: fire
> water
> eddying insolitude: air
> forest

Chapters alternate between a prose chapter, whose title typically includes a colon, and a photograph chapter, whose title is a single word. Perusing the Contents, one may observe that the four classical elements of earth, water, air, and fire are featured. In assessing how the Black Rock has been perceived, Paul Starrs assembled a fascinating archive of twenty maps of the region dating back to 1863. His commentary on these maps constitutes a thumbnail history of cartography as well as an argument that maps represent not objective reality but the map-maker's perception, perceptions that evolve through time.

We have already noted how Goin's essay "A Good Place to Make Love" contributes poetic insight. Poet Gary Snyder has also written about the Black Rock Desert and lines from his poem "Finding the Space in the Heart" are printed as a companion to Goin's opening triptych, discussed above.[2] A brief sample from that poem echoes the catalog of absences enumerated by emigrant J. Quinn Thornton, but finds those absences liberating and the Black Rock playa to be the quintessential space for meditation, the place where Snyder concludes his epic poem cycle *Mountains and Rivers Without End*, emptiness being the ultimate destination of a Buddhist seeker:

> Off nowhere, to be or not be,
> all equal, far reaches, no bounds.
> Sound swallowed away,
> no waters, no mountains, no
> bush no grass and
> because no grass
> no shade but your shadow.
> (Gary Snyder, "Finding the Space in the Heart,"
> *Mountains and Rivers Without End*, 151–52)

Sharing the sensibility of Snyder's poem, Goin's photographs in *Black Rock* could be thought of as visual haiku. The "Playa" chapter features images made between 1985 and 2002 on and of the playa. Photographs of a pristine, blank expanse are juxtaposed with images showing vehicle tracks—receding into the distance, forming a circle, or creating line art. There are also performance pieces of solitary human figures standing on and leaping above the playa and staged scenes, such as a lone patio umbrella casting a simple black octagonal shadow on the beige lakebed or an unattended campfire on the treeless playa, charred burning boards propped up in a triangle. The chapter entitled "Light" premiers an unfurling sky drama of ever-changing illumination, with scenes including a time-lapse, night image of Venus setting; moonrise over the Black Rock; a pink-tinted distant storm; soft-purple evening light; a variegated cloud canvas at dusk; the last rays of sun brightening shrubs in a rocky canyon; or the artificial lights and dusty air of Black Rock City at Burning Man. By assembling this assortment of landscape photographs in a gallery entitled "Light," Goin alerts the reader to notice—to be aware of, to prize—the special qualities and dynamism of light itself in the Black Rock Desert.

Understanding spirit of place in the Black Rock must take into account the people, activities, and human traces that pepper that space. The middle three photograph chapters in *Black Rock*—"Drama," "Signs," and "Gerlach"—detail the desert's human element, including portraits of Burners, rock hounds, hunters, and ranchers; photographs of environmental artworks, a memorial, petroglyphs, a license plate collection, and an American flag made of ribbons tied to a chain-link fence; and a typology of vernacular building fronts in the small town of Gerlach. Many of these images reference the history of art and photography; for example, the light-delineated buildings reference Walker Evans's similar technique when he photographed in the American South. Together, these chapters remind pilgrims to the wild that the Black Rock has a human history and ongoing human element.

Returning to the classical natural elements, Goin's photograph chapters next explore "Earth" and "Water." "Earth" takes inventory of the range of rock formations, patterns, and colors in the Black Rock Desert and surrounding country. There are images of the famed black rock itself, a couple of whitish, eroded rock formations, and a hillock of red volcanic rock. There are several photographs of sensuous hills and mounds, reminiscent of Goin's early images in the Mesabi Range in Minnesota. And throughout this chapter runs an enigmatic motif of a human eye hidden in the picture, such as in the shadows of the windows of a distant cabin or a faint petroglyph in Little High Rock Canyon. The subtle message pertains to perception, to having eyes to *see* what is in view and a mindset to appreciate that which is often overlooked or misunderstood. "Water" may seem to be an oxymoron in a desert, but this chapter celebrates scattered oases in the form of hot springs, lakes, reservoirs, creeks, a geyser, mist, and mud puddles. Precious for its scarcity, water animates these images, which contrast the sparkle, reflections, and pellucid aquamarine of water pockets with the arid austerity of the surroundings. A number of the photographs picture people relishing the sacred waters—swimming, bathing, floating, coating themselves in mud, and gazing like Narcissus into the mirror of the water surface.

Like "Water," the final chapter, "Forests," may seem out of place in a book about the Black Rock Desert, whose signature element is the bare playa. However, for Goin, the spirit of the Black Rock country includes the surrounding hills, canyons, and mountains, where habitat can change dramatically at different elevations. It is magic to camp in a mountain meadow by a brook near an aspen grove and peer outward from that verdant refuge to behold the vast playa below. A variety of shrubs and trees appear in this gallery—sage, juniper, poplar, elderberry, and apple—attesting to the Black Rock's biodiversity; but the star of the series is the aspen, a photogenic, smooth, white-barked tree with trembling leaves. Goin's aspen photographs create a catalog of the many moods and guises of this Black Rock life form, changing appearance according to the season—green leaves in spring and summer, yellow in autumn, bare in winter—or time of day or night, revealing the imprint of a historically heavy snow year in a stand of crooked trunks, and exhibiting the marks of past human visitors who left behind arborglyphs.

Peter Goin has this to say about his Black Rock photographs: "Within the arid, barren Black Rock, a deep-rooted spirit of life exists. These photographs are merely artifacts of the passage of a memory in time, a brief spiritual reference to a lyrical knowledge, cycling back upon itself" (156).

■

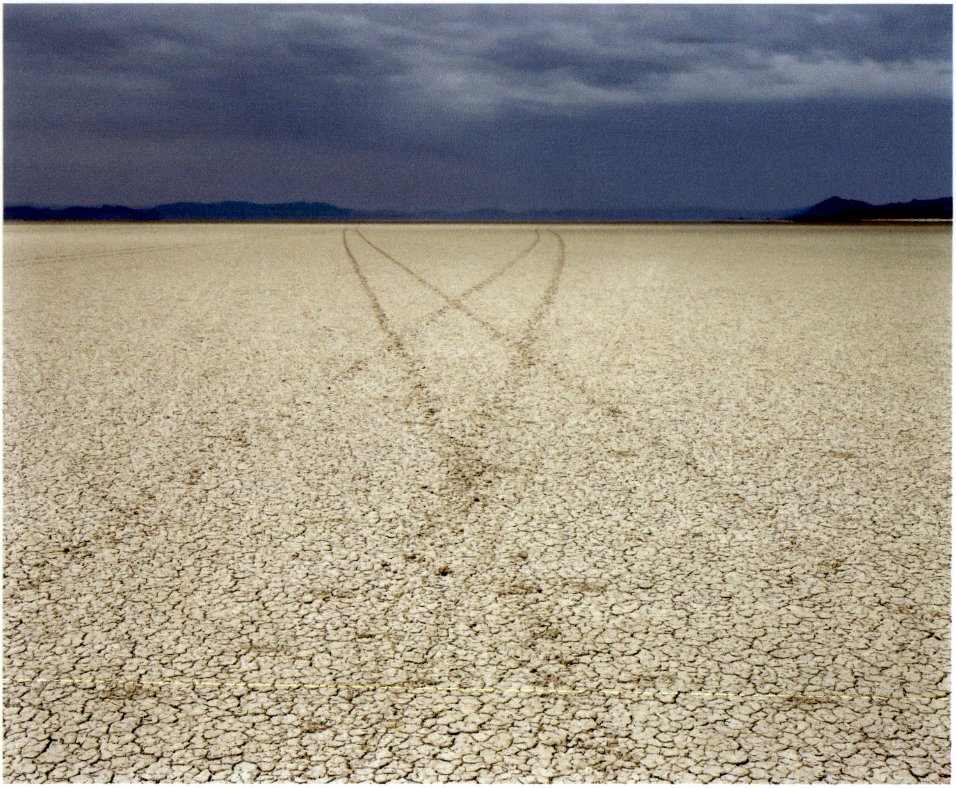

16.2 *Intersecting tracks*, 1998.

Without vehicle tracks, this image could be timeless, picturing what the world must have been like when dinosaurs roamed, speaking to the fundamental power of global atmospheric conditions. There is something engaging about this amorphous sense of space and time. Where is time? There is no time. There's the time of the light. The time of the wind. Two sets of vehicle tracks in four hundred square miles of playa reference human presence and insignificance. The geometric pattern made by the tracks could be seen as a diamond, not quite a star, perhaps a pair of pointed ears of a fox or the devil. The horizon in the distance creates the illusion of perspective and distance, and yet, graphically, the lines made by the tracks draw the surface up flat and make it a construction, and the dark storm almost seems to be propped up by the tips of the tracks. Peter notes that some people write about the Black Rock Desert like it is a tabula rasa, but his view is that "it's only a blank slate to those who will not see. There's a lot going on there, and it's a different realm of discovery."

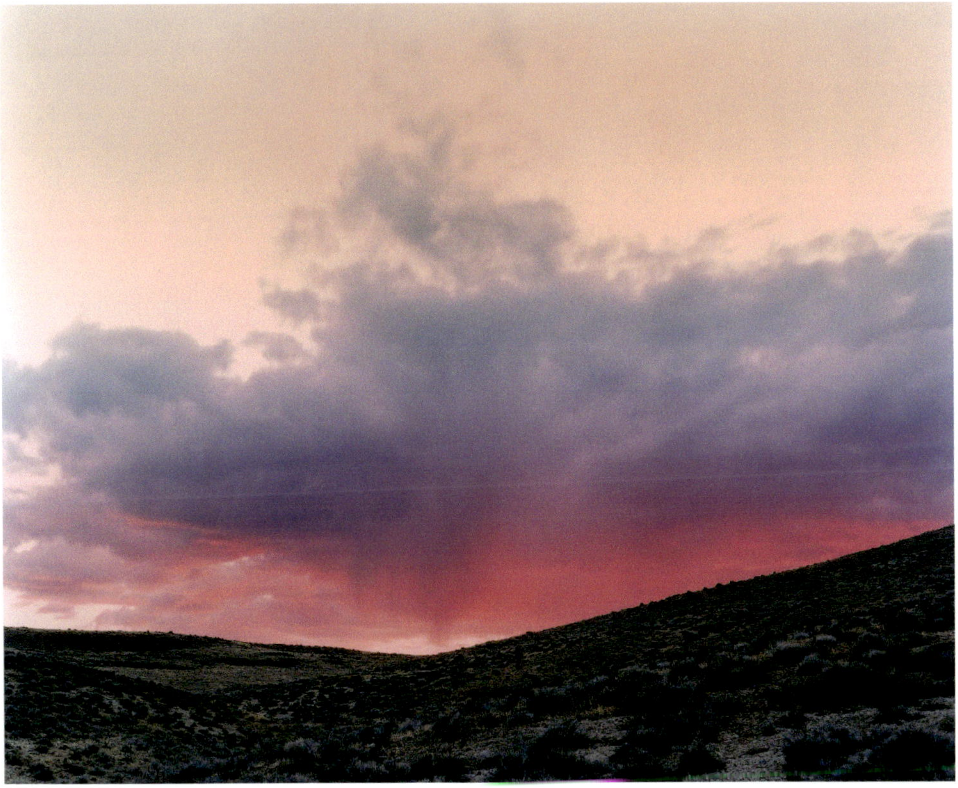

16.3 *Last weather of the day*, 2003.

An excerpt from Goin's "A Good Place to Make Love" reads:

> During the summer months, thunderstorms come and go, rolling across the expanse, often more threatening than precipitating. In the grand scheme of the Black Rock, these moments are temporal, reminders that there exists somewhere else an atmosphere of regularly unpredictable weather, inclement and variant.
>
> *(154)*

Located in the chapter entitled "Light," this image suggests that in the Black Rock, spirit of place encompasses not just land but also firmament. In fact, the composition of this image—the land in shadow, occupying just a narrow strip in the shape of a wide-angle V—diverts the viewer's attention heavenward to the painterly blend of pink, salmon, gray, purple, and peach. Goin's image alludes to photographer Alfred Stieglitz's *Equivalents* series of cloud photographs, which pushed photography into abstract art by obscuring the photograph's referential content, making the photograph an artifact equivalent to a higher expression in the fine arts such as painting. As is true of many of Goin's photographs, this image references the history of art and has philosophical and theoretical significance that may elude the viewer who sees it merely as a nice sunset.

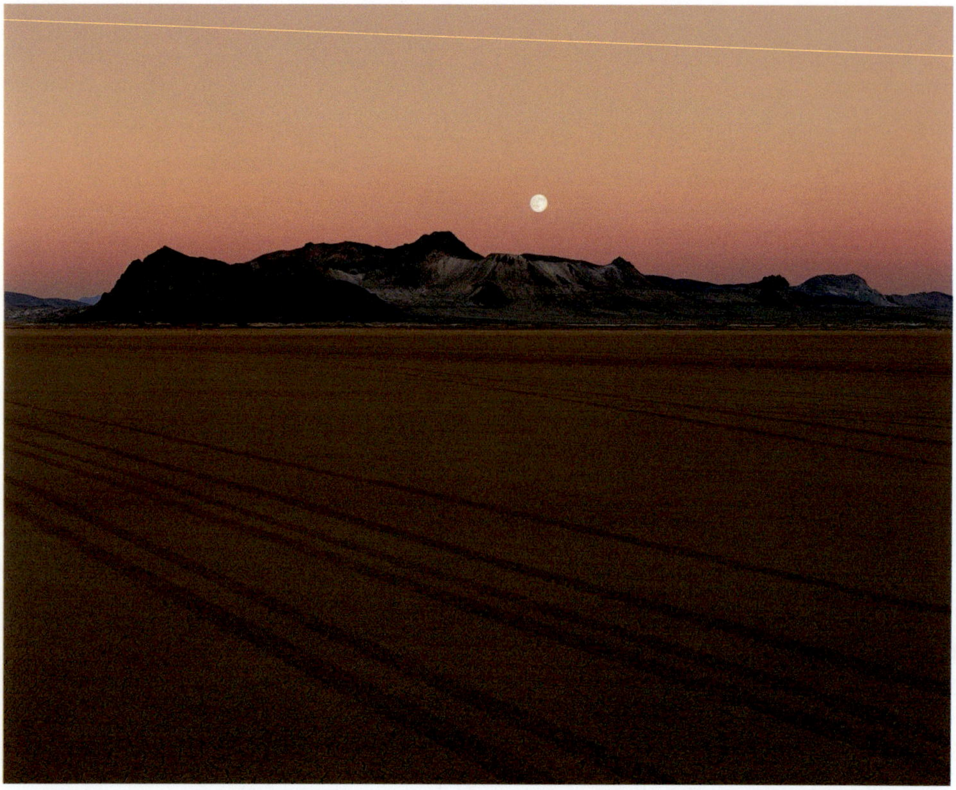

16.4 *Moonrise over Black Rock*, 1990.

Those familiar with the history of photography will recall Ansel Adams's popular 1941 black-and-white photograph *Moonrise, Hernandez, New Mexico*. Indeed, Goin had that image in mind when he camped on the Black Rock playa in 1990, staying an extra day for the moon to be full. Although Peter used a 4x5 camera—before Photoshop—people have accused him of faking it. There is something about this picture that doesn't look real. Nevertheless, although composed with forethought and craftsmanship, this photograph of the rock outcropping named "black rock" is authentic. Although their traces have been erased by time and overwritten by automobile and truck tracks, emigrants sighting on the black rock may have crossed over the spot where Peter Goin camped. This image contains a second art reference—Bertolt Brecht's *Waiting for Godot*. In Brecht's play two men stand by a lamppost on a bare stage, waiting for Godot or God. Brecht's existential drama is about the human condition of waiting. However, as Goin sat on the playa waiting for the moon to rise, waiting became *being*, as he experienced a theatrical event taking place right before his eyes.

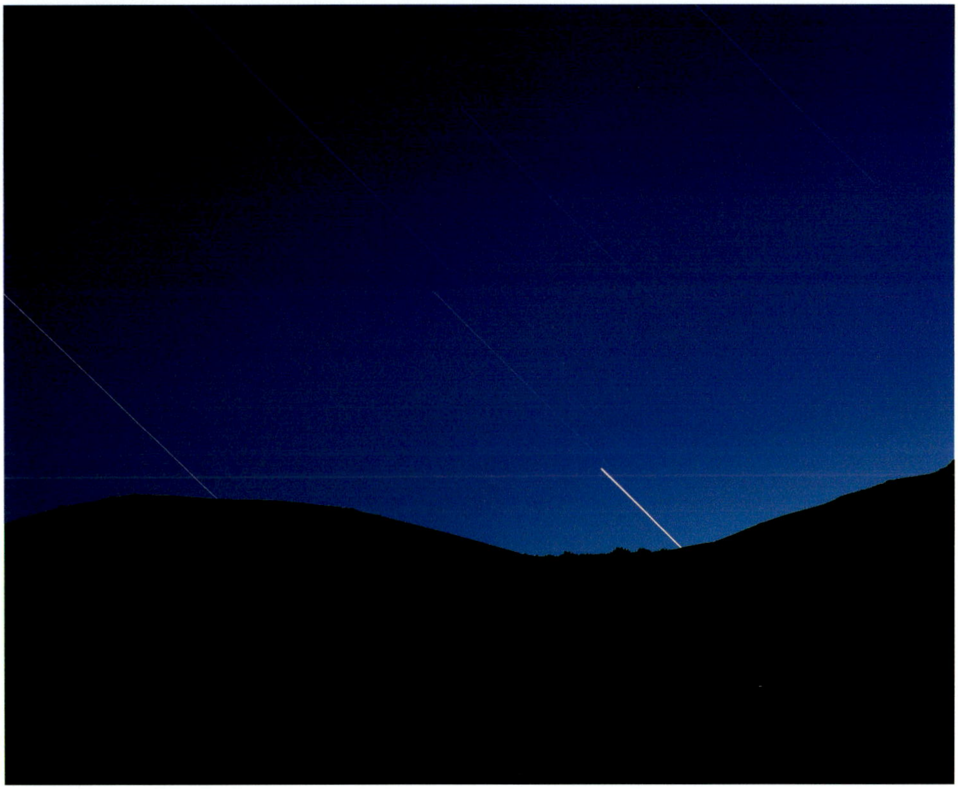

16.5 *Venus setting*, 2002.

This night image occurs in the *Black Rock* chapter on "Light," an intriguing placement since the photograph was made by time exposure on a pitch-black, moonless night. In many places in the United States, light pollution at night would prevent the foreground from being so featurelessly black. The purity of that blackness is integral to spirit of place in the Black Rock as are the starry nights, which awaken our atavistic inner selves to gaze skyward as humans have done for millennia. The long exposure allowed Goin to record the passage of Venus in the sky from its exposure at dusk to its setting. For Goin, Venus is a signifier of Nevada, a way of orienting oneself. Another essential Nevada element is the sharp horizon line. In the East, the horizon is usually obscured by trees, but in Nevada the sky line is as crisp as kirigami. Like the previous two images, this photograph references art history, in this case, Claes Oldenburg's jumbo-sized public sculptures of normally small objects, for example, a badminton birdie, a teaspoon, a clothespin. Peter imagines that the track of Venus could inspire a Nevada sculpture of an enormous neon tube. The line of Venus's setting also references the artist's simple act of making a mark.

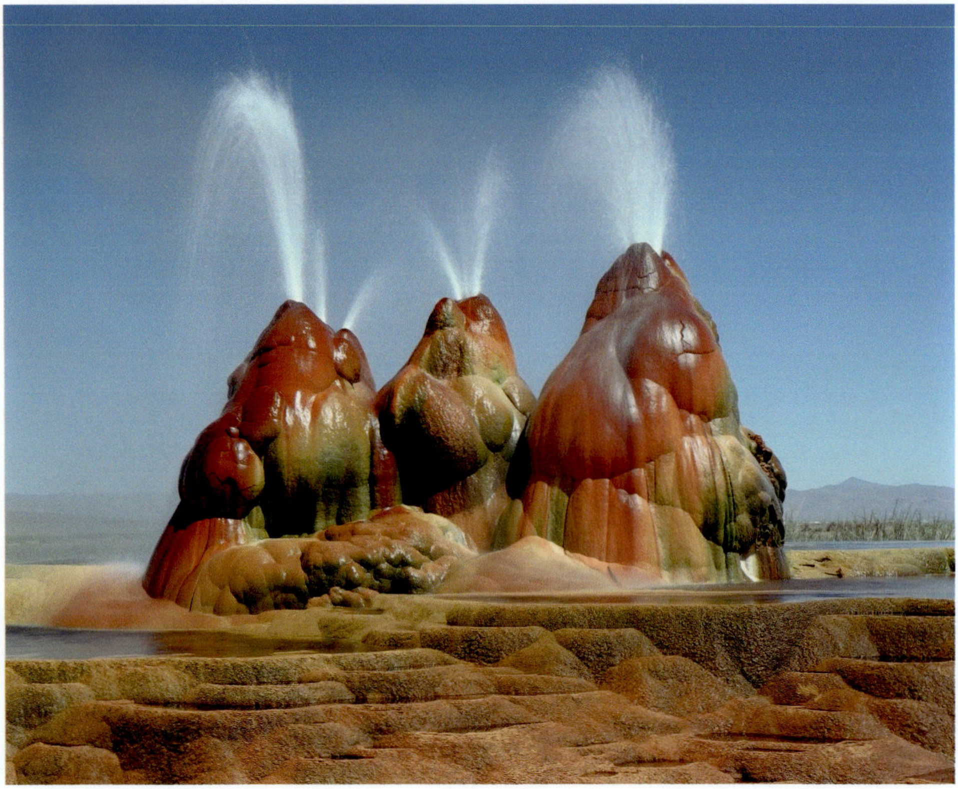

16.6 *Fly Geyser*, 2002.

Yes, this is a real photograph and these colors are natural. However, Fly Geyser, on the western edge of the Black Rock Desert, about twenty miles north of Gerlach, is not itself natural. In 1964 a geothermic energy company drilled a test well at Fly Ranch and hit two hundred-degree water. Although hot, the water was not hot enough for their purposes, so they sealed the well and left. Apparently, the seal leaked, and a multi-spouted geyser formed, gradually building up a surrounding mound made of dissolved minerals, including calcium carbonate and silica. Thermophilic algae create the green and red hues. In terms of Goin's work, Fly Geyser fits the *humanature* theme of natural-looking, human-altered landscapes. In 2002 Goin got permission from the landowners to access the geyser and make this photograph, which was published on a Black Rock poster. Since that time, many people have contacted him, wanting to visit the site and make a copycat photograph—such is the way that images circulate and create desire. According to the *Geology Page* website, the Burning Man Project purchased the 3,800-acre Fly Ranch with geyser in 2016 for $6.5 million and since 2018 has offered limited public access.

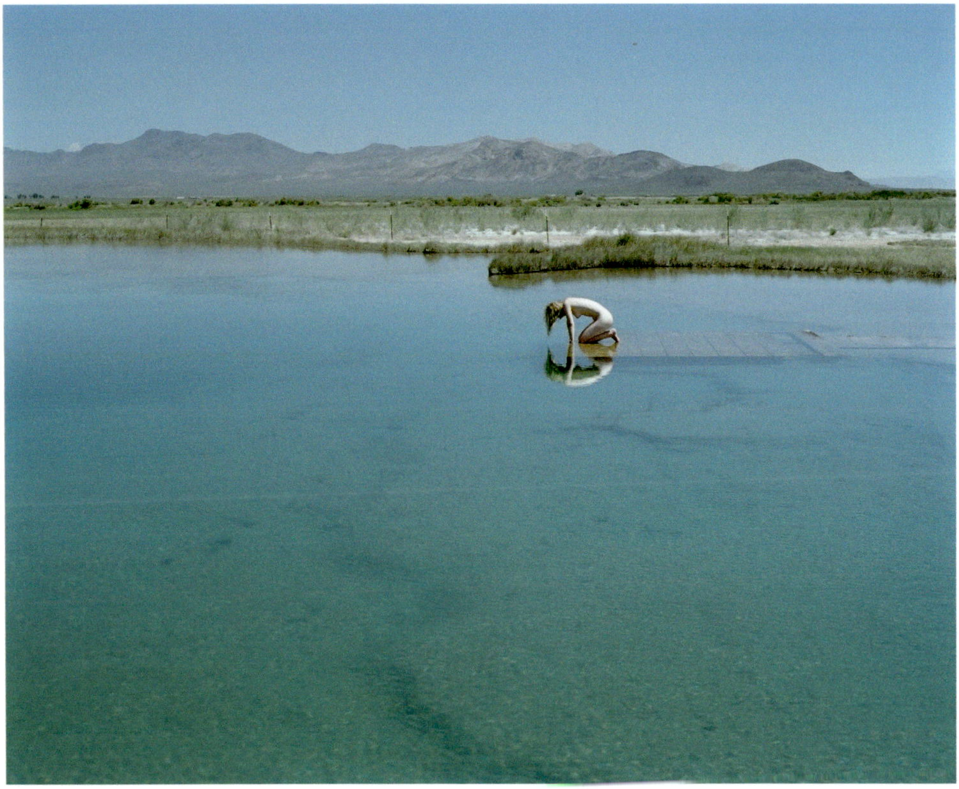

16.7 *Swimmer leaning into reflection, hot springs pool near Fly Geyser,* 2002.

Minerals in the water near Fly Geyser create aquamarine travertine pools, which are pleasantly warm, fed by hot springs. There is something about this remote oasis in the desert that makes people want to strip down and get naked. Of this image, Goin says:

> In our culture, we're kind of hyper-focused on sexuality, but the photograph is really more about the reflection and about the circular shape that the reflection creates in that image. It's also about the history of art and the history of representation, a direct reference to a particular type of Edward Weston photograph where a woman is floating on her back in a pool. One of the great traditions in the history of art is portraying a woman bathing. If you look at Degas' paintings, a woman is getting in or out of a bath. There's something about that process of having that contemplative moment. There's a certain kind of beauty in that moment, and it's not meant to be eroticized.

A sensuous, aqueous experience is not what one would expect from a desert, but it is part of the spirit of the Black Rock.

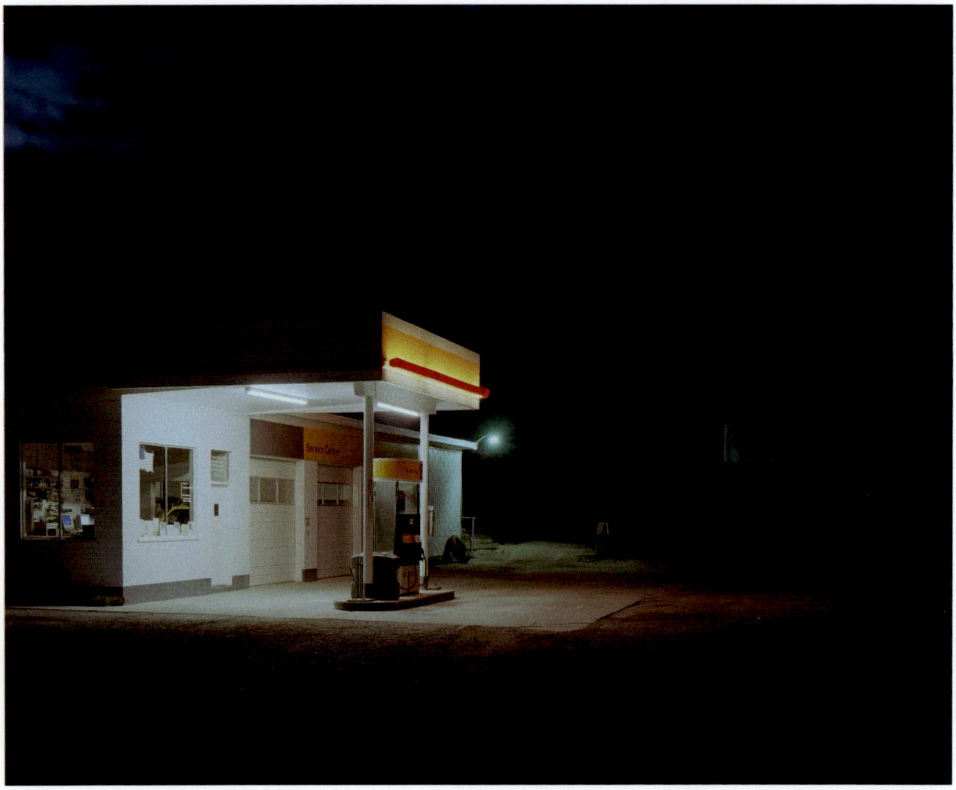

16.8 *Shell station*, 2004.

This image of a Shell Station in Gerlach at night should trigger memories of a famous American painting. Edward Hopper's 1942 *Nighthawks* depicts people seated in a cheap diner at night. *Nighthawks* is Hopper's most famous painting, admired for the way it expresses the loneliness and isolation of a big city and for how skillfully it captures the effects of artificial light. Art critics speculate that Hopper may have been inspired by Van Gogh's 1888 painting Café Terrace at Night. Here, Goin adds his artwork to an ongoing artistic thread, thereby assuming photography's status as a fine art. Goin's image of the Shell Station also references Edward Ruscha's 1960s pop art paintings of gas stations and commercial graphics, ultimately a comment on American culture. In terms of spirit of place of the Black Rock, the importance of this humble gas station cannot be overstated. This Shell Station could spell the difference between a safe trip and being stranded in the American outback with an empty tank. If the preceding image of a travertine pool represents a literal oasis, what we have here is an oasis of light in the existential desert of night.

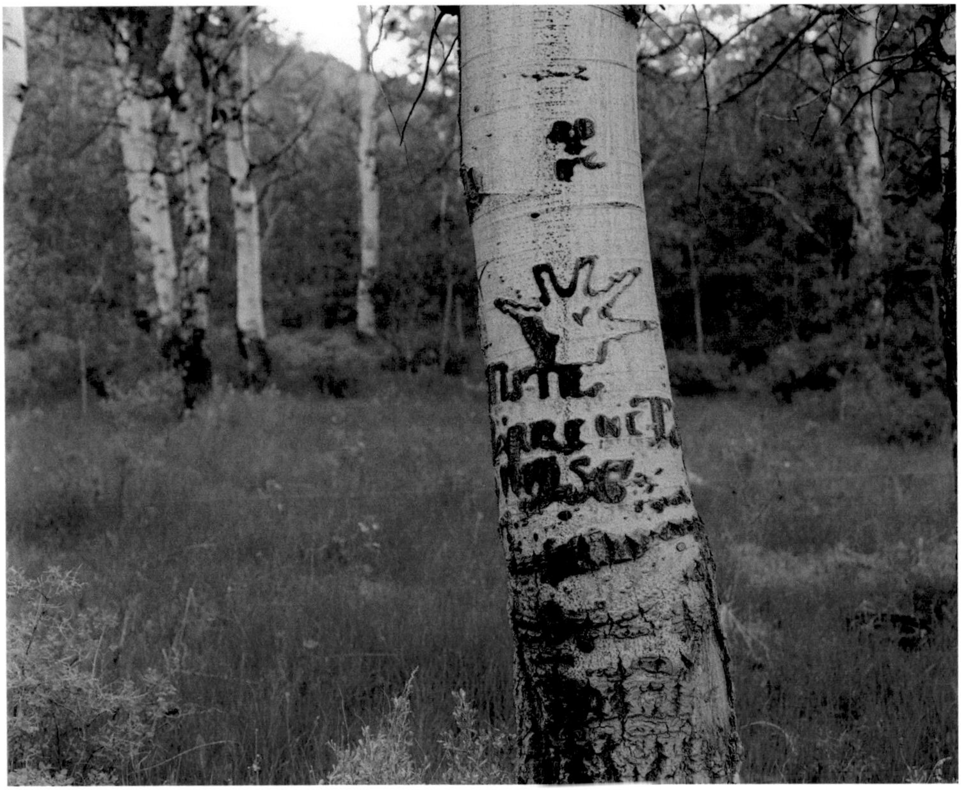

16.9 *Hand arborglyph, Summer Camp Canyon, south of Mahogany Creek, Black Rock Desert*, 2003.

In the first decades of the twentieth century, Basque sheepherders cut designs into the smooth, white trunks of aspen trees, leaving folk art groves sprinkled in the high country throughout the American West, particularly Nevada. Arborglyphs last only as long as the aspen tree lives, approximately one hundred years, so it is literally a disappearing art form, which photographs can preserve. Peter remarks:

> This tendency to mark our landscape with our identities and with our images, there's something elemental about the human need to do this. Carvings of images add complexity to our sense of understanding about the landscapes we inhabit. We are participating and interacting with the landscape. This is a way of marking a landscape which is very atavistic.

The Black Rock is not just the playa, but includes the surrounding Basin and Range country, where forested alpine environments exist, and this image represents that diversity. In the fine arts, there is a conversation about the artwork revealing the hand of the artist. But how does a *photographer* show his hand? Goin's image proffers a playful answer.

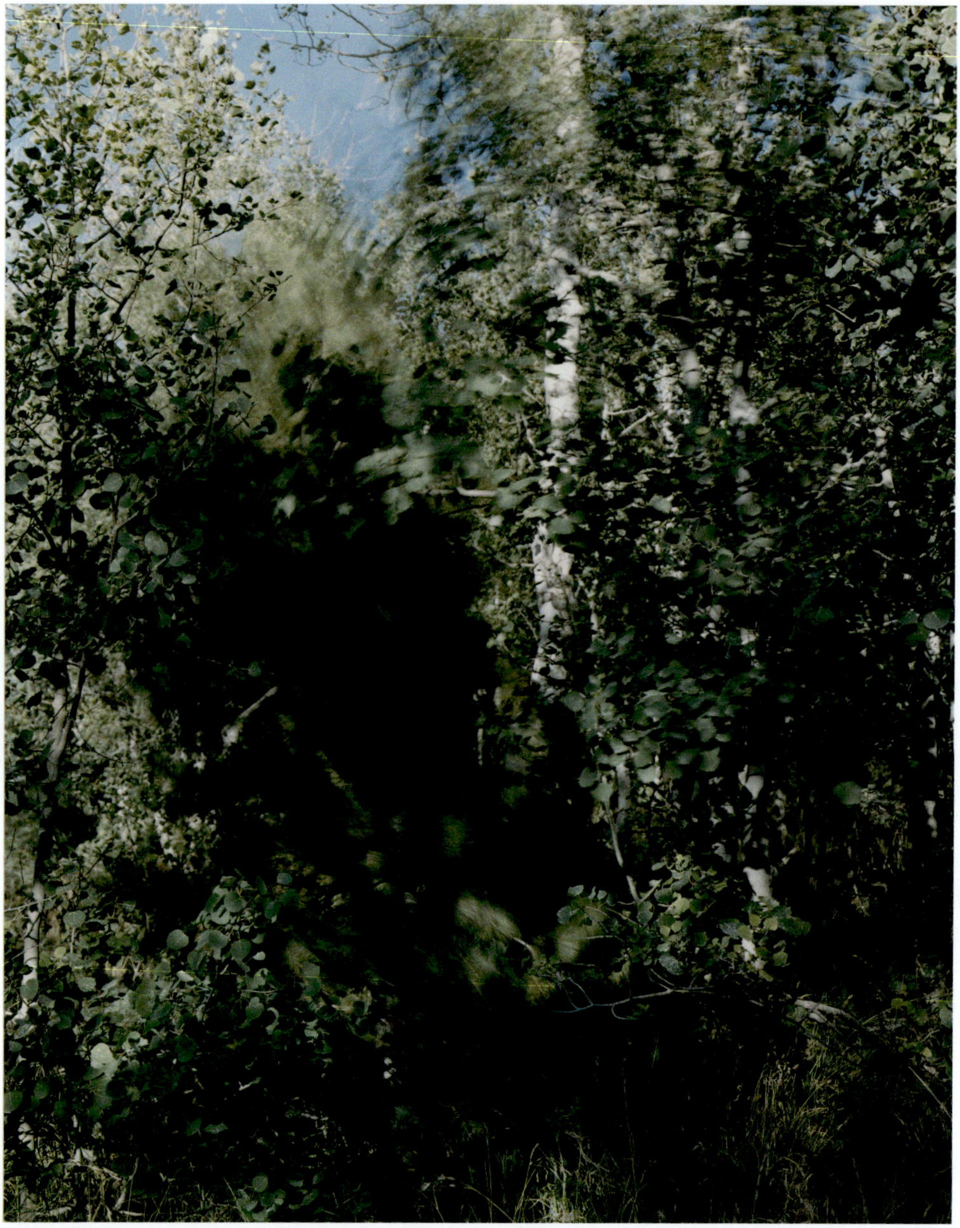

16.10 *Wind through trees, Pine Forest Range*, 2002.

This image complicates our understanding of the Black Rock. In this dense stand of aspen trees, the view is confined to a green foreground bereft of horizon. The edges of the photograph are in focus, but part of it is out of focus, as if our vision is going bad. An element of the Black Rock country is wind, but how does a photographer picture that which is invisible? Goin was camping in the high country near Blue Lakes, and he noticed a small whirlwind stirring up a few trees but not others. Today, one might whip out their smart phone, snap a photo burst, and select the best image. Back then, Goin mounted his 4x5 camera to a tripod and clicked the shutter repeatedly, using a lot of film, serendipitously capturing a whisper of wind. Conceptually, this photograph suggests how elements merge together—wind and weather and place. This interfusion happens all the time, but we rarely think about it. In terms of the history of art, this image references the evolution of photography away from realism and pictorial representation and into abstraction.

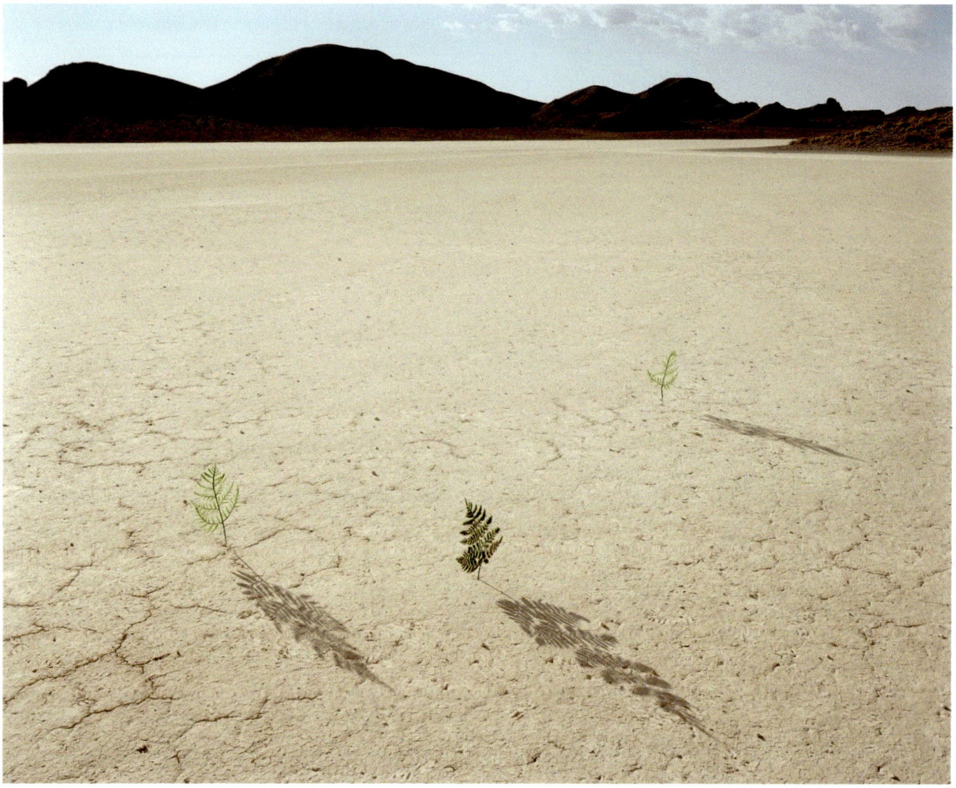

16.11 *Shadows and faux ferns*, 2004.

Chapter 11, "Humanature," included an image of a patch of plastic ferns decorating a forest floor of fallen brown leaves (refer to Figure 11.2). That image commented on synthetic nature, human-manufactured products designed to simulate nature. Here, the valence is different. Like the participants of Burning Man, Goin is interacting with playa as a canvas that invites him to make art. The out-of-place ferns, the novelty of green, and the play of shadows add a human touch of whimsy and dry humor to this inhuman environment. This photograph also subtly references time. Implied is geologic time, the photograph made during the relatively hot and dry Holocene, but alluding to the cooler, wetter Pleistocene, when abundant plant life flourished in the Black Rock and when this now-dry lakebed lay at the bottom of Lake Lahontan. Also evident is historical time, namely the modern period when plastic ferns are a mass-produced commodity and when modern art such as this photograph is de rigueur. And there is solar time of day, late afternoon, when the east face of the background hills lies in shadow and when little ferns cast long shadows.

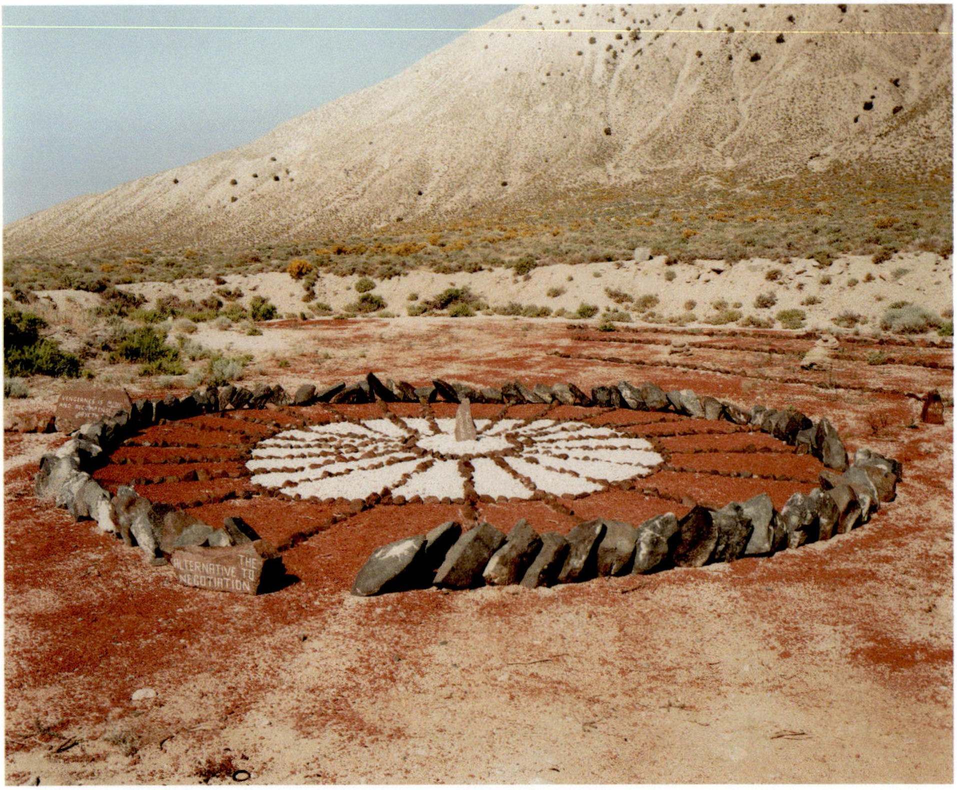

16.12 *"Ground Zero." Outsider art of DeWayne Williams*, 1994.

Goin observes that there is something about the human spirit that compels us to make a mark. The next chapter, "Ancestral Artisans," traces that universal human impulse back to aboriginal petroglyphs and pictographs. Here, Goin classifies modern-day geoglyphs by DeWayne Williams at the western edge of the Black Rock Desert as "outsider art." Outsider art is most often made by unemployed males—retired, laid off, or injured—primarily from the working class, who have always defined their self-value by producing something. DeWayne Williams, who served in the Marine Corps in World War II, settled in Gerlach in 1970 and spent the next quarter century until his death in 1995 etching rocks and arranging them into monuments and sculptures along a dirt spur road off State Route 34. Peter likes to think of Williams's work as a kind of "vernacular funkadelic street poetry" in the desert. This art piece of Ground Zero is an anti-nuclear statement that aligns well with Goin's *Nuclear Landscapes* work. In 2016 Goin collaborated with poet Gary Snyder to produce *Dooby Lane*, a book that celebrates Williams's glyphic writing of life and spirit.

Notes

1 By invitation from series editor Gregory McNamee, poet and essayist William L. Fox teamed up with photographer Mark Klett to produce a slim black-and-white volume entitled *The Black Rock Desert* (2002) for the University of Arizona Press's Desert Places series. As Fox explains in the Preface, they found themselves interested in "the function of metaphor in space and time." By their reckoning, metaphor is what it takes to get over "cognitive dissonance—the confounding of our perceptual expectations in an isotropic space, where features are uniformly distributed in all directions" (xiii). Their book is a valuable complement to the comprehensive treatment of perception by Goin and Starrs in *Black Rock*.

2 Gary Snyder and Peter Goin are good friends off the page as well as in print. Each has perennially gravitated to the Black Rock Desert as an essentially spiritual place. They collaborated on a book entitled *Dooby Lane* (2016), whose subtitle reads, "Also Known as *Guru Road*, a Testament Inscribed in Stone Tablets." The book attempts to preserve in print a mile-long outsider art installation on the southwestern edge of the Black Rock Desert, where a resident of the nearby town of Gerlach created his own message to the world. The artist, DeWayne Williams, who suffered from cancer, chiseled, etched, and painted approximately 450 different names and inscriptions, each on its own red-brown volcanic rock with a flat face like a tablet. In 1988 Gary Snyder happened upon Guru Road and recorded the inscriptions present at that time—concise nuggets such as "Never Say Die," "Don't Panic," "It's Your Trip," and "Believe in Yourself." Snyder's field notes are printed in *Dooby Lane*, along with a transcript of a conversation that he and Goin had about the site. Goin's color photographs document most of the sayings and sculptures along Dooby Lane, a valuable record given that even words set in stone are mutable—stone weathers, vandals steal, visitors augment, and mudslides displace.

Works cited

Burning Man. Burning Man Project, 2019, burningman.org/event/brc/. Accessed 6 June 2019.

"Fly Geyser, Nevada, USA." *Geology Page*. 20 March 2019, www.geologypage.com/2019/03/fly-geyser-nevada-usa.html. Accessed 14 June 2019.

Fox, William L., and Mark Klett. *The Black Rock Desert*. University of Arizona Press, 2002.

Goin, Peter, director and producer. *In Search of Ritual: The Burning Man*. Channel 5 Public Broadcasting, 1993.

Goin, Peter, and Gary Snyder. *Dooby Lane*. Counterpoint, 2016.

Goin, Peter, and Paul F. Starrs. *Black Rock*. University of Nevada Press, 2005.

Ramos, Julie. Interview with Peter Goin. *Écoplasties: Art et Environnement*, edited by Nathalie Blanc and Julie Ramos. Manuella Éditions, 2010.

Thornton, J. Quinn. *Oregon and California in 1848*. 2 vols. Harper and Brothers, 1849.

17

ANCESTRAL ARTISANS

Coming after *Black Rock*, Peter Goin's next book project, *Nevada Rock Art*, anticipates a conclusion to his Nevada work. By 2009, when *Nevada Rock Art* was published, Peter was fifty-eight and had lived in Nevada for twenty-five years. He had been making photographs for more than forty years, and, as he puts it, "I was beginning to envision that there's a beginning, a middle, and an end to my scholarly and creative expression of landscape studies." In terms of his Nevada work, Peter was beginning to think in terms of a concluding chapter. *Nevada Rock Art*, which documents Nevada's petroglyphs and pictographs, is Goin's gift to Nevada. He envisions it as a book about hope.

Nevada Rock Art defines petroglyphs as "intentionally created designs on rock surfaces. They reflect incising, pecking, carving, and abrading using a variety of tools made principally of stone" (19). Makers of petroglyphs typically choose a rock surface that has patina, "the dark coating found on exposed rock surfaces, principally in arid environments" (20). Chipping down through the patina exposes lighter colored rock below, and thus the lines of a Nevada petroglyph tend to show up light against a darker background. More vulnerable to weathering, *pictographs* are painted designs on stone, using colored pigment. Pictographs that survive the test of time are usually located under an overhang, protected from sun and rain. Given that the Nevada desert is so often perceived as an empty quarter or barren wasteland, it is astounding that there are some 1,200 petroglyph sites in the state of Nevada, harboring tens of thousands of individual petroglyphs, some of which may be 15,000 years old.

What is it about petroglyphs that might appeal to Peter Goin and inspire a sense of hope? Well, as he observes, "there's some hope in realizing that there are common elements in our humanity that endure." Chief among these is the universal human impulse to make a mark. Peter explains:

> With *Nevada Rock Art*, what I was really interested in doing was beginning to explore the concept of this historical urge to leave a mark that isn't a mark about destruction or isn't a mark of meanness or of cruelty. It's more about this artisanal need to make a mark and to represent some kind of metaphoric understanding of ourselves.

DOI: 10.4324/9781003212607-17

Recall that Peter's earlier projects documented landscapes forever altered by the nuclear era as well as landscapes gutted by large-scale mining. In that vein, Goin was bearing witness to the lasting effects on the landscape of America's warrior and consumer culture. By reaching much farther back in time to our ancestors thousands of years ago, *Nevada Rock Art* taps into a more hopeful and ongoing human legacy of art and creativity. Hopeful as it is meant to be, *Nevada Rock Art* is nevertheless controversial. This chapter explores how Peter became involved in this project, why this work is important, and what complexities and conflicts arise when we—through Goin's lens—contemplate petroglyphs, apparently simple designs carved into rock.

In doing fieldwork for such projects as *Nuclear Landscapes*, *A Doubtful River*, and *Black Rock*, Peter had many times encountered petroglyphs. That in itself speaks to his powers of observation and spirit of receptivity, because, as anyone who goes in search of petroglyphs will tell you, they are oftentimes hidden in plain sight. One can stand a foot from a petroglyph and fail to see it. In one instance, when Peter spotted a pictograph of a human hand on the Nevada Test Site, his Native American guide told him that petroglyphs will only reveal themselves to you when you are ready to see them.

Knowing of his Nevada landscape work, a couple of board members of the recently formed Nevada Rock Art Foundation (NRAF) invited Peter to lunch. The NRAF was established in 2002 by Shari Chase and Dr. Alanah Woody as a 501(c)(3) organization "that actively promotes the protection of prehistoric rock art in Nevada and surrounding areas." The Foundation "implements its mission through programs that improve archaeological knowledge of rock art and raise public awareness of rock art's heritage significance." The Foundation trains volunteers to record petroglyph sites with the rationale that:

> although it may not be possible to physically preserve all rock art sites, they can be saved for posterity by recording them in the field, and these data can be used for education, research, publication, analysis and legal protection for years to come.

"The NRAF envisions a future in which the people of Nevada and the Great Basin, value, protect, and are knowledgeable about the characteristics of Nevada rock art sites; and where research into Nevada's rock art attains international significance" (*Nevada Rock Art Foundation*).

Over lunch, the NRAF board members asked Peter if he would be interested in doing a book on Nevada rock art. Peter has never viewed himself as a photographer for hire. In order to agree to undertake such a project, he needed to feel convinced that the work was congruent with his own mission. He had already been thinking about a broad-based understanding of his own career, and he concluded that a book on Nevada rock art would allow him to reflect a part of who he is, which is essentially of an optimist. He confides:

> I'm a positive person. I am addicted to hope. And I think that I probably have a larger addiction to hope than might be totally healthy. Yet I think if you look at the history of artists there's always been a little bit of delusion of hope that allows us to create these artifacts of beauty that tell a story of our humanity.
>
> You know, of course, Buddha said, "yes, we will destroy ourselves. So now that we know this, go plant a tree." It's such a great way to live. And so, in that way, less literally but more philosophically or organically, I wanted to start working on projects that represented or incorporated on some level a vein of hope.

Peter agreed to do the book, understanding that his role was to be that of photographer. Ultimately, however, he also became the fundraiser and even played a central role in publishing the book.

Peter photographically documented Nevada petroglyphs in more than one hundred locations throughout the state, respecting the NRAF's mission to record rock art sites. For every photograph, Peter recorded GPS coordinates, and his set of geo-located images are archived in the University of Nevada Special Collections Department for use by qualified researchers. In this practical sense, Peter's work is a contribution to Nevada, as he had envisioned, providing professional documentation, using a 4x5 camera with architectural control (which can compensate for perspectival distortion to produce an accurate record) to create a historical record for posterity. Furthermore, Goin's methodology of photographing petroglyphs in context—including landscape features in the frame rather than cropping the image to show only an isolated glyph or panel—means that his photographs can be used as legal documents in a court of law to prosecute cases of petroglyph theft or vandalism.

Personally, Goin found the project a lot of fun, as it gave him a reason to go on a Nevada treasure hunt, in search of *in situ* artifacts. The book project also earned him special permission to visit restricted areas and access to knowledgeable guides, including anthropologists, local Indian guides, and Bureau of Land Management personnel. What he didn't realize was how politically complicated this project would be.

One politically charged issue that Goin encountered concerns different understandings of who is "native" and the presumed rights associated with that status. He tactfully broaches this issue in his introduction to *Nevada Rock Art* by telling a story of taking his eldest daughter, who was four or five at the time, to an indigenous prayer ceremony on the campus of the University of Nevada, Reno. The shaman offered blessings and poetry in his native tongue. And then, scanning the assembled crowd, he summoned Peter's daughter to come to him, whereupon "he waved feather arms over her head and along her body, dangled noisemakers, and then ended his performance by pronouncing, in perfect English, 'Now you are a Native Nevadan'" (19). On the drive home, Peter's daughter asked, "Dad, wasn't I a native Nevadan before the blessing? . . . Dad, what I meant to say was, I was born in Nevada. Don't I belong here too?" (19). The question of who belongs and who is native can be a divisive one, as Peter explains in conversation:

> To say that you belong in this landscape *less* because you're Caucasian than, say, a Paiute is a statement that is consistent with the divisions of ethnic identity, ownership, and heritage. It's a separating device rather than inclusive device. And that's really one of the things I'm trying to argue against.
>
> What's native? What's exotic? Where did the aboriginals originate? I'm not going to get into origin stories. I'm just saying if you want to trace the DNA lineage, where do people originate? It fits within the same philosophical debate over the role of exotic species in a biotic community. There's a tremendous romance for the purity of native species. And so we look at exotic species as interlopers, as invaders, as something that upsets this delicate balance that we presume existed before we were here. And there's a certain kind of arrogance in environmentalism in this debate, because who is to say exactly what is native and what is exotic?

Bioregionalist thinkers propose the radical notion that one can "become native" to a place by learning its plants, animals, soils, weather patterns, topography, and ecology and by harmonizing one's lifestyle to fit into and live sustainably in that place. In such a view, being "native" has less to do with bloodlines or birthplace than it does with attitude and behavior.

Debates about who is native and who belongs carry policy consequences. For example, some Native American groups wish to limit access to petroglyphs to tribal members. It is, however, a matter of conjecture whether the makers of petroglyphs that are thousands of years old are of the same lineage as the historically recent indigenous tribes in the area. Peter's view—shared by the UN in its world heritage designation—is that the makers of petroglyphs are common ancestors to all of humanity. As he opines in conversation:

> To say that these will be forever removed from our opportunity as citizens of this earth to visit these sites I think is shortsighted and more a function of guilt over the way we've historically treated aboriginal societies, rather than embracing these as evidence of the human spirit and of the ability of humanity to express and represent itself in graphic images that are more hopeful than they are violent.
>
> We are also part of this ancestral evolution. We might say, "Well, my individual ancestry is from northern Europe," or wherever you want to say; and while a Paiute would define their ancestry differently, ultimately, we share common essential qualities, which is we are human. And we share certain needs and desires for community, for procreation, for those aspects that define us as humans, and that transcends those differences.

In his essay for *Nevada Rock Art*, Goin states his position publicly: "Petroglyphs and pictographs are creative expressions that bind ancestral energies to us all. Listening to the voices of creative energy is not an exclusive privilege but a universal right, for we are human, as were they" (24). Another controversy is that certain interest groups wish to forbid photographing petroglyphs, and there are efforts to prevent, legally, any documentation of these sites. As Peter notes, "There are a number of the tribes and others who feel that photography is a way of appropriating or violating the spirit of a place." For his part, Goin believes that photography has the potential to enrich a sense of place and honor that spirit.

One thing is indisputable. Nevada's wealth of petroglyphs proves that the state's undeveloped lands are not "virgin wilderness," but, in fact, landscapes with a human history. When exploring the Nevada outback, some people enjoy fancying that they are the first person to ever step foot where they are. But a more accurate way of thinking is that wild Nevada is an open-air *museum* and that hikers are walking across evidence and artifacts of peoples who have lived there for thousands of years before recorded history. The landscape is alive historically, if only one knows how to read the clues. As Goin writes:

> Encountering a glyph—versus the more egocentric 'discovering'—is a unique experience that alters the nature of walking, of history, of time. The landscape is now written with a text whose origin may be indecipherable. . . . Petroglyphs and pictographs are conclusive evidence that humans as creative beings have been part of the landscape for thousands of years.
>
> *(21)*

In *Humanature* Goin explored the idea that humans are part of the landscape, integrated into the natural world. There too always surfing the edge of what is nature and what is human, he was interested in human-made marks on the landscape. In *Nevada Rock Art*, he projects that interest backwards in time.

If landscape is a museum, one might wonder who curates it and according to what principles. These concerns undergird Goin's book *Time and Time Again*, and we shall now make a brief side trip away from *Nevada Rock Art* to review that book, as it too investigates traces of indigeneity on the land. Coauthored with art and cultural critic Lucy Lippard, *Time and Time Again*'s subtitle is *History, Rephotography, and Preservation in the Chaco World*, published by the Museum of New Mexico Press in 2013. Lippard explains in the introduction that "This book is a visual and textual speculation on the manner in which concepts of history and time affect the way we see the ruins, cultures, and landscapes" of Chaco Canyon, New Mexico, and Mesa Verde, Colorado (9–10). The origin of the project was a conference that Goin participated in in 2000, organized by the Center of the American West on "Chaco, Mesa Verde, and the Confrontation with Time." To prepare his presentation for that conference and for ten years thereafter, Goin located historical, black-and-white photographs of these two sites and rephotographed the same views in color to document change through time.

The relationship to time at these ancient Puebloan sites is complex. Different ways of thinking about and measuring time—astronomical, historical, cyclical, nomadic—reflect fundamentally different world views, and employing one dating system tends to occlude other ways of experiencing time and the past. Just as a photograph creates an artifact of a place arrested at a given moment on the time continuum, so too restoration of these sites pins them to a particular moment in time. As Goin observes:

> From a conceptual point of view, restoration is traveling back into time, to a place where the purity of a site's architectural integrity is presented as a representation of an ideal. . . . Ultimately . . . a site is reconstructed within a presumptive visual context, specifically the romance of the 'ruin'.
>
> *(124)*

Both Chaco Canyon and Mesa Verde are *humanature* landscapes, managed and curated by the National Park Service to protect ancestral artifacts, strive for archaeological accuracy, and enable visitation.

Compared to Chaco and Mesa Verde, Nevada petroglyphs lie under the radar, unseen by most people. The vast majority of Nevada's rock art sites are not marked or restored. Rather, they are subject to sun, wind, and weather and are vulnerable to defacement by vandals. Instead of arrested decay, we have ongoing evolution. Goin's photographs in *Nevada Rock Art* document petroglyphs at one moment in time, the first decade of the twenty-first century. And whereas Chaco and Mesa Verde are preserved as monuments of the past, Nevada rock art, at least in Goin's view, is a tradition that is still alive. As he explains in conversation:

> we look at petroglyphs and we say, 'Oh, prehistory.' No, it is simply prehistory that continues. So, we have to look at that, and a kind of undercurrent narrative of *Nevada Rock Art* is that it continues to this day. In this book, I include photographs of

> contemporary petroglyphs that are as complex and as difficult to understand as those that clearly predate us and our civilization by centuries.

"We feel compelled to make a mark," Goin contends:

> One of the universal impulses of human nature is to express ourselves creatively. In marking the landscape, we establish a presence. We leave behind a variety of different references and symbols of our existence, proclaiming, 'I am here; therefore, I am.'

In Goin's view, one of the essential characteristics of human social identity is to be of a place, or to claim a place, or to engage with a place. Evincing an interactive relationship to place, petroglyphs are an enduring reflection of the human spirit.

Nevada petroglyphs can seem inscrutable and mysterious—a puzzle without an answer key, a coded message without a codebreaker. Anthropologists debate the meaning of petroglyphs, some scholars hypothesizing that certain glyphs are tied to shamanistic rituals, while other glyphs—depicting bow and arrow or spear and hunted animal—are related to hunting practices. Experts read certain petroglyphs as maps, other glyphs as solar calendars, and others as indicators of water. Although we may not definitively know the *meaning* of petroglyphs, we can still connect with them on an emotional level. We are human. The makers of petroglyphs were human. Goin writes, "Connecting to an object or glyph made a thousand years ago is a reach across time, sparking imagination and a visceral recognition deep within the soul" (23).

If petroglyphs are a form of enduring artistic expression still discernable if not fully decipherable by people living hundreds, thousands, and even ten thousand years later, what artworks do people of today create that will endure far into the future? So much is up to chance. In the realm of print, it is generally the case that the fewer copies of a work that are printed, the more highly they are valued and the longer they are likely to be preserved. Goin had experienced this phenomenon himself with his 1987 book *Tracing the Line*. He kept one personal copy, and the other fourteen copies are preserved in the best archives in the nation. It is perhaps this logic that led Goin to conceive of publishing *Nevada Rock Art* as a limited edition, fine art book. One thousand signed and numbered copies were printed. This beautifully designed, weighty book measures 12¼ inches wide, 11½ inches tall, and 1⅛ inches thick, is printed on acid-free paper—Chinese Matte Art, 157 gsm weight—and slip-cased in arrestox b in Paprika. Of the book's 252 pages, 165 are devoted to the Rock Art Gallery of Goin's photographs. Hence, even if a petroglyph is lost to weathering, theft, or vandalism, and even if future access to it becomes restricted, this petroglyph will be preserved and available for contemplation by future generations in *Nevada Rock Art*.

Nevada Rock Art is the inaugural publication of the Black Rock Institute Press, founded by Peter Goin and his *Black Rock* coauthor and University of Nevada, Reno, colleague Dr. Paul Starrs. Because Goin co-directed the press, he was able to exercise greater-than-usual control over the final product, and *Nevada Rock Art* helped bring attention and future projects to the Institute. According to its website, the Black Rock Institute, based in Reno, Nevada:

> is a nonprofit, tax-exempt 501(c)(3) organization founded in 2009, that fosters public interest and understanding of landscapes throughout the American West. Projects and

> research supported by the Black Rock Institute focus on the diverse and complex reality of western landscapes, which may include sense of place, lyrical interpretation of the nature of mountains and aridity, inquiry into demographic and historical change in the West, or examination of the material reality of physical landscapes.

It is no small feat to publish a book, let alone create a dot-org institute and publishing house from the ground up, let alone do so while fully employed as a university professor and single parent of two children. However, because during these years the University of Nevada Press had made cutbacks and was turning away worthy projects, Goin felt strongly that another venue for Nevada writing needed to exist, and he was able to draw upon his previous experience in the book industry to make it happen. Following *Nevada Rock Art* (2009), the Black Rock Institute Press published *Bombast: Spinning Atoms in the Desert* (2010) by Michon Mackedon, former member of the Nevada Commission on Nuclear Projects, and *Death After Life: Tales of Nevada* (2015) by "Nevada scholar-resident-writer-academic William A. Douglass" (*Black Rock Institute*).

The text of *Nevada Rock Art* includes a preface and introductory essay by Peter Goin; a scholarly review essay on changing archaeological theories concerning petroglyphs by Dr. Angus R. Quinlan, who was at that time executive director of the NRAF, and Dr. Alanah Woody, NRAF's founding director; a rumination by cultural geographer Paul F. Starrs on petroglyphs, perception, and failures of perception; and an epilogue by Bureau of Land Management archaeologist Mark Boatwright on petroglyph conservation issues in Nevada. Goin's preface, "Between a Rock and a Hard Place," characterizes the book as "a study of Nevada's petroglyphs and pictographs laid out for serious students of the archaeological record," then broaches the restricted access controversy, and introduces the contributors (9). His essay, "If Only Lines Could Talk," takes a more lyrical and personal approach, beginning retrospectively by revisiting the questions that have sparked his decades-long photographic journey, musing on "how way leads on to way," to lift a line from Robert Frost's "The Road Not Taken." Goin writes, "With no real rest for days, driven by intangible questions, and compelled to finally finish this quest, my journey seems endless" (14). Without explicitly naming the nine- or ten-book projects that preceded *Nevada Rock Art*, Goin nevertheless alludes to each by theme and guiding inquiry. Key questions include: What does a line imposed on a landscape create? How does a photographer photograph that which cannot be seen? How are photographs and time interrelated? How does a visual artist communicate the insight that nature is a social construct? And, overarching virtually all of Goin's work, what projects must be undertaken to responsibly bear witness to our era and its consequences?

In contrast to those who regard petroglyphs as random doodling or graffiti, Goin contends that the makers of petroglyphs are artisans, that is, skilled at a craft. If you have ever tried to make a petroglyph, you will understand the labor, time, and skill required. As he puts it, "a petroglyph requires planning, scripting, and committed completion" (20). Whereas archaeologists may strive to "decode" their meaning, Goin is more interested in appreciating petroglyphs as art. He points to modern art and abstract photographs as having taught us that *form* can convey psychological meaning even when a work's content is non-representational. On the basis of form, we can emotionally connect with the work. Goin employs the metaphor of text to help viewers apprehend a petroglyph's graphic elements:

> The edges, the density, the pattern and angle of sculpted strikes function . . . much as a font may emphasize the title of a book or the mood of a story. The negative spaces and image/position relationships function as letter kerning and type leading; the rhyme and sequence function as specialized punctuation, creating subtle complexity in a compound sentence.
>
> *(23)*

One thing that intrigues Goin when he thinks about petroglyphs in their landscape context is that aboriginal artisans appear to have been integrated into the social fabric of their societies. In contrast, in his view, our own society seems to take some pleasure in excising artists and considering the artifacts that they create nonessential. In conversation, Goin speaks forcefully on this issue:

> In this contemporary era, we tend to devalue the creative acts of children and of adults by ostracizing the products and the artifacts of that expression into sanitized white walls and institutions that we call museums, and to consider artistic production a non-essential commodity. I would argue that it is a fundamental essential commodity. Art builds community, it builds history, it reinforces society. That we underestimate art is more a function of our market economy than it is of the importance of these things. But what are the objects that will endure? The artisan object can last thousands of years, and these will be the objects by which we're defined.

Perhaps the closest modern-day descendent of prehistoric petroglyphs is public art—open-air sculptures, monuments, murals, and architecture. Such pieces may be viewed by anyone in the vicinity, any day. Parents take their children to see artworks that they themselves may have seen when they were young. Pedestrians hurry past a colorful mural one day, absorbed in their own thoughts, and the next day suddenly stop to notice it. Sculptures and buildings become landmarks, and people orient themselves by them—"turn right just before you get to the statue of the giant butterfly." Public art is art that lives among us. Goin has contributed to the public art of Reno. His photographs of Lake Tahoe, printed as large outdoor banners, have welcomed travelers at the Reno-Tahoe International Airport. He and his colleague Scott Hinton participated in the redevelopment of downtown Reno by designing inlays of Nevada cattle brands, set into the concrete platforms at three city bus stations, a contemporary analog to ancient petroglyphs. Goin's artist's statement reads:

> I believe that the graphic quality of these [cattle] brands is engaging and dynamic, and their history in folklore and legend made them an ideal element for brass and stainless steel imbeds into concrete. Passengers stand, wait for the bus, and often look to the ground for obstacles, for directions, for safety . . . and to have a surprisingly unique part of Nevada's history embedded beneath their feet might provide a moment that can best be described as a celebration of lightness of being. Perhaps, for a second or more, a passenger can enjoy a visual surprise that connects them to the history of Nevada, and beyond.

In the Washoe County Regional Transportation Commission's statement about public art, listen for a thousand-year-old echo: "Integrating art helps build a sense of pride, ownership, and identity with the surrounding community" (*RTC Public Transportation*).

As we now consider Goin's photographs of petroglyphs in *Nevada Rock Art*, try thinking of them as public art, art that their cultures lived with. These designs are so well integrated into the landscape that if the readers of *Nevada Rock Art* blur their eyes or stand back, the petroglyphs disappear, and what one sees is a photograph of rocks, cliffs, canyons, mountains, and desert vegetation. Upon closer scrutiny, however, one notices fine examples of the wide range of petroglyph designs characteristic of the Great Basin. There are representational human-like "anthropomorphs," animal-like "zoomorphs," insect forms, snakes, fish, plants, shields, and *atlatls* (spear throwers). And there are "reductive rectilinear abstract designs, curvilinear designs, geometric motifs—especially zigzags—cross-hatching, wavy lines, parallel lines, rakes, bisected circles, concentric circles, dots, scratched dendritic and grid patterns, and spirals" (21).

Intriguingly, *Nevada Rock Art* also includes photographs of ancient-looking petroglyphs that nevertheless depict recent history—a man in a cowboy hat carrying a rifle, a horse with a western saddle, a wagon, a frame house—as well as obviously more recent glyphs that incorporate English words, names, initials, and dates. Some petroglyph panels are very crowded, suggesting that multiple generations of artisans added their work to an existing site and embellished or overwrote earlier glyphs. Because some petroglyphs show up best in sun while others are most visible in shade, for each image Goin had to patiently wait for the best lighting conditions. While waiting, he could look across the landscape to distant hills or basin and range, imagine being transported back in time hundreds or thousands of years, see the light that they might have seen, breathe the air, wonder where the next hunt or the next campsite might appear, and be absorbed in that quintessentially human activity—making a mark.

After paging through the photograph gallery and illustrated essays of *Nevada Rock Art*, one is amazed to realize that Nevada is such a graphically busy state. "Where *are* all of these petroglyphs?!" the reader may wonder, "Where can *I* see them?" *Nevada Rock Art* won't tell you. It is not a field guide. As Goin forewarns in the preface, "The photographs in this volume are geographically shuffled, mostly to prevent unscrupulous access. Accordingly, they are provided without map coordinates, captions, or other textual identifiers" (9). Without photograph captions or even labels, the reader can only make guesses about the age, subject, significance, season, time of day, and, of course, location of a given petroglyph in *Nevada Rock Art*. The book tantalizes by offering an unmarked gallery, presented as timeless art. Provocatively, *Nevada Rock Art* whets the reader's appetite to see these petroglyphs in person, and then counterpunches with the message, "ACCESS DENIED." The petroglyph gallery in this book follows suit, providing only simple, descriptive titles and generalized locations for the photographs. In this sense, the reader experiences the complexities that Goin stumbled into when he agreed to undertake this project. If you are determined see petroglyphs in the field, join the NRAF and get involved. For now, contemplate the photographs below and imagine that you are in the picture. The captions provide a glimpse into Goin's mind as he beheld these sites and composed these images.

■

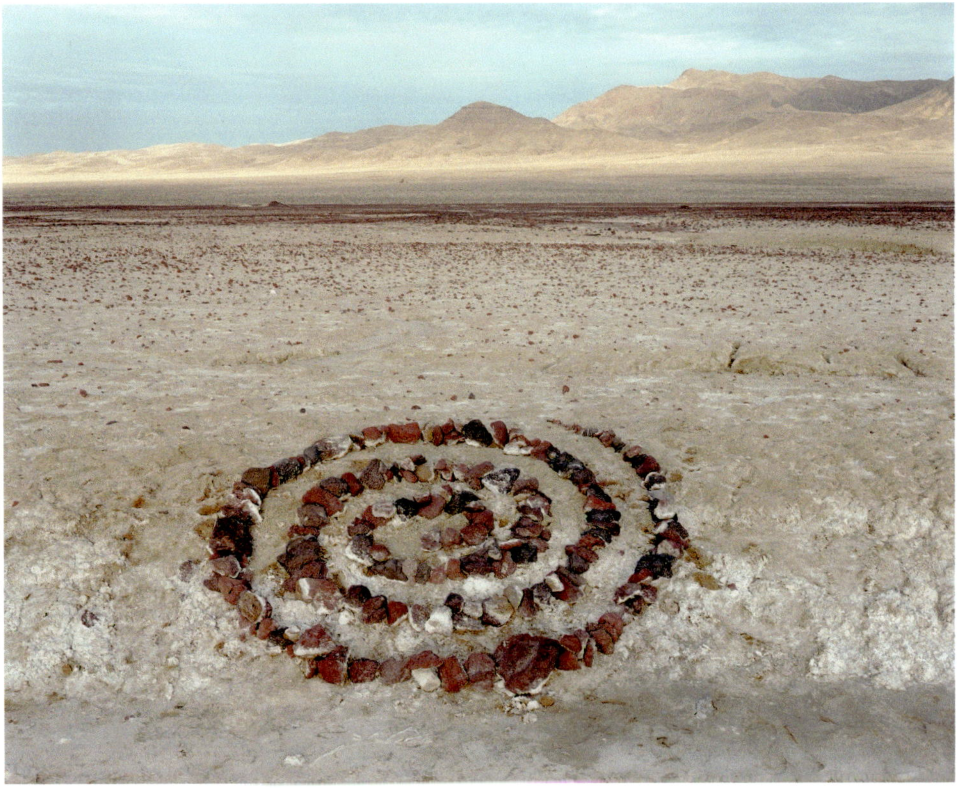

17.1 *Spiral, alkali flat, southeast of Fallon*, 1999.

The last image of the previous chapter was a modern-day geoglyph installation by DeWayne Williams entitled "Ground Zero." Williams's sculpture intimates that Nevada rock art is not a bygone art form, but is still being produced. Indeed, Goin's scholarly opinion is that Nevada rock art is an ongoing tradition, a position buttressed by his photographs of modern day petroglyphs, pictured later in this chapter. Aboriginal people also made *geoglyphs*—the artful placement of rocks—that survive to this day. How then is one to date a geoglyph? In this case, the location of this spiral geoglyph—along Highway 50, southeast of Fallon—is a giveaway, as that stretch of road is highly decorated with contemporary rock designs by passersby who stopped their cars to add their own creation to the roadside gallery. Peter Goin himself constructed this geoglyph in 1999, inspired by a Hopi belt buckle with a spiral motif that his wife, Chelsea, had given him. In making this spiral, Peter engaged with the landscape anonymously, as did aboriginal artisans. In *Nevada Rock Art*, the provenance of Goin's art piece is not revealed, and Peter thinks of it as a Hitchcock-like performative cameo appearance in his own book.

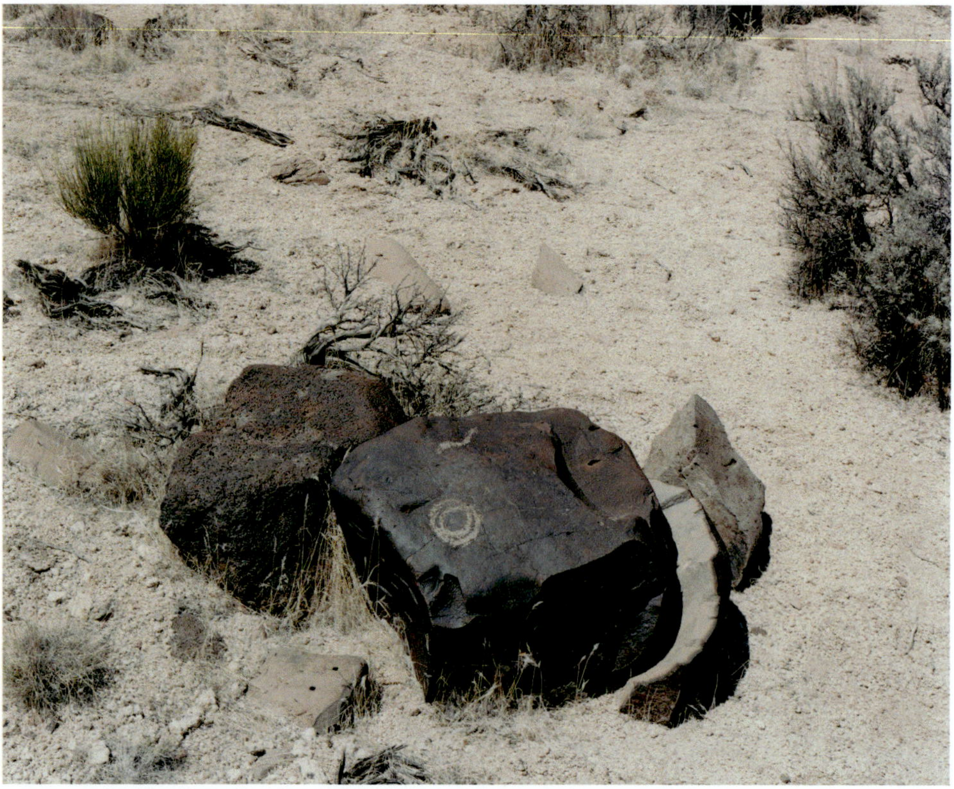

17.2 *Double circles, Nevada Test Site*, 2007.

This photograph shows the dark patina that forms over millennia on some of Nevada's rock surfaces. Scratching through the patina layer exposes lighter, unweathered rock below, forming a petroglyph, which is subject to continued weathering. The degree of weathering is a clue to the age of a petroglyph. In this example, the host rock on which the concentric circles are drawn is located in a circle of rocks twelve to fifteen feet in diameter that were intentionally moved there. The circular geoglyph mimics the shape of a semi-nomadic aboriginal dwelling and the opening in the circle of rocks faces east, directly in line with sunrise. One speculation is that this petroglyph marks a home site. Once you become attuned to such patterns in the landscape, you begin to notice other oddities, such as a mound with lime and other nonlocal rock that may indicate the site of a kiln. That this possible home site occurs on the Nevada Test Site, itself marked with human-made circles, is an irony, in that the modern circles are atomic bomb craters, too radioactively contaminated to linger in, let alone dwell in.

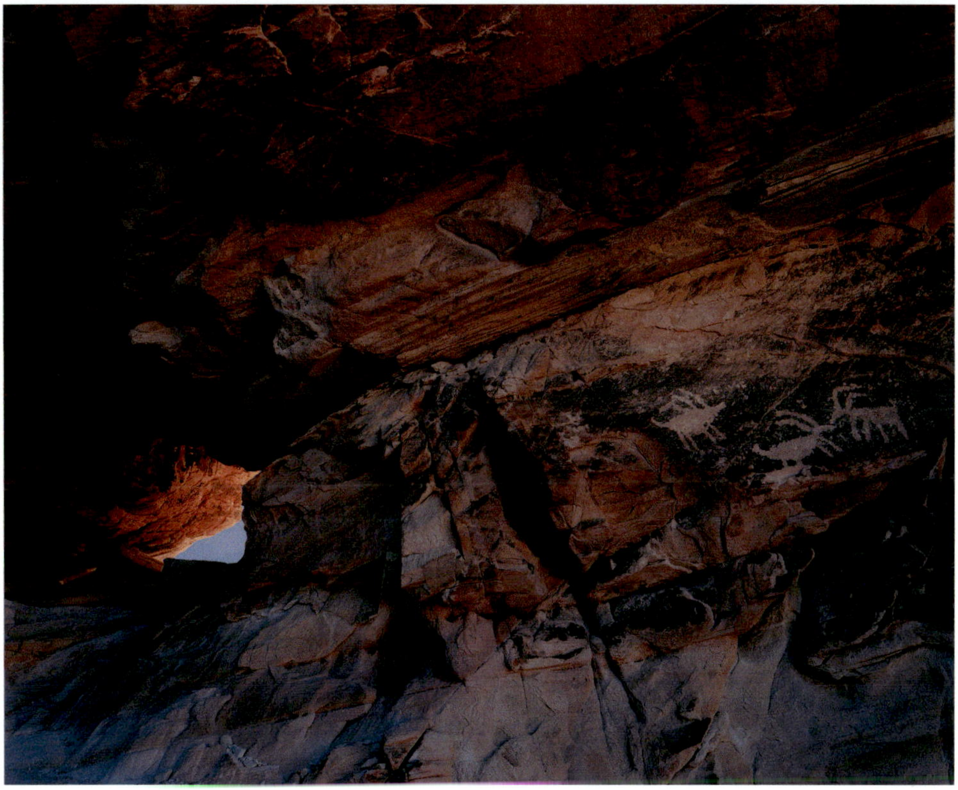

17.3 *Animals on wall, southern Nevada*, 2004.

These animals are just above a high ledge that is shaded in the afternoon. The site would have been an ideal place for aboriginal hunters to sit and survey the landscape, waiting for game to pass by. Petroglyphs such as these, with well-defined edges and a carefully considered design, take a long time to make, perhaps days or weeks. In these glyphs, note the positions of the legs, the parallel lines, the attentiveness to detail. The interaction of the two animals at right may recall a narrative. Such sites probably indicate an area where people lived or at least spent significant time. The artisans who made the petroglyphs were thus integrated into their societies in a way that modern-day artists are not, at least as Peter sees it. These representational animal petroglyphs give us a sense of how people at that time perceived the landscape and what elements were important to them.

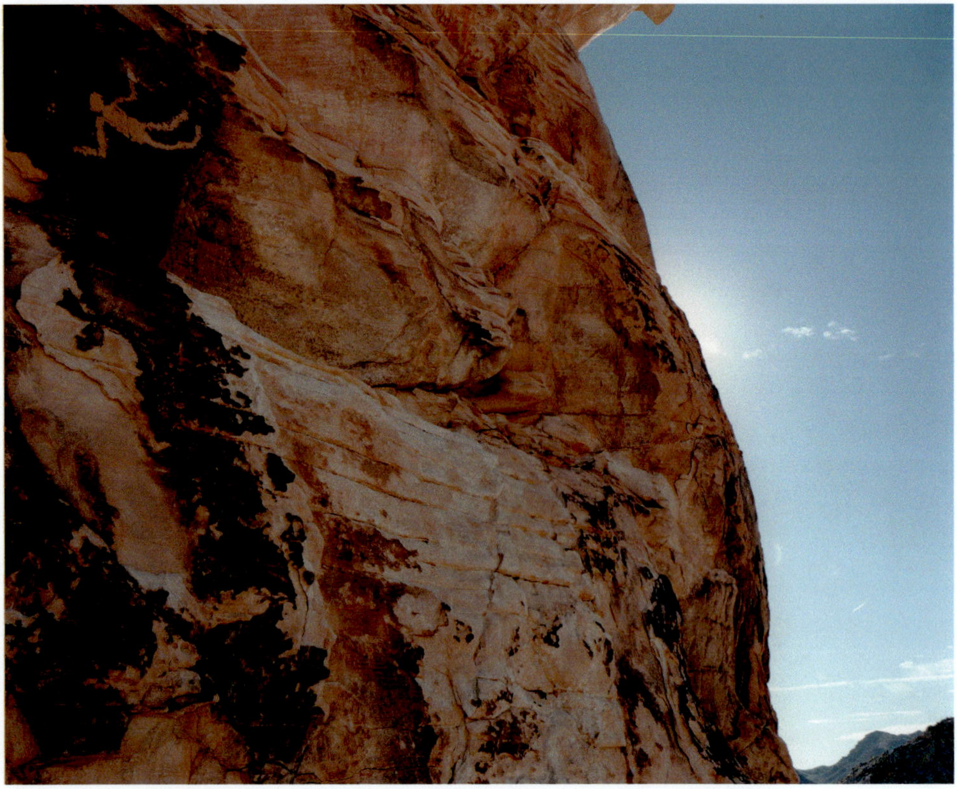

17.4 *Falling figure*, 2004.

This petroglyph, which figures as a design element of this book's frontispiece, is popularly known as the "falling man" petroglyph. It is positioned high up on a cliff face. To fall from the cliff edge above or the narrow ledge just below this petroglyph could mean death. Whether it be a memorial to a friend or a caution to children, this figure surely signifies a fall from height. Because the petroglyph is above the reach of an average adult, Peter speculates that the maker of this petroglyph may have stood on a ladder placed on the ledge. A ladder would have allowed the artisan to reach the dark patina overhead, an ideal canvas for the design. Even to petroglyph experts, "falling man" stands out as special—its narrative still seems intelligible to us today, and its design seems uncannily modern. Is this glyph two hundred years old or two thousand? We do not know. For his part, Peter felt precarious making this image. He was perched on that same narrow ledge, where a fall would be dangerous. Using a 4x5 camera with tripod required that he crouch under a black cloth, which was unnerving, especially as he was hurrying to make the image before the sun peeked around the cliff edge.

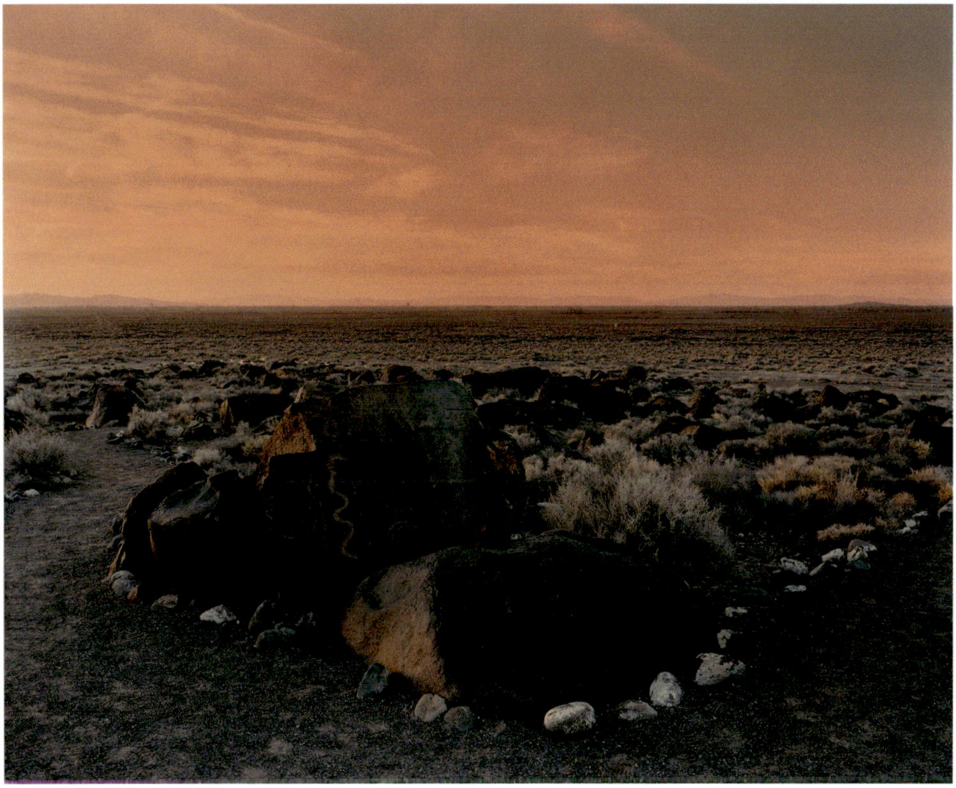

17.5 *Grimes Point*, 2004.

The orange sunset lighting in this photograph is intensified by smoke from wildfires at the time, adding a note of the temporal, the momentary, and the meteorological circumstance to a subject thousands of years old on the same palate—point and counterpoint. This landscape composition reveals the spectacular context of this petroglyph, an awareness that would be missing if Goin had focused solely on the rock art designs in isolation. In ancient times, today's sagebrush plain was a vast lake and marsh, rich habitat for birds, fish, reptiles, and humans. In the foreground, note the path bordered by whitish rocks and—in the middle ground—the glint off the horizontal row of parked vehicles in the left center of the image. This snakelike petroglyph is located at Grimes Point, Nevada, on Highway 50, east of Fallon. Grimes Point was inhabited by Native Americans more than ten thousand years ago, as evidenced by the radiocarbon-dated skeleton of the Spirit Cave Mummy found near this site, which scientists determined to be 10,600 years old. Visitors today can view examples of petroglyphs along a short, self-guided interpretive trail, originally constructed by the Youth Conservation Corps and named Nevada's first National Recreation Trail in 1978.

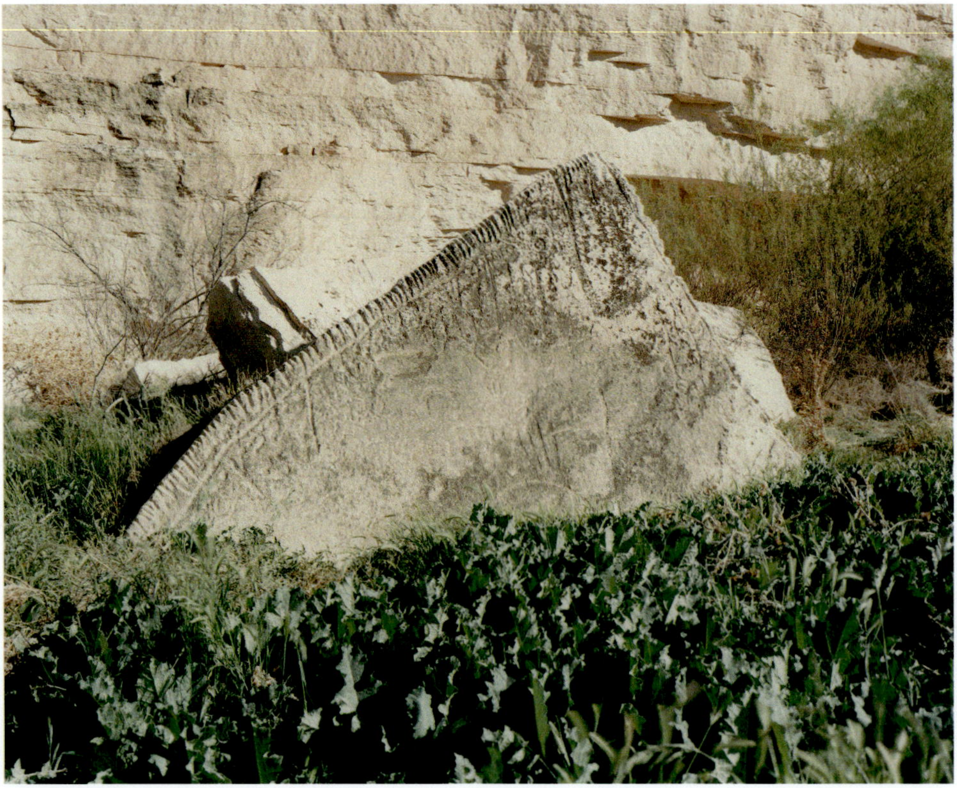

17.6 *Notched triangular rock*, 2007.

Wallace Stevens' poem "Thirteen Ways of Looking at a Blackbird" offers multiple ways of perceiving a blackbird, none definitive, each adding a facet of significance to a common bird that may very well have gone unremarked were it not for the poem's making it an object of contemplation. Similarly, Goin's image invites us to notice a rock and contemplate it from multiple conceptual angles. One facet is the fieldwork that Goin undertook to make the image. He hiked through a canyon, aware that people walked in that very place thousands of years ago. Another facet is the vegetation, mostly invasive, a subtle reference to the debate about what and who is "native." Were aboriginal nomadic cultures "invasive?" What about the meaning of the notches along the edge of the rock—calendar? seasonal timepiece? Then, there is the varying depth and degree of weathering of the markings, likely indicating a progression of narratives and interaction that may continue to this day. Also, why *this* rock, why *this* space, which lies at the junction of three canyons. Whereas scientists aspire to draw conclusions, petroglyphs teach us to live with uncertainty.

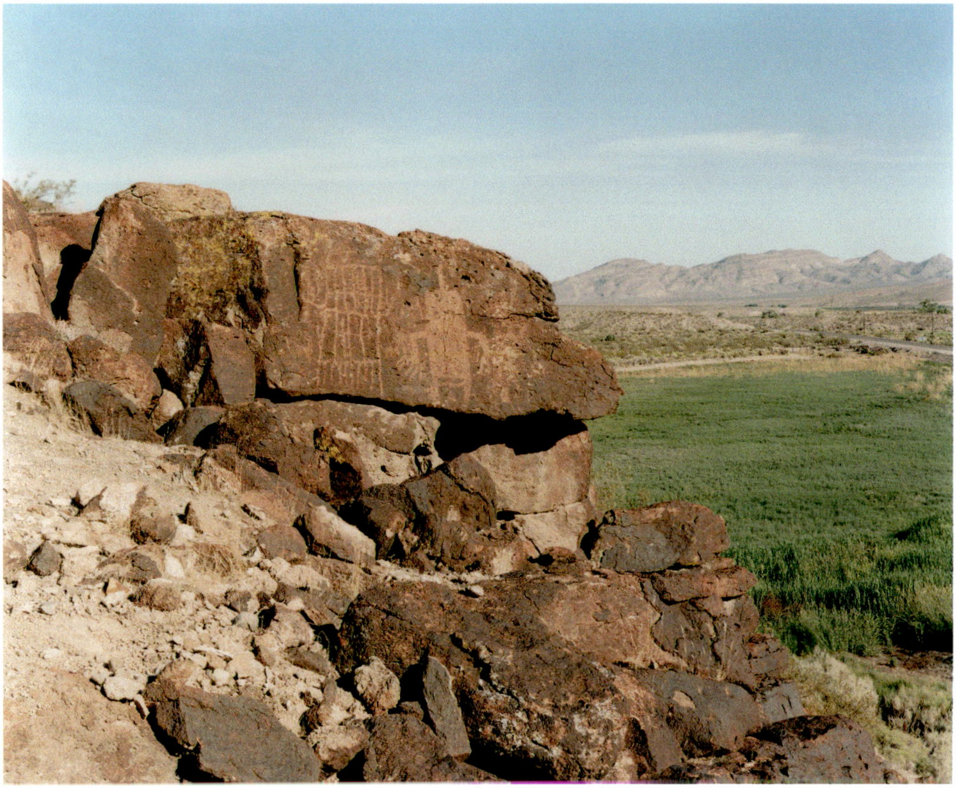

17.7 *Pahranagat Man*, 2007.

Here is a rectangular bodied, human-like figure with a spike sticking up out of its head and with splayed fingers. Resembling a video game character or space-age robot, this style of anthropomorph, which recurs near the Pahranagat National Wildlife Refuge in southern Nevada, has been given the name Pahranagat Man. "Pah" is the Paiute word for water, and naturally occurring springs water the verdant alfalfa field in this photograph, forming vivid juxtaposition of modern agriculture and ancient desert rock art, as if we are looking at not just two different cultures but two different climate regimes. In her essay "Fossil Water" about the Pahranagat Wildlife Refuge, Diana Kappel-Smith writes that according to experts in desert hydrodynamics, "the water that fills Pahranagat now is from rain that fell twelve thousand years ago." It is called "fossil water." The rectangular grid to the viewer's left of Pahranagat Man invites speculation—rug? weaving? map? game? Grids are a common petroglyph design in Nevada, an informational or organizational system whose meaning is open to interpretation. Peter appreciates that he was able to photograph this petroglyph, because public access to this site may soon be restricted.

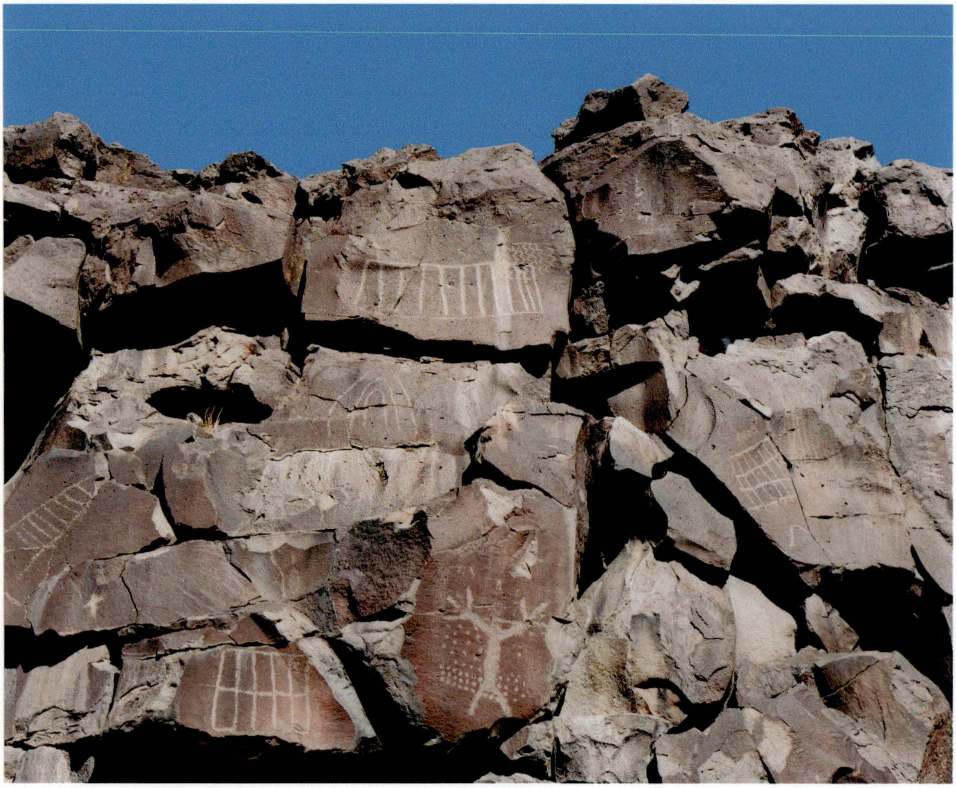

17.8 *Lagomarsino Canyon*, 2004.

Among petroglyph enthusiasts, Lagomarsino Canyon in northern Nevada is a favorite destination, although one must have a good map and 4-wheel-drive vehicle to reach it. Approximately two thousand petroglyphs decorate an extended, basalt cliff band and talus slopes below. The endpapers of *Nevada Rock Art* display a collage of thirty photographs Goin made at Lagomarsino, representing a stunning variety of geometric and representational designs and clearly evincing a long time span, including faint marks almost erased by time; bolder, fresher glyphs that show up vividly against the dark patina; and even a recent petroglyph of the capital letter "A." The image of the "A"—a subtle allusion to Hawthorne's *The Scarlet Letter*—is a potentially controversial photograph, as it may seem to countenance defacement. Nevertheless, the "A" glyph supports Goin's argument that petroglyphs are "prehistory that continues" into the present. One original interpretation of the Lagomarsino grid designs, made by Peter's wife, Chelsea, a scholar of material culture and hand-woven, decorative rugs, is that grids represent the tradition of weaving, including patterns of warp and weft, loom shuttle, and fabric. This revolutionary theory recognizes the role of women in ancestral societies and petroglyph design.

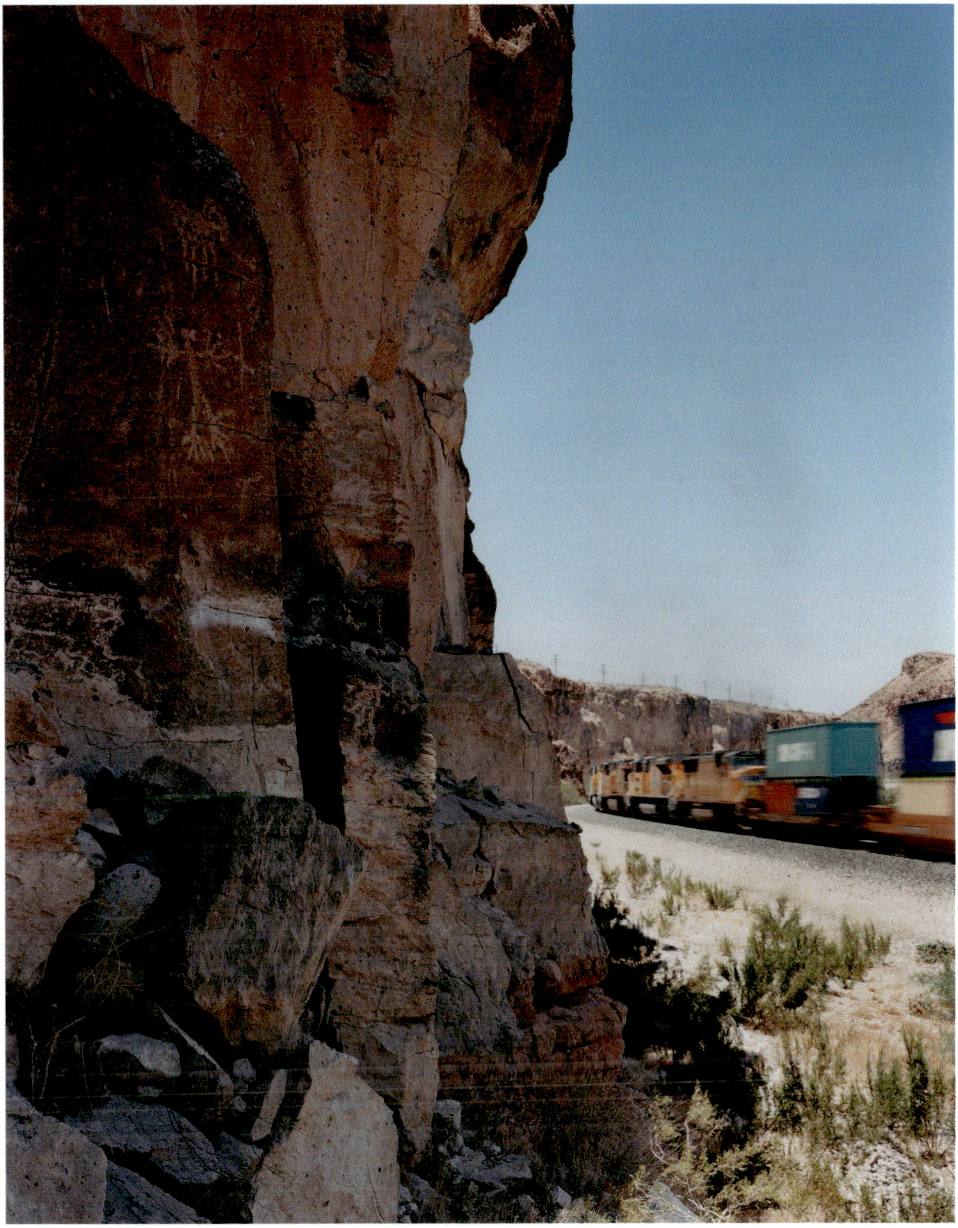

17.9 *Petroglyph, railroad, Clark County*, 2007.

In this chiaroscuro, the ancient petroglyph in shade contrasts with the blindingly bright incursion of a speeding train and row of power poles, striking examples of the long history of human marks on the landscape. Amateurs call this anthropomorph a shaman, but that reading, and the assumption that petroglyphs served shamanistic rituals, is speculative. Making this photograph was not for the faint of heart. As Goin recalls, it was about 110 degrees, and he and his assistant walked about a mile from their cars and then waited for about three hours for a train. The assistant suffered heat exhaustion and thought that Peter must be crazy to do this kind of work. Compositionally, the lines in this image—horizontal rock seam at left blending into rimrock at right, and the irregular vertical line of the cliff edge against the sky and ground—divide this image into quadrants, forming a grid that subtly references the grid designs of some petroglyphs. Note, too, how the 4x5 camera renders both foreground and distance in sharp focus, against which the blurring of the train signals speed and the smoke from its exhaust smudges the sky.

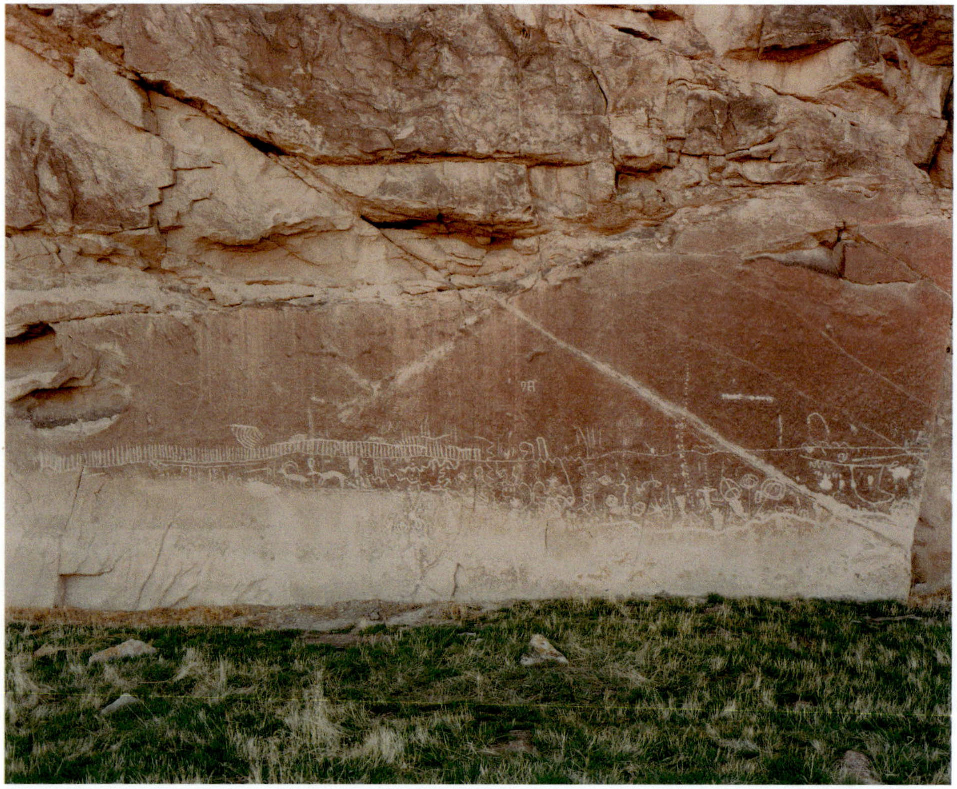

17.10 *Petroglyph panel, southern Nevada*, 2005.

To Goin, this busy petroglyph panel represents the cacophony of life, equivalent to the hustle and bustle of aboriginal experience. Whether the makers intended it or not, this panel seems to imply a narrative. Hatch marks along the horizontal line may indicate chronology. The washed out lower petroglyphs suggest that sometime after the petroglyphs were made, the water level rose and scoured the rock face, possibly depositing rocks and sands that effectively raised the ground level, which may explain why the contemporary carvings—"'78" and "MIKE DAVIS"—are higher up than the prehistoric panel. Goin suggests two interesting ways to think about this image. First, perhaps this panel correlates to sound, analogous to a musical score. Second, note how the image is framed such that the diagonal seams and scratches on the cliff form a giant "X." In black-and-white film photography, photographers drew a big black "X" across negatives they did not want printed. Metaphorically, the "X" on this image signifies its inscrutability to present-day viewers—conceptual access denied.

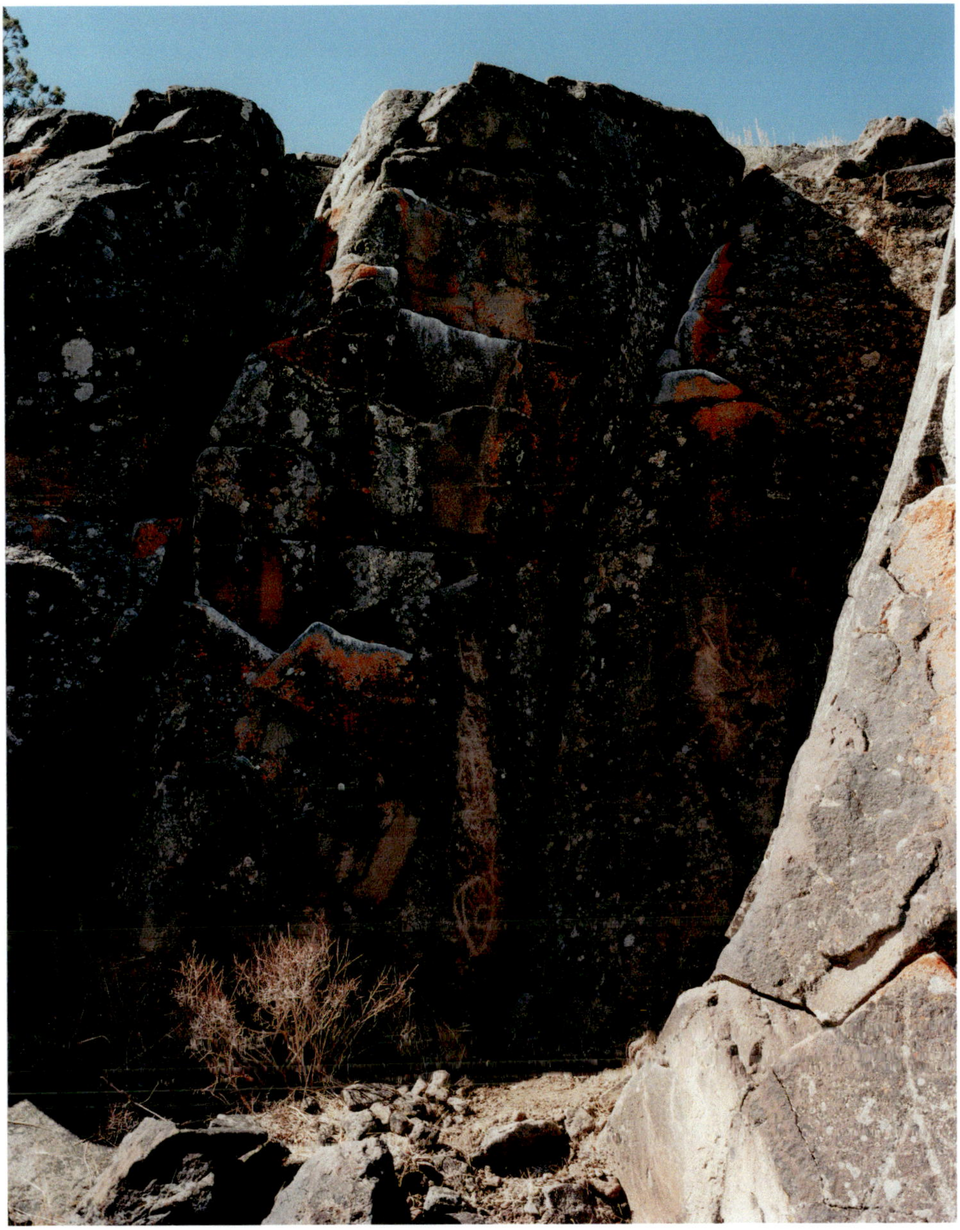

17.11 *Petroglyph fish near Massacre Lake*, 2007.

Can you spot the large petroglyph of a fish, hanging upside down from its tail? Is it a large petroglyph of a small fish or a life-size petroglyph of a big fish? Probably the former, a three- to four-foot-long graphic equivalent of campfire tales of the big fish that got away. This petroglyph is located near Massacre Lake in northern Nevada. The whitewash on the rock outcroppings indicate bird perches, a clue that the Massacre Lake environs is a good source of food for wildlife. The petroglyph itself is a curiosity. Its style seems oddly modern, yet the weathering it has undergone suggests that it is quite old. One theory is that this glyph was made when much of Nevada was covered by a vast inland sea known as Lake Lahontan, which supported very large fish. This rock would have been near the shore. In 2019, Massacre Rim was designated a Dark Sky Sanctuary by the International Dark Sky Association, only the fourth sanctuary of its kind in the United States. Skies unpolluted by artificial light are rare and remotely located, providing a glimpse of how the night sky and landscape would have looked to the maker of this petroglyph.

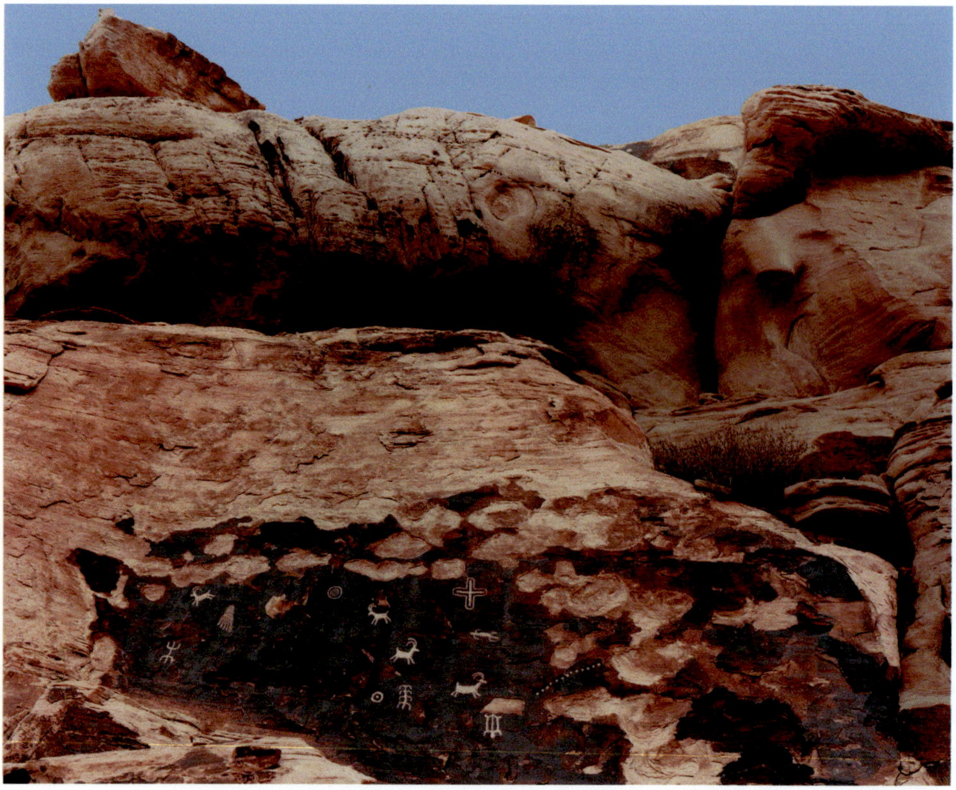

17.12 *Gallery of Petroglyphs, Clark County*, 2007.

Among the archaeologists Goin accompanied to this well-known popular site, there was considerable discussion about the mix of periods and potential time frame for this panel. Known as the "perfect panel," its quality is among the highest of all Nevada petroglyph sites. The individual petroglyphs on this canvas seem relatively isolated from one another, and their differing design styles, depths of engraving, and degree of weathering may point to different origins. However, the archaeologists that Peter spoke with had another theory. To them, the high craft of these petroglyphs—super clean edges, beautifully drawn shapes, consistency in craftsmanship—suggest that they were made by a single, highly skilled artisan, who may have intended this to be a "greatest hits" panel, reproducing the best petroglyphs in one place, akin to a collection in a museum. Bighorn sheep are a prevalent petroglyph in northern and southern Nevada. It's interesting how abundant sheep are in the petroglyph record, yet how uncommon it is today to see real bighorn sheep. Were they once more abundant? Hunted to near extinction? Today the Nevada Department of Wildlife has a program to maintain or enhance desert bighorn populations through transplantation and habitat restoration.

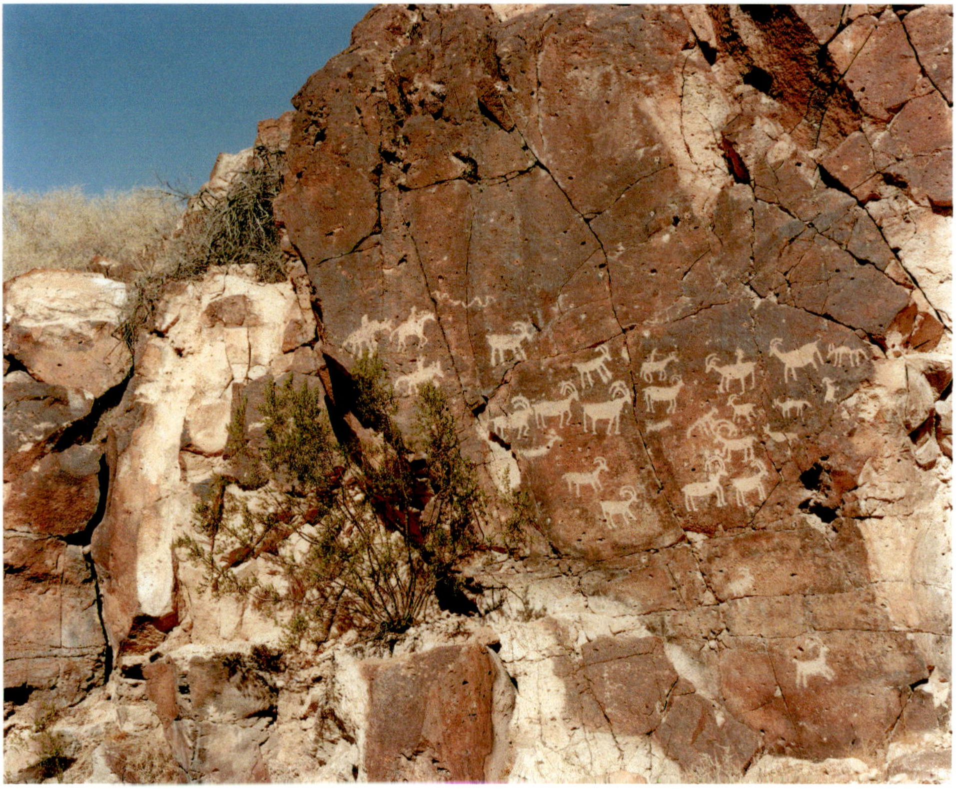

17.13 *Petroglyph panel, Clark County near Lincoln County Line*, 2007.

More sheep. Lots of sheep. Big ones, little ones, facing right, facing left. While at first glance this petroglyph panel may seem prehistoric, a closer look tells a different story. On the left is a cowboy on a horse, accompanied by two other mounted companions. This petroglyph, which is on private property, must have been made sometime in the last two hundred years! On the far right are two small, skinny animals with long tails that appear to be dogs. At top middle there is at least one other mounted rider. What we have depicted here may be cowboys and their dogs rounding up a herd of sheep. The fact that many of the sheep have clearly drawn feet and some have bent knees indicates a different style from the ancient glyphs of sheep. Because the sheep do not look panicked, this petroglyph may tell us that the sheep were at least partially domesticated. This panel reinforces Peter's contention that petroglyphs are a continuing tradition, even if our sense of time prevents us from realizing it.

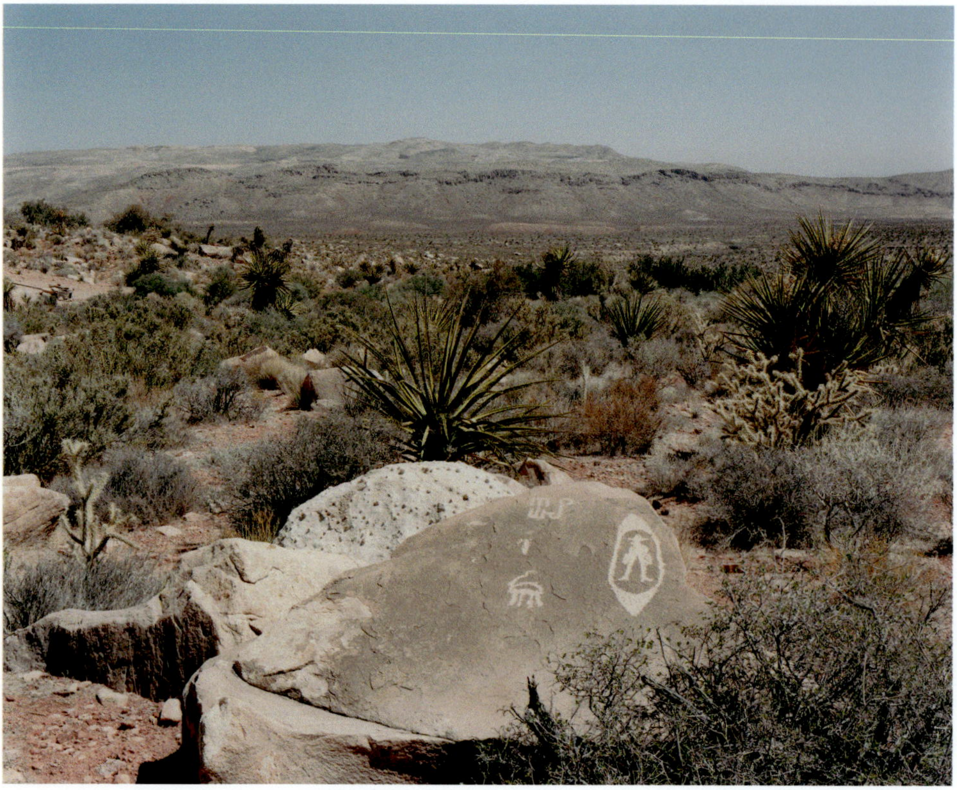

17.14 *Petroglyph, Cottonwood Valley, southern Nevada*, 2007.

Continuing to mount evidence for the ongoing tradition and recent dates of some Nevada petroglyphs, Goin includes this photograph of a man wearing a hat and boots. The capsule around the man is probably meant to be a boat. If you look closely, you will see that the man appears to be missing an arm. Indeed, this petroglyph has been defined as depicting Major John Wesley Powell, the one-armed Civil War veteran who, in 1869 with nine men and four small wooden boats, floated the Colorado River through the Grand Canyon, believed to be the first ever to make the trip. What looks to be a cattle brand symbol at the top of this rock may be a cavalry marking. Imagine hiking around in the wild Mojave Desert country and coming across this petroglyph. It would be a shock—so unexpected and seemingly out of place. Like the jar placed upon a hill in Tennessee in Wallace Stevens' poem "Anecdote of the Jar," this petroglyph is an artifact that totally changes one's sense of the surrounding landscape, making the wilderness sprawling around it "no longer wild."

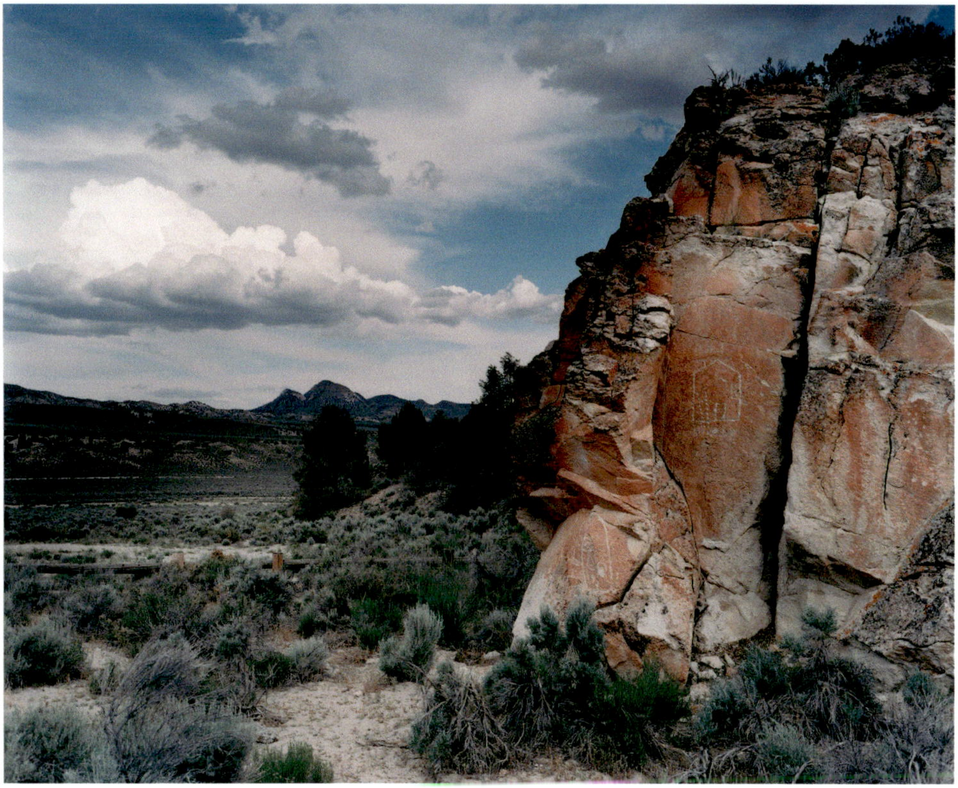

17.15 *House petroglyph, near Moores Station*, 2005

This petroglyph of a house supplies conclusive evidence that petroglyphs continue to be made in modern times. Although some Nevada petroglyphs have been documented to be more than ten thousand years old, clearly others are of recent vintage. This petroglyph is located near Moores Station, which is near the ghost town of Warm Springs, at the junction of US Route 6 and State Route 375. This petroglyph may even depict Moores Station itself, a two-story stone building that was established in 1875 as a stagecoach stop. A nearby spring makes this spot a green oasis in the desert. Given that Goin intends *Nevada Rock Art* to be a book about hope—about marks on the landscape that endure, reflecting a creative rather than destructive spirit—this concluding image is not without irony. In 1968 the Atomic Energy Commission conducted an underground atomic test near Moores Station, known as Project Faultless. Prior to the test, due to concerns that the historic building would be damaged, Moores Station was reinforced with wooden beams on the outside, held together with steel rods. Although Moores Station is now a private residence, the reinforcements are still visible today, enduring marks of the Atomic Era.

Works cited

Black Rock Institute. Black Rock Institute, 2018, https://blackrockinstitute.org/. Accessed 24 July 2019.

Goin, Peter. *Nevada Rock Art*. Black Rock Institute Press, 2009.

Goin, Peter, and Lucy Lippard. *Time and Time Again: History, Rephotography, and Preservation in the Chaco World*. Museum of New Mexico Press, 2013.

The Nevada Rock Art Foundation. Nevada Rock Art Foundation, 2016, nvrockart.org. Accessed 24 July 2019.

RTC Public Transportation. Regional Transportation Commission of Washoe County, 2019, https://www.rtcwashoe.com/public-transportation/about-transit-centers/. Accessed 24 July 2019.

18

NEW NATURE

Although their publication dates are just a year apart, *Nevada Rock Art* (2009) and Peter's next project—with Paul Starrs—*Field Guide to California Agriculture* (2010) represent a big leap in the space-time continuum. There is, of course, the crossing of state lines for the fieldwork, Nevada in one case, California in the next. Timewise, the rock art project preserves prehistory, while the California agriculture guide is steeped in the present. Petroglyphs last ten thousand years or more; agricultural products spoil in days or weeks. People used Stone Age tools to create petroglyphs versus sophisticated technology to grow modern-day crops. Small-scale hunter/gatherer societies give way to mass-market, consumer culture. *Nevada Rock Art* is a hardcover, slip cased, limited edition Fine Art book, while *Field Guide to California Agriculture* is available as a handheld paperback. The color palette of *Nevada Rock Art* is subdued, spare, and nearly monochromatic—earth tones, subtle variations of brown and tan—while the feast that is *California Agriculture* explodes in vibrant, bold colors—red! orange! yellow! green! purple! blue!—a cornucopia of chromatic effulgence, indulgence, and exuberance. The tone of the two books oscillates from reverential in the first to humorous in the next. These two projects, for which Goin did the fieldwork back and forth during the same period of six or seven years, might be thought of as the yin and yang of his psyche.

Despite their striking differences, *Nevada Rock Art* and *Field Guide to California Agriculture* are each significant waypoints on Goin's career-long quest to investigate and document *humanature* in America. Both petroglyphs and agriculture reflect human engagement with nature and result in human alteration of landscapes. By this time a mature and accomplished artist, Goin became increasingly able to articulate how his varied work—spanning almost twenty books—forms an integrated whole. Consider his 2018 artist's statement in *Colors of California Agriculture*, the theme-based companion to *Field Guide to California Agriculture*:

> Peter Goin produces photographs dealing with the long-term impact of culture and technology on the environment. Using the language of the Fine Arts, Peter's work

DOI: 10.4324/9781003212607-18

> integrates history, architecture, urban planning, and the social and political sciences in an interdisciplinary manner.
>
> (Colors *214*)

When one considers that agriculture occupies more than half the area of California, one begins to realize the extent to which California's environment has been altered by humans. As Paul Starrs writes, "From the air, California seems like the empire of a creator bent on constructing the world's most cruelly demented jigsaw puzzle, a vast agriscape cast in shades of green, tan, emerald, and gold" (*Field* 3).

"Agriscape" is Peter and Paul's invented word. Without a doubt agriculture is a ripe field for *humanature* exploration. Still, how did Goin, living and working in the Silver State, become involved in tilling the fertile soil of the Golden State? This project was the brainchild of Paul Starrs. During their collaboration on *Black Rock*, Goin and Starrs had already begun to talk about a next project. Starrs, whose home is in El Cerrito, California, commuted to Reno to teach geography courses at the University of Nevada. On the drive from El Cerrito to Reno and back again, Starrs sped by many agricultural landscapes, particularly in the California Delta and Sacramento Valley, a patchwork of fruit and nut tree orchards, rice fields, and row crops, such as beans, asparagus, corn, and strawberries. As a geographer who had spent much time traveling with UC Berkeley Professor Jim Parsons and with fellow geographers around parts of California, Starrs was intrigued by these land uses and had begun thinking about a field guide. Furthermore, Starr's wife, Lynn Huntsinger, Professor of Environmental Science, Policy, and Management at UC Berkeley, taught a course on "Follow Your Food," which included field trips to farming and livestock operations and farmers markets, culminating in a student-prepared dinner at the Huntsinger-Starrs residence. As Starrs recalls, "This all finally jelled: Here was a great story, visual and geographical, waiting to be told. . . . There was no field guide to agriculture, certainly not for California with its universe of crops and products" (*Colors* 195). Entrepreneurially, Starrs discovered a niche that the two of them could fill. Fortuitously, an editor who was influential in the Field Guide Series at the University of California Press encouraged him to pursue the idea.

Peter readily assented to the idea, immediately grasping its relevance to his *humanature* studies and eagerly anticipating the adventures that fieldwork would bring. Peter's roots in agriculture and California run deep. His pioneering ancestors arrived in Oregon in 1853 by wagon train, and his family took up permanent residence in California in 1919. Peter's father picked lemons during the summer while attending UC Berkeley. And Peter himself lived in San Francisco from 1978 to 1984. Furthermore, the time seemed ripe for a book on California agriculture. Alice Waters' Chez Panisse, the Berkeley restaurant established in 1971 to specialize in locally sourced organic food, had begun a California cuisine fad and jumpstarted a foodie culture. Michael Pollan's books on food and agriculture topics were hot best sellers. Starrs observes in the early pages of *Field Guide to California Agriculture* that although less than one percent of Californians are involved in ag production:

> In an ever more urbanized society, there is quite noticeably a growing attention to agriculture, food, and domesticated animals. It may seem contradictory, but unfamiliarity can breed respect, not obliviousness, spawning a sense of remorse: 'I should know more about food and its production than I do'.
>
> (Field *9)*

Enter Paul F. Starrs and Peter Goin. As Goin explains in *Colors of California Agriculture*:

> When we started our multi-year journey, our premise was, and remains simple: As a rule, the generally urban population no longer knows much about agriculture, which is the predominant use of California's private land (and, given ranching, no insignificant part of its public land). Where does the food we eat come from? Who grows the crops? Who raises the animals?
>
> (Colors *200–01)*

Peter and Paul aimed to provide answers. Into the fray they jumped, going first to the 2006 World Ag Expo in Tulare, in the heart of the San Joaquin Valley and at the center of California agriculture. Peter remembers with a smile:

> On a warm February California day, we entered the fairgrounds, instantly confronted with acre upon acre of green tractors, earth-mover tourist wagons, red tractors, decorated dairy cow statues, antique irrigation pump-fountains, blue tractors, Gophinators, cowhides on racks, trimmers, tillers, and dump trucks. And more upon more, naturally . . . a feast of agricultural innovation and mechanization, an industrial harbinger for solving every problem by primary-colored machines.
>
> (Colors *196)*

As Peter and Paul's education progressed, they found themselves in awe at the sheer size and diversity of California agriculture. In describing the topic, they appeal to words such as *kaleidoscope, cornucopia, compendium, complexity,* and *conundrum*. Agriculture is "big business in scale, in profits, and in prominence" (*Field* 4). In 2007, near the beginning of their investigations, California agriculture was better than a $37 billion activity—wholesale—on 75,000 farms and ranches. For diversity, scale, and quality, California is without equal. California agriculture exists at every step along a production scale, "from intensive and hand-raised crops with limited outputs," catering to the "locavore" movement, "to agribusiness growers whose gross production expenses are in the tens of millions of dollars" (*Field* 10). Goin notes that:

> California's varied ecosystems allow for agricultural production on a massive scale. The state produces 450+ commodities, accounting for more than a third of the country's vegetables and two-thirds of the country's fruits and nuts. Milk and cream, grapes, almonds, cattle and calves, lettuce, strawberries, pistachios, tomatoes, walnuts, and broilers comprise the top-ten valued commodities, and this doesn't even include the most probable number one crop, cannabis.
>
> (Colors *201)*

And Starrs marvels that "Within the hundred million acres of California is more concentrated and diverse agricultural bounty than any place else in the world" (*Colors* 16).

Philosophically, Peter finds the subject of California agriculture appealing as a manifestation of *humanature* and exemplar of what he calls "the new nature." Recall that Goin's 1996 book *Humanature* documented instances of human-designed landscapes that people nevertheless experience as nature—a manufactured swamp, a planted forest, grass overgrowing a mine tailings hill. These human-altered environments that we experience as natural, constitute, in effect, a "new nature." Rather than celebrating so-called "pristine" nature and "virgin" wilderness, Peter exposes such places as always already human constructions through management policies; fire regimes; human interventions such as trails, signage, and use; the presence of "invasive," "non-native" species; devastation resulting from the nuclear era; and the effects of anthropogenic climate change. And rather than believing that a person can experience nature only in "unadulterated" nature, Peter embraces the experiences to be found in urban nature—a backyard wildlife corridor, for example—parks, and reclaimed landscapes. In Peter's view, these unabashedly human-architected natural spaces are to be commended and enjoyed rather than condemned as somehow tainted, inauthentic, or impure. They are the *new* nature. Although agriculture makes no pretense to recreate a "natural" environment, the ever so highly humanized landscapes of California agriculture have become, in effect, components of this new nature. As Peter explains in conversation:

> If you look at agricultural environments and you walk through them, it's oftentimes the closest urban opportunity for the solace of open spaces. We have a different response driving down a concrete jungle than we do driving through agricultural fields. Our sensory input is different. We can smell the orange blossoms. We can absorb the moisture of irrigation. We interact with agricultural environments differently than we do city streets. If you walk through agricultural fields, surprisingly, it's remarkably similar to taking a walk in the woods. As we begin to embrace agriculture and absorb its rewarding aspects—the solace of it, the labor of it, the circumstances of its liveliness—we experience some of those similar feelings, and we're comfortable within that built environment because of the sights and smells and the life within, the teeming life within those rows.

In his *humanature* work, Peter often seeks to draw attention to that which is typically unseen and correct common misconceptions. In agriculture, one common misconception concerns scale. Peter notes that our concepts are outdated: "We're still caught in the sentimental concept that agriculture must be 160 acres, and we cling to the Jeffersonian ideal of creating an agrarian democracy that will propel idealism and purity and innocence to the global stage. It's naive." The fact is that economies of scale dictate that to feed the millions of people who depend on California agriculture requires that farming methods be bigger, quicker, speedier, and more efficient than ever before. Essentially, to feed the world requires that the Industrial Revolution transform agriculture. And it has. The problem is that most people never see the process. "We're not seeing what is really involved in this huge industrial culture to generate foods that we consume to sustain ourselves," Peter points out. And our imaginary or concerted naiveté to the consumption of food extends to animals. "When people think of agriculture, they think of crops that are grown," he continues, "but they don't realize that agriculture includes animals that are raised." Animals provide a variety of different foodstuffs, including meat, milk, cheese, butter, eggs, and even Jell-O. And animal husbandry takes place on as grand a scale as crop growing; however, most of the feeding and all the slaughter of animals takes place out of sight of the general public. As Peter would learn, some aspects of the animal food industry, poultry production in particular, are even off-limits to the most diligent of photographers.

Goin's California agriculture photographs not only picture individual fruits and vegetables, they also document the practices of modern-day corporate agriculture—endless parallel rows of crops, enormous machines, cadres of farm laborers at work, irrigation canals and sprinkler systems, pesticide crop dusters, research stations, and computers. Peter accepts this new reality—the garden writ large—and investigates it with curiosity and an open mind, aware not only of problems but of successes. In fact, one of the big problems that he documents is not the growth of agribusiness, but rather the growth of cities and suburbs that pave over fertile soil with concrete, asphalt, and new housing developments. But even as urban sprawl gobbles up arable land, the people who live in cities and suburbs must eat. The relationship of city to farm and field is succinctly set down by Starrs in the preface of *Field Guide to California Agriculture*. He quotes author Paul Richardson, who posits that "Cities represent demand. The countryside represents supply." Starrs adds:

> That point may seem obvious—but if so, why does such apparent separation exist? As authors, we hope this book will begin the suturing of the food-to-table world back into an elegant whole . . . and in the process again merge city and countryside.
>
> (Field *xxii)*

One step in reconnecting city folk with their agricultural roots is to teach them the names of the crops. Peter observes in conversation that:

> one of the classic definitions of a connection to a place is the ability to name the trees in a place we live. And most people can't. And if you look at agriculture when you're driving down the road, and you see something, most often you say, 'Well, what is that? I have no idea.'

Starrs and Goin's *Field Guide* teaches readers what grows where and why, "making sense of a world we perceive but don't necessarily understand in California agriculture" (8). Peter believes that this kind of knowledge helps to sustain the human spirit, which requires a community of all kinds of influences. In effect, the *Field Guide* introduces city folk to members of the biotic community that feeds them, helping not only to mend the city-country split but to heal the body-mind bifurcation as well. As The Bancroft library curator Jack Von Euw writes, the *Field Guide*, "replete with clear descriptions and vivid visuals of greenhouses, fields of soybeans, spinach, and asparagus rescues me from drowning in my own urban ignorance" (*Colors* 13).

Given the extent and complexity of California agriculture, doing the photography for the *Field Guide* was destined to be a monumental undertaking. In fact, Goin and Starrs made more than two dozen trips spanning more than seven years, logging thousands of miles in Paul's Toyota Rav4. Most of the field excursions were made together so that Peter and Paul could compare notes and challenge one another on the ground. Both coauthors are fluent in Spanish, the idiom of much California farm labor, and one of Paul's responsibilities was to engage the workers in conversation long enough for Peter to make photographs of the crops. After that, as Paul recalls fondly:

> Peter and I reunite, finish the conversation, and head to the next dairy, the next field crop, the next farm, a previously unseen variety or festival or taste. This is a voyage of many stops, many crops, many themes—and exposure to a goodly number of California dreams.
>
> (Colors *195)*

Field Guide to California Agriculture was the first of Peter's projects to be conducted entirely with a digital camera. All told, he made approximately 1500 images, about ten percent of which were published in the *Field Guide*. Since by virtue of genre the *Field Guide* restricted

Goin to mostly declarative photographs of the "agri" side of agriculture, Goin and Starrs then expanded their revelation of California agriculture to a large format, Fine Art, encore volume, *Colors of California Agriculture*, featuring some one hundred of Goin's images that explore the "culture" side of agriculture.

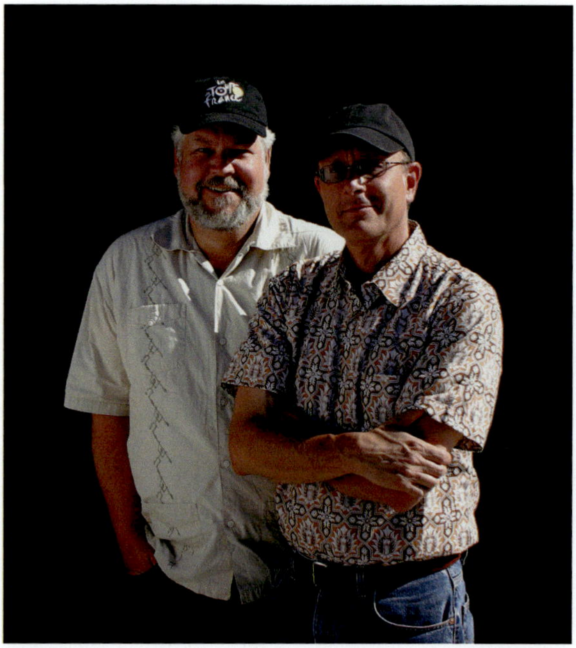

18.1 *Paul F. Starrs and Peter Goin* (left to right, respectively). Photograph by Jean Dixon, 2005.

Lest the reader garner a false impression of relentless toil and martyrdom, rest assured that Peter and Paul's road trips were very fun. The two distinguished professors hit it off famously and cruised in high spirits most of the time, as one can intuit from the "About the Authors" photograph of the pair.

In this picture, both men are handsomely tanned from being outside, and they are dressed for fieldwork, sporting baseball caps and short-sleeved, buttoned shirts. But there the resemblance begins to unravel as their complementary characters are revealed. Paul's slightly rumpled shirt is untucked and his hand rests casually in his pants pocket. Peter's crisp shirt is neatly tucked in to his belted jeans, and his arms are crossed in front of his body. Slightly to the side and behind Peter stands the bearded, genial figure of Paul, a self-described "man-mountain," looking like Santa Claus on summer vacation, smiling openly, with a twinkle in his eyes as if he has just cracked a joke or is about to hatch one. Peter, in contrast, is clean shaven and offers a Mona Lisa, close-lipped smile, peering at the camera from behind slightly darkened glasses. Whereas Paul appears open, approachable, and jovial, behind Peter's cooperatively pleasant demeanor one senses intensity, privacy, purpose, and just a smidge of mirth under wraps. As Paul good-naturedly recalls in conversation, Peter "flogs" his collaborators, but in a good way because things get done, adding, "Peter needs someone who will give him shit and feed him tequila at the end of the day." For his part, Peter quips of Paul that "he's an adolescent male who happens to be in a late-fifty-year-old man's body."

On the road, the Goin-Starrs partnership blossomed into mutual respect and brotherhood. Both are sons of US diplomats and spent their formative years living abroad in multiple countries, giving them a third-culture, outsider-insider, comparative vantage point on the United States. Both owe an intellectual debt to UC Berkeley's foundational work in cultural geography, Peter having been influenced by John Brinckerhoff Jackson's *The Necessity for Ruins*, and Paul having earned a PhD at UC Berkeley, roving the agricultural landscapes of California with his Cal mentor, Professor James Parsons. The bonhomie between Peter and Paul influenced how they saw what they saw when they looked at California agriculture. To wit, they delighted in roadside humor and in instances of what they called "agri-porn," two themes that Peter explicitly develops in his photographs for the project. When road-tripping, kitsch, folk culture, roadside signs, puns, the unexpected, the voluptuous, the grotesque, and the puzzling can all seem funny.

One hilarious sight during their fieldwork occurred when Peter was photographing at the Johann Dairy near Fresno, where he and Paul witnessed a carousel milking machine. To their astonishment, cows vied to be next in line to ride the merry-go-round milker:

> In a large staging area, they jostle and edge each other, half-ton milk engines, entering the chute. Round they go in a full circle, only to be forced to back out by a cold-water spray directed at their face. Without the startling surprise, they wouldn't leave the comfort of reducing their milk load and riding the ride. . . . Cows and cats, silage and backhoes . . . all, together, in concert like a rural opera, except to say that all the singers are low mooers.
>
> *(Goin,* Colors *198)*

Another roadside curiosity was a field of green plants, above which poked occasional wooden stakes; on top of each stake was impaled a desiccated and droopy red bell pepper. Peter remarks dryly, "Although the piked-pepper action was repeated through this field, the meaning is unknown," an unsolved mystery of California agriculture (*Field* 91). Peter's fondest memory is of a sign that he photographed along Highway 99 near Dairyville. Framing a big, round, red tomato painting was the words, "HOME GROWN TOMATOES" (see Figure 18.2). For Peter, the sign symbolizes "the pride of agriculture, the can-do spirit of farmers and ranchers, the tenacity of the agricultural enterprise, the visual richness of a fecund tomato, and the simple beauty of *what is real*" (*Colors* 202). Peter instantly thought of an esoteric reference—the lyrics of a Guy Clark song:

> Homegrown tomatoes, homegrown tomatoes
> What'd life be without homegrown tomatoes?
> Only two things that money can't buy
> That's true love and homegrown tomatoes.
>
> (Colors *202–03)*

When Peter showed Paul the photograph, unprompted, Paul broke out in song and Peter joined in. Peter smiles, "It was one of those rare moments of two guys and two minds. He gets it."

One doesn't expect a nature field guide to be humorous, but *Field Guide to California Agriculture* is undeniably witty. Admittedly, at times the humor veers a bit off-color, as when Goin's photograph of a cabbage looks like a Georgia O'Keeffe painting of a cabbage that viewers tend to perceive as a vagina, or when his photograph of two dangling kiwi fruits resembles a scrotum. Even though he created the photographs, Peter implicates Paul, saying:

> my coauthor is quite proud of his adolescent brain, and he has a tendency to go to the phallic references sooner than I do. He loves the idea of inventing words, and we took the prefix *agri-* and put it in front of all kinds of things, agritourism, agriporn.

Jests aside, Peter avers that he has tapped into a universal human tendency:

> Our minds do this. We think in sensual terms. Our desire is paramount in the way that we function. I didn't purposefully say I'm going to seek out eroticized nature, but if it appeared, I wasn't going to reject it.

In the history of art, eroticized fruits and vegetables appear frequently, sensuality being an overt theme of much art.

While agriporn lurks as a muted motif in *Field Guide to California Agriculture* and *Colors of California Agriculture*, the sensuality of agriculture surfaces as a dominant theme, and Peter has much to say about it. First of all, when we travel or experience any kind of agricultural environment, whether it's our own garden or flowers in our home or plants, we associate smells, colors, shapes. Our experience is sensual. Secondly, agriculture is linked to our relationship with food. Peter observes in conversation that:

> We find pleasure in food. We find pleasure in preparing food. We find pleasure in the way food looks. And so, if you were to encapsulate this project in one sentence, it would be, 'The sensuality of the new nature emerges in California agriculture.'

Philosophically, Goin wanted to reunite the Cartesian split between the mind and the body by bringing together ideas of the sensual, the eroticized, and the experiential—sight, sound, smell, touch, and taste. As a critic of American culture, Peter laments that:

> We fail to understand the sensuality of the food we eat. When we look at a pear hanging from a tree, it's pretty spectacular. And it translates in the way we shop: We want our apples to be nice and polished… no blemishes. Shopping for food, therefore, is in a subliminal way, a sensualized path to nurturing. When you look at the whole sequence of how we treat foods, it's very tactile, it's very sensual. The premise of the California agriculture project is that we fail to recognize and identify with our own intrinsic relationship to sensuality.

As Peter developed his approach to interpreting agriculture sensually, color became an active verb in his visual language. Although Goin is a stickler for documentary accuracy, and he did not Photoshop any images to intensify the colors, he nevertheless employed a richer, more saturated color scheme for the California agriculture books than for any of his other books. Challenging David Batchelor's premise in *Chromophobia* that Western cultural and intellectual thought is chromophobic—afraid of corruption and contamination through color—Goin asserts that we embrace color in our daily lives in a variety of different ways. But what methodology could he forge to represent the wide range of colors that agriculture exhibits? His strategies included selecting naturally vibrant subjects; photographing collections of items—fruits, vegetables, and brightly clad farmworkers—to exhibit a medley of colors; favoring primary colors and contrasting colors; creating compositions built around color (see the image of a blood orange slice on a purple, nitrile glove; Figure 18.6); and waiting for weather conditions and times of day that made colors pop. Last, in a brilliant, abstract art piece, Goin extracted representative colors from the produce, products, and paraphernalia of agriculture to form two adjacent grids of brightly colored squares; each grid of eighty squares is precisely arranged in ten rows and eight columns, partitioned by pure white borders (see Figure 18.18). The squares and corresponding identification key are arranged alphabetically, from alfalfa flower to zucchini blossom, and include such elements as blueberry and bok choy, flame grapefruit and free-range fryer, latex glove and leeks, pomegranate and portable toilet, roma tomato and rubidoux pummelo, and white potato and worker housing. This artwork entitled "Colors of California Agriculture," measures forty-four inches × sixty-nine inches in the exhibition print and exults in the sensuality of color. Peter touts it as a proposal for fashion choice colors rooted in agriculture.

As intellectual laborers, Peter and Paul's fieldwork yielded not beans or broccoli but books. The first, *Field Guide to California Agriculture* (2010), published in the University of California Press's prestigious Natural History Guide series, is a tour de force. Conventional field guides identify and interpret the flora and fauna of "wild" nature, whereas a field guide to agriculture radically posits that we need guides to the "new" nature, that is, humanized landscapes. Whereas most field guides target readers who are on foot and able to closely inspect individual plants and animals, *Field Guide to California Agriculture* assumes that users of the guide will be driving by in a vehicle. As Starrs writes, *Field Guide to California Agriculture* is "designed to be on your dashboard . . . for a road-based adventure," and it will have succeeded "if through time this book is hard worn by being drawn repeatedly from car side pockets and seatbacks, unholstered like a gunfighter's six-shooter" (9, 412). Starrs and Goin's award-winning *Field Guide* has been called a "mini-encyclopedia," "an empirical study," and "a primer in California agriculture," "brimming with information" that gives readers a means for understanding what they see. The guide begins with essays on the historical, environmental, and sociocultural contexts of California agriculture, and then moves to an inventory of crops and animal products. Goin's photographs picture individual fruits and vegetables as well as representative agricultural California landscapes. In addition, a gallery entitled "The Paradox and Poetics of Agriculture," many of whose images are featured below in this chapter, offers a more overtly artistic viewpoint. As Peter confirms in conversation, "we're reinterpreting the concept of the field guide intellectually. We are redefining or challenging it to be less mundane, factual, and encyclopedic and more lyrical, interpretive, and expressive." These strategic choices enliven *Field Guide to California Agriculture*, a refreshing departure from the "kill it, pin it on a board, and label it" fare of guides wedded to rigid classification systems.

Colors of California Agriculture, published by The Bancroft Library of the University of California, Berkeley, in 2018, develops the lyrical, interpretive, and expressive angle more fully than is possible even in a redefined field guide. This visually sumptuous volume pairs a remarkable collection of historical photographs, illustrations, maps, and documents from

The Bancroft's extensive archives with Goin's color photographs of the same or similar topics, deftly related, explained, and expounded by Starrs' expert commentary. As Bancroft director Elaine C. Tennant writes of the book:

> The merry earnestness of the Goin photographs presented with their specific historical counterparts and the Starrs descriptions of both are an invitation to gasp at the beauty and diversity of the agricultural bounty of the state, to laugh out loud at some visual jokes, and to think hard about the next hundred years of agriculture in California.
>
> (Colors *9)*

Jack Von Euw, Bancroft curator, aptly praises Goin for being "intimately interested in what is in front of him. In doing so, Goin can take the seemingly ordinary and present us with something extraordinary while preserving the documentary aspect of the photograph" (*Colors* 15).

Prints of Goin's photographs were on display in the "Colors of California Agriculture" exhibit at The Bancroft Library in 2013, and he and Starrs donated their California agriculture materials to The Bancroft, catalogued as the *Peter Goin and Paul F. Starrs California Agriculture Archive, 2004–2013.* This archive consists of their notebooks, atlases, fieldwork guides, drafts, and 250 working prints. The *Peter Goin Digital Photograph Archive: The Colors of California Agriculture* contains approximately 1500 digital files, in Tag Image File Format. And the *Portfolio of Photographs of California Agriculture by Peter Goin* contains sixty-four digital archival pigment prints in a clamshell case. These archives augment major California Agriculture collections in The Bancroft dating back to 1803, including photographs by Carleton Watkins, Alice Iola Hare, Charles B. Turrill, Dorothea Lange, and Ansel Adams.

The images below, culled from the California agriculture archive, incline toward Goin's more lyrical images. Featured are a number of close-ups, views of extended hands holding food-in-the-raw, and a miscellany of roadside signage. In this gallery, be alert for art history references, as food—still life studies, cornucopias, banquets, gardens, peasants in the field—crops up repeatedly in the artistic tradition, and Peter is joining that pictorial feast. As well, enjoy the humor; Peter's "merry earnestness" adds a dash of piquant seasoning to a subject that might otherwise be bland. Finally, know that each image is a portal to larger conversations on complex topics, such as color, sensuality, eroticism, industrialization, land use, suburban sprawl, the new nature, labor, and the American dream.

■

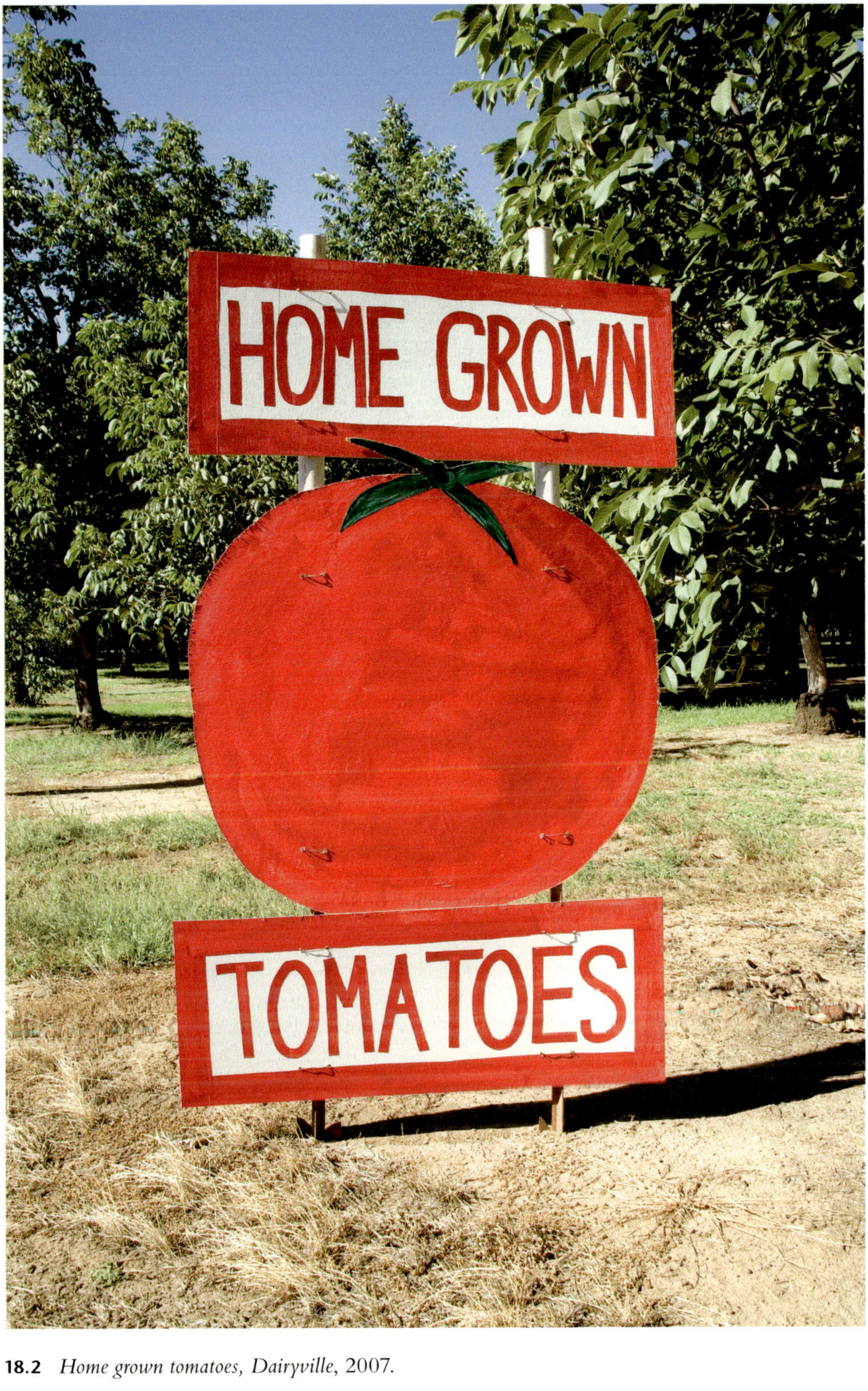

18.2 *Home grown tomatoes, Dairyville*, 2007.

Intriguingly, Goin chose to photograph a sign advertising tomatoes. Why? Signs mediate our relationship to the real. Signs reflect our identities and transform our understanding of a place. They allow us to embrace the simulation of nature. Signs express the "culture" embedded in agriculture. This sign for "Home Grown Tomatoes" at a roadside fruit stand in Dairyville encodes what the tomatoes for sale represent. "Home Grown," as opposed to "factory farmed" invokes positive associations we have of home—family, hearth, Mom, nostalgia, security, love—and of "home grown" produce—garden, flavorful, juicy, fresh, healthy, limited quantities. When you purchase a tomato from this vendor, you are buying into meaning not just matter.

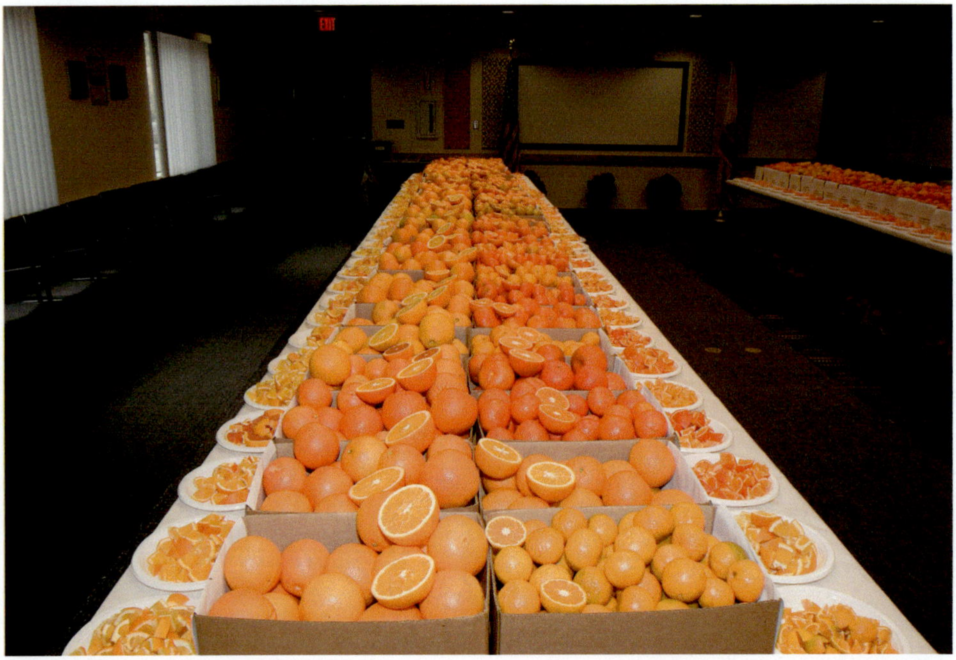

18.3 *Annual growers' citrus tasting, Lindcove Research Center, University of California*, 2006.

In his essay "Visualizing Agriculture," Goin writes:

> Paul and I traveled to the Annual Growers' Citrus Tasting Event conducted by the University of California, at the Lindcove Research and Extension Center. . . . Walking into the room, the citrus smell boosted our body's production of serotonin, a hormone that makes you feel happy. This might be true or not, but we were content just to see so many experimental citrus varieties. There were Ceridwen Navel, Atwood Navel, and Leng Navel oranges. Limonette DeMar lemons, Vaniglia Sanguigno oranges, Yuzu, Mary Ellen sweet limes and Millsweet lemons, Cadenera sweet oranges, Thornless Mex limes, Flame grapefruit . . . There were many more, and while I want to list them all, suffice it to say that the table itself reminded me of Leonardo da Vinci's *The Last Supper*. . . The plethora of citrus made it another day at the circus.
>
> (Colors *198)*

There is a huge industry for citrus, but a fickle public keeps growers guessing. Special varieties of citrus are experimentally developed for color, sweetness, drought-resistance, freeze-resistance, insect-resistance, shipping hardiness, and shelf life. Culturally, we associate oranges with health and sunshine, and we traditionally give oranges as gifts during the holiday season.

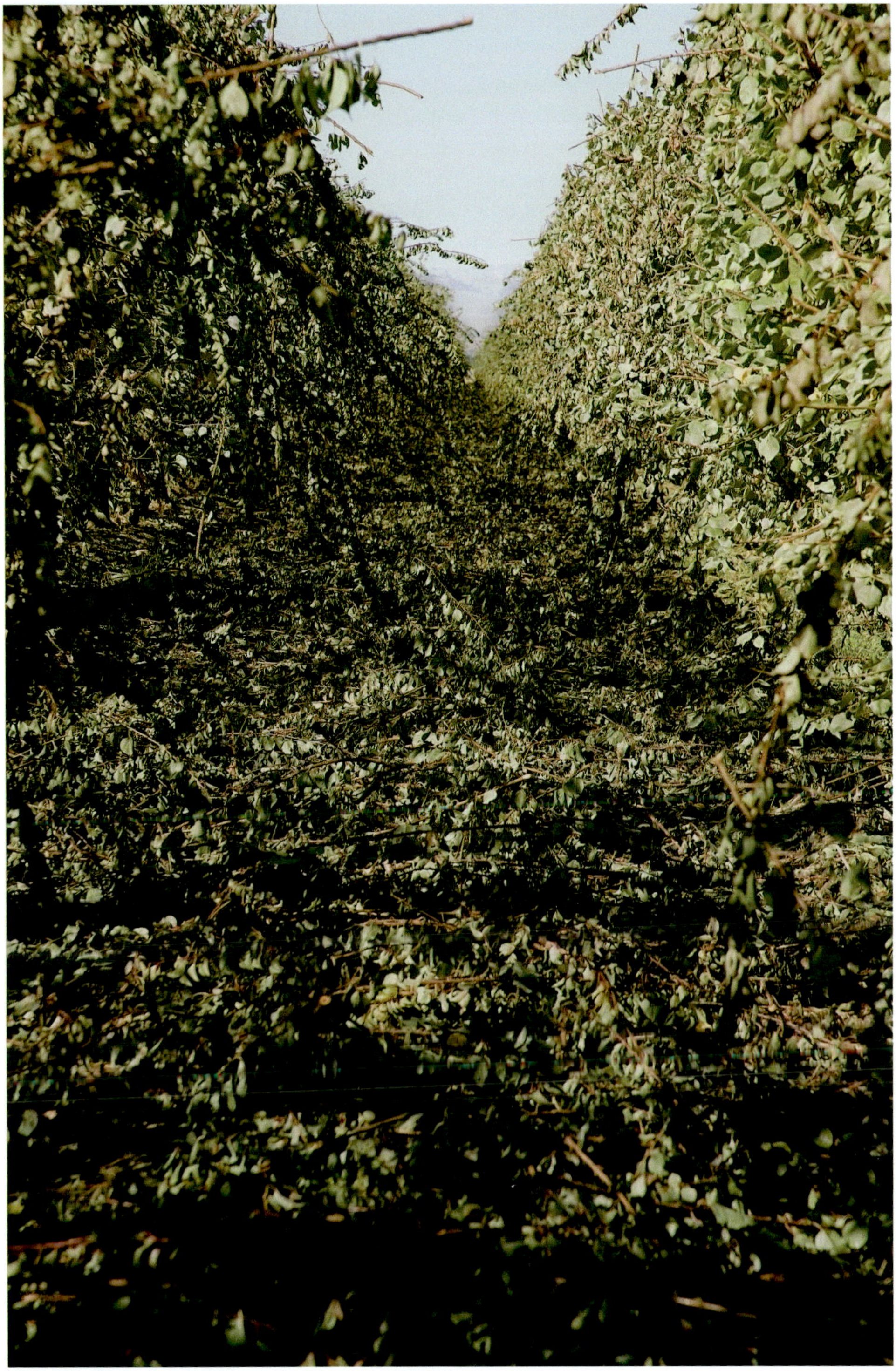

18.4 *Topping and tree-trimming after harvest, Stanislaus County*, 2007.

The rectilinear lines, receding perspective, and rank growth in this image resemble Goin's Erie Canal photographs of vines overgrowing the walls of a decommissioned concrete canal—technology being reshaped by nature. But in this California fruit orchard, we witness the opposite dynamic: Trees reshaped by technology. Pictured here is a phase in large-scale agriculture, namely the topping and tree-trimming after harvest to render next year's fruit crop easier to mechanically harvest. This image documents the evolution away from labor-intensive hand-picking to mechanized agriculture on a grand scale, where a single person driving an enormous, expensive machine can harvest hundreds of acres of monoculture crop. As Peter explains, "It's a two-part process to design machines that can harvest, but then alter trees to be more conducive to mechanize. It's not just changing the machine, it's also changing the tree, sometimes by genetic engineering." In an homage to abstract expressionism Goin designed this image to be green all the way down to the ground, visually flattening the space and creating abstract geometric shapes. The contradiction between the flattened space and perspective control is confusing and provokes an intellectual discussion about what is nature and what is this kind of regulated, machine-produced, boxlike hedge.

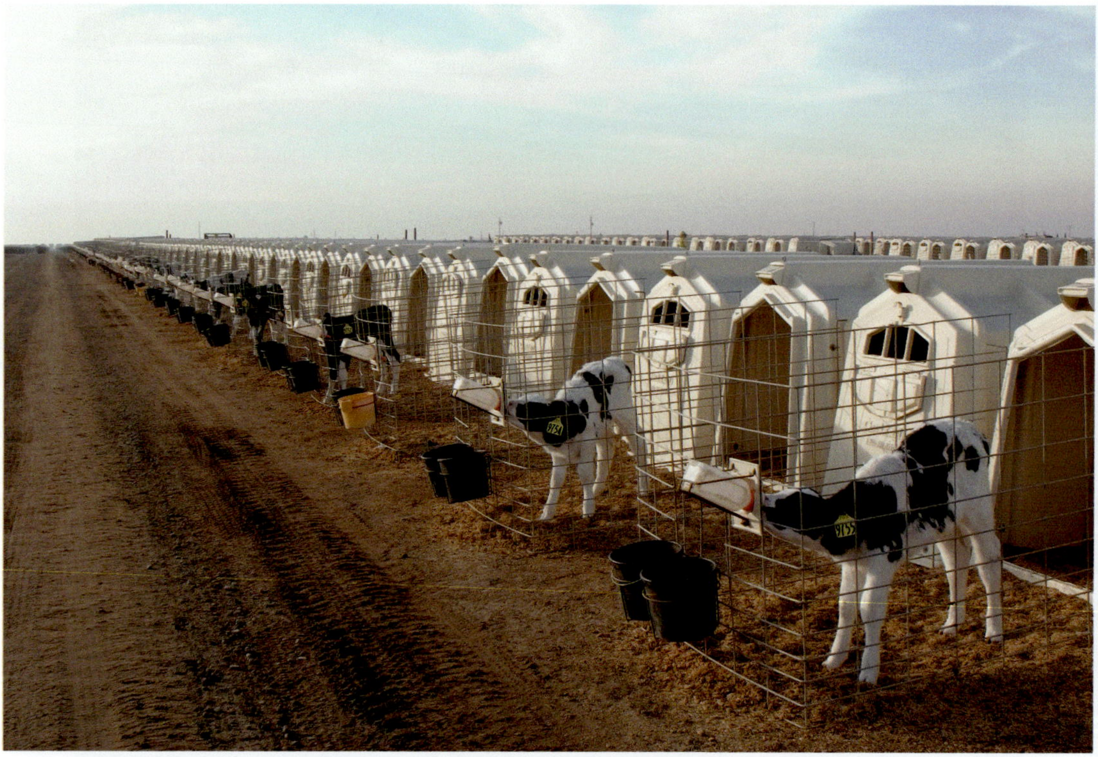

18.5 *Row of calves, bottle-feeding, Johann Dairy, Fresno,* 2006.

Dairy is the number one agricultural industry in California. But we don't often think of it in the terms this photograph proposes. We don't often think of the economy of scale of the manufacture of animals—the businesslike management of sentient creatures. Our agricultural empire is reminiscent of a widget factory, producing orderly iterations of a standardized consumable, each with a serial number. This dairy calf feeding operation at the family owned Johann Dairy in Fresno is not the largest of its kind, but it is considered one of the most humane. Each calf enjoys a large hut, outdoor space, and bottle service. Goin's photograph takes advantage of perspective and a second row of huts to suggest infinity, calf-rearing ad infinitum. In this image, the rows of pens extend from close-ups at right to infinity at left, directing our reading of the cornucopia-shaped view from right to left. Milk cartons and dairy billboards picture happy cows grazing in verdant fields.

Peter editorializes, "I really enjoy the Garden of Eden worship that we engage in, that kind of Christian expectation of paradise. But look up the etymological origins of the word 'paradise.' It is a walled enclosure where animals were kept."

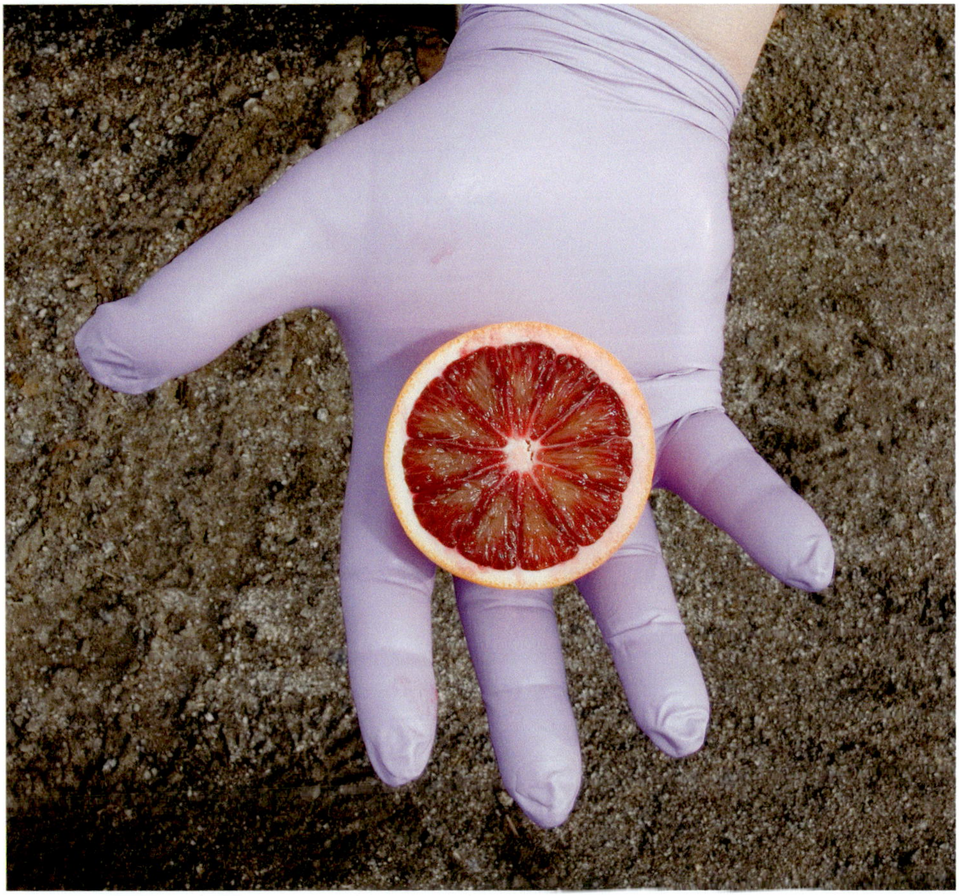

18.6 *Moro blood orange, annual growers' citrus tasting, Lindcove Research Center, University of California*, 2006.

Color is the theme of *Colors of California Agriculture*, and this image of a sliced-open blood orange on a purple Nitrile glove triggers an ocular orgasm, so to speak. Using erotic language to describe the moist, juicy interior of a blood orange flirts with Goin and Starrs' idea of agriporn. As Peter says, "the concept of a blood orange is interesting if you think about sensuality of agriporn. Blood is really red. It's ruby red. Lipstick is red." As shoppers, we are finicky about fruits. We won't buy anything with a blemish on the peel, even though the blemish will not affect the taste of the fruit and even though the peel will be discarded. We are seduced by appearance, especially color. A supermarket that sells boutique blood oranges will likely display a sample cut open, so shoppers will be lured by the novel color palette of the fruit. As humans, we tend to classify things, often by color. The first thing a classifier does is show you what the object is. The concept of showing is a repeated theme throughout Goin's work, illustrated here by an artificially colored hand showing us the nature of plenty.

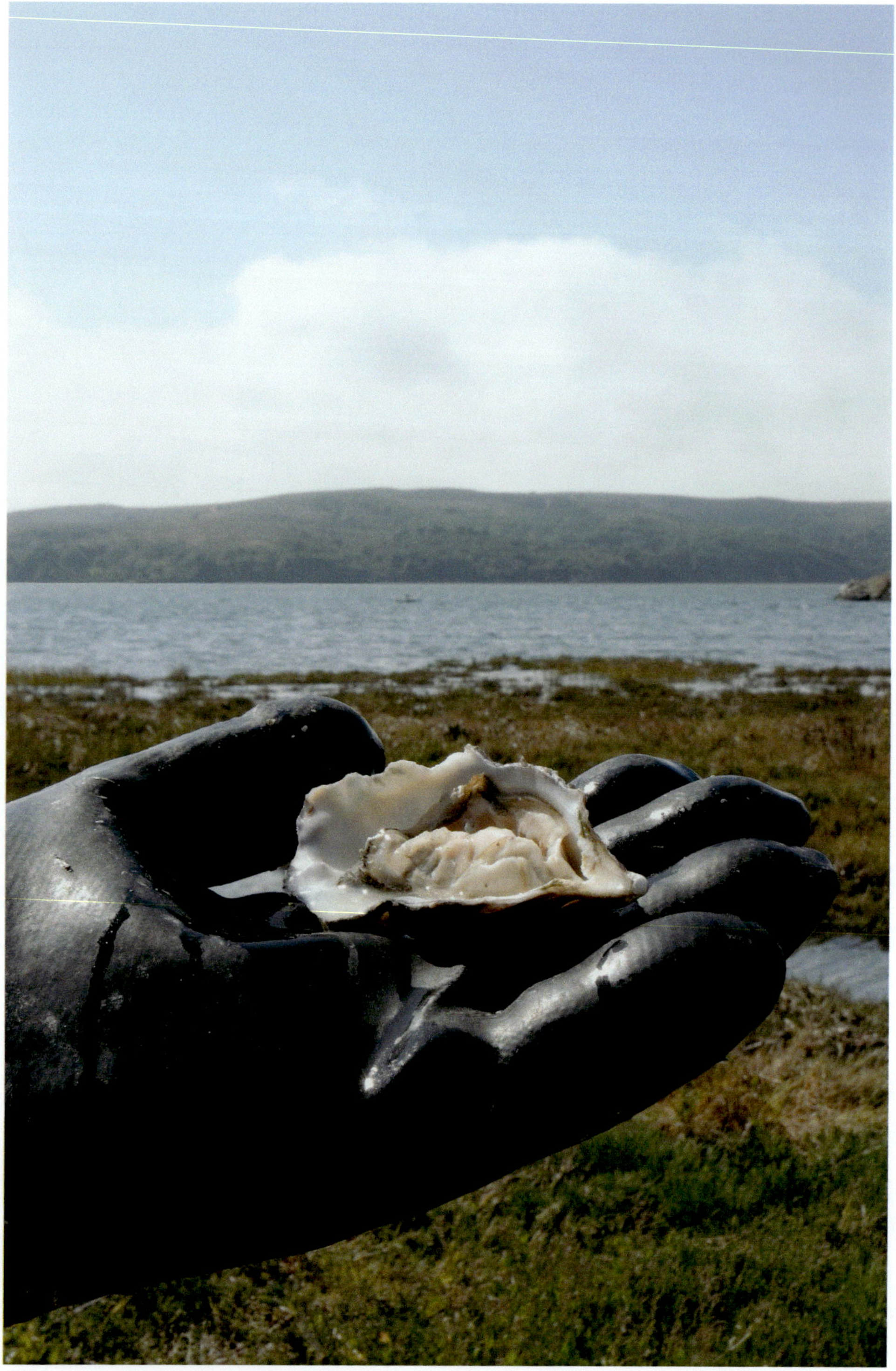

18.7 *Oyster in black rubber glove, Hog Island, Tomales Bay*, 2006.

Oysters are not commonly associated with agriculture, yet they are cultivated, harvested, and sold to eat. The specimen in this photograph is a *Crassostrea gigas*, the Pacific oyster, farmed by the Hog Island Oyster Company. On Goin's mind when he made this image was Botticelli's fifteenth-century painting *The Birth of Venus*, which depicts the goddess Venus emerging naked and fully grown at the seashore, standing on a seashell, a symbol in classical antiquity for a woman's vulva. Ever since Roman times, oysters have been considered an aphrodisiac. As one blogger writes in a post entitled "The Erotic Oyster," "As the oysters lie there, raw, moist, tender and glistening, in their half-shell, waiting to be devoured, it is easy to conjure up erotic associations . . . 'eat oysters and try *not* to think of sex'" (Moore). In this image, the gloved hand offers the oyster, a gift from the sea. We cannot tell the age or gender of the person wearing the glove, and this ambiguity tends to universalize the representation of the hand, the hand being one of the most universal artistic elements through time.

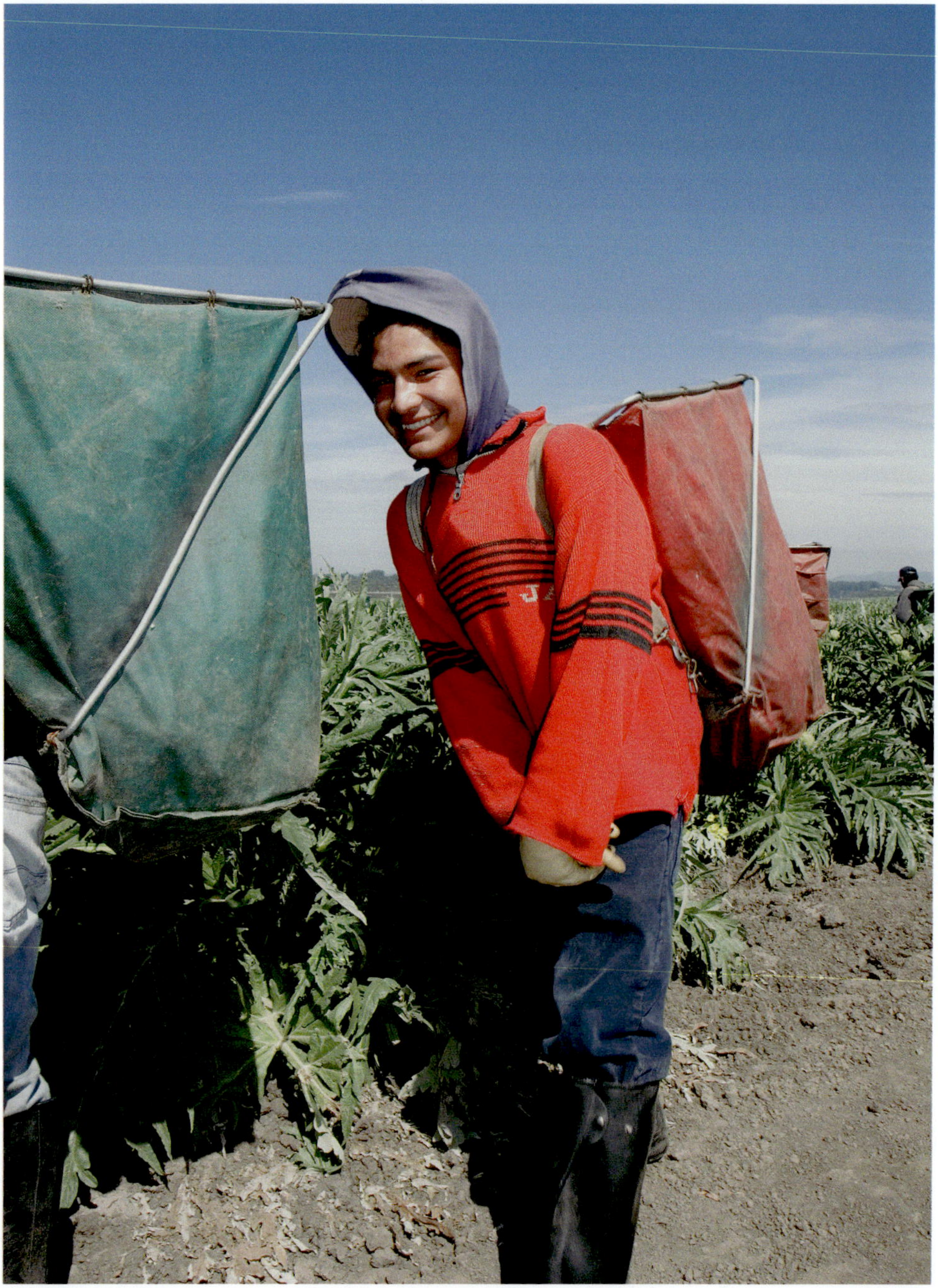

18.8 *Young man in red sweater, bin of artichokes*, 2007.

How to photographically represent the hotly contested issue of migrant farm labor in California agriculture? That complex topic could fill a book. Goin here chooses to tell a positive story. This young man, harvesting artichokes near Castroville in Monterey County, clad in red, white, and blue, will earn in one day what it would take two months to make in Mexico. As Goin muses:

> We so often want to talk about the exploitation of the Mexican farmworker. Nothing that I'm doing here refutes any of those arguments; however, we oftentimes don't seem to understand the basic human need to want to work, to consume food, to nourish ourselves, to take care of our responsibilities, to participate in the social contract. It's a more complicated narrative. Just as we can express our concerns over the exploitation of the farmworker, we diminish them as humans by isolating their experience to that one understanding. These are individuals. They have lives and loves and families. Here's a small slice of something that you wouldn't necessarily expect given the stereotypical representation of a Mexican farmworker.

In this single image, Goin combines respect for people who work with themes of youth, innocence, purity, color, sensuality, and fecundity.

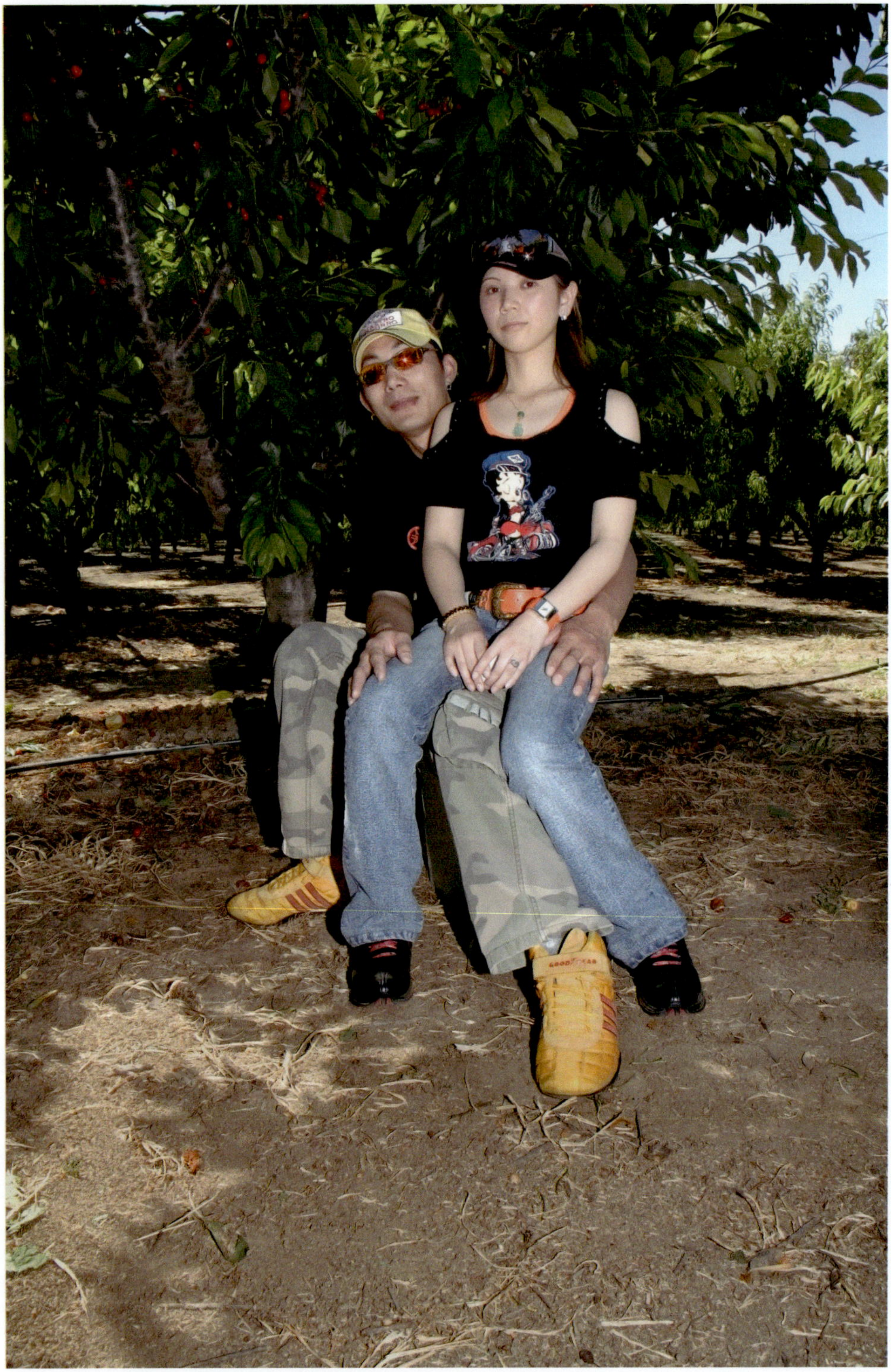

18.9 *Delon and Eva at U-Pick, Brentwood*, 2007.

"Our concepts of agriculture are outdated," Goin insists, "Agriculture has changed." U-pick fruit orchards and berry farms are one example of how growers are adapting to changing markets. In the cherry orchard pictured here, people from the city pay money to have an experience of agriculture. Agritourism. The young couple in this portrait are visiting Bacchini's Fruit Tree U-pick farm with their family of Philippine heritage. More interested in romancing than in picking cherries, this couple readily agreed to be photographed, and they posed themselves—the man sat on a crate and the woman sat on his lap. Goin delights in the cultural references at play in this image—t-shirts, youth, Betty Boop, and the red of the cherries reflected in the young man's sunglasses. "There's something about the whole human relationship to fruit," Goin observes, "We think of fruit as a dessert, as a reference to pleasure in some way, shape, or form. And of course, metaphorically, cherries have slang references in western culture." Here in the "new nature," California agriculture represents fecundity, fertility, sensuality, health, and healing by getting back to the earth.

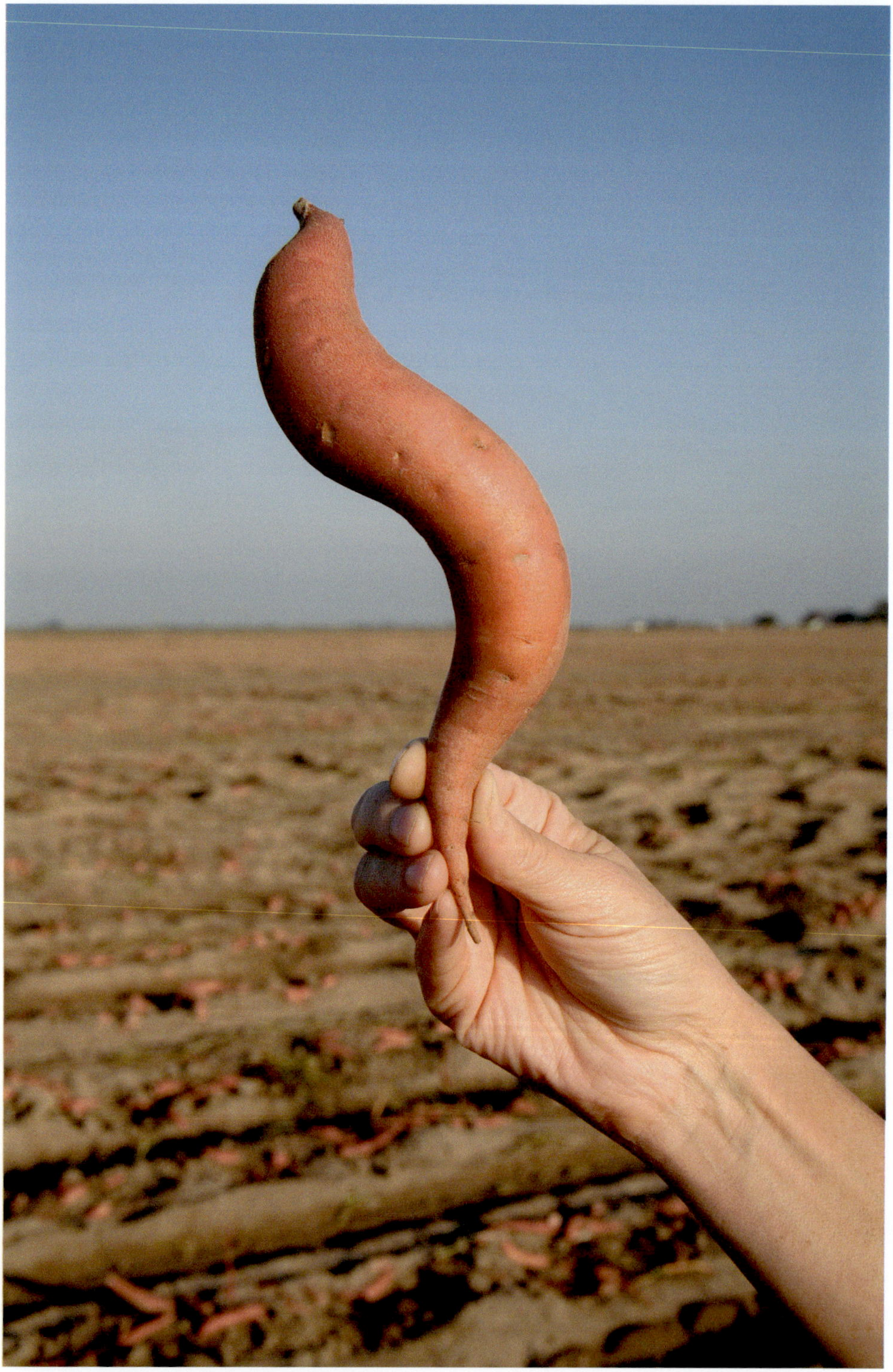

18.10 *Sweet potato, Merced County*, 2007.

How would you caption this image? What stands out? Centered and sharply focused is a close-up of a curvaceous, brownish orange sweet potato, held up torch-like by a human hand, as if this venerable vegetable is intended to be gazed upon as found art, admired for its abstract, sculptural form. The hand itself appears to belong to a white, middle-aged adult with clean, smooth skin and well-groomed fingernails, likely one of the urban, field guide authors or their companion rather than a farmhand. The blurred-focus backdrop reveals row upon row of unearthed sweet potatoes awaiting harvest and a horizon line that includes a thin fringe of trees and dwellings capped by a diffuse blue sky. The context of industrial-scale agriculture contrasts with the sensuality of the individual sweet potato. Compositionally, the diagonal line made by the arm and hand and the curved shape of the sweet potato intersecting with the horizon line creates a visual illusion of curved space. Goin's portrait of a sweet potato is an environmental still life that pays homage to the great history of still life in art, in which flowers, fruits, and vegetables sometimes carry symbolic meanings. "I'm participating in that tradition in a kind of playful, trickster way," Goin smiles, "Look at your food. Food is beautiful."

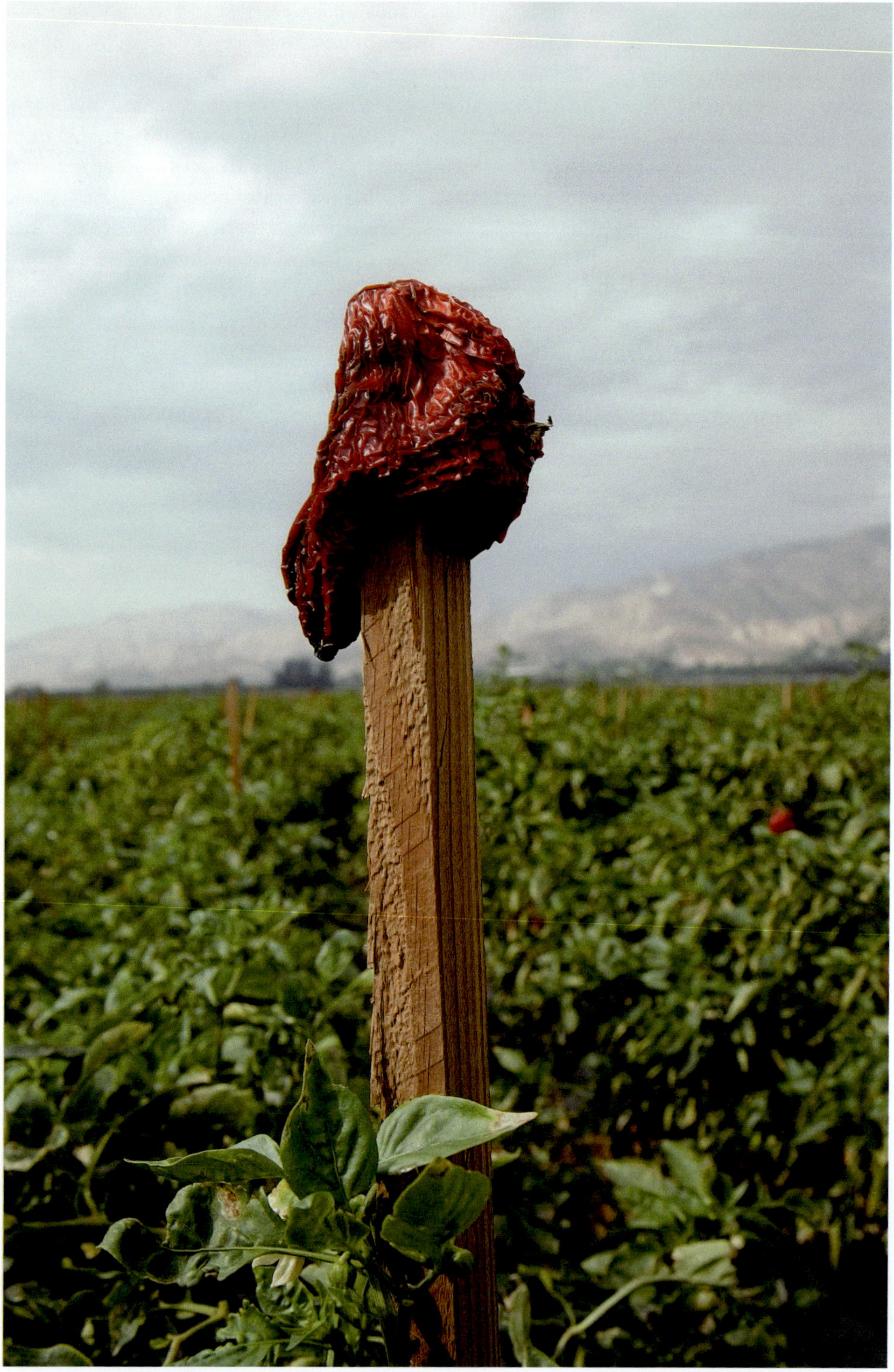

18.11 *Bell pepper on wooden stake, Ventura County*, 2007.

Continuing in the still life tradition, this image of a shriveled red bell pepper drooped atop a rough wooden stake in a green field presents an enigma. Why would someone put a pepper on a post? In fact, Goin noticed dozens of pepper-topped stakes sprinkled throughout this field. Why? he wondered. Crop identification? Bird deterrence? Method of propagation? Agritourists frequently encounter things they don't understand, a measure of the cognitive gap between city and country folk. While *Field Guide to California Agriculture* illuminates many of these mysteries, the case of the perched, parched bell peppers remains unsolved. No one Peter asked, even an ag station supervisor, could tell him. Visually, Goin finds this photograph alluring—the red, the green, the high contrast. "It's red. It looks like meat on a stick. And it's agriculture," he says, "To me this image is about this richness." If you read this image as phallic, then you, too, have been bitten by the agriporn bug.

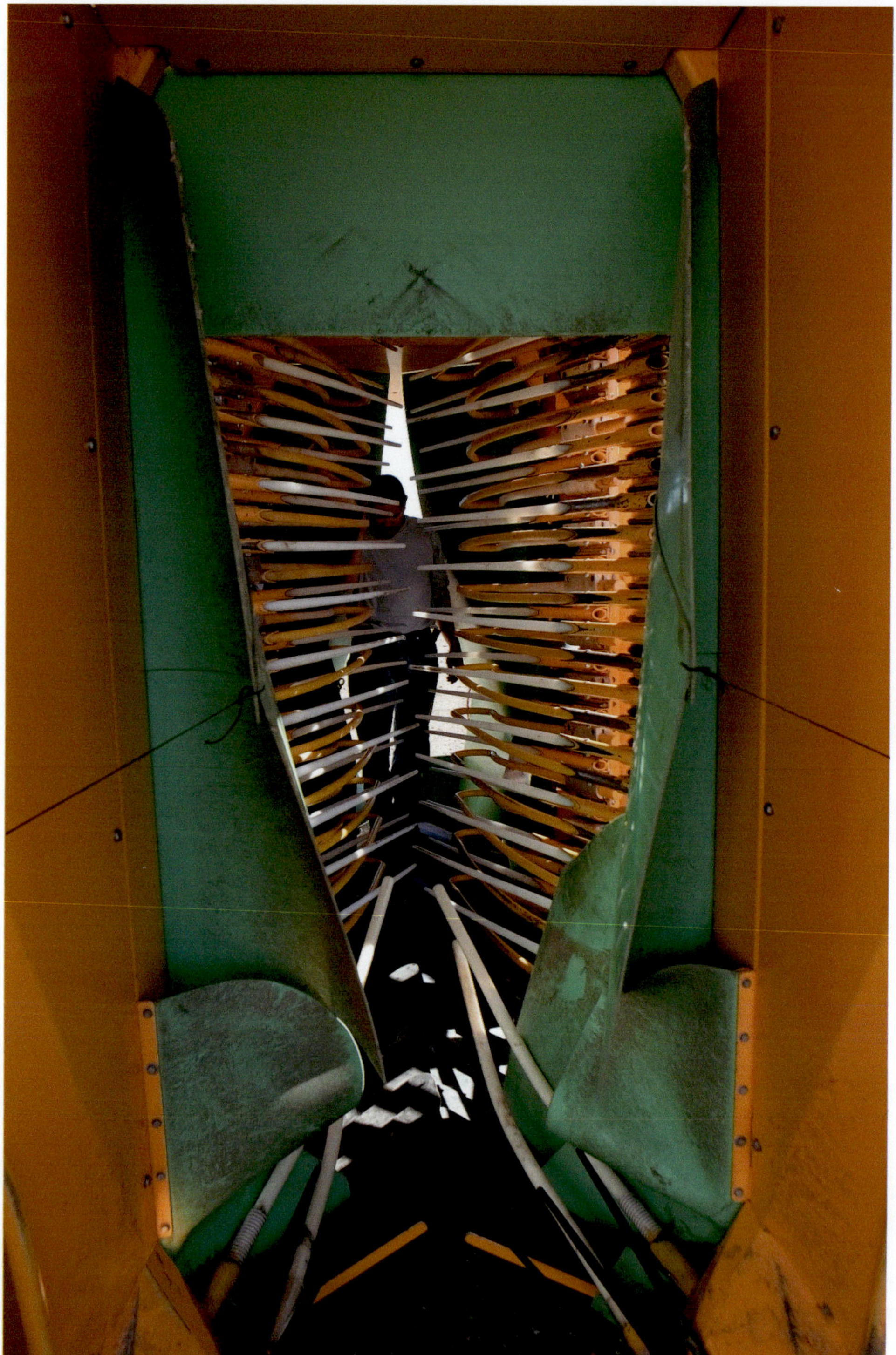

18.12 *Olive harvester, interior,* 2009.

A recurring motif in Native American stories is the toothed vagina of a virgin girl; breaching it preoccupies wily coyote. In this photograph, a postmodern, industrial toothed vagina has swallowed up a man and imprisoned him. But no, wait! This is not art, this is ag. As Starrs explains:

> At the California Olive Ranch near Gridley, a Gregoire Model 167 is adapted to harvest ultra-high-density plantings of olives, grown on trellises, that are planted with up to 700 trees per acre—a marked contrast to the traditional olive tree density of 20–40 stems per acre. What seem to be scary 'teeth' are actually nylon straight rods designed to penetrate deeper into the olive canopy to knock loose up to 95% of the olives, far more than might be gleaned by other mechanical harvesters. In a few seconds, a tree can be stripped of fruit; it would take a human picking crew a quarter hour or more, working as a team, to get the same results—and these days, such a labor supply is notoriously difficult to find and hire.
>
> *(Colors 165)*

Goin makes the point that "there is a great deal of inventiveness in agricultural harvesting, planting, and management."

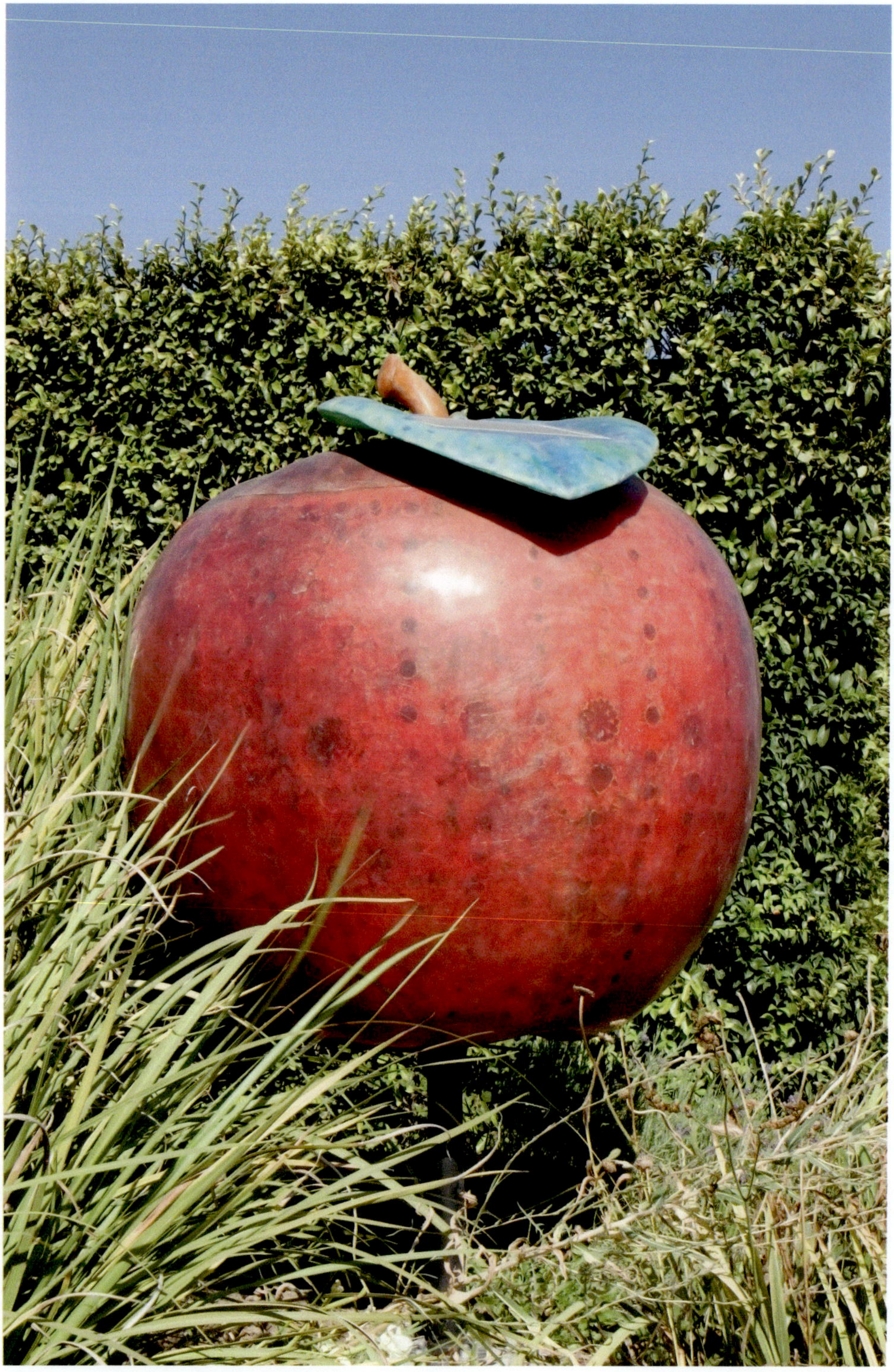

18.13–18.16 *Roadside typology of faux food*, 2006–2008.

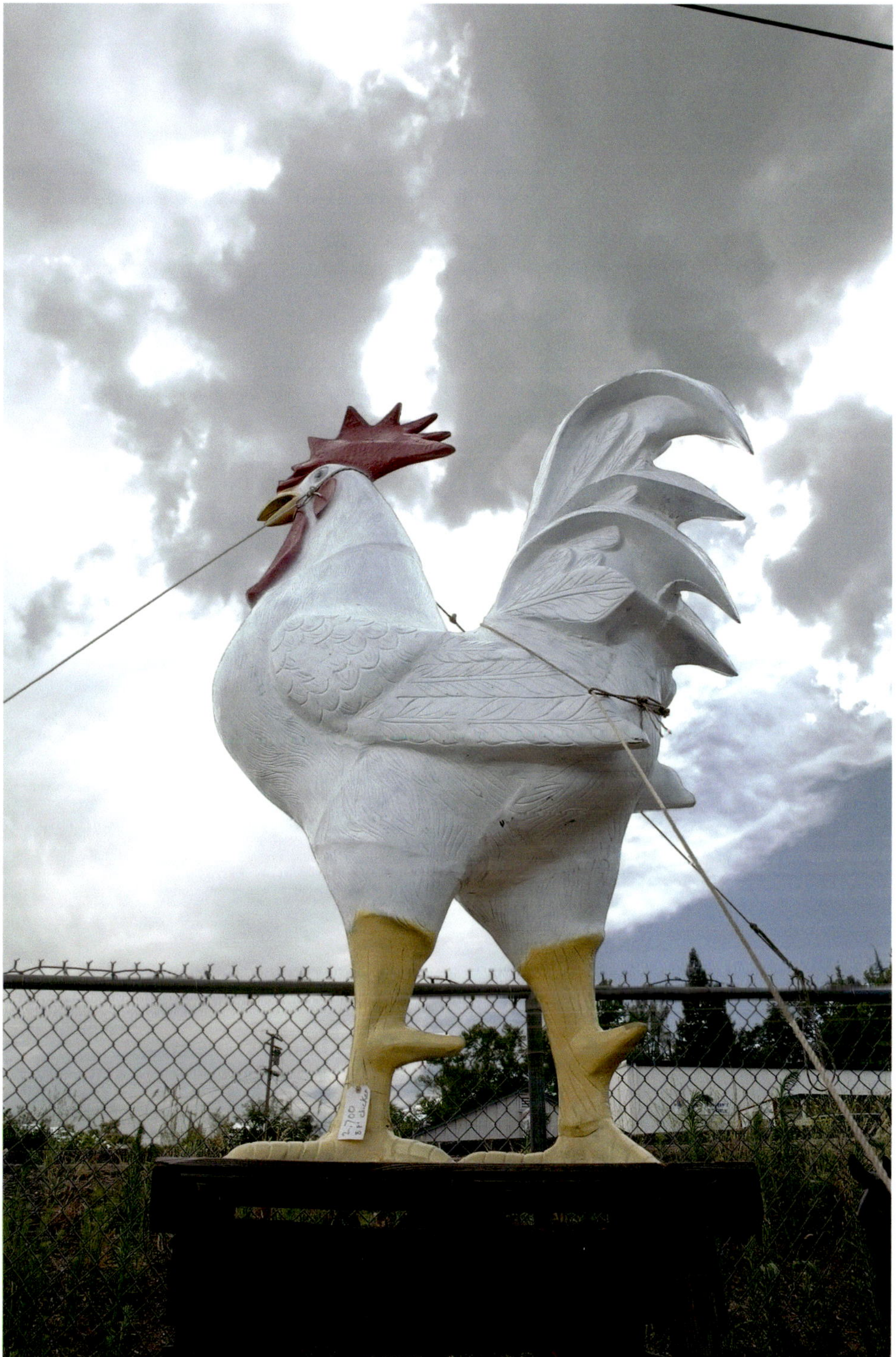

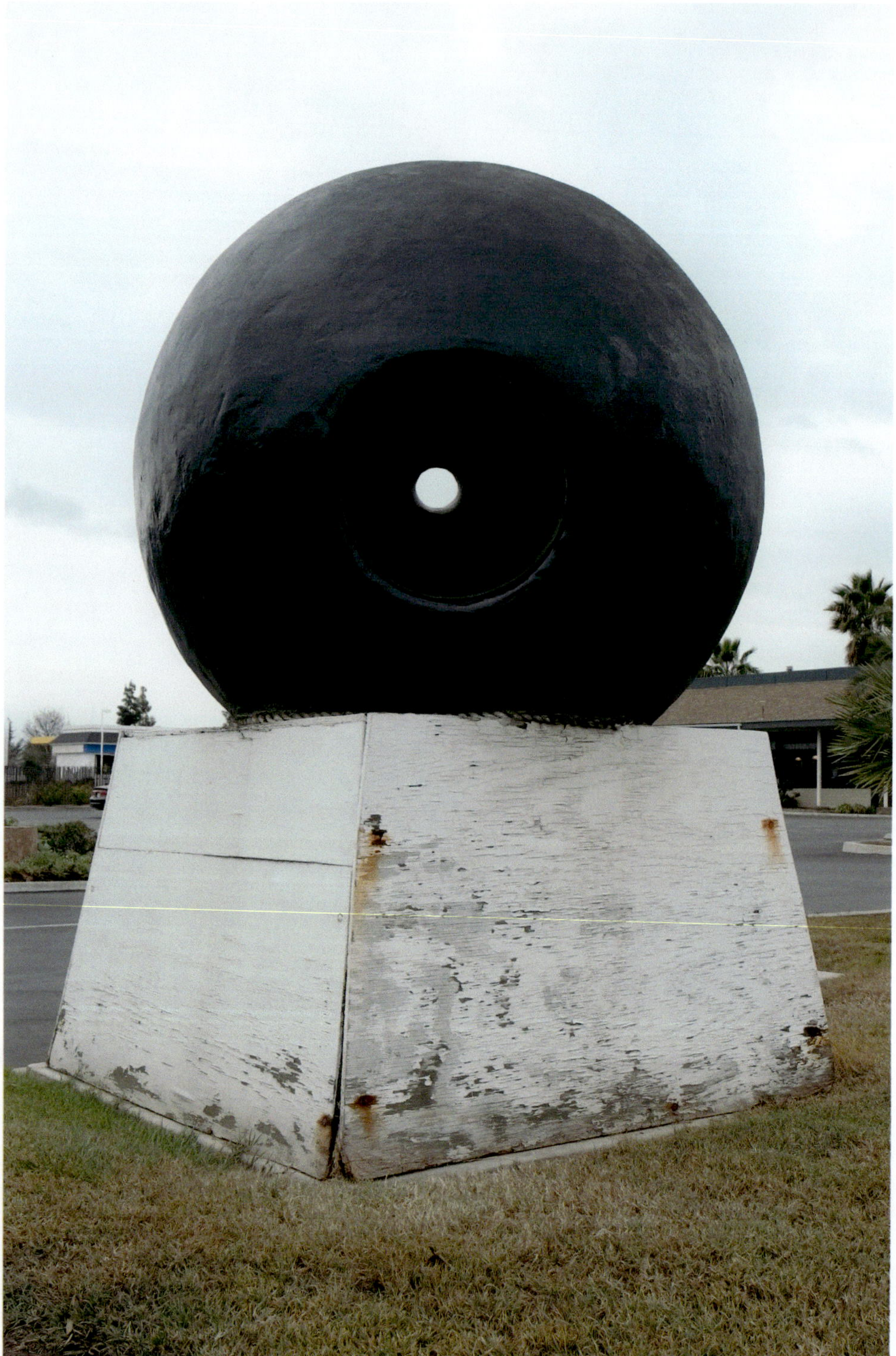

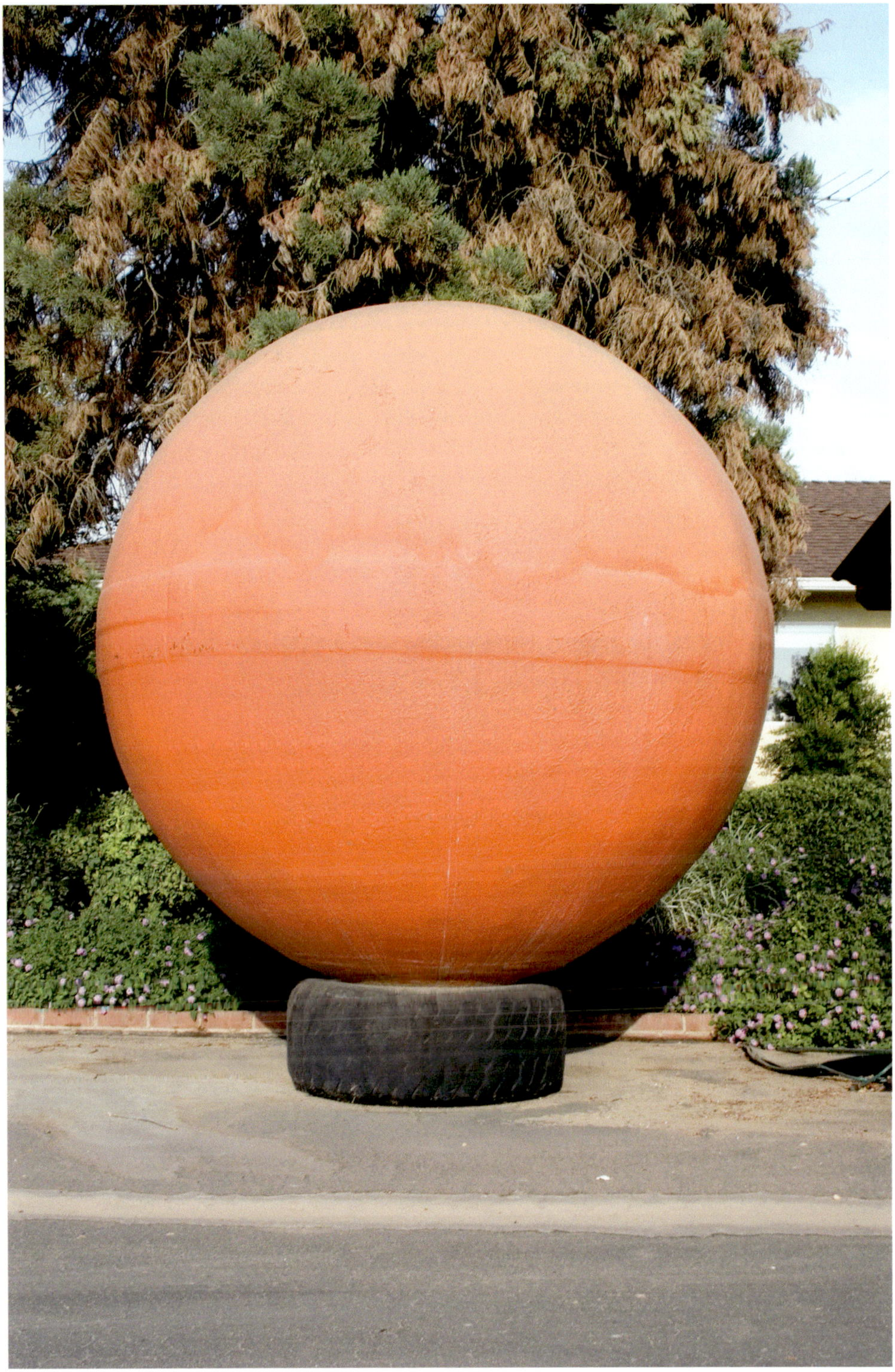

German artists Bernd and Hilla Becher became known for their typologies, grids of black-and-white photographs of varying examples of a single type of industrial structure, such as the blast furnaces of steel mills. As a Museum of Modern Art exhibit of their work explains, "The rigorous frontality of the individual images gives them the simplicity of diagrams, while their density of detail offers encyclopedic richness" (Galassi). In this grid of supersized replicas of California agricultural products, Goin invites us to think about how places take pride in and identify with their prime agricultural crops. The red apple resides next to a bank in Sebastapool, a town historically known for its apple harvest; the black olive adorns a parking lot in front of an Olive Garden restaurant in Lindsay, which claims to be the "Olive Capital of the World"; and the fiberglass orange sits atop a surplus tire beside a fruit stand southeast of Fresno in Tulare County. Goin's catalog of kitsch references to California agriculture explores our tendency to make monuments to what matters to us. Even on the urban fringe, agriculture can matter to people. As he notes, "these vernacular signs represent not just a particular industry, but a commitment to the ideology of farming, to nutrition." Because poultry producers strictly prohibit photography, we include the rooster model to symbolize California's poultry industry.

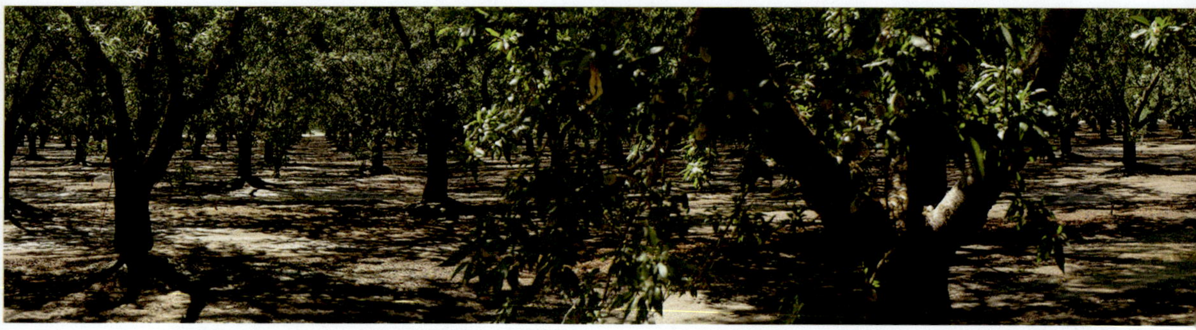

18.17 *Almond grove, Stanislaus County*, 2008.

A four-page spread in *Colors of California Agriculture*—displayed as a large, 24" × 139" panorama in the 2013 exhibit of Goin's photographs at The Bancroft—gives viewers the feeling of walking by the orderly rows of an almond grove, the neat rows of trees ticking past one's peripheral vision like the beat of a metronome. Conceptually, this photograph is about the process of walking. Notice the visual confusion that results from incorporating multiple vantage points in a single image, creating a disorienting sense of motion. To make this image, Goin laboriously stitched together eleven separate photographs so precisely that the seams disappear. He explains that the Almond Grove panorama "is an intellectual recontextualization of the multiple vantage points that historical painters have used to represent with mathematical accuracy the depth and dimension of a two-dimensional surface." He references David Hockney's book *Secret Knowledge: Rediscovering the Lost Techniques of the Old Masters*, which investigates the painting techniques of artists such as Caravaggio, Velázquez, and da Vinci, concluding that they used optics, lenses, and the *camera obscura* to render accurate perspectives and often multiple perspectives in the same painting. Goin's images often allude to scholarship, although that dimension of his work is frequently overlooked.

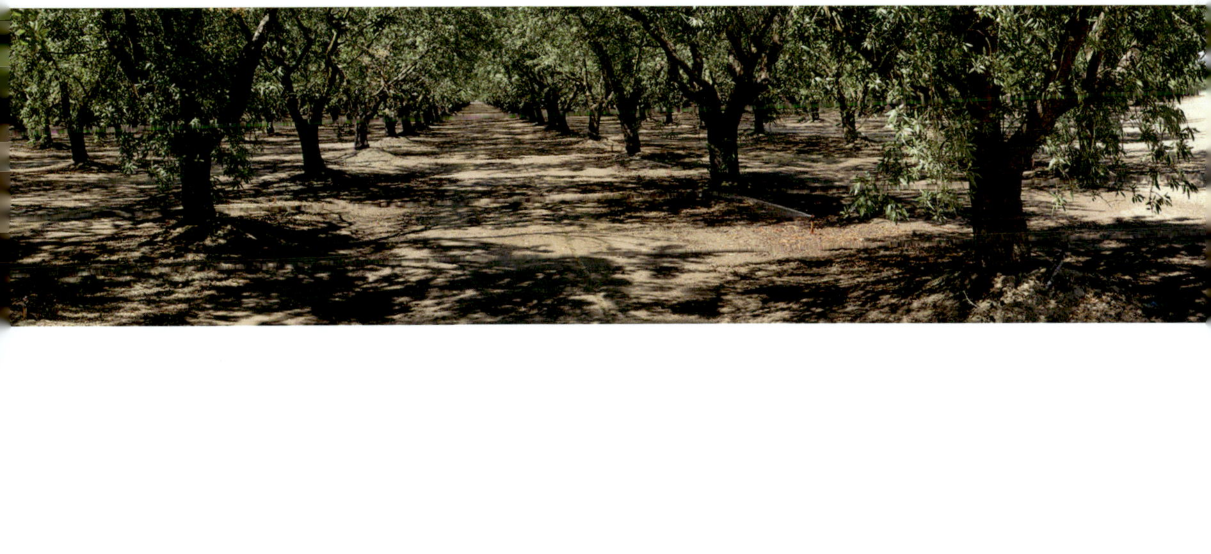

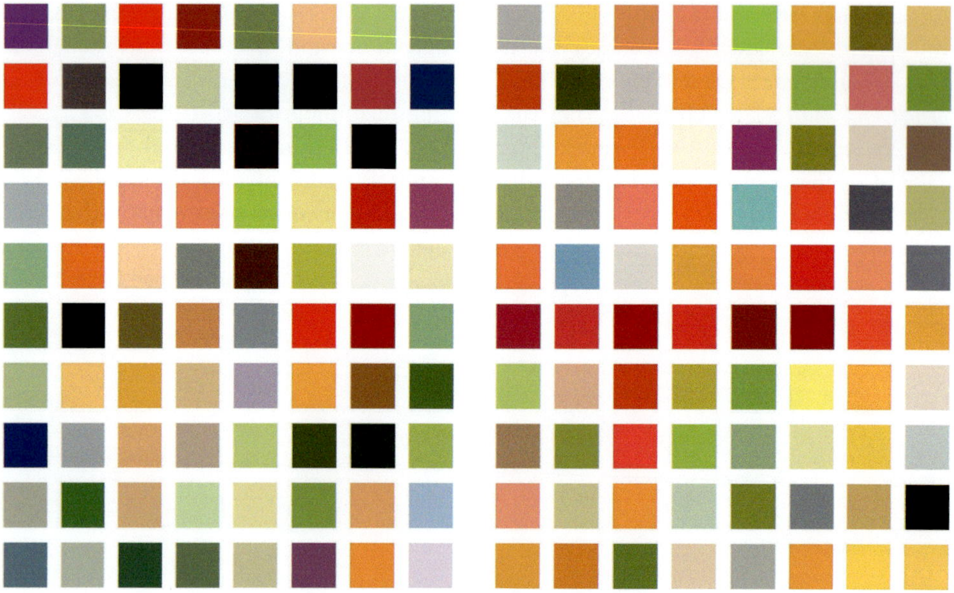

18.18 *Colors of California Agriculture*, exhibition print made in 2010.

This abstract, modern art piece, measuring with borders forty-four inches × sixty-nine inches in the exhibition print, is Goin's creative interpretation of California agriculture. To him, agriculture embodies the sensuality of the "new nature" of human-made landscapes. John Muir famously wrote:

> Climb the mountains and get their good tidings. Nature's peace will flow into you as sunshine flows into trees. The winds will blow their own freshness into you, and the storms their energy, while cares will drop off like autumn leaves.

Goin discovered that similarly uplifting, rich sensory experiences can be had amidst agricultural fields, a hopeful idea that he imparts via vibrant, saturated color photographs. The 160 different colors in this grid, extracted from digital files made between 2009 and 2010, can be seen in California agricultural products and equipment. Try relaxing your eyes and look at this image with a soft focus. Can you see how the color squares seem to pulse with kinetic energy? Do you begin to see little gray boxes where the white lines intersect? "How do we get people to use their brains differently?" Goin asks. One way is through optical illusions, such as this image, which heighten visual awareness and mindfulness. Goin's work aims to explode old stereotypes and catalyze a fresh look at agriculture.

Works cited

Galassi, Peter. "Bernd and Hilla Becher: Landscape/Typology." *MoMA*. The Museum of Modern Art, 2019, https://www.moma.org/calendar/exhibitions/95. Accessed 27 August 2019.

Goin, Peter, and Paul F. Starrs. *Colors of California Agriculture*. The Bancroft Library, University of California, Berkeley, 2018.

Moore, Nigel. "The Erotic Oyster." 17 December 2011 post. *The Oyster's My World*. The Oyster's My World, 2019, https://theoystersmyworld.com/2011/12/17/the-erotic-oyster/. Accessed 27 August 2019.

Muir, John. *Our National Parks*. Houghton Mifflin, 1901.

Starrs, Paul F., and Peter Goin. *Field Guide to California Agriculture*. University of California Press, 2010.

19

ARTIFACTS OF THE FUTURE

Our narrative—and Peter Goin's photographic career—began with a journey that Goin and his graduate school friend Robert Cree made in 1977–1978. The young adventurers went on the road to explore Mexico and Central America. In Tulum, Mexico, Peter made a series of photographs of ancient Mayan ruins, struck by the realization that nature was vigorously overgrowing and in some cases obliterating human architecture. In his words, architecture was "melting back into nature." This theme of the dynamism and visual evidence of the human-nature interface would propel Goin on the career-long investigation that we have chronicled. The theme is explicitly articulated in Goin's 1996 landmark book *Humanature*, a philosophical and photographic manifesto that reveals how the spaces we experience as nature invariably bear the signature of human manipulation. Every realm of the natural world—trees, beaches, land, rivers, and wildlife—reflects a degree of human management however covert; conversely, many overtly human-made structures—zoos, mines, and dams—create landscapes that on some level simulate or revert to nature. Humans and nature are not separate entities, Goin vehemently asserts, but always already enmeshed. In 2016, approximately forty years after Goin's Mayan images and twenty years after *Humanature*, Goin coauthored a book on Lake Powell entitled *A New Form of Beauty: Glen Canyon Beyond Climate Change*, a final destination for our narrative if not for Goin's ongoing work. Whereas the Mayan series was steeped in antiquity, *A New Form of Beauty* carries Goin's perdurable humanature theme into the future, an unknown and some would say alarming future, in which anthropogenic climate change will alter landscapes; in fact, already *is* altering them.

Just as the photographs collected in Goin and Starrs' *Black Rock* spanned some twenty years, the Lake Powell photographs likewise represent more than a dozen trips made over an extended period, 1987–2014. Peter first became interested in Lake Powell in the context of water in the arid West. Having moved to the high desert environment of Reno in 1984 to begin his teaching and research career at the University of Nevada, Goin soon became drawn to inland waters, intrigued, as he puts it, by "how the absence of water makes its own appearance." He began spending time at Lake Tahoe, the Truckee River, and Pyramid

Lake. He laughs, "If a moth is attracted to flame, I'm certainly attracted to water." Fittingly, *A New Form of Beauty* is dedicated "To the waters of the West, and to knowing, experiencing, and visualizing this most essential compound of life."

It was only a matter of time before Goin became curious about Lake Powell, a colossal human-made lake on the border of Utah and Arizona, created by damming the Colorado River in 1963 and submerging Glen Canyon. Depending on the lake level, which fluctuates, this two hundred-mile-long reservoir has approximately 1960 miles of shoreline, making it the second-largest reservoir in the United States—after Lake Mead—and one of the largest created environments on the planet. As Peter recalls, "I had been interested in Lake Powell but hadn't explored it, and so I thought, 'Go look.'" He and his friend and old college roommate Dennis Albrecht made their first of many trips to Lake Powell in 1987. Despite enduring almost continuous lightning and thunder storms that year, Goin found Lake Powell immediately engaging, and he wanted to keep going back to photograph. As he recalls in conversation, for him Lake Powell:

> satisfied that need for the hike, the solace, the swim, the view, the water. I love the view of water and the way that it at times appears like a huge enormous ocean of mercury, and the waves and reflections of light on the water. It's a very engaging light environment—the physical quality of reflected light.

In the "Photographer's Note" for *A New Form of Beauty*, Goin summarizes his varied experiences at the lake over the years:

> I've traveled Glen Canyon and Lake Powell for more than twenty-five years, first from Hite then Bullfrog, Dangling Rope, and Wahweap; I've driven roads from Page to Glendale and Kanab to Cedar City, from Winslow to Hurricane. I've photographed repeatedly throughout the canyons and byways within the Escalante River corridor. . . . I've made no fewer than fifteen research trips, the last five times with the University of Nevada, Reno's photographic research vessel, a seventeen-foot aluminum boat outfitted for photography. I've experienced fear and doubt, joy and sadness, light and darkness within the sublime landscape of Lake Powell.
>
> *(143)*

Before detailing how Goin's conceptual understanding, imaginative response, and photographs evolved over the course of his immersion in what he calls "the conundrum of Lake Powell," let us briefly establish the origins and history of the place as Goin presents that information in the introduction to *Humanature*, where he points to Lake Powell as an example of a human-created natural landscape. Reflecting the pervasive management of nature in the United States, the 1500-mile, mighty Colorado River is restrained by the Glen Canyon and Hoover dams, which create the huge reservoirs known as Lake Powell and Lake Mead. The human effort to control nature has resulted in the complete harnessing of the Colorado River, diverting it through long siphons into the water supply reservoirs of Phoenix, Las Vegas, San Diego, and Los Angeles, desert megacities whose populations exceed the capacity of local water sources. The flow and volume of this once-wild river are managed according to "the peaks and valleys of urban hydro-electric power demands, while entitlements are challenged through the justice system" (14). Built in 1963 near Page,

Arizona, Glen Canyon Dam was the crowning jewel of Floyd Dominy, commissioner for the US Bureau of Reclamation from 1959 to 1969, who wrote in his book *Lake Powell: Jewel of the Colorado* that "multitudes hunger for a lake in the sun" (quoted in *Humanature* 14). In that spirit, the Glen Canyon National Recreation Area was established by the US Congress on October 27, 1972, to "provide for public outdoor recreation use and enjoyment of Lake Powell" ("Park Statistics"). Lake Powell's water is defined in two ways, Goin says: One is a resource—a commodity—and the other is a recreational space.

In the 1960s, the flooding of Glen Canyon for contemporary use was highly controversial, as the canyon was very beautiful and had tremendous aboriginal value, including petroglyphs that would be drowned. According to an article in *High Country News*, "The Sierra Club's first executive director, David Brower, considered the dam's construction a personal failure and spent the rest of his life advocating for its removal" (Langlois). Nevertheless, as Peter Friederici writes in *A New Form of Beauty*, "a rising reservoir is a tool of forgetting, wiping out the signs of what it destroys below, each stepping-up of the water level a new status quo and another opportunity for yet more growth" (76). After the dam, he notes:

> for most of those who came to visit this new recreational mecca, it was as strange to contemplate the time before the reservoir's creation as it is to imagine life before one's own birth: Something must have been happening here then, yes, but it was so confounding to really try to understand what it was like that it was easier—safer perhaps—to pretend it never did.
>
> *(76)*

Although the text of *A New Form of Beauty* discusses the building of Glen Canyon dam and the inundation of the canyon, visually the book enacts the very amnesia that Friederici diagnoses. Whereas several of Goin's other books—*Stopping Time, Time and Time Again*, and *A Doubtful River*—juxtapose historical images with Goin's contemporary rephotographs to document change, intriguingly, he chose to omit historical images of Glen Canyon from a book whose subtitle *Glen Canyon Beyond Climate Change* references that very canyon. Why?

When Glen Canyon dam was completed, the impounded waters of the Colorado River began creeping up the canyon walls for the first time ever, eventually forming an enormous lake where there was once a network of red sandstone desert canyons, canyons carved by water action eons earlier. The Ice Age over eons formed prehistoric Lake Lahontan in the Great Basin, but the Industrial Age formed Lake Powell, which filled to its engineered maximum capacity in only twenty years. The question for Goin was what story to tell. In his view, the controversy over the dam had been well documented. The elegy over the loss of Glen Canyon had already been sung in words and images, and were he to feature before-and-after photographs of Glen Canyon, he would simply be retelling that familiar tale and fixating on that singular moment in time. Peter was drawn to Lake Powell itself. As he says, "We need to tell the history now, too. And the history now is of this lake. This lake is Glen Canyon now." Independent of Goin, Peter Friederici, a professor in the School of Communication at Northern Arizona University, was similarly drawn to explore the lake, in his case by kayak, explaining, "Whether you saw Lake Powell as a shining success, or as an abomination that never should have been born, you couldn't deny the sheer actuality of its presence" (78). It would take a quarter century of regular visits for Goin to receive from Lake Powell the story he wanted to tell and to team up with Friederici to produce the book *A New Form of Beauty*.

Over the years, Goin's work at Lake Powell evolved from an initial interest in making performative pieces, to an awareness of Lake Powell as *humanature* writ large, to a planetary perspective of Lake Powell as an index of climate change. Desert lakes are difficult to photograph. It is hard to embrace that elemental landscape in a formal rectangle and to deal with the technical challenges of photographing in that environment. Goin writes, "My camera has been absorbed in whirlwinds, slipped down the angle of repose into a watery grave, dangled above cliffs, and swayed by the swells of passing houseboats" (145). Peter's original creative response to Lake Powell was to experiment with performative pieces, several of which appear in *A New Form of Beauty*. Poetically, the light at Lake Powell is itself performative, and Peter thinks of his early images as recording light memory in the prehistoric sense of light in landscape. A stunning example of light memory is the first photograph in the book—in a gallery entitled Light—which pictures a sandstone rock overhang exquisitely lit from below by sunbeams reflected off a pellucid shallow pool at its base. If you've ever been mesmerized by the moving light patterns on the bottom of a swimming pool in the sun, just imagine that light dance taking place overhead on a rock ceiling of orange, tan, and white. Peter fondly recalls long hours on the "lazy river of time" watching the ever-changing light show:

> The theatrics of the performance of light's actions throughout the day is spectacular. You can literally just sit there and watch the light, because it's constantly on the move and measured by the canyons. The light is coming up or moving down or shifting or dropping or highlighting or bouncing off the water, and the water is moving. If you're willing to step outside the mechanics of your own mind's debate and just pay attention to the landscape, the light will entertain you, and you literally can sit and watch the light for hours.

Goin videotaped a continuous loop of light reflecting off the water at Lake Powell. He calls the reflections "dancing sparklies," and his video celebrates water as the spiritual center of human experience. "Even though Lake Powell is a human-constructed environment, you still have the wind and the waves," Peter points out, "Water is water, whether it's impounded or whatever."

With the assistance of his friend Dennis, Goin also experimented with human performance pieces. As Goin recalls:

> I would have my assistant go out. We would have a fire or a lantern or sparklers or some kind of light, flashlights even, and I would make 4 × 5 photographs of his shadow and reflection on the wall as a way of using light to mimic pictographs. To me, a *light pictograph*, which is actually a phrase I use, was intriguing.

Making these performance photographs, many of which involved multiple exposures, was a way of creating experiences using the landscape as a stage.

Although he enjoyed making the performance pieces, Goin never envisioned them as building toward a book. At best, he thought that they might become a portfolio of a dozen or so prints, and that might be the full product of his visual scholarship at Lake Powell. However, Lake Powell continued to draw him, and as his visits compounded, the personal assertiveness that fueled his performative work quieted as he became increasingly receptive to a lyrical, melodic narrative asserted by the place itself. As he read more about the origins and history of the lake, the seeds for his book *Humanature* were planted. In that book, however, Lake Powell functioned more as catalyst than primary subject. Goin's 1987 black-and-white quadriptych of Lake Powell at ninety-nine percent full pond figures in the book's introduction, but *Humanature*'s chapter "Dams" focuses on smaller, shrinking reservoirs in drought-stricken California, Idaho, and Nevada.

Like other reservoirs throughout the arid West, the lake level of Lake Powell fluctuates. Friederici writes, "Rise and fall, rise and fall—this was the quotidian life of a working reservoir in the modern American West" (78). When water rises, land disappears; when water recedes, land reappears a lighter color than before, scrubbed clean by the action of the water, leaving a so-called bathtub ring that defines the historical high-water mark. The level of Lake Powell constitutes a humanature system, where the natural processes of snowpack melt, rainfall, and runoff determine the amount of inflow, while human control of water releases at Glen Canyon dam for power generation, flood control, municipal, industrial, and agricultural water use, and Lake Mead equalization accounts for outflow rates, in concert with the natural processes of evaporation and seep. The US Bureau of Reclamation notes that:

> Each year, the lake level increases between May and July from runoff followed by a decrease in lake level throughout the remainder of the year, leaving a visible ring along the shoreline. The amount of fluctuation varies each year but during extended periods of drought, Lake Powell's elevation could drop by more than 200 feet below its maximum elevation.
>
> (*"Glen Canyon Unit"*)

Peter's visits to Lake Powell between 1987 and 2014 corresponded with a generally declining lake level. To give a sense of the actual numbers, in 1983, after a flood, Lake Powell reached an all-time high level of 3708 feet above sea level. In 1987, when Peter's visits began, the lake level reached 3698 feet at the year's maximum. Since Peter's visits began in 1987, the lake reached a low water record of 3555 feet in 2005, a drop of 143 feet in eighteen years. In 2014, when many of the photographs for *A New Form of Beauty* were made, the low water level for that year stood at 3574 feet. Five years later, in 2019, the year's water level minimum was lower still, at 3568 feet (lake level statistics are from *Lake Powell Water Database*). Essentially, in the almost thirty years that Peter made visits to Lake Powell, he witnessed a highly visible decline in lake level; 140 feet of once-submerged shoreline were now exposed to view, creating a totally different experience. *A New Form of Beauty* features two, fold-out panoramas, high- and low water bookends. The high-water panorama, in the front matter of the book, is a compilation of the 1987 quadriptych that appeared in black-and-white in *Humanature*, now seamlessly stitched together and printed in color. This inviting image of Forgotten Canyon depicts the wind-rumpled surface of a large body of water, lapping up against beautiful red-rock formations under a pleasant, cloudless blue sky. There is not a tree or anything green in sight. The scene is elemental: Rock and water. Hidden coves and waterways extend beyond the borders of the image, making the viewer yearn to be on a boat and go exploring. The low water panorama of Moqui Canyon, made in 2014, opens the Low Water gallery toward the end of the book. From a distance, this long picture appears to be abstract art, consisting of four horizontal color bands of roughly equal height: Pale blue sky on top, rust-colored canyon wall next, chalk-white canyon wall below that, and slate-blue water on the bottom. Beautiful. Disturbing.

For Peter, the story that Lake Powell was telling him required an unfolding sense of patience to discern. It is a story of the drying up of a huge desert lake, a tale of geological-scale change taking place in a single human lifetime. Climate change. Not climate change as a potential threat for a vaguely distant future, but climate change happening *now*, right before our eyes. Peter Friederici concurs, writing that *A New Form of Beauty* is about "the vanishing of the second-largest artificial lake in America in the face of the new, potent phenomenon we call climate change" (3). He explains that many factors are responsible for decreasing water levels of Lake Powell:

> Increased temperatures. Diminished mountain snowpacks. More dust. Faster runoff. A longer growing season. More human pressure on all sorts of water sources. Taken together, all these changes tell us that the Colorado River is 'dammed at the source,' as Bill McKibben presciently put it in his 1989 book *The End of Nature*.
>
> *(83)*

McKibben's best-selling book about climate change is an apt reference, also cited by Goin in *Humanature*. Scientific models attribute climate change to increased carbon emissions from industrial civilization. Carbon emissions form a "greenhouse" layer in the atmosphere, trapping heat that results in hotter global temperatures and megastorm systems. Essentially, climate change is *humanature* of the planet, and we are living in the Age of the Anthropocene, an epochal time when human actions are defining a new geological era.

Composed to ponder the coming of fall, but equally apropos of a falling lake, Robert Frost's poem "The Oven Bird" poses the question "what to make of a diminished thing." Whereas the majority of songbirds are most vocal in spring, ovenbirds sing loudly in mid-summer when other birds have tapered off. The repetitive song of the ovenbird, which sounds like "teacher, teacher, teacher," signals that leaves are old; that flowers are mostly done; that highway dust is over all; and that it will soon be fall and then winter. The sonnet concludes:

> The bird would cease and be as other birds
> But that he knows in singing not to sing.
> The question that he frames in all but words
> Is what to make of a diminished thing.

Peter Goin, in contemplating a drying lake, becomes the oven bird of lyrical documentary photography. Just as the oven bird knows "in singing not to sing," Goin knows that his fine art photography need not—indeed, should not—be conventionally beautiful as he creatively responds to and interprets Lake Powell. In *A New Form of Beauty*, Friederici writes of climate change:

> Maybe what we need to do, then, is to embrace this new ambiguity, to accept that we are as gods but far from omnipotent, that we are rather cocreators, that we are

> as much nature as what we once labeled as nature because it seemed outside ourselves, but that with this new promotion comes new responsibility that might truly be labeled what has become one of the most clichéd words in American English: *Awesome.*
>
> *(131)*

Accepting our new role as cocreators will not be easy, Friederici warns. It will require recognizing that "unbridled use of new powers without responsibility results only in disaster" (131). And it will entail inventing a new form of beauty. Goin agrees with New Topographics photographer Robert Adams that each generation must redefine beauty for itself. Goin adds, "part of the human condition is to find beauty in our circumstances." Finding beauty in Lake Powell, Friederici writes, will be a matter of:

> finding poetry in a new mud flat where speedboats once raced—or in the muddy, debris-filled broth caught in the branches of drowned cottonwood trees after a wet winter. There will not be much easy majesty to our way of seeing. . . . We will have to gain a layered appreciation of complexity, and of flux.
>
> *(131)*

As a student of Zen Buddhism and as a man whose wife died suddenly in a car crash, Peter has learned the lesson of impermanence. He titled his photographer's note for *A New Form of Beauty* "Nothing Lasts Forever." In this metaphysical light, Glen Canyon and Lake Powell become teachers of impermanence. (Recall that the ovenbird cries, "teacher, teacher, teacher.") As Peter observes in conversation, "It's not a static place. Glen Canyon is changing, has always been changing, and will always continue to change. So, within that context, how can we find meaning?" J. B. Jackson's *The Necessity for Ruins* has intrigued Peter for a long time in its argument that we need ruins to connect us meaningfully to the past. Ruins are tangible remnants of human history, visible artifacts of a human story left for future generations to decipher. Peter's creative insight at Lake Powell is that:

> New ruins are being made. There are new stories to be told. When I go to the lake what I see is evidence of people at the lake. I see evidence of people who have lived a life. It's almost as if, in this really bizarre, ironic way, I'm walking on *The Planet of the Apes*—which of course was filmed at Lake Powell. It's not a post-apocalyptical world, but it's a post-climate change world. We're so overwhelmed by the history of the damming of Glen Canyon and the recreational industry of it that we're missing that interpretation, which is the generating of ruins, generating a changing landscape in our consciousness, realizing how the landscape reflects change.

In his Glen Canyon/Lake Powell photographic investigations, Peter began thinking of his project as future archaeology, documenting the remains of our present civilization. Just as we value petroglyphs, arrowheads, and pottery shards as extant traces of aboriginal cultures, so too the miscellaneous detritus exposed by Lake Powell's declining waters. Peter opines, "are the objects that will define the future of this place. So, I'm traveling to the future, looking back to right now. It's a new form a beauty. It's Glen Canyon beyond climate change."

Were it not for the book's title phrase—*A New Form of Beauty*—most readers would not readily classify Goin's Lake Powell photographs as beautiful. For conventional beauty, one might expect to see Kodachrome blue skies and stunning vistas. Instead, the images document a human-altered landscape. In the Artifacts gallery, for example, the viewer encounters a decidedly unromantic, visual inventory of modern-day ruins—a submerged green plastic chair; an exposed anchor chain; a makeshift danger buoy dangling in the air from a dead snag; the bent rods of a shade structure "skeleton"; another skeleton of a camp chair sans fabric; a half-buried cooler (playfully labeled, "Pseudolith"); a weathered cardboard rocket launcher; stranded fishing lures wedged in trees; a boat ladder without a boat; an exposed sunken cruiser; a dirt-filled shoe embedded like a fossil in the strata of a dirt embankment; and a motorboat high and dry on a rocky shelf above the lake (oxymoronically labeled "Suspended and sunken boat").

The "Flora and Fauna" gallery might likewise strike one as dystopic. In a gallery of that name, one would normally expect to be treated to beautiful photographs of native plants, birds, and wildlife in their natural habitat. Instead, this gallery features invasive plants—Russian thistle, aka tumbleweeds—and non-native fish—a swirling cauldron of carp at the Hall's Crossing Marina. Also pictured are dead and designer fish—fish skeletons, fish carcasses, a fish pattern on the fabric of a pair of mud-encrusted board shorts. This gallery also pictures graffiti carved into and scrawled on rocks (whimsically labeled, respectively, "Contemporary vernacular petroglyphs" and "Naming rock"). As one can see from the tricksterish photograph labels, Goin encourages readers not to be repelled by but rather to engage with these contemporary ruins imaginatively. He employs humor in both subject and labeling to break us out of the familiar responses of guilt, regret, and denial, and instead, to arouse our curiosity, inventing new paradigms for engagement, meaning, and beauty. While the images at first glance may appear to feature the anti-aesthetic, further contemplation and an open mind render them curious and playful, framing the present as future history. Peter explains in conversation:

> Just because we can find beauty in a transformed landscape does not intrinsically enable or justify the transformation of that landscape. The landscape has been changed and that's what we have to deal with, for better or for worse. You have to go in with your eyes wide open. But the fact is that the world changes in a whole variety of ways, and we adapt to those. We are beauty seekers. We are meaning seekers. We are order seekers. Those are intrinsic to the human spirit.

The human imprint is subtler in the gallery Low Water and the beauty more accessible, although discomfiting when one realizes the root causes of low water, that is, anthropogenic climate change and water overconsumption linked to population growth and unsustainable lifestyles. In this gallery, one finds scenic natural vistas of magnificent cliff faces, enticing coves, sensuous rock formations, and natural arches, all marked by a lovely off-white stripe between the water below and the reddish sandstone above. Ah, one remembers with a sigh, the band where water used to be. . . Also pictured are the evocative, sculptural forms of dead tree trunks and branches, forming black or white silhouettes depending on the light, perfectly mirrored on still water, or arranged into intriguing trails of driftwood in the desert. The story of these trees haunts the formally beautiful images—trees once alive and growing in the canyon, then killed and forgotten by the canyon's filling, now reappearing as ghosts from the past.

As we now prepare to contemplate selected images from the Lake Powell project, the lens of *visual limnology* will helpfully illuminate the nature of the work. Limnology is typically defined as the scientific discipline that studies the physical, chemical, and biological components of inland water ecosystems, such as lakes, reservoirs, rivers, and streams. Goin's long-standing interest in Water in the West has led him to photograph the same ecosystems that scientific limnologists study, and people's eyebrows raise when he introduces himself as a visual limnologist. Granted, he does visual scholarship and produces factually accurate documentary photographs. Yet there is no denying the creativity of his images. This paradox brings us full circle to one of Goin's favorite quotations, which defines documentary photography as "deliberately wrought visual poetry disguised as plain, prosaic fact" (Thompson 12). Peter Goin's documentary photographs *express the experience of the lyrical.* He understands that the Cartesian split between science and art has made it very difficult for people to embrace the creative interpretation of fact. Yet, in the past, in the work of Leonardo da Vinci, for example, "creative expression has existed side by side with scientific inquiry" (144). Goin argues that "Verifying *only the facts* denies us our heritage and our senses as together the processes of inquiry—fact and feeling—are integrated however the boundaries are defined or defended" (144). When you are puzzled by a photograph that defies traditional standards of beauty, you are being given a gift. You are being challenged to *see* things in a new way—and to remarry fact and feeling in your own consciousness.

■

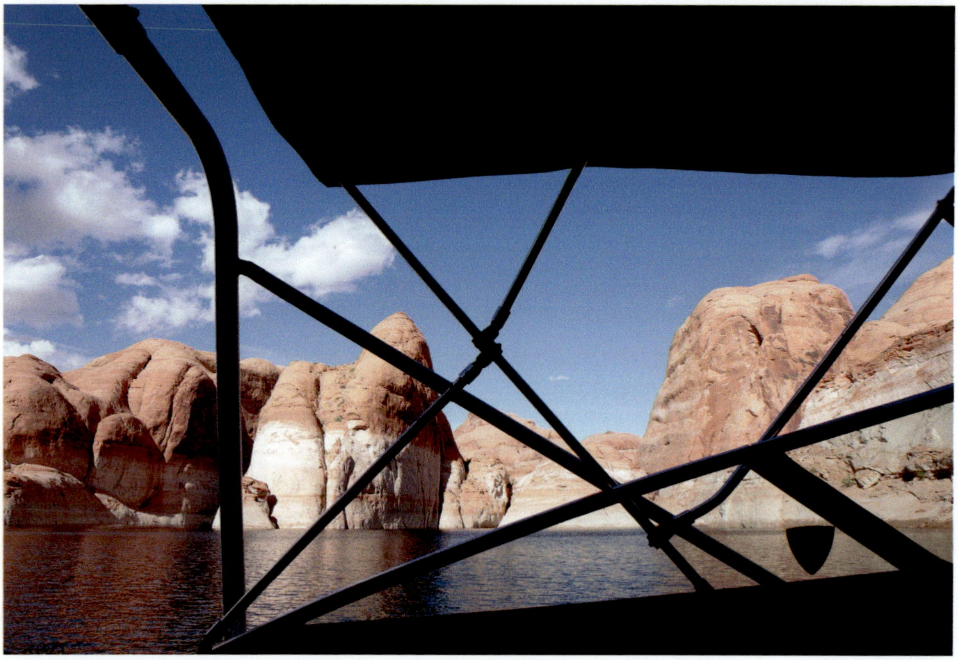

19.1 *Bimini-framed landscape, Anasazi Canyon*, 2010.

This view of a canyon is interrupted by the black supports of the Bimini top of a boat, competing for conceptual space and superimposing formalist design elements onto the documentary image of Lake Powell at low water. The Bimini top draws attention to the mediated view and the act of making the photograph only possible from a vantage point on the water. The image is framed by the means of transportation used to experience it, a principle of viewing the landscape that goes back to the days of horse and buggy. The angles and shapes of the Bimini top break up the landscape and fracture it, deconstructing it into necessary elements. As viewers, we tend to reconstruct that landscape in our mind's eye. Yet the reality is it is a very fractured landscape. Goin explains that Lake Powell is a human-made lake that we experience as natural:

> 'Reservoir' sounds so mundane. We would prefer to define our experience lyrically as a place of beauty, of light and sun and canyon walls. And so, we're always in this border between the lyrical and the mundane, the fractured and the composed, the fully constructed and the abstracted.

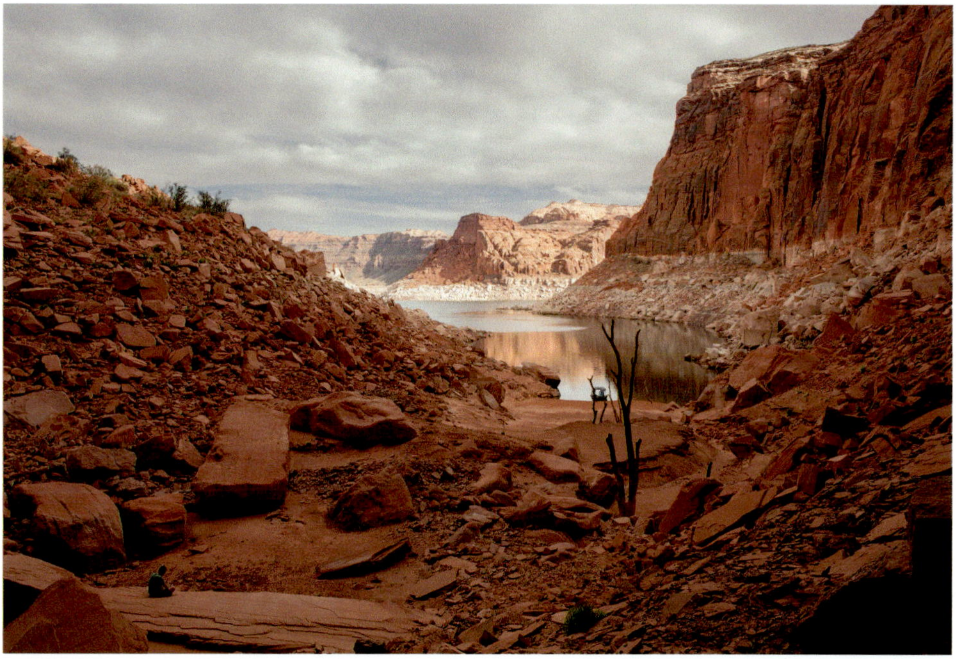

19.2 *Low water line, Good Hope Bay (distant view)*, 2014.

This distance view of a remote bay pays homage to nineteenth-century landscape photographs and paintings that evoke the grandeur of the American West. In order to convey the sublimity of those landscapes, artists would include a human figure in the scene, dwarfed by the scale of the surrounding mountains and valleys. Here, in that tradition, Goin's associate is pictured in the lower left in quiet contemplation. Referencing another artistic tradition, in this case Alfred Hitchcock, who made cameo appearances in his films, Goin includes his boat, looking very tiny at the end of the bay—the same boat whose Bimini top loomed large in the previous image. Scale is an important conceptual aspect of this view, scale not just of Lake Powell, but of forces that are beyond any of us individually. "Even collectively," Goin notes, "we seem unable to really come to terms with concepts of the human role in climate change." Climate change is a subtle theme of this image, hinted at by references to the sublime and, of course, by the low water ring that circumnavigates the photograph. In recent history, the vantage point from which this photograph was made would have been underwater.

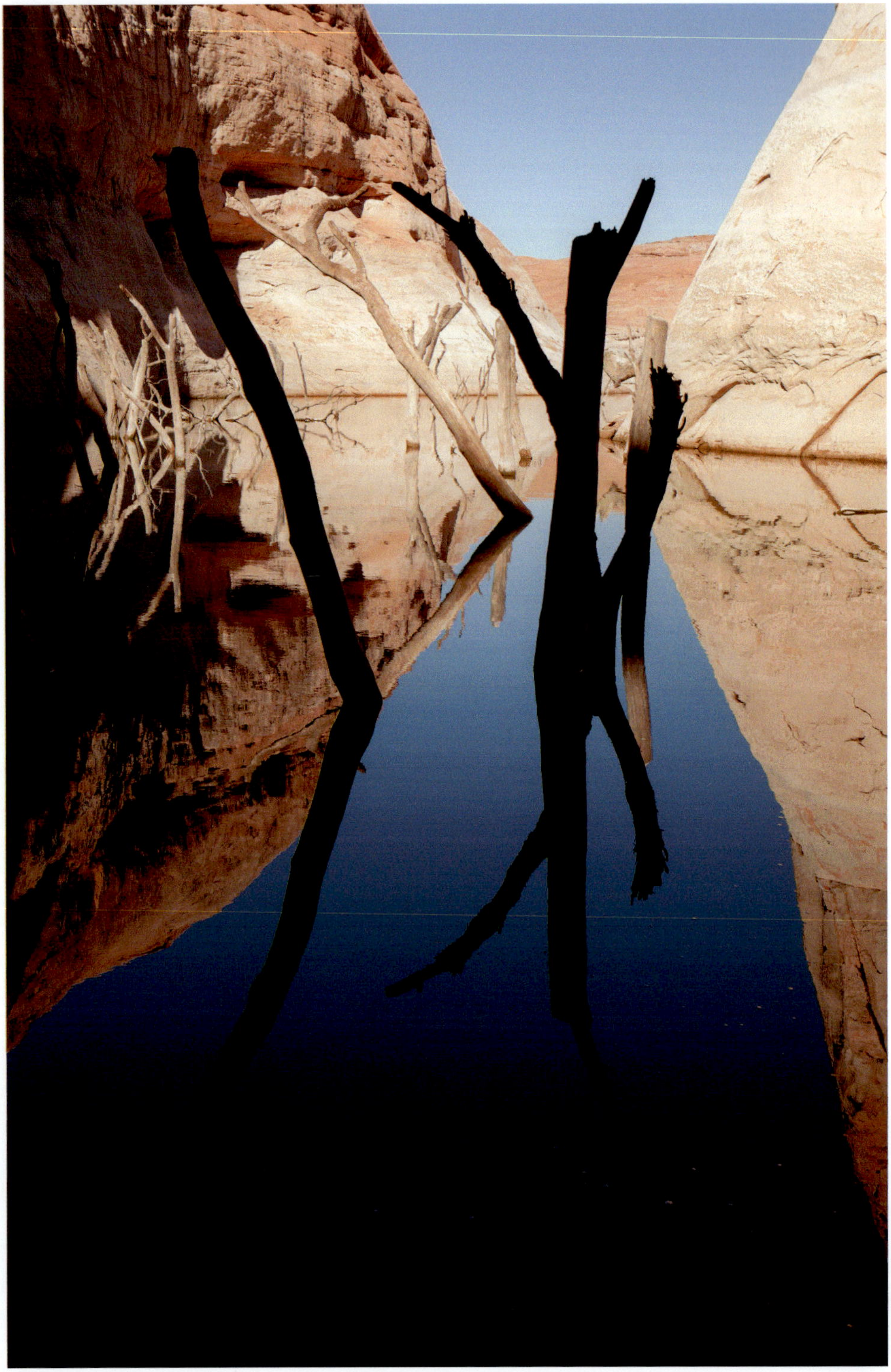

19.3 *Mirrored low-water landscape, Lake Canyon*, 2013.

Perfectly still water in a side canyon of Lake Powell mirrors the landscape so seamlessly that the vertical tree trunks appear to be continuous, like stitches binding air and water. Employing a Lake Powell color palette of blue, orange, white, and black, this photograph works as abstract art, forming an X that divides blue triangles on top and bottom and orange triangles on the sides. Not necessarily implying a sexualized interpretation but allowing for that reading, Goin positioned the boat and his camera such that one of the black tree trunks exactly intrudes into the V of the landscape. Resembling Egyptian hieroglyphics, the trees as a set stand as humanature sentinels, growing naturally in the canyon before human-caused flooding submerged and drowned them, now again exposed to the air due to a declining lake level tied to a concatenation of natural and human forces. Visually, the effect of the trees is to flatten the space into a vertical plane, an abstract concept of spatial complexity that reflects Goin's lyrical interpretation of the confusion we experience when we deal with the climate change debate.

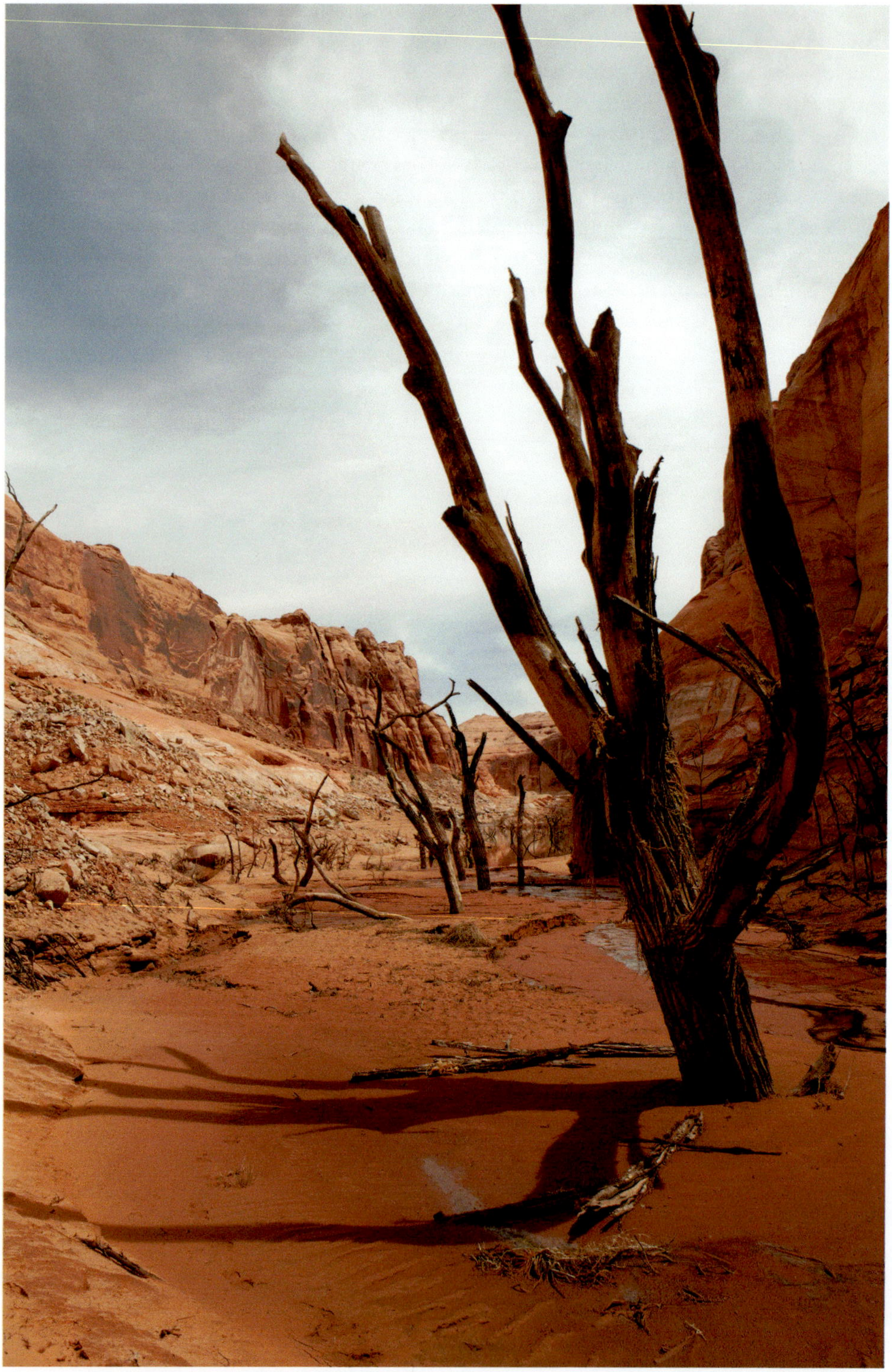

19.4 *Dead tree, shadow, Iceberg Canyon*, 2014.

Is this a movie set? If so, it's not a romantic comedy. As Goin explored the canyons of Lake Powell, he rarely felt like he was in a prehistoric landscape more than this one. He recalls:

> I think it's the nature of the trees, the shadows, the sand landscape. It just seemed like dinosaurs roamed here. If I were a billionaire, I would find some way to put some kind of huge robotic dinosaur in there just for the pure entertainment of it.

Like the previous image, this one bespeaks landscape change. Once a riparian habitat, shaded by tall trees and green vegetation and resounding with the trills, chirps, and calls of busy birds, this habitat was destroyed by the arrival of Lake Powell. Now the absence of water has transformed that landscape into a more desolate environment. Whereas the previous image flattened the space, this one manipulates light and shadow and foreground and background to create depth, as if you could walk into the scene. This photograph recalls stereoscopic views popular in the late nineteenth and early twentieth centuries, where a pair of nearly identical images viewed through a stereoscope created an illusion of three-dimensionality.

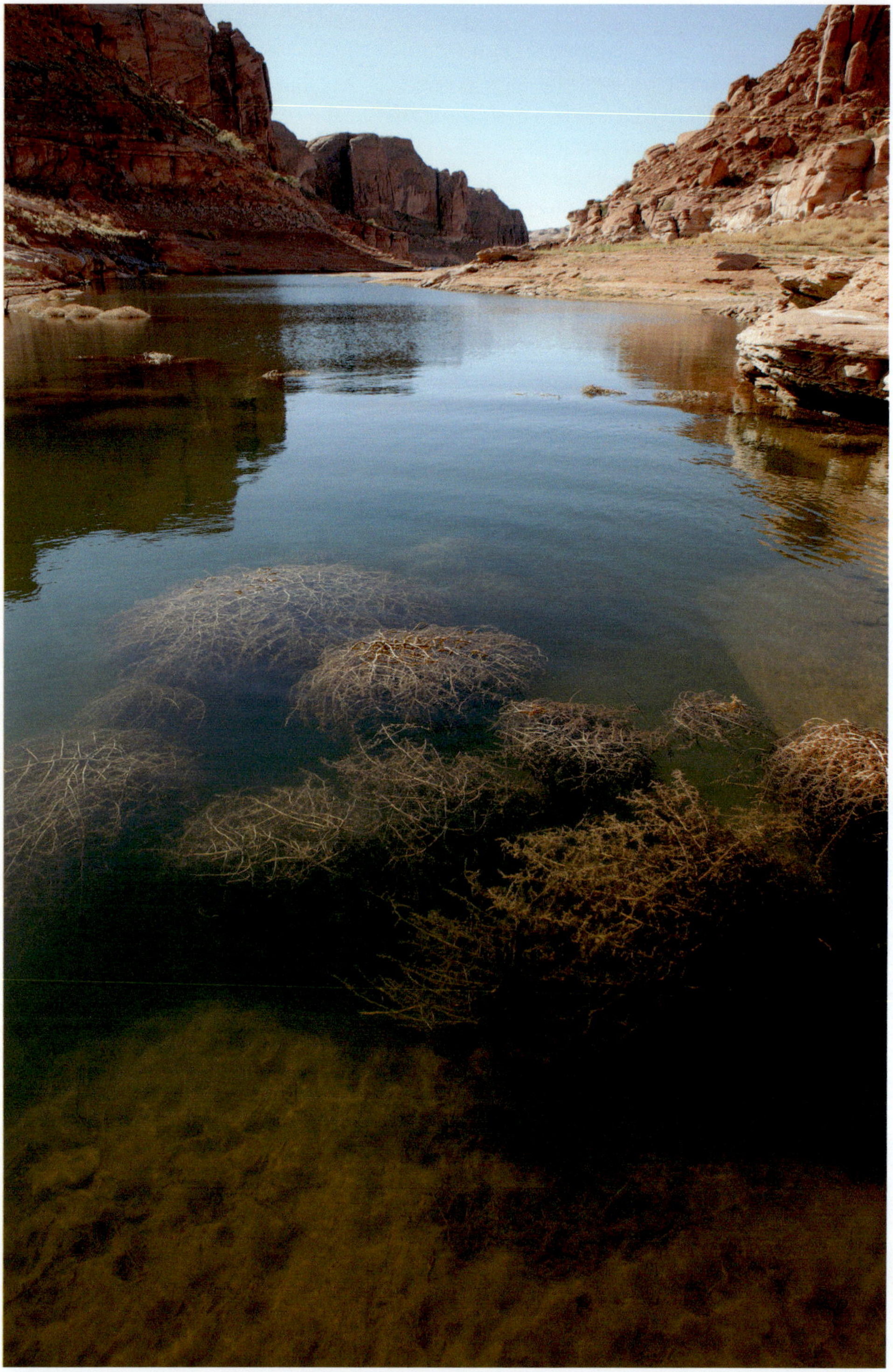

19.5 *Submerged tumbleweeds, Knowles Canyon*, 2013.

When regrowth occurs after disturbance, the pioneering new plants often are not the same species as those that were obliterated. In this low water view, tumbleweeds are colonizing the shoreline, whose banks were denuded of vegetation when Lake Powell was created. An iconic species in the arid West, the tumbleweed is an invasive, thorny annual forb believed to have been brought to the United States in contaminated flax seeds by

Russian immigrants to South Dakota in 1873 ("Russian Thistle"). Goin, who questions the notion of "native" species, delights in the irony that a symbol of the American West is an exotic species. This photograph of submerged tumbleweeds strikes one as odd since tumbleweeds are not an aquatic species. Tumbleweeds spread their seeds when the mature plant breaks off at the ground and tumbles along, blown by the wind. Presumably, the submerged tumbleweeds in the foreground were blown into the water, captured in the nook, and covered by water. *A New Form of Beauty* celebrates the elements of earth, wind, water, and fire. Wind, being invisible, is difficult to photograph. However, tumbleweeds imply wind, and the photograph becomes an artifact of the evidence of forces that are still operating.

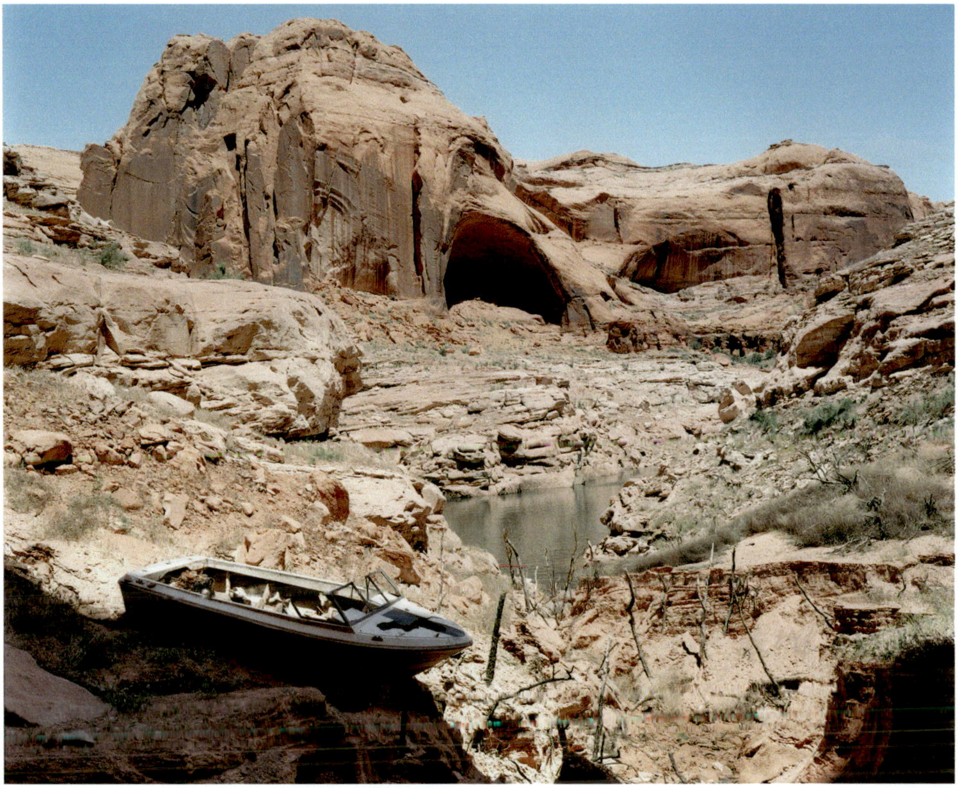

19.6 *Suspended and sunken boat, Ribbon Canyon*, 2005.

Goin's description of this image—"suspended and sunken boat"—is a paradox. Boats are supposed to float on the water. But this boat is described as being in two places at once—both *above* and *below* the water—neither of which is *on* the water. The riddle is solved when one adds the dimension of time and recalls the landscape history of Ribbon Canyon. At one time, when the waters of Lake Powell were high, Ribbon Canyon was a cove, where this boat capsized and sank. When the lake level dropped, as evidenced by the lighter strata of exposed rock, the boat remained in place but became high and dry, an example of contemporary ruins. In every ruin there lies a story, and one wonders about the buried narrative of this boat. Almost no one sinks a boat on purpose. Goin lingered at this site for several days to observe the movement of the light, making the photograph when the boat cast a dark shadow, mimicking the dark shadow of the arched cave in the background, creating a sense of mythology of the sunken boat. Curiously, when Goin returned in a later year to rephotograph, the boat was gone.

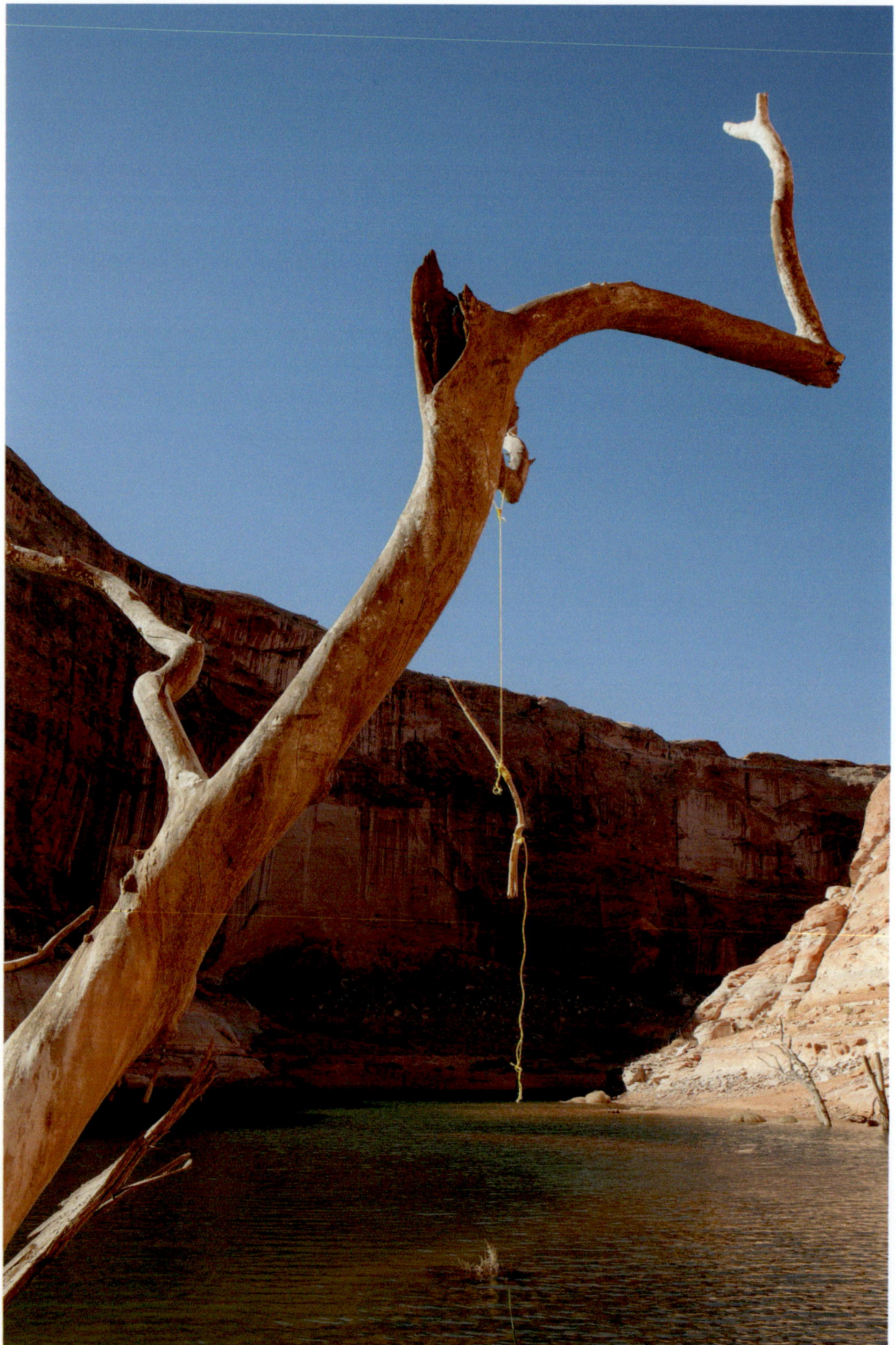

19.7 *Makeshift danger buoy, high above low water, Forbidden Canyon*, 2013.

As in the previous image, Goin's description—"buoy, high above low water"—presents a conundrum. This makeshift buoy, tied to a tree branch, was originally intended to float on top of the water to warn boaters of the submerged tree. Now, of course, the buoy hangs down from that branch, a reminder that the entire tree used to be underwater. Just as water displaced air when Glen Canyon Dam was built, air now displaces water, attesting to the dynamic instability of the reservoir. From a composition standpoint, the color yellow is associated with danger (recall the yellow of nuclear contamination signs), creating a note of unease on this blue sky day in Forbidden Canyon; the place name "Forbidden" likewise arouses a twinge of fear. Emotional tension is intensified by the juxtaposition of calming elements—rippling water, blue sky, sunlit shore—with ominous ones—dead tree snag, noose-like suspended rope, and dark background shadows. Also of note is the way Goin composed the view such that the diagonal vector of the tree trunk directs the viewer's attention to the arm-like limb, which bends upward at the elbow as if to fit obediently within the confines of the photograph.

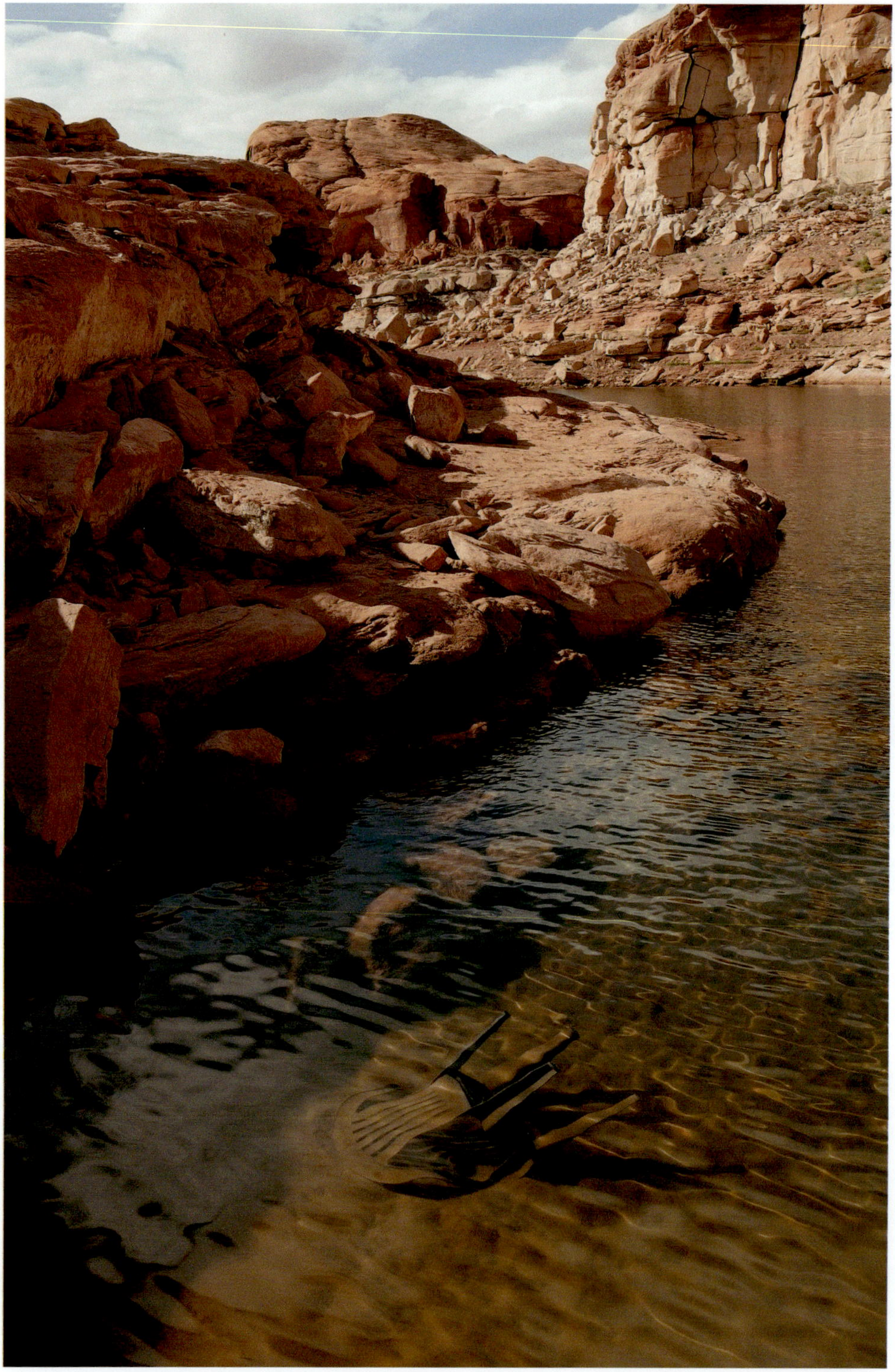

19.8 *Submerged green chair, Cedar Canyon*, 2014.

This photograph of a submerged chair echoes a previous image of submerged tumble-weeds. In each case, objects barely visible below the water's surface appear weirdly out of place, normally found on land not underwater. The tumbleweed is a humanature plant, whereas this chair is a manufactured product; however, both images suggest incipient fossils. Fossils form when objects get submerged, are covered by layers of silt, compressed by the accumulated weight of eons of deposition, and left entombed in rock, awaiting discovery by future paleontologists and archaeologists, who use fossils to reconstruct the past. A fossilized lone plastic chair would yield information about human use of this place. Absent evidence of dwellings and roads, this portable chair might indicate leisure, recreation, accessibility by boat, a fossil fuel economy (plastic is made from oil), and mass-produced, factory-made goods. Those who know Goin's work will recall green plastic chairs appearing in other contexts throughout his oeuvre, a sign of their ubiquity in our culture. While some readers may regard this submerged chair as ugly and unfortunate litter, despoiling the natural beauty of the canyon, Goin enjoys the humor and the implied story, and he questions how "natural" this place is anyway.

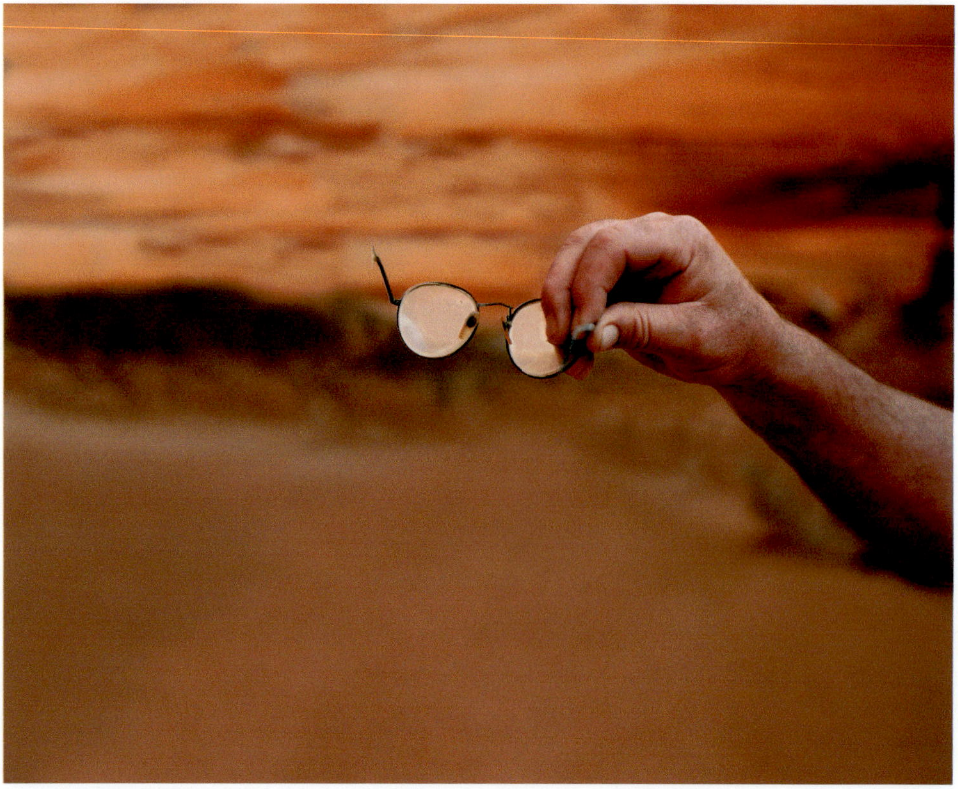

19.9 *Found eyeglasses, Iceberg Canyon*, 2004.

A motif throughout Goin's work is the human hand, offering up a taken-for-granted object for contemplation. This photograph of found eyeglasses is the first photograph in the Artifacts gallery in *A New Form of Beauty.* We normally define artifacts as human-made objects of cultural or historical interest. We dispose of garbage in landfills; we study artifacts. But really, the difference between them is perceptual—one person's trash is another's treasure. The Artifacts gallery features images of found objects that most people would dismiss as litter. But, if seen through a different "lens"—as the eyeglasses hint—the detritus documented by Goin in this gallery reveals quite a lot about American culture at Lake Powell in the 1980s through the 2010s. As he explains in conversation, a photograph itself is an artifact, so this image represents an artifact of an artifact and suggests the concept of generating artifacts. This image also references the Bauhaus photographers of the 1920s and the 1930s, who oftentimes made photographs of glasses and patterns of light and shadows as refracted through lenses. Photography is about the view and the vision, and Goin's photograph of eyeglasses is a self-referential allusion to the history of photography itself.

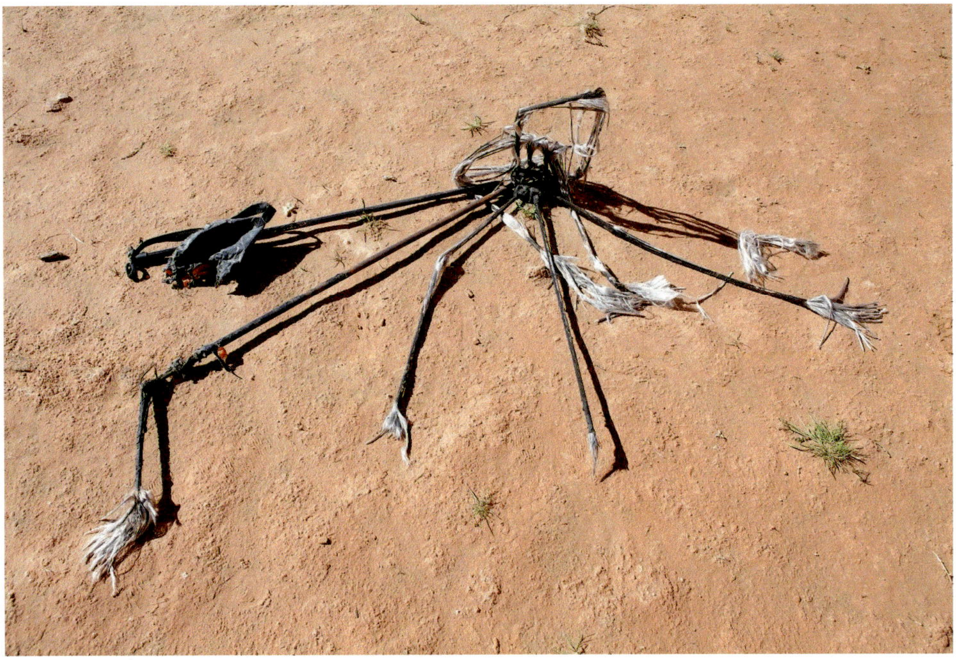

19.10 *Camp chair skeleton, Forbidden Canyon*, 2013.

Giant arachnid? Bionic creature from a sci-fi horror movie? Mechanical toy? The first question this photograph raises is: *What are we looking at*? We assume that photography connects to reality. Here, the viewer is like an archaeologist who discovers a curious artifact and must figure out what it was used for by a vanished race. Goin has revealingly labeled this image "Camp chair skeleton," which helps us notice the manufactured joints and make sense of the frayed shock cord tails. "Skeleton" is Goin's whimsical tip that we are welcome to imagine these chair remains as belonging to a deceased vertebrate. No matter what we must face, be it a pile of trash in the wilderness or alarming statistics about climate change, this image playfully reminds us that, as Goin says, "You have to laugh because life is just way too serious." In terms of composition, Goin could have positioned his camera directly overhead to hide the shadows, but he chose to make the photograph from an angle where the shadows stand out and animate the scuttling spider, implying motion in the same way that lines drawn behind a comic book figure indicate speed.

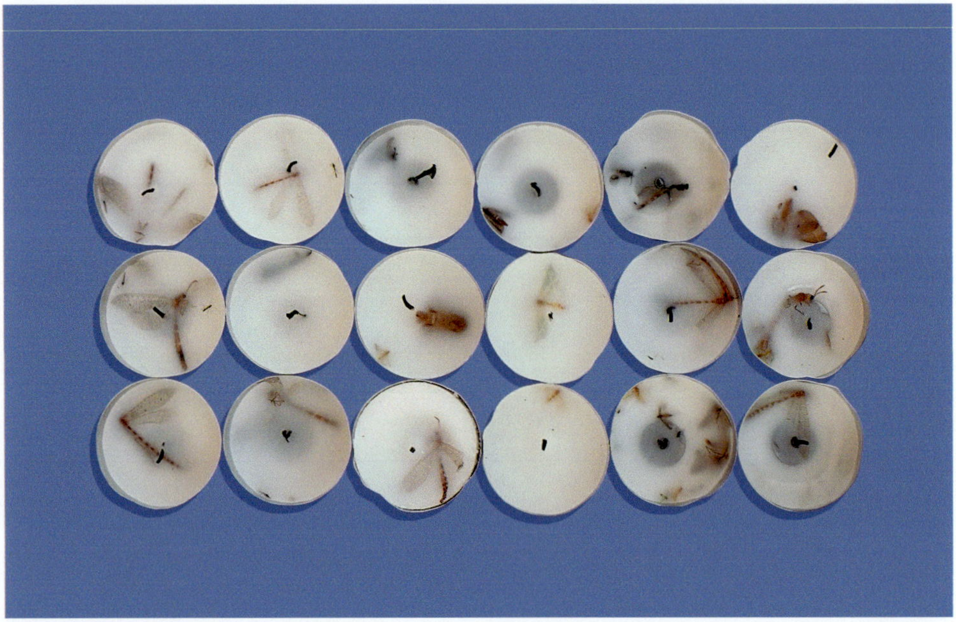

19.11 *Candle wax, insects*, 2011.

Leading off the Flora & Fauna gallery in *A New Form of Beauty* is a work of modern art—eighteen mottled white circles positioned in a six-by-three rectangle against a periwinkle background. Goin arranged the used tea candles that he burned at night on his rented houseboat, each of which attracted bugs that committed unintentional suicide in the melted wax, which hardened when it cooled, preserving them like prehistoric insects. The geometric regularity of this display mimics the orderly taxonomic system scientists use to categorize species. In this image Goin is satirizing traditional methods in biology, in which to classify an insect, you have to first kill it and pin it to a board. Performative elements of this photograph include the wax museum memory of watching insects put on a nightly show on the houseboat. Goin also performed his own magic in manipulating the original image, in which the candles were arranged on the bumpy lid of a dirty cooler. Goin digitally removed that background and replaced it with a blue-violet color that he extracted from a digital photograph of the lake's surface. Hence, the periwinkle color alludes to Lake Powell even though the image itself is abstracted from place.

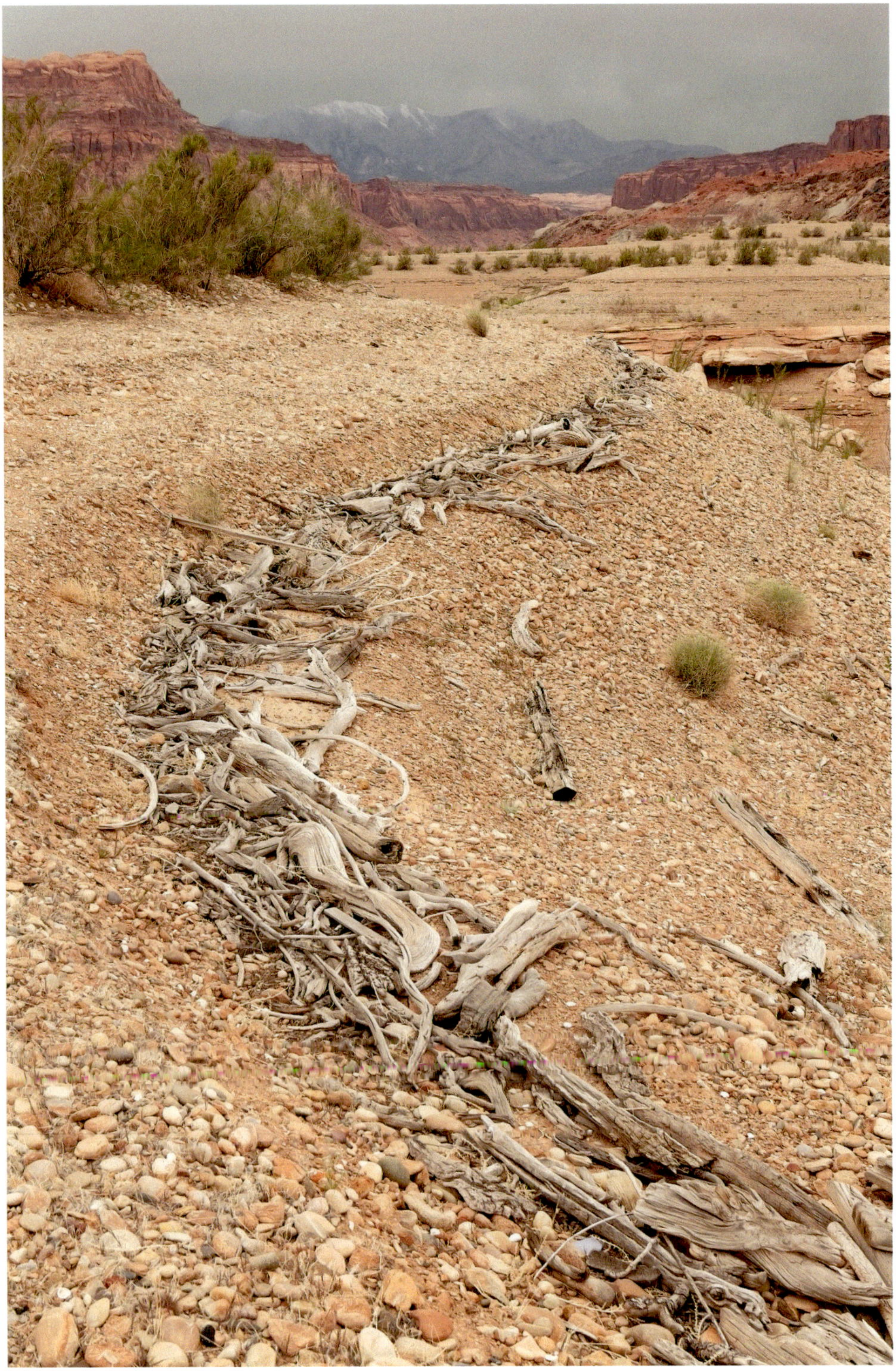

19.12 *View from a driftwood beach, farthest upriver point reachable by boat at Lake Powell,* 2014.

In the Low Water gallery, this photograph speaks to the conundrum of Lake Powell. Just as Goin's book *Black Rock* documented remnants of the presence of water in the Black Rock desert, so too does this image of a driftwood beach evoke the absence of the former presence of water. But this view of Lake Powell is ironic in a way that the Black Rock series was not.

Glen Canyon was carved by the erosive history of water itself. Then came the artificial, human-made reservoir that, when full, formed an enormous lake. In decline, the lake left remnants of driftwood, weirdly out of place in a desert landscape. It's more than an anomaly. It's an ironic view of a place whose identity is now defined by the declining presence of water. In *Nuclear Landscapes* Goin photographed radioactivity, which is invisible. Here he is photographing a lake that is invisible. An admirer of French farce, Goin creates an absurd image in which a line of driftwood along a dry shore symbolizes the absence of water in a reservoir that is supposed to be about the sustaining value of a "bank"—a bank account of water.

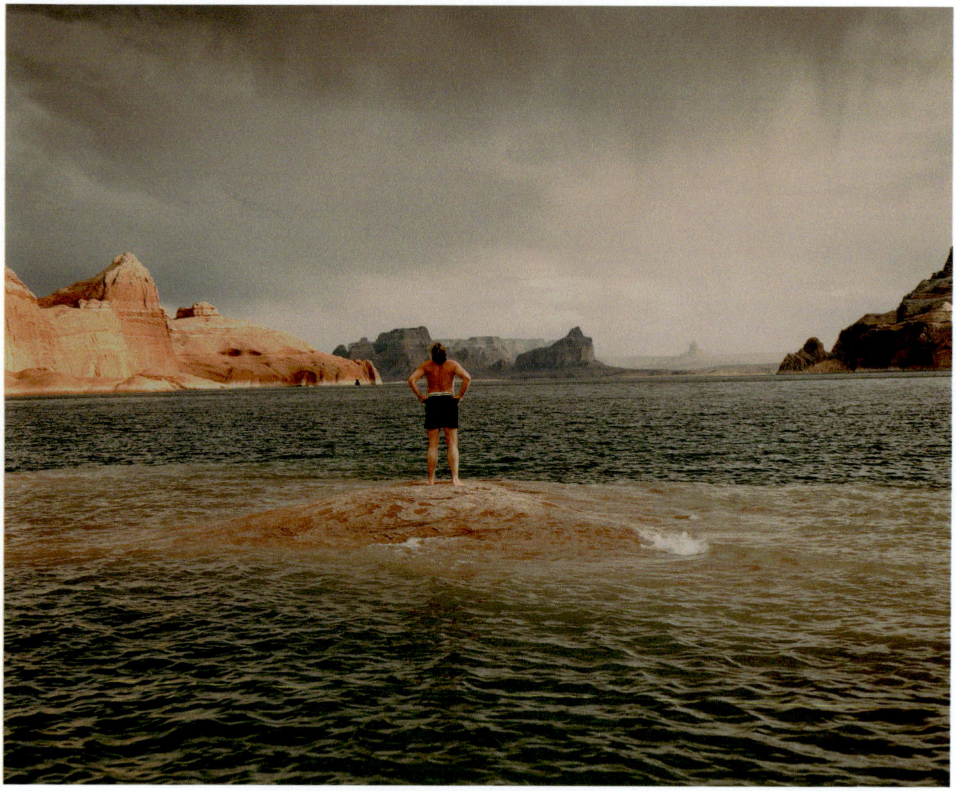

19.13 *Impending storm near Gunsight Pass, Padre Canyon*, 1996.

Whereas the previous image, made in 2014, documents the absence of water, this orchestral, Wagnerian image, made in 1996, conveys the magnificence and power of the huge lake at near full capacity. In the grand tradition of nineteenth-century landscape paintings that express the epic scale of the American West, this photograph evokes a sense of the sublime. The impending storm, water movement, infinitely receding background, and solitary human all evoke the power of nature. Peter's assistant, Dennis, swam out to this rock to pose for the photograph and then began anxiously looking for lightning. Made with a 4x5 camera, this image is composed such that the flesh-colored, illuminated rock formation directs the viewer's attention to the human figure, through whose bent arms the horizon can be seen intersecting his torso, while the triangular shape of his arms mirrors the negative space of his hands. This photograph appears in the Flora & Fauna gallery of *A New Form of Beauty*. We normally don't think of humans as fauna, but in this landscape the human body stands out against the elements, as if the human evolved from the water to crawl up on the rock and then stand erect.

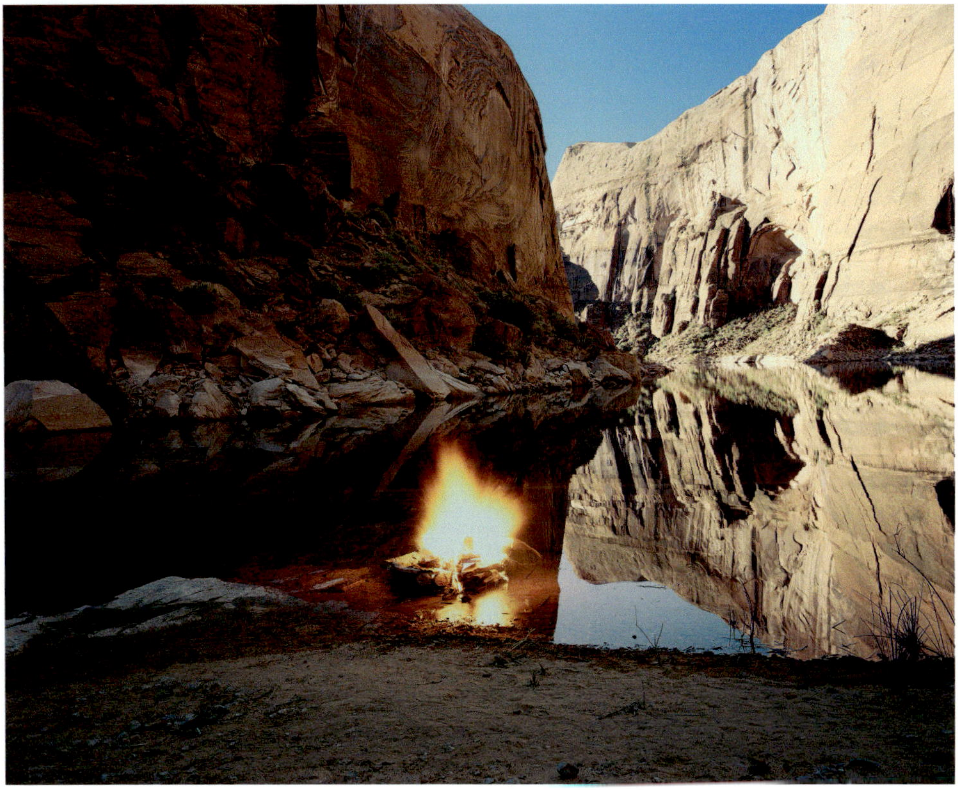

19.14 *Fire in water, Sevenmile Canyon*, 1999.

Goin and his friend Dennis set up camp on a beautiful beach. When they woke up in the morning, the water line had crept up. They lit a fire for warmth and after breakfast noticed that the water level now engulfed the fire ring. The slow exposure of this photograph reflects the passing of time and the transition from night at left to day at right. The reflected landscape looks like a kaleidoscope lit by the flare of the fire, an amalgam of the basic elements—earth, air, fire, and water. For Goin, this image is about the rhythms of the lake:

> It's about the rising of the water, the lowering of the water, the merging of the elements, the quiet. The whole experience of just sitting there in profound quiet, which is something a lot of people don't ever do. They need noise around them. You see people at Lake Powell who have the stereo blasting and the motors roaring. They're going around in circles like some kind of fish disease that they've caught that they have to go in circles all the time. Yet, when do they sit and listen? I think the listening is really very important.

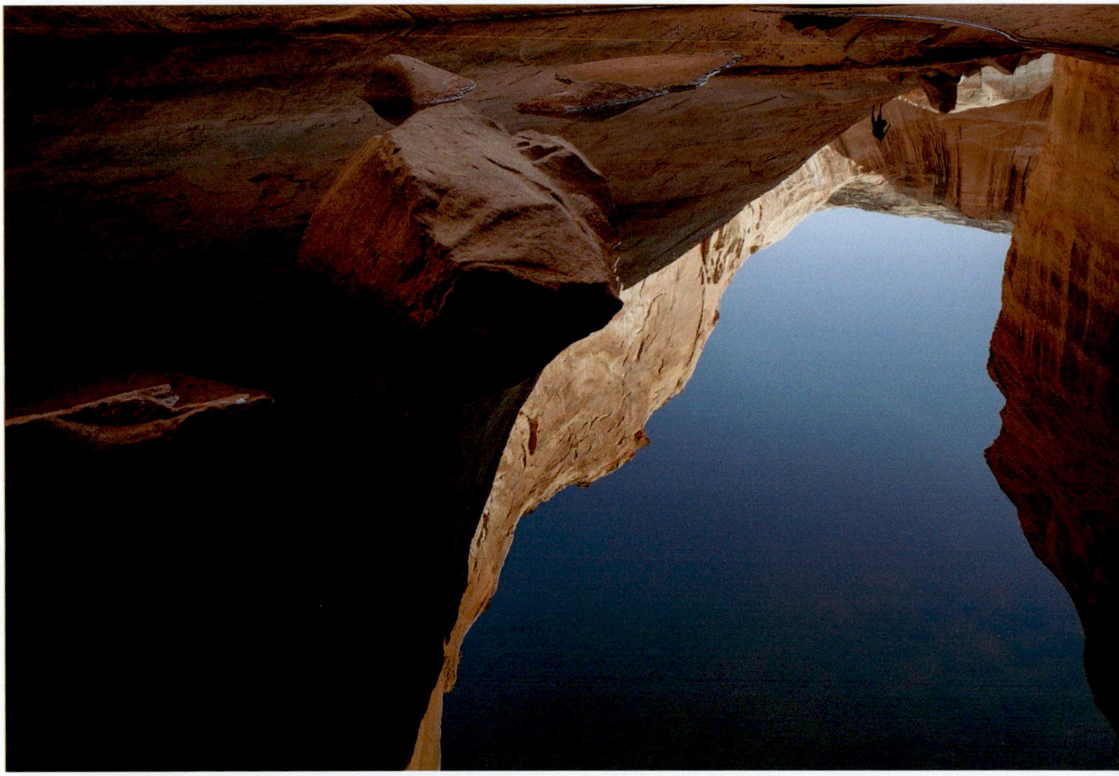

19.15 *Reflections, figure for scale, Crystal Canyon*, 2014.

It is easier to make sense of this image if you rotate it 180 degrees so that the tiny human figure dangling in the upper right stands erect in the lower left. Once you take some time to study the photograph as printed, however, you will begin to see pale blue lines that denote the edges of the mirror-smooth water surface, marking off a few exposed rocks and a narrow margin of beach at the very top. The poetic effort here is to think of Lake Powell in an inverted sensibility. As Goin explains:

> If you fill up a canyon, you're inverting it. And that process transforms the landscape. It also recreates the landscape within a context of functionality. And this is where I think sometimes we resist the human-made landscape. With some exceptions such as churches, we deny our own ability to create spiritual places. We define ourselves as spoilers. We don't think of ourselves as *creating* wild landscapes.

While this beautiful photograph does not necessarily endorse the damming of Glen Canyon or the profligate water and power consumption that led to its construction, in yin-yang fashion the image's elegant simplicity coupled with visual confusion does suggest that reality is complex.

Works cited

"Glen Canyon Unit." *Reclamation: Managing Water in the West*. US Department of the Interior Bureau of Reclamation, 2019, https://www.usbr.gov/uc/rm/crsp/gc/. Accessed 2 October 2019.

Goin, Peter. *Humanature*. University of Texas Press, 1996.

Goin, Peter, and Peter Friederici. *A New Form of Beauty: Glen Canyon Beyond Climate Change*. University of Arizona Press, 2016.

Lake Powell Water Database. Summit Technologies, Inc. from Data Courtesy of US Bureau of Reclamation, 2019, http://lakepowell.water-data.com/index2.php?as_of=1987-01-29. Accessed 2 October 2019.

Langlois, Krista. "Down with the Glen Canyon Dam?" *High Country News*, 4 September 2017. High Country News, 2019, https://www.hcn.org/issues/49.15/rivers-down-with-the-glen-canyon-dam. Accessed 2 Oct. 2019.

"Park Statistics." *Glen Canyon National Recreation Area*. National Park Service US Department of the Interior, 2019, https://www.nps.gov/glca/learn/management/statistics.htm. Accessed 2 October 2019.

"Russian Thistle." *The Great Basin and Invasive Weeds*. Utah State University, 2019, https://www.usu.edu/weeds/plant_species/weedspecies/russianthis.html. Accessed 16 October 2019.

Thompson, Jerry L. Essay. *Walker Evans at Work: 745 Photographs Together with Documents Selected from Letters, Memoranda, Interviews, Notes*. By Walker Evans and Jerry L. Thompson. Harper & Row, 1982.

20

EPILOGUE

Goin home

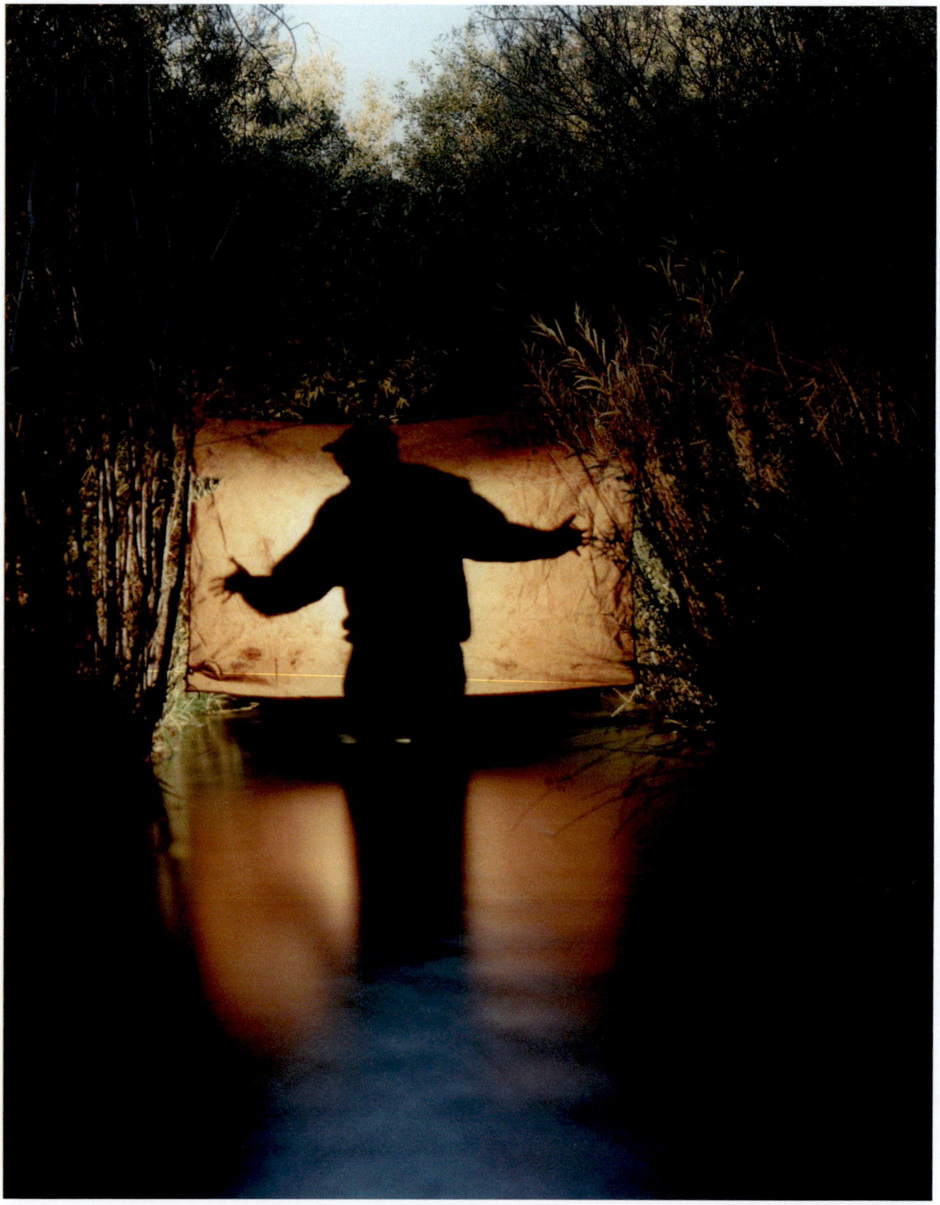

20.1 *Shadow figure.*

DOI: 10.4324/9781003212607-20

Final exam: What would you say about this photograph? Would you be able to identify the artist? Hints: Peter Goin typically wears glasses and a baseball cap. As a child, Peter and his siblings enjoyed watching shadow puppet shows in Indonesia. He published a book on Nevada rock art, which includes pictographs and modern-day petroglyphs. He made performative photographs as a way to creatively engage with the landscape at Joshua Tree National Park and Lake Powell. Peter Goin has been drawn to arid waters for much of his career. He has photographed vegetation reclaiming Mayan ruins and overgrowing the Erie Canal. The compositional device of a cross can be found in many of Goin's projects, including his US-Mexico border survey and post-mining landscapes book. Peter Goin's *Nuclear Landscapes* evokes the energy of that which cannot be seen. Peter describes his photographs as "portals to conversation."

Peter Goin and the Photography of Environmental Change has explored and interpreted Peter Goin's art. His has been a decades-long photographic investigation, beginning roughly in 1977 with the Mesoamerican ruins work and continuing up to the time of this writing, 2021. Forty-four years dedicated to craft, answering the call to bear witness. However, before closing the covers of this book and sending Peter and camera off the page and into the future, it seems important to acknowledge that since 1984, Goin has pursued his photography as one strand in the more encompassing career of being a professor of photography and time-based media in the Department of Art of the University of Nevada, Reno (UNR). While some photographers work unrelated day jobs to support their art and others manage to earn a living by selling their photographs, Goin has enjoyed the luxury—and shouldered the responsibilities—of an academic position, which has provided a steady paycheck and included "research and creative activity," that is, photography, as part of his job description but has also required other kinds of work. A typical assignment for a tenure track professor in the Department of Art at UNR is forty percent teaching, forty percent research/creative activity, and twenty percent service. This study, then, has so far examined only forty percent of Goin's professional life. Teaching, administration, and service complete the picture of his distinguished career and reveal core personal motivators of love, ambition, and a sense of duty.

"I love being a teacher," Peter effuses, "I love the students that we have. They're so fantastic—their sense of hope. You just want to be there for them." For Goin, love manifests as work. In his career at UNR, he has taught beginning through advanced photography, color photography, digital photography, large format and architectural photography, videography, and the history of photography. Special topics have included "the book," photography and culture, experimental processes, and documentary practice. Goin's teaching has extended to internships, directed readings, and graduate seminars. Anonymous student evaluations on *Rate My Professors* website characterize Professor Goin as "well prepared for class," "beyond knowledgeable," "awesome," "passionate," "funny and intelligent," "smart and engaging," and "extremely helpful." They comment, "He makes learning about photography fun, but challenges you to open up and be a risk taker"; "Not only does he get you to make better pictures, but he also helps you think about the world around you"; and "Be prepared to be pushed. He hates slackers." Goin's strong teaching has been formally recognized: He earned the college of Arts & Science Alan Bible Teaching Excellence Award in 2005 and the university Outstanding Undergraduate Research Faculty Mentor Award in 2013.

While Peter strives to provoke students' interest and make every classroom come alive, his true passion is to get students outside. His field courses entail an enormous amount of planning and have put him in positions of extraordinary responsibility, not only for the students' learning but for their life and limb. In 2006 and again in 2008, Goin taught courses at the

Universidad de Iberoamerica in Puebla, Mexico, under the auspices of UNR's University Studies Abroad Consortium (USAC). He taught another USAC course in Viterbo, Italy, in the summer of 2015. He has also run his own international field workshops for UNR students, one to the Pacific Coast of Mexico in March 2010 and several others to Oaxaca, in 1999, 2004, and 2013. Here is how Goin described his Oaxaca mini-session in the UNR course catalog:

> The program of study will focus on themes such as a "sense of place" by visiting unique and little-known sites within a three-hour drive of Oaxaca. Site photography, digital printing, basic portfolio construction, and classroom discussion and critique will be investigated. The goal of this project is to introduce, thereby significantly improving, the concept of transcultural visual practice. After an 11-day field workshop in Oaxaca, the remainder of the mini-session will be conducted in UNR's photographic laboratories for processing, printing, and preparing the portfolios for publication and exhibition at a local non-profit, the *Holland Gallery,* and at the Department of Art Galleries South. Workshop participants will create mural panels focusing on the narrative as a conceptual element.

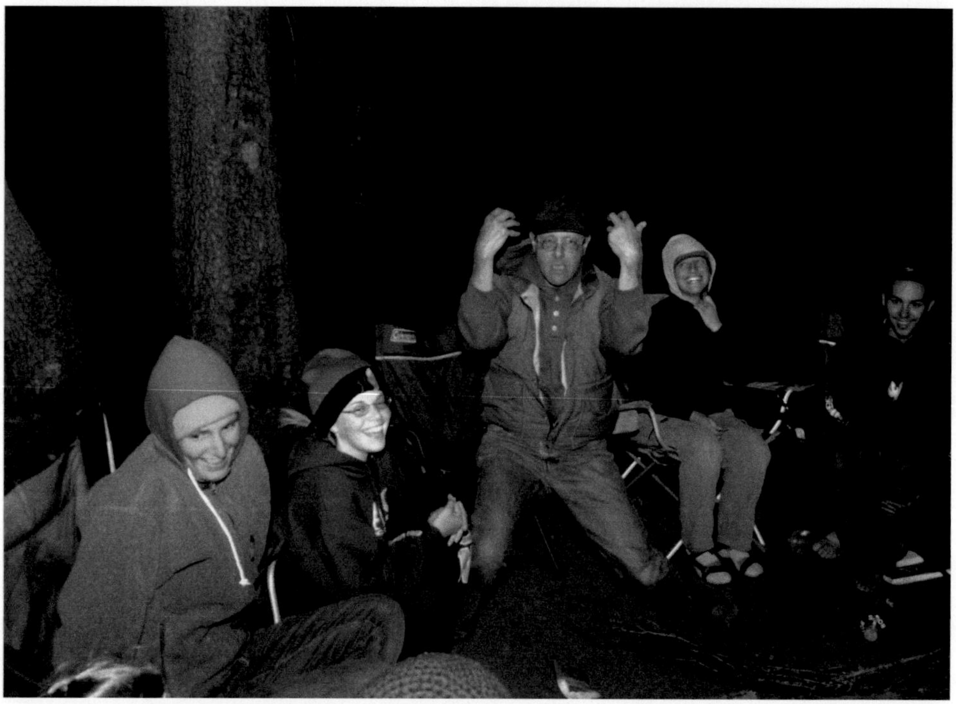

20.2 *Professor Goin entertaining his photography students around the campfire on a week-long, horseback field session in the Hoover Wilderness, California*, summer 2003.

Closer to home, Goin has run both field courses and optional field trips for his UNR photography courses, including trips via horseback in the Sierra Nevada and Ruby mountains and camping trips on Nevada's Black Rock Desert. In Oregon, he co-taught a course with Ursula LeGuin in 1997 for the Oregon College of Art & Craft on "Visual Narrative and the Photographic Book." In 2017 Goin participated in a conference roundtable conversation for the Regional Society for Photographic Education on "Getting Students Out of the Classroom," and his courses have been featured in UNR's alumni magazine *Silver & Blue* under the titles "A Dirt Floor Classroom" and "Once Upon a Time in Mexico." As is true for most of Goin's photographic projects, he finds that teaching in the field fuses work and play. For the students, Goin's field courses must surely stand out as one of the highlights of their college years.

Goin's commitment to teaching has made him an ambitious program builder, who has done yeoman work behind the scenes. He has applied for and won Instructional Enhancement Grants to plan new courses and revamp labs as the evolving medium of photography advanced to color printing and digital technologies. He facilitated the hire of his former student Scott Hinton, who went on to earn an MFA from San Jose State University and is currently employed at UNR as the Department of Art's Coordinator of Digital Labs and Faculty Research. Hinton has helped Goin develop a robust photography program within the Department of Art and has also collaborated with Goin on grant-funded projects as well as assisted him in the field. Peter Goin has borne his share of administration and service duties, not necessarily his favorite assignments, but handled with integrity, perseverance, and attention to detail. At UNR, he has served on the Center for Advanced Studies Executive Committee, the Editorial Board of the University of Nevada Press, and numerous award committees. He chaired the Department of Art for four years, from April 2011 to June 2015.

As one measure of Goin's success at UNR, consider his streamlined promotion path—hired as an assistant professor in 1984, promoted to associate professor and tenured 1989, and promoted to full professor 1992. He has been awarded some of the university and the state's highest awards, including the University of Nevada Board of Regents Creative Activities Award (1995), the university Foundation Professorship (1996), and the Governor's Millennium Award for Excellence in the Arts (1999–2000). We need not belabor this litany of professional achievements and distinctions nor delve into every category let alone every line item on Goin's sixty-two-page, single-spaced academic resume, but they do reveal a dimension of Peter Goin not apparent in his photographs alone.

Preserving, cataloguing, and archiving his work has been an ongoing commitment for Goin. It is a tedious process with no immediate returns, but Peter is in it for posterity. Collections of Goin's images, some with project notes, are available for future scholars at The Bancroft Library in Berkeley, the Center for Creative Photography in Tucson, the Library of Congress in Washington, DC, Special Collections at Duke University in Durham, the Nevada Museum of Art, and the UNR's Special Collections and University Archives Department. When he began making photographs and for much of his early career, Peter had no idea that anyone might one day care about his work. "I thought that my medium was doing my books," he explains, "and that's my contribution. I didn't think anybody would care about the photographs necessarily or the archive materials. I just assumed that I would have a bonfire on the Black Rock Desert." Nevertheless, something about his nature caused him to save field notes, maps, and research findings in boxes, which he kept and hauled along on his various moves. It was in talking with his friend Gary Snyder, a nationally distinguished poet, when Peter began to think differently about legacy. Snyder chided Goin for his false modesty and counseled him to document and save all of his work. Snyder himself had apparently been doing so since the age of twenty-one. Having been taken to task, Goin developed a cataloguing system for his photographs, quipping that "A negative without a number is a citizen without a country." A photograph labeled MA4, for example, stands for "Miscellaneous, A-series, number 4," while a photograph labeled CK13 indicates color film, series K, image 13. Each image also has an identifying caption, saved in a master database. While Goin's is an idiosyncratic numbering system, when someone requests permission to reprint any of his thousands of photographs, Goin can put his hands on the original negative or digital file within five minutes.

Goin has done extensive research of his own in private and publicly available collections to locate historical images of Lake Tahoe, Chaco Canyon, Mesa Verde, and other places he has photographed. As such, he has come to regard archives as "soulful repositories of place history." "History is continually written and rewritten," he explains; however, future generations' ability to write history relies on the availability of documents and visual artifacts. In the case of photographs:

> What is lost, or denied due to institutional barriers, or leads an image to dissolve because of the vagaries of decay and time (especially with color film), denies our sense of ourselves, as history is relegated to the remaining images that are publicly available, digitally scanned, usable, and mostly free.
>
> *(Goin,* Colors *196, 202)*

As this observation suggests, "archiving" one's photography involves much more than dropping off unsorted boxes of papers to a library. Goin and his paid assistants have located, organized, catalogued, cleaned up blemishes, digitally scanned, and printed hundreds of his photographs on archival paper. He has become self-reflective about the value of preserving his own work in archives, musing:

> The object that we leave behind is our communication across time. If we ever want to talk to people after we no longer exist, one of the best ways to do it is to leave behind a mark, to leave behind a legacy, to leave behind those objects that are preserved. That's why I work so hard to put my work into archives, so that others can look back and say, "well, what was it like in Nevada from this one view?"

Peter Goin and the Photography of Environmental Change began by noting that a retrospective of Peter Goin's photography is overdue. True, but a retrospective is also premature, for Goin's work/play continues. When he is not archiving past work, Peter is making new images and dreaming of future projects. Here is a sneak preview of work in progress and books in the making. In Goin's 1992 Lake Tahoe book *Stopping Time* he made a rephotographic survey of Lake Tahoe, publishing fifty-five pairs of then-and-now photographs documenting how the Tahoe Basin landscape evolved over time. He subsequently produced four more spin-off books on specific places and facets of Lake Tahoe's history for Arcadia Publishers' "Images of America" series, slim volumes that present history through historical photographs and educational captions. In doing the research for these Lake Tahoe projects, Peter discovered many more historical images than he could reprint in that limited format, images that he felt needed to be published to amplify the historical record and make it publicly available.

In the meantime, he acquired a boat—aka the "photographic research vessel"—that made it possible for him to study and photograph the shore of Lake Tahoe from the water, which was, in fact, the vantage point of many of the historical photographs. For Peter, the view of Lake Tahoe from a boat changed everything. It was an epiphany. Keen to continue and in a sense complete his earlier rephotographic survey, Goin has embarked on a more comprehensive follow-up volume to *Stopping Time*, this one entitled *Re-Visiting Lake Tahoe*, slated to include more than 285 photographic pairs.

One other new Lake Tahoe project was published in fall 2021, this one a bit more philosophical than the rephotography books. In *The Nature of Lake Tahoe: A Photographic History, 1860–1960* Goin explores how Lake Tahoe came to be the place we understand it to be through the visions of photographers from 1860 to 1960, the black-and-white era. Goin's central thesis is that Lake Tahoe is a place defined in part by the visual record. *The Nature of Lake Tahoe* features approximately 185 black-and-white images by others and approximately 40 color photographs by Goin, representing his visual interpretation of those historical narratives.

After that? What does Goin's photographic future portend post Tahoe? "I'll probably always create photographs," he predicts, "but books are marathons. How often do we run a marathon?" Even though he envisions an end to book-making, nevertheless, Peter does have one more book project. The book would be a coffee table book of panoramic photographs of one hundred lakes in Nevada. "That will be my swan song," Peter says, "plus, it will give me a swan song with people I've known, whom I can invite to come with me. I've done all the research, and I plan to move at my own speed." Most people will be surprised that there are *one hundred lakes*?! in Nevada, statistically the most arid state in the United States and written off by many as a barren wasteland. Where *are* all of these beautiful lakes, readers may wonder. Peter Goin knows. Because Goin has gone there.

Having experienced an international childhood and itinerant youth, Goin has found a home in Nevada, a seemingly unlikely place to sink roots, but one that inspires his loyalty. He points out that:

> Nevada is populated by transplanted individuals. If you look at the vast majority of people who call themselves Nevadans, they almost all came from somewhere else. When I came here in 1984 the population of Reno was barely over 100,000. It's the funniest thing—I'm more rooted here than the vast majority of people who live here.

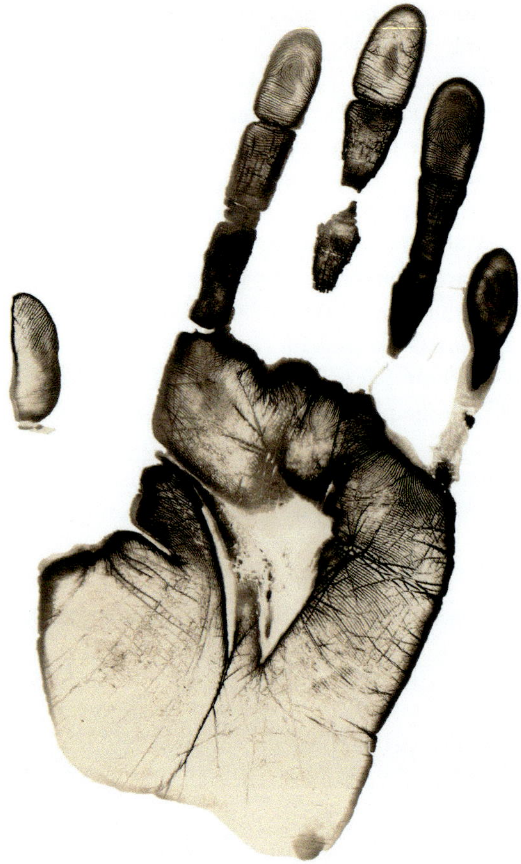

20.3 *Hand of the artist.*

Goin has been an advocate, interpreter, and spokesperson for the Great Basin, explaining, "It is a landscape that has been ignored and derided and misunderstood and diminished and rejected, and I think that there is something that needs to be said and written about here." Devoting more than three decades to doing just that—and shaping his way of life to harmonize with aridity—Peter Goin has achieved his own high standard to become *of* a place. He has made a mark.

Works cited

Goin, Peter, and Paul F. Starrs. *Colors of California Agriculture*. The Bancroft Library, University of California, Berkeley, 2018.

"Peter Goin." *Rate My Professors*. Cheddar, Inc., 2019, https://www.ratemyprofessors.com/ShowRatings.jsp?tid=148567. Accessed 5 December 2019.

APPENDIX A

PETER GOIN'S BOOKS, ARCHIVES, AND COLLECTIONS

Books

Arid Waters: Photographs from the Water in the West Project, edited by Peter Goin, 1992.
Atlas of the New West, with William E. Riebsame, James J. Robb, Patricia Nelson Limerick, and Charles Wilkinson, 1997.
Black Rock, with Paul F. Starrs, 2005.
Changing Mines in America, with C. Elizabeth Raymond, 2004.
Colors of California Agriculture, with Paul F. Starrs, 2018.
Dooby Lane, with Gary Snyder, 2016.
A Doubtful River, with Robert Dawson and Mary Webb, 2000.
Emerald Bay and Desolation Wilderness, 2018.
A Field Guide to California Agriculture, with Paul F. Starrs, 2010.
Humanature, 1996.
Illustrations for *Death After Life: Tales of Nevada,* by William A. Douglass, 2015.
Lake Tahoe, 2005.
Lake Tahoe: A Maritime History, 2012.
The Nature of Lake Tahoe: A Photographic History, 1860–1960, 2021.
Nevada Rock Art, 2009.
A New Form of Beauty: Glen Canyon Beyond Climate Change, with Peter Friederici, 2016.
Nuclear Landscapes, 1991.
Re-Visiting Lake Tahoe, 2022.
South Lake Tahoe: Then & Now, 2010.
Stopping Time: A Rephotographic Survey of Lake Tahoe, with C. Elizabeth Raymond, 1992.
Time and Time Again: History, Rephotography, and Preservation in the Chaco World, with Lucy Lippard, 2013.
Tracing the Line: A Photographic Survey of the Mexican-American Border, 1987.

Artist books

Material Culture, *Signifying the Moment*, and *The Vicarious and the Visual*, three artist books produced for an installation at the Southeast Museum of Photography, Daytona Beach, FL, 2000.

Book chapters

"Atomic Pop," in *Bombast: Spinning Atoms in the Desert*, by Michon Mackedon, 2010.
"Magical Realism: The West as Spiritual Playground," in *Western Places, American Myths*, edited by Gary J. Hausladen, 2006.
"The Nuclear Past in the Landscape Present," *Atomic Culture: How We Learned to Stop Worrying and Love the Bomb*, edited by Scott C. Zeman and Michael A. Amundson, 2004.
"Nuclear Temples," *Resisting the Nuclear: Art and Activism Across the Pacific,* edited by Elyssa Faison and Alison Fields, forthcoming.

Archives

The Bancroft Library, University of California, Berkeley
Center for Creative Photography, University of Arizona, Tucson
Library of Congress, Prints and Photographs Division, Washington, DC
Nevada Museum of Art, Reno
Rubenstein Library, Duke University, Durham, NC
University of Arizona, Tucson, University Libraries, Special Collections
University of Nevada, Reno, Special Collections and University Archives Department

Collections

American University, Washington, DC
Amon Carter Museum of Art, Fort Worth, TX
Atomic Photographer's Guild Permanent Collection, Canada
Baltimore Museum of Art, Baltimore, MD
Beinecke Rare Book & Manuscript Library, Western Americana Collection, Yale University, New Haven, CT
Bibliotheque National, Paris, France
Birmingham Museum of Fine Art, Birmingham, AL
Boston Museum of Fine Arts, Boston, MA
California Museum of Photography, Riverside, CA
California State Library, Sacramento, CA
Center for Documentary Studies at Duke University, Durham, NC
Comstock Bank Collection, NV
Corcoran Gallery of Art, Washington, DC
Dominion Resources Art Collection, Virginia Commonwealth University, Richmond, VA
El Paso Museum of Art, El Paso, TX
Graham Nash Collection, Los Angeles, CA
Humanities Research Center, University of Texas at Austin, TX
Jordan Schnitzer Museum of Art, Eugene, OR
Los Angeles County Museum of Art, Los Angeles, CA
MacArthur Foundation, Chicago, IL
Minneapolis Institute of Art, Minneapolis, MN
Mint Museum of Art, Charlotte, NC
Museet for Fotokunst, Odense, Denmark
Museum of Fine Art, Houston, TX
Museum of Modern Art, New York, NY
National Museum of American Art, Washington, DC
New York Public Library, New York, NY
Newmont Mining Corporation, Carlin, NV

Palm Springs Museum of Art, Palm Springs, CA
Phoenix Arts Commission, Phoenix, AZ
Portland Art Museum, Portland, OR
Princeton University Art Museum, Princeton, NJ
Reno City Hall Collection, Reno, NV
Riverside Museum of Art, Riverside, CA
San Francisco Museum of Modern Art, San Francisco, CA
Snell & Wilmer Law Offices Collection, AZ
Southeast Museum of Photography, Daytona Beach, FL
Special Collections, University of New Mexico, NM
Special Collections, University of Southern California, Los Angeles, CA
Special Collections, University of Utah, Salt Lake City, UT
Tennessee State University Collection, Nashville, TN
University Art Museum, California State University, Long Beach, CA
University of Iowa Museum of Art, Iowa City, IA
University of Nevada, Reno Academy for the Environment, Reno, NV
University of Wyoming Art Museum, Laramie, WY
Walter Cuchine Collection, Eureka, NV
Washington State Public Art Collection, Seattle, WA
Whitney Museum of American Art, New York, NY

APPENDIX B

KEYS TO READING THE VISUAL LANGUAGE OF PHOTOGRAPHY, BY PETER GOIN

The three keys shared below present a methodology for reading the visual language of photography. These keys may be used to analyze and interpret any photograph by any photographer from any era. These lists originated as lecture outlines for college courses in photography that Professor Goin taught at the University of Nevada, Reno. For this appendix, we have developed Goin's original outlines to work as stand-alone guides, independent of Goin's in-class commentary. Admittedly, an entire book could be written about any one of these keys. However, there are advantages to a stripped-down, concise checklist of points to consider when analyzing a photograph. Therefore, in the interest of improving visual literacy, we offer the following tools:

Elements of composition

This key introduces photographers to the many variables available to them when making a photograph, variables they should bear in mind when composing an image. Skilled photographers consciously manipulate these variables for effect, and their choices constitute a stylistic approach that, if repeated, results in an artistic "signature." This list may also be used by viewers to analyze and interpret a photograph. Essentially, Elements of Composition is a concise guide to knowing what to *look for* when you look at a photograph, namely how to "read" the image. Once you are aware of the compositional choices that the photographer made, the next step is to think about the *effect* of those choices and the reason the photographer may have made them.

Ten points of critical analysis

The ten points in this key will help the viewer move beyond an "I like it" or "I don't like it" response to a photograph and understand an image as an element of a larger narrative. While the Elements of Composition list is a guide to analyzing the formal properties of a photograph, this Ten Points list will help you move beyond formalist analysis to produce a *critical* reading. This process considers the photographer's frame of reference and point of view and alludes to a continuing conversation with previous works in the visual art tradition. Critical analysis also recognizes that how a photograph is interpreted depends on *who* is making the interpretation—what their cultural biases are—and *in what context* they encountered the photograph—historical archive? newspaper? family album? art museum?

Seven layers of meaning

This key is for the more advanced viewer of photography, who understands that a photograph is layered with meaning and is not just always about the visible subject. The seven layers of meaning detailed here form a taxonomy that may help a viewer to categorize an image. In some images, one of these layers will predominate, while in others, multiple layers may be combined in a variety of ways, and not all layers will be included in all images. The point of this key is to emphasize that a photograph is more than just *technically* layered; it has levels of complexity. There are different *kinds* of photographs, and a full analysis of a photograph must go beyond formalist description and critical analysis to considerations of genre.

Elements of composition

Photography is a language, and the variables below function as the grammar. How these variables are used—and in what priority and order—is the equivalent of photography's syntax. Analyzing a photograph using this key will produce a reading that focuses the primacy of form over content.

1. Compose the image as you would a sentence, minding syntax, built from myriad elements, such as line, form, resolution, shape, symmetry, exposure, and color.
   The variables contained within a photograph are emphasized, diminished, altered, varied, among other approaches, and the combination of the variables creates the composition. And yes, the subject of the photograph is a variable.
2. Be aware that the eye is directed to the lightest part of the composition.
   Our visual acuity naturally seeks out light, and where a beam of light projects across an image, the highlight attracts the viewer's gaze.
3. Consider light as a verb. Notice shadows.
   Often, beginning and self-trained photographers think of light as "that which illuminates" rather than a compositional tool for creating and dividing space. Directional lighting offers a myriad of possibilities for recreating a composition, and the technique of chiaroscuro derives specifically from the use of the camera obscura and harsh shadow compositions.
4. Understand the effect of lines → diagonal lines are active while horizontal lines are passive.
   A line that divides diagonally implies action, as if the viewer's eye slides down the angle of repose. Movement is associated with angles, while a horizontal line implies a static landscape or a distant edge. The horizontal line conforms to the grid of the rectangle, while a diagonal line divides and intersects the grid.

5. Evaluate and condense the linear elements into geometric shapes (triangles and circles are more visually dynamic than squares).
   A traditional drawing technique is to create abstract forms from a still life, reductively defining objects and elements in space as active geometry. These geometric shapes are the building blocks of design, not just compositions in photography.

6. Understand and engage the angle of view, frame of reference, and point of view.
   The angle of view is the position from which the photographer made the image. Sometimes that view is directly at the photographer's standing position, but often it is altered, such as with Bauhaus photographers who experimented with looking up to the structure of a composition, or down from up-high, or at an angle that defies what has become a traditional grid-like compositional structure. The frame of reference includes the cultural point of view, which calls for unique visual interpretations. For example, some people view the spatial environment by looking at in-between spaces, while others look at the objects that exist within that space. The point of view is the maker's opinion—what the photographer felt or thought about what was portrayed.

7. Remember the rule of thirds → foreground, middle, and high ground.
   In the rule of thirds, there are two vertical lines that break the image up in to three sections vertically and two horizontal lines that break the image into three sections horizontally. The dominant subject typically occupies two-thirds of the image area.

8. Pay attention to the subject and its relationship to the background.
   A common occurrence by photographers is that they neglect to notice that a photograph essentially compresses the viewed world into a two-dimensional composition, flattening the scene. As such, the background intersects the foreground, and how the background functions in relationship to the foreground subject is a variable. Failure to pay attention to the interaction of background and foreground subject may result in unintended effects, such as a tree or other object appearing to grow out of a person's head.

9. Pay attention to negative space (shadows, spaces between defined spaces).
   Negative space is the space contained within and surrounding the subject of an image. By itself, negative space can have a shape and can be used as an integrated element in a visual composition. Notice the shapes that negative space creates.

10. Pay attention to the relativity of color.
    A basic principle of color theory is that color is relative. How one color makes another color appear lighter, darker, more intense, or even entirely different depends on the surrounding matrix of color or colors. Color unto itself is rarely static, but is more often transformed by its relationship to the tones, hues, and saturation of surrounding color.

11. Pay attention to the importance and utility of text.
    The visual world has a plethora of text everywhere, and when a photograph is made, the "text" of the word elevates to a visual element. The evolution of the alphabet of font

styles reveals how important and meaningful text design can function, but text unto itself can be employed as a sign, symbol, and/or a reference to something other than the pure translation of the literal word.

12. Use visual metaphors.
    A photograph is often more than the simple portrayal of the subject itself. A likeness is a feature of the essentialism of the photograph, but visual metaphors are an evolution of a more complex visual syntax. For example, a photograph of a dead bird may be a visual metaphor for DDT; a distant horizon may communicate loneliness; a corsage of wilted roses may code for a failed relationship; and directional light from above may imply some form of celestial purity.

13. Negotiate motion.
    During the 19th century, there was a period of time when it was considered bad form to include blurred objects; simply, those were errors. However, once motion was negotiated, the concept of the passage of time expanded the visual language of photography beyond the fragment of the exposure itself.

14. Understand and negotiate differential focus.
    Pictorialism used differential focus to represent the "artist's hand." No longer are photographic artists expected to produce images with only a sharpened, edge-to-edge focus. A range of focus, or a centrally focused point in an image with blurred distance or surroundings, allows the maker to direct the viewer's eye.

15. Know that you have the option of acting directorially → creating an image that would not otherwise exist.
    Acting directorially, a phrase introduced by critic Allan Coleman, implies that the photographer created the scene as well as the resulting photograph. Acting directorially is a form of pre-visualization, of theater. The camera need not be limited to documenting the observable world; it can be a tool for creative expression, as when a photographer arranges objects prior to photographing them or literally invents a scene. For example, in Cindy Sherman's *Film Stills*, Sherman dresses and poses herself to resemble generic female characters from the 1950s and the 1960s films.

16. Introduce scale in relationship to repetition.
    Repetition implies a form of emphasis, almost like saying the same thing multiple times. It can be a creative device, such as Sandy Skoglund's constructions that include multiples of goldfish hanging from the ceiling. Scale is a standard technique in graphic design, used to exaggerate depth and distance.

17. Understand that a photograph is flat; therefore, negotiate 3D → 2D.
    The world is multidimensional, but a photograph, most often, is a flat object with a printed representation of that which was included in the frame. The illusion of 3D is often readily apparent, as viewers recognize the illusion of depth and perception. On the flip side, photographers can diminish the inherent perception of depth, intentionally creating a flat look by, for example, removing any horizon lines. This process of flattening the visual field has been very popular in compositional design during late Modernism.

18. Engage multiple perspectives within the same image.
    Painters such as Jan van Eyck and Caravaggio used the camera obscura to focus on multiple sections of a scene and then paint the resulting whole view, but within the painting each angle of view was a separate perspective. David Hockney used this concept in his Polaroid multiples, emphasizing certain features of the subject by exaggerating or filtering the angle of view.

19. Previsualize = anticipate before making the image, both technically and conceptually.
    Previsualization is related to the process of making an image. Often, a photographer will see a scene that they want to record, but that scene, in real time, has already vanished. One of the key conceptual developments in photography is *The Decisive Moment*, encapsulated in the photographs of Henri Cartier Bresson. Prepare to create the image before the scene departs, before the moment arrives.

20. Present the invisible = allude to that which you cannot literally see.
    An essential aspect of defining photography is that it is deeply connected to the "real"— to that which exists in the observable world. However, photographs record emotions, dangerous gases, smells, radioactivity, and even viruses. The challenge is to determine the most appropriate combination of variables so that the viewer understands that which is unseen.

Ten points of critical analysis

These points push interpretation of a photograph beyond a purely formalist analysis by taking into account the historical and cultural context of the photographer and the interpreter.

Definitions

Frame of reference: Generally agreed upon cultural rules. Frame of reference refers to pervasive, generally accepted assumptions (e.g., reading left-to-right, culturally determined distance, identifications of gender).

Point of view: Specific opinion or bias. Point of view refers to a specific and possibly oppositional (and conscious) stance, opinion, or statement. It is common for the photographer's point of view to be more specific and obvious than the frame of reference.

Critical thinking: A process of thought and critical discourse, whereby both frame of reference and point of view are exposed and challenged.

1. Using the Elements of Composition key above, consider this question: What variables has the photographer used, and how, to create the specific image? Variables range from time, light, subject, angle of view, to scale, space, size, tone, and frame, among many others. Consider, for example, the role of the subject within the image. Is it presented as a likeness, symbol, or element within a narrative device?

2. What is the photographer's frame of reference? Frame of reference implies not just where the photographer "stood" but more specifically, what cultural values does the photograph represent? For example, a male-dominated culture may spawn photographs that objectify women, presenting them as objects of the photographer's consumptive gaze.

3. What point of view and biases did the photographer incorporate into the image? Point of view extends beyond implicit biases, which tend to reflect cultural frames of

reference, to the opinion of the artist. In other words, what is the maker's opinion of the subject? In terms of biases, it is helpful to understand how the camera can transform relationships, understanding that a camera is a tool that has historically been used to objectify others.

4. What is the role of the photographer in the making of the image—participant, observer, directorial? A participant is involved in the scene portrayed; an observer is literally watching from afar, not within the midst of the event or scene; and the directorial photographer created the scene or actions. Consider also the role of the photographer in the construction of the final image. For example, was the image Photoshopped? "Construction" refers to the design and structure of the final object.

5. How is text used? Throughout the history of photography, photographers have tried to combine the written word with the image. In many cases, the words supplement the image as captions, poetry, or text, while in other cases the photographs complement the text. There are also photographs that include the word within the image. Notice how these words are used (e.g., do they have a purely textual role or do they also play a visual role), how they are incorporated within the image, and how that incorporation changes over time. Also, note how text can alter the context of the image. What effect, for example, does removing the text have on the image?

6. How does context define meaning? If the context changes, the meaning of the photograph most probably changes. For example, a photograph displayed in a fine art museum will have a different meaning than the same photograph printed in a government report. Also, over time the cultural context for viewing photographs changes. There are generally three levels of context: The photographer's intent, which is possibly vague and not very dominant or determinant; the distribution system, including the requisite audience; and changing interpretations (i.e., how it was intended to be seen, how it was critically seen, and how it is seen now).

7. What are the elements that distinguish the identity of the photographer? Do photographers possess a specific "style" and approach, much like a signature? The assemblage of variables, if used consistently, especially subject, lighting, and compositional structure, can reveal a style.

8. Are there particular themes that are specific only to photography? Fashion in technical style is as prevalent in photography as in any other medium. Notice how prevailing fashions affect the way photographers choose to represent their subjects, how certain conventions are repeated, and how themes are generated. Photographs are descendants and connected by a compositional lineage to previous images, although, sadly, many enthusiastic photographers are unaware of the visual history of the medium. Not many universities teach the history of photography!

9. How does any particular photographer contribute to the development of our collective vision? The corollary to this question is: do you perceive the accelerating "literacy of vision?" Think about how the language of photography is changed and expanded by the work of a particular photographer. Our visual sophistication is continually emerging in complexity and range of styles. No longer are photographs "faithful witnesses." The visual language is evolving, and it is a dynamic, enriching process, in which photographers experiment with how the variables are employed in the construction of the image.

10. Why and when is photography an art? This question goes back to context, and the easy answer is that "A photograph is art whenever a curator hangs it on a gallery wall." But the more complicated response involves the difference between attempting to make a likeness without emotional shading or perspective—in other words, a fact-based document—and using the medium as a means of personal expression.

Seven layers of meaning

Beyond the overt subject of a photograph and beyond its formal and critical aspects, awareness of genre and genre mixing illuminate a photograph's significance and help us to place it in a tradition.

1. Scientific.
   Scientific photography is fact-based only, without emotional shading. The photograph functions as verifiable evidence, such as, for example, an image produced by electron microscopy or the photograph made for a driver's license. The medium is not used as a means of personal expression; in fact, the individual as maker is not introduced into the image.
2. Vernacular.
   A vernacular photograph is a common, everyday photograph. Vernacular photographs embrace language or words that derive from a place or region, and subjects or compositions that are native or peculiar to popular taste. To take the example of architecture, "vernacular" refers to the characteristic style of houses of the ordinary people of a particular region. Examples of vernacular photographs include snapshots, the photographs within family albums, and most vacation photography.
3. Documentary.
   A documentary photograph may appear to be a fact-based, prosaic document; however, a documentary photograph departs from a scientific photograph, in that, to quote critic Jerry L. Thompson, it is "deliberately wrought visual poetry." As such, it *does* possess emotional shading. For example, the Depression-era images of the Farm Security Administration's documentary photographers, including Dorothea Lange and Walker Evans, are simultaneously factual and lyrical.
4. Simulacra.
   Simulacra and hyperreality denote a representation that maintains its own status as an object and is not reducible simply to the thing it represents. A simulacrum becomes a "sign" or "signifier" in its own right, whose connection to the world outside the image is fluid, not fixed. The idea of the simulacrum emphasizes that the sign—that is, the photograph—is a metaphor or simile, in other words, a semblance not a copy. Plato spoke of two kinds of image-making. The first is a faithful reproduction, an attempt to precisely copy the original. The second is distorted by intent so that the copy appears accurate. An example is Michelangelo's *David* that was sculpted larger at the head than the feet so that viewers on the ground would see it correctly. A simulacrum is not a copy of the real, but it becomes "real" in its own right, paving the way for a pretense or perversion of reality, depending on the point of view. An example is the idealized

versions of public figures, including actors and fashion models, who through Photoshop become hyperreal.

5. Abstract.
   An abstract composition moves away from the recognizable subject into pure elements of relative color and design. An abstract image is easily recognized, as its lack of figurative structure provokes the viewer to ask, "what is that?" Consider the work of Aaron Siskind, whose flat picture planes, shallow depth of field, and focus on surface textures of mundane objects such as tree bark and walls create a flatness consistent with the painterly works of Mark Rothko and Franz Kline.

6. Conceptual.
   A conceptual photograph is primarily a device to reflect an idea conceived in the mind. The photograph, which may be figurative or abstract, derives from a mental image that can be an underground symbol unto itself, as, for example, in the gay community, a calla lily codes for the idea of male sexuality. Often, this style of photography is an allusive style characterized by ambiguous metaphors and a combination of elements signaling that thought rather than the perceptible subject of the image is the fundamental subject of the photograph. A classic example is when Alfred Stieglitz photographed Marcel Duchamp's urinal, a readymade sculpture for an exhibition in New York.

7. Transcendent.
   A transcendent photograph extends and exceeds the usual limits of recognizable visual knowledge. The composition may contain a subject, broadly interpreted, which is not normally perceived in the human experience. Some consider this step vague or obscure, principally because the representation could exceed the limits of experience, lying beyond comprehension or understanding. A transcendent layer might be suggested by a photograph of an afterlife, a spirit, or a future that is only imagined, not yet realized. Surreal photographs embrace transcendency by their very nature. Examples include Tommy Ingberg's *Crow*, which resembles a Rene Magritte painting, and Dimitri Daniloff's *PS1 Supermarket*, which depicts a headless shopper with an empty supermarket cart, poised to make a selection from well-stocked shelves of human heads.

INDEX

Note: *Italic* page numbers refer to figures and page numbers followed by "n" refer to end notes.

(FOLLOWING PAGES) *The Pursuit of Happiness (Mengejar Kebehagian)* is a large, wall-sized narrative photogram 16.5 × 24 feet. The process of making a photogram involves laying objects on light-sensitive paper, exposing it to light, and then developing through normal chemical processes. The objects project shadows of white or colored light, creating the photogram. In the history of photography, the photogram has been used experimentally to design and articulate abstract forms. Rarely, however, has the photogram been used to tell a story. Goin's *The Pursuit of Happiness* is composed of 120 assembled photogram panels; each panel can be read individually or collectively, as a group. The overall image is an implied, if not sequential, narrative. This work reflects the design and concept of Indonesian shadow puppets; instead of a screen, photo paper is used. The "puppets" are paper cutouts, and the "sets" are made of paperboard. *Mengejar Kebehagian*, the Indonesian phrase for "Pursuit of Happiness, "is part of Indonesian culture and is included in Indonesia's constitution.

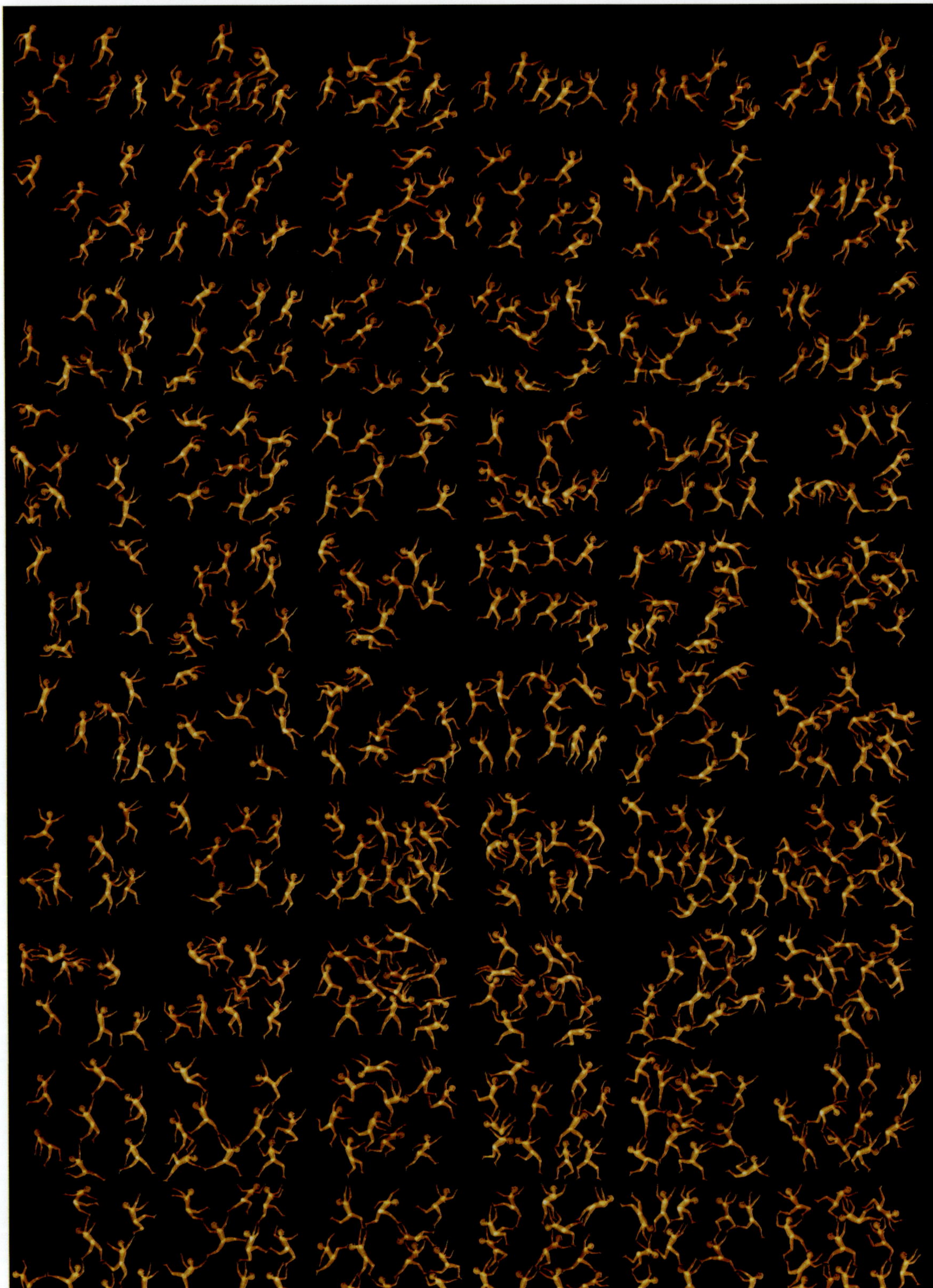

The Pursuit of Happiness (Mengejar Kebehagian), 2007

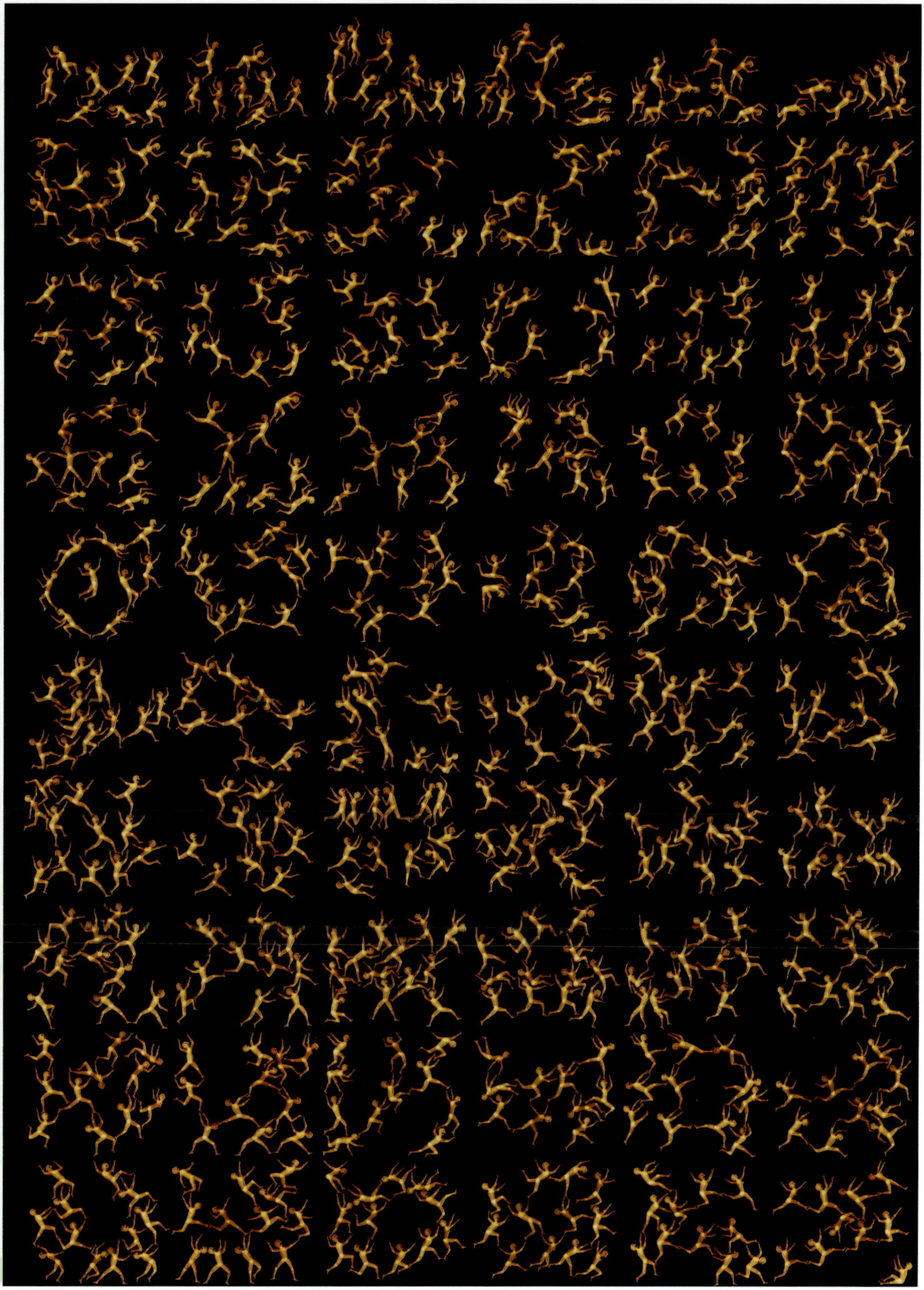

For Product Safety Concerns and Information please contact our EU representative GPSR@taylorandfrancis.com Taylor & Francis Verlag GmbH, Kaufingerstraße 24, 80331 München, Germany

Batch number: 10733627

Printed by Printforce, the Netherlands